Applications
of Optical Fourier Transforms

CONTRIBUTORS

SILVERIO P. ALMEIDA

WALLACE L. ANDERSON

F. M. M. AYUB

H. BARTELT

S. K. CASE

P. S. CHEATHAM

P. DAS

R. HAUCK

GUY INDEBETOUW

MARVIN KING

SING H. LEE

JAMES R. LEGER

ROBERT O'TOOLE

WILLIAM T. RHODES

BAHAA E. A. SALEH

A. A. SAWCHUK

R. R. SHANNON

HENRY STARK

T. C. STRAND

Applications
of Optical
Fourier Transforms

Edited by *HENRY STARK*

DEPARTMENT OF ELECTRICAL, COMPUTER, AND SYSTEMS ENGINEERING
RENSSELAER POLYTECHNIC INSTITUTE
TROY, NEW YORK

1982

 *ACADEMIC PRESS*

A Subsidiary of Harcourt Brace Jovanovich, Publishers

New York   London
Paris   San Diego   San Francisco   São Paulo   Sydney   Tokyo   Toronto

ACADEMIC PRESS, INC.
111 Fifth Avenue, New York, New York 10003

United Kingdom Edition published by
ACADEMIC PRESS, INC. (LONDON) LTD.
24/28 Oval Road, London NW1 7DX

Library of Congress Cataloging in Publication Data
Main entry under title:

Applications of optical Fourier transforms.

    Includes bibliographical references.
    1. Optical transfer functions. 2. Fourier
transformations. 3. Optical data processing.
I. Stark, Henry, Date.
TA1632.A68   621.36'7   81-20509
ISBN 0-12-663220-0   AACR2

PRINTED IN THE UNITED STATES OF AMERICA

82 83 84 85   9 8 7 6 5 4 3 2 1

To Anna N. Stark

Contents

Chapter 11   **Statistical Pattern Recognition
Using Optical Fourier Transform Features**

HENRY STARK AND ROBERT O'TOOLE

Chapter 12   **Incoherent Optical Processing**

H. BARTELT, S. K. CASE, AND R. HAUCK

List of Contributors

Numbers in parentheses indicate the pages on which the authors' contributions begin.

SILVERIO P. ALMEIDA (*41*), Department of Physics, Virginia Polytechnic Institute and State University, Blacksburg, Virginia 24061

WALLACE L. ANDERSON (*89*), Electrical Engineering Department, University of Houston, Houston, Texas 77004

F. M. M. AYUB[1] (*289*), Department of Electrical, Computer, and Systems Engineering, Rensselaer Polytechnic Institute, Troy, New York 12181

H. BARTELT (*499*), Physikalisches Institut der Universität Erlangen-Nürnberg, Federal Republic of Germany

S. K. CASE (*499*), Department of Electrical Engineering, University of Minnesota, Minneapolis, Minnesota 55455

P. S. CHEATHAM (*253*), Pacific-Sierra Research Corporation, Santa Monica, California 90404

P. DAS (*289*), Department of Electrical, Computer, and Systems Engineering, Rensselaer Polytechnic Institute, Troy, New York 12181

R. HAUCK (*499*), FB7, Universität Essen GHS, Essen, Federal Republic of Germany

GUY INDEBETOUW (*41*), Department of Physics, Virginia Polytechnic Institute and State University, Blacksburg, Virginia 24061

MARVIN KING (*209*), Riverside Research Institute, New York, New York 10023

SING H. LEE (*131*), Department of Electrical Engineering and Computer Sciences, University of California, San Diego, La Jolla, California 92093

[1] Present address: Advanced Technology Center, Bendix Corporation, Columbia, Maryland 21045.

JAMES R. LEGER[2] (*131*), Department of Electrical Engineering and Computer Sciences, University of California, San Diego, La Jolla, California 92093

ROBERT O'TOOLE (*465*), Department of Electrical, Computer, and Systems Engineering, Rensselaer Polytechnic Institute, Troy, New York 12181

WILLIAM T. RHODES (*333*), School of Electrical Engineering, Georgia Institute of Technology, Atlanta, Georgia 30332

BAHAA E. A. SALEH (*431*), Department of Electrical and Computer Engineering, University of Wisconsin, Madison, Wisconsin 53706

A. A. SAWCHUK (*371*), Image Processing Institute, Department of Electrical Engineering, University of Southern California, Los Angeles, California 90007

R. R. SHANNON (*253*), Optical Sciences Center, University of Arizona, Tucson, Arizona 85721

HENRY STARK (*1, 465*), Department of Electrical, Computer, and Systems Engineering, Rensselaer Polytechnic Institute, Troy, New York 12181

T. C. STRAND (*371*), Image Processing Institute, Department of Electrical Engineering, University of Southern California, Los Angeles, California 90007

[2] Present address: 3M Company, St. Paul, Minnesota 55144.

Preface

Several years ago I was attending an image processing conference near Monterey and, being in the enviable position of not having to give a paper, I could both concentrate on what was being said and enjoy the magnificent Monterey scenery without feeling the trace of anxiety that normally goes with giving a paper. In those days there was quite a debate raging on whether image processing should be done with optics or computers. One evening, while thinking about the papers that I had heard, it occurred to me that if the opticists had, so to speak, an ace in the hole it was their ability to generate a high resolution, two-dimensional Fourier transform using their own version of an FFT computer, namely a lens. Moreover, this "computer" could do it at the speed of light, at extremely high resolution, and do it as quickly for a 10^6-pixel image as for a 100-pixel one. It was then that I became interested in producing a book that would stress the applications of optical Fourier transforms. In retrospect, perhaps the Monterey scenery and the fine California wine combined to generate uncharacteristic euphoria and ambition in me. Frankly, I don't know. But I did raise the issue with colleagues and they thought such a book a fine idea. Academic Press showed interest and, perhaps even before the effect of the last bottle of the Cabernet Sauvignon wore off, an agreement was signed and there I was, an editor.

Leaving myself out, I do feel that these pages are the work of an unusually talented group of co-authors who are among the most active contributors to the field. Readers will find in this book some of the extraordinary achievements of Fourier optics. Not all, perhaps not even all of the most important applications are to be found here. Unfortunately, time and circumstances did not permit the inclusion of applications of Fourier optics to medical imaging, diffraction-limited imaging through a turbulent atmosphere, and other timely topics. Nevertheless, what is included is impressive and furnishes convincing evidence that Fourier optical systems, in their broadest sense, and expecially when backed up by computers, constitute an elegant and useful technology.

Now I would like to go over briefly the contents and organization of this book. In Chapter 1: *Theory and Measurement of the Optical Fourier Transform*, I review the Fourier transform property of a lens and discuss some of the care that must be taken if the light observed in the focal plane of a lens is indeed to represent the spectrum of the input. Optical power spectral estimation is discussed along with conventional and some unconventional "windowing" considerations required for stable and accurate spectral estimates. In Chapter 2: *Pattern Recognition via Complex Spatial Filtering*, S. P. Almeida and G. Indebetouw discuss the theory and applications of complex spatial filters: how they can be made scale and rotation invariant; and how they can be applied, in signal detection, character recognition, water pollution monitoring, and other pattern recognition problems. Over 280 references are given.

The theme of pattern recognition is continued by W. L. Anderson in Chapter 3: *Particle Identification and Counting by Fourier-Optical Pattern Recognition*. After discussing the theoretical foundations of the method, Anderson furnishes a careful computation of the statistical characteristics of the Fourier irradiance pattern and shows how the inverse scattering problem, that is, obtaining the particle distribution from the spectrum, can be solved with the help of the celebrated Gauss–Markov Theorem. Hybrid methods for realizing the technique are thoroughly reviewed.

J. R. Leger and S. H. Lee continue the discussion of pattern recognition as well as the more general signal processing problem in Chapter 4: *Signal Processing Using Hybrid Systems*. The emphasis is on hybrid systems, that is, those eclectic systems that combine the best of optics, analog electronics, and digital computers to solve problems. Included here is an attempt to answer the question, "Which parts of an information processing computation are best done by optical systems, and which parts should be done by electronic ones?" Also included here are two topics of particular interest in hybrid system design: sampling theory with attendant aliasing considerations, and optical–electrical interconnections. Examples of working hybrid systems are given to illustrate the design techniques unique to this approach to signal processing.

Perhaps one of the earliest and most successful applications of Fourier optics was to the problem of radar signal processing. M. King, who is associated with a laboratory that pioneered in the use of coherent optical systems to extract range-Doppler information from radar echoes, reviews that activity in Chapter 5: *Fourier Optics and Radar Signal Processing*. Included here is a discussion of pulse-Doppler and chirp signals and of how these can be processed optically, optical processing for synthetic aperture radar, and recent progress in the field.

Despite progress with other media, the single most important medium for storing signals in optics is still photographic film. In Chapter 6: *Application of Optical Power Spectra to Photographic Image Measurement*, R. R. Shannon and P. S. Cheatham discuss their experimental investigation that involves photographic image evaluation by measuring the optical power spectra of signal and noise. Their work shows that the signal-to-noise power spectrum is a useful and reliable measure of the information content of photographic film and is useful in image quality determinations.

A newer technology, whose integration with Fourier optics can enable all kinds of useful operations such as light modulation, spectrum analysis, correlation and convolution, and others, is the phenomenon of ultrasound with surface confinement, or SAW (for *surface acoustic waves*). A thorough discussion of SAW devices that includes their underlying physics, modes of operations, and their integration and application in Fourier optics is furnished by P. Das and F. M. M. Ayub in Chapter 7: *Fourier Optics and SAW Devices.*

The signal processing systems described in Chapters 1–7 share the property of space invariance. Most signal processing systems are in fact, by design, space invariant. But not all systems share this property; for example, space invariance is not a property of ultrawide-angle lens systems or lenses exhibiting coma. For such systems, a lateral shift anywhere in the input plane does not result in a corresponding lateral shift anywhere in the output plane. In Chapter 8: *Space-Variant Optical Systems and Processing*, William T. Rhodes discusses such systems and shows how space-variant systems can be designed to carry out useful functions, for example, the realization of an operation such as the superposition integral.

Broadening the arena of Fourier optics still further, A. A. Sawchuk and T. C. Strand, in Chapter 9: *Fourier Optics in Nonlinear Signal Processing*, discuss how nonlinear systems can usefully be applied in Fourier optics. How do Fourier transforms enter into nonlinear systems? In some systems the Fourier transform is integral to the operation of the nonlinearity (for example, halftone techniques). In other systems, the Fourier transform is an important adjunct in a composite nonlinear system (for example, a system with input–output nonlinearities suffering from intermodulation noise). The authors describe a number of nonlinear systems and components (for example, the variable grating mode device) and show how they can be used in many applications, including digital logic.

Is there a link between optical Fourier transforms and the human visual system? This is the subject of Chapter 10: *Optical Information Processing and the Human Visual System*, by B. E. A. Saleh. The author discusses how Fourier methods are used to study the transmission of spatial information through the human visual system and reviews how coherent techniques are used in vision research, for example, the use of speckle to measure the dioptics of the eye and the application of spatial filtering techniques to pictures used in psychophysical studies.

In Chapter 11: *Statistical Pattern Recognition Using Optical Fourier Transforms*, R. K. O'Toole and I again pick up the theme of pattern recognition first introduced in Chapter 2. We consider two well-known pattern recognition problems that have received wide attention in the digital signal processing community. We concentrate on two questions: (i) can features based only on the optical Fourier transforms lead to a high probability of correct classification? and (ii) can a hybrid, that is, optical–digital, pattern recognizer, using only the Fourier irradiance spectrum, do as well as sophisticated all-digital routines based on co-occurrence gray scale statistics? As readers will find out, the answer to both questions is yes.

A discussion of Fourier optics presupposes coherent illumination, but does it need to? This is what H. Bartelt, S. K. Case, and R. Hauck (BCH) set out to explore in the last chapter: *Incoherent Optical Processing*. This chapter, like the ones by Rhodes and by Sawchuk and Strand (RSS) expands the realm of Fourier optics. But whereas RSS expanded the class of spatial systems, BCH expand in the time domain to include incoherent illumination. Starting with a telecentric optical system illuminated by an arbitrary source, they furnish a general theoretical model of the propagation of light fields from input to output. The model is then used to compute the fields when the source is restricted in size or bandwidth. They show that spatially incoherent optical systems can perform convolutions linear in intensity, that spatially coherent but temporally incoherent systems can be used in scale-multiplexed matched filtering (useful when the shape of the object is known but the size is not), and that such a system can be used in numerous other applications such as pseudocolor encoding and signal-to-noise improvement.

HENRY STARK
Troy, New York

Acknowledgments

In my role as editor of this book, I wish first of all to thank my co-authors who did such an excellent job with a minimum of exhortation. So many people and agencies are responsible for creating a book such as this that it is really impossible to thank all of them individually. However, I do wish to single out individuals and organizations who played a vital, continuing role. Special thanks are due to the following: Rosan Laviolette for her typing and cooperation; the administration of Rensselaer Polytechnic Institute for permission to take the time to complete this project; the various funding agencies, especially the National Science Foundation, the Army Research Office, the Air Force Office of Scientific Research, and the Office of Naval Research for their continued support; the highly competent staff at Academic Press; and finally my wife Alice whose patience and support were much appreciated.

Chapter 1

Theory and Measurement of the Optical Fourier Transform

HENRY STARK

DEPARTMENT OF ELECTRICAL, COMPUTER,
   AND SYSTEMS ENGINEERING
RENSSELAER POLYTECHNIC INSTITUTE
TROY, NEW YORK

In a book devoted to applications of optical Fourier transforms (FTs), it somehow seems fitting that we discuss the celebrated Fourier transform property (FTP) of a lens in the first chapter. The material is covered in several places (see, for example, Refs. [1–3]), so that we shall not give a very detailed and rigorous treatment here. As we shall see, the FTP of a lens is easily established from the diffraction integral that describes the propagation of monochromatic light in free space. The diffraction integral is the single most important object in the study of Fourier optics, and virtually all optical phenomena can be explained mathematically in terms of it. Surprisingly the diffraction integral can be easily derived from Fourier transform considerations alone plus certain notions such as plane waves and complex amplitudes.

The FTP of a lens is usually derived assuming an idealized lens between object and back focal plane (BFP) and—although sometimes not stated—the validity of the stationary phase approximation for the Fourier transform configuration. The diffraction integral is then applied twice—from the object to the lens and from the lens to the BFP. When vignetting† is ignored, the light amplitude in the BFP is then shown to be, up to a constant and a quadratic phase factor, the classical two-dimensional Fourier transform (2-DFT) of the transmittance of the diffracting object. The validity of this

---

† The phenomenon of vignetting is discussed in Section 1.3.

1

result depends on the object being illuminated by a plane wave, although it is more generally true that—subject to the constraints listed above—there exists a Fourier transform relation between the light amplitude adjacent to the object and the light amplitude in the BFP. In the following few sections we shall derive the FTP of a lens and see under what conditions a true Fourier transform is obtained.

1.1 PLANE WAVES

In scalar diffraction theory, we are concerned with solutions to the scalar wave equation

$$\nabla^2 u(\mathbf{r}, t) = \frac{1}{c^2} \frac{\partial^2 u(\mathbf{r}, t)}{\partial t^2}, \qquad (1.1\text{-}1)$$

where ∇^2 is the Laplacian operator, $u(\mathbf{r}, t)$ is the scalar field (or a single component of a vector field), $\mathbf{r}$ is the position vector, t is time, and c is the speed of light. For monochromatic illumination, a solution of Eq. (1.1-1) is

$$u(\mathbf{r}, t) = a(\mathbf{r}) \cos[2\pi v t + \phi(\mathbf{r})], \qquad (1.1\text{-}2)$$

where $a(\mathbf{r})$ is the time-independent amplitude, v is the frequency in hertz (Hz), and $\phi(\mathbf{r})$ is a time-independent phase. It is customary to write Eq. (1.1-2) as

$$u(\mathbf{r}, t) = \text{Re}[a(\mathbf{r})e^{-j\phi(\mathbf{r})}e^{-j2\pi v t}] \equiv \text{Re}[U(\mathbf{r})e^{-j2\pi v t}], \qquad (1.1\text{-}3)$$

where $U(\mathbf{r}) \equiv a(\mathbf{r}) \exp[-j\phi(\mathbf{r})]$, is commonly called the complex amplitude and satisfies the Helmholtz equation

$$(\nabla^2 + k^2)U(\mathbf{r}) = 0. \qquad (1.1\text{-}4)$$

The constant k in Eq. (1.1-4) is called the wave number and is related to the free-space wavelength λ according to

$$k = 2\pi/\lambda = 2\pi v/c. \qquad (1.1\text{-}5)$$

All linear operations involving $u(\mathbf{r}, t)$ can be done using only $U(\mathbf{r})$; real solutions, i.e., those whose form is as in Eq. (1.1-2) or linear combinations thereof, can always be obtained from $U(\mathbf{r})$ by the formalism of Eq. (1.1-3).

From now on, unless otherwise stated, we shall assume that all representations of the field $u(\mathbf{r}, t)$ will be through the artifice of the complex amplitude.

A. Propagation of Plane Waves

A unit amplitude plane with wave vector $\mathbf{k}$ is described by a complex amplitude

$$B(\mathbf{r}) = e^{j\mathbf{k} \cdot \mathbf{r}} \qquad (1.1\text{-}6)$$

and corresponds to a real wave of the form

$$b(\mathbf{r}, t) = \cos(2\pi v t - \mathbf{k} \cdot \mathbf{r}). \tag{1.1-7}$$

In a linear homogeneous medium the wave vector has magnitude $|\mathbf{k}| = k$, and its direction is in the direction of propagation of the wave. The complex amplitude is constant over points $\mathbf{r}$ such that $\mathbf{k} \cdot \mathbf{r} = c_0$, a constant. But $\mathbf{k} \cdot \mathbf{r} = c_0$ is the equation of a plane—hence the term plane wave. The surface of constant phase is normal to k, the direction of propagation. Letting $\mathbf{k} = k_x \mathbf{i}_x + k_y \mathbf{i}_y + k_z \mathbf{i}_z$, where $\mathbf{i}_x, \mathbf{i}_y, \mathbf{i}_z$ are unit vectors parallel to the Cartesian coordinate axes, we can define a set of direction cosines α, β, γ and a set of angles $\hat{\alpha}, \hat{\beta}, \hat{\gamma}$ according to

$$\alpha \equiv \cos \hat{\alpha} = (\mathbf{k} \cdot \mathbf{i}_x)/k = k_x/k,$$

$$\beta \equiv \cos \hat{\beta} = (\mathbf{k} \cdot \mathbf{i}_y)/k = k_y/k, \tag{1.1-8}$$

$$\gamma \equiv \cos \hat{\gamma} = (\mathbf{k} \cdot \mathbf{i}_z)/k = k_z/k.$$

We note from Fig. 1.1-1 that $\hat{\alpha}, \hat{\beta}, \hat{\gamma}$ are the angles between $\mathbf{k}$ and the Cartesian unit vectors $\mathbf{i}_x, \mathbf{i}_y, \mathbf{i}_z$, respectively. Since $\mathbf{k} \cdot \mathbf{k}/k^2 = 1$, $\alpha^2 + \beta^2 + \gamma^2 = 1$ and we can write Eq. (1.1-6) as

$$B(x, y, z) = e^{j(k_x x + k_y y + k_z z)} = e^{jk(\alpha x + \beta y)} e^{jk(1 - \alpha^2 - \beta^2)^{1/2} z}, \tag{1.1-9}$$

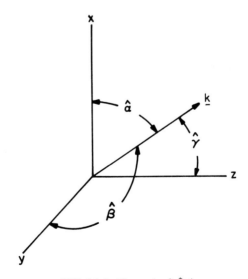

FIG. 1.1-1. The angles $\hat{\alpha}, \hat{\beta}, \hat{\gamma}$.

where we have used the fact that $\mathbf{r} = x\mathbf{i}_x + y\mathbf{i}_y + z\mathbf{i}_z$. If we adopt the notation $B(x, y, z) \equiv B_z(x, y)$ [e.g., $B(x, y, 0) \equiv B_0(x, y)$], we note that

$$\frac{B_z(x, y)}{B_0(x, y)} = \exp\left\{j2\pi\left[\frac{1}{\lambda^2} - \left(\frac{\alpha}{\lambda}\right)^2 - \left(\frac{\beta}{\lambda}\right)^2\right]^{1/2}z\right\} \equiv H_z\left(\frac{\alpha}{\lambda}, \frac{\beta}{\lambda}\right), \quad (1.1\text{-}10)$$

where $H_z(\alpha/\lambda, \beta/\lambda)$ is the free-space transfer function that describes the propagation of a plane wave with direction cosines α, β, γ over a distance z. It is useful to define the Cartesian spatial frequencies u, v, w by

$$u \equiv \alpha/\lambda, \quad (1.1\text{-}11a)$$

$$v \equiv \beta/\lambda, \quad (1.1\text{-}11b)$$

$$w \equiv \gamma/\lambda. \quad (1.1\text{-}11c)$$

In terms of these spatial frequencies, the free-space transfer function can be written as

$$H_z(u, v) = \exp[j(2\pi/\lambda)(1 - \lambda^2 u^2 - \lambda^2 v^2)^{1/2}z]. \quad (1.1\text{-}12)$$

Propagation requires that $u^2 + v^2 < 1/\lambda^2$. Otherwise, the exponent in Eq. (1.1-12) is real, and evanescence of the waves occurs. Our interest is in the propagation condition only.

1.2 THE DIFFRACTION INTEGRAL

Let us consider a wave traveling essentially along the z-axis and incident on an xy-plane at $z = 0$ (Fig. 1.2-1). The Fourier transform enables us to

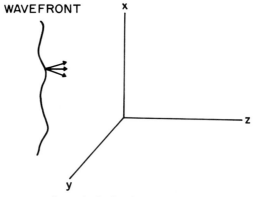

FIG. 1.2-1. A wave traveling principally along the z-axis, incident on the xy-plane at $z = 0$. The wave may be viewed as a superposition of a large number of plane waves, each with a different set of direction cosines.

represent such a wave in terms of an infinite number of infinitesimal plane waves according to

$$U_0(x, y) = \int_{-\infty}^{\infty} \int_{-\infty}^{\infty} A_0(u, v) e^{j2\pi(ux+vy)} \, du \, dv, \qquad (1.2\text{-}1)$$

where $U_0(x, y)$ is the complex amplitude across the xy-plane at $z = 0$ and

$$dU_0 \equiv A_0(u, v) \, du \, dv \, e^{j2\pi(ux+vy)} \qquad (1.2\text{-}2)$$

represents a plane wave of infinitesimal amplitude $A_0(u, v) \, du \, dv$ traveling with direction cosines

$$\alpha = u\lambda, \qquad \beta = v\lambda, \qquad \gamma = \sqrt{1 - (u\lambda)^2 - (v\lambda)^2}. \qquad (1.2\text{-}3)$$

Similarly, the complex amplitude in a transversal xy-plane a distance z away can be decomposed into infinitesimal plane waves $dU_z \equiv A_z(u, v) \, du \, dv$ $\times \exp[j2\pi(ux + vy)]$ according to

$$U_z(x, y) = \int_{-\infty}^{\infty} \int_{-\infty}^{\infty} A_z(u, v) e^{j2\pi(ux+vy)} \, du \, dv. \qquad (1.2\text{-}4)$$

From Eqs. (1.1-10) and (1.1-12), we can relate dU_z to dU_0 according to

$$dU_z = H_z(u, v) \cdot dU_0,$$

from which it follows that

$$A_z(u, v) = A_0(u, v) \exp\left[j\frac{2\pi}{\lambda}(1 - \lambda^2 u^2 - \lambda^2 v^2)^{1/2} z \right]. \qquad (1.2\text{-}5)$$

Our aim is to relate $U_z(\cdot)$ to $U_0(\cdot)$. To this end we use the forward Fourier transform to write

$$A_0(u, v) = \int_{-\infty}^{\infty} \int_{-\infty}^{\infty} U_0(\xi, \eta) e^{-j2\pi(u\xi+v\eta)} \, d\xi \, d\eta. \qquad (1.2\text{-}6)$$

We can now relate the field $U_z(\cdot)$ to the field $U_0(\cdot)$ by using Eq. (1.2-6) for $A_0(u, v)$ in Eq. (1.2-5) and then replacing $A_z(u, v)$ in Eq. (1.2-4) by the right side of Eq. (1.2-5). The result is

$$U_z(x, y) = \int_{-\infty}^{\infty} \int_{-\infty}^{\infty} d\xi \, d\eta \, U_0(\xi, \eta) G(x - \xi, y - \eta), \qquad (1.2\text{-}7)$$

where

$$G(x - \xi, y - \eta) \equiv \int_{-\infty}^{\infty} \int_{-\infty}^{\infty} e^{jk[1 - (u\lambda)^2 - (v\lambda)^2]^{1/2}z} e^{j2\pi[u(x-\xi)+v(y-\eta)]} \, du \, dv. \quad (1.2\text{-}8)$$

Equations (1.2-7) and (1.2-8) furnish the solution to the diffraction problem. They allow the complex amplitude in a transversal plane at z to be expressed in terms of the field of a diffracting aperture in the plane $z = 0$. The only remaining problem is the simplification of Eq. (1.2-8). Fortunately this can be done rather easily, as will now be demonstrated.

In Eq. (1.2-8) let $u = \rho \cos \phi$, $v = \rho \sin \phi$, $x - \xi = r \cos \theta$, $y - \eta = r \sin \theta$; then

$$G(r \cos \theta, r \sin \theta) = \int_0^\infty \int_0^{2\pi} e^{jk(1 - \lambda^2 \rho^2)^{1/2}z} e^{j2\pi r\rho \cos(\theta - \phi)} \rho \, d\rho \, d\phi$$

$$= 2\pi \int_0^\infty e^{jkz(1 - \lambda^2 \rho^2)^{1/2}} J_0(2\pi r\rho)\rho \, d\rho \equiv \tilde{G}(r). \quad (1.2-9)$$

With the change in variable $t \equiv 2\pi\rho$, Eq. (1.2-9) can be rewritten as

$$\tilde{G}(r) = \frac{1}{2\pi} \int_0^\infty e^{-z(t^2 - k^2)^{1/2}} J_0(tr)t \, dt. \quad (1.2-10)$$

This integral may be evaluated from a result given by Erdélyi [4, formula 52, p. 95], namely,

$$\int_0^\infty J_0(bt)e^{-a(t^2 - y^2)^{1/2}}(t^2 - y^2)^{-1/2}t \, dt = e^{-jy(a^2 + b^2)^{1/2}}(a^2 + b^2)^{-1/2},$$
$$\hspace{8cm} (1.2-11)$$
$$\text{arg}(t^2 - y^2)^{1/2} = \pi/2 \quad \text{if} \quad t < y.$$

If we differentiate Erdélyis' formula with respect to a and then make the associations $a \equiv z$, $y \equiv -k$, $b \equiv r$, we obtain

$$\tilde{G}(r) = \frac{e^{jk(z^2 + r^2)^{1/2}}}{j\lambda\sqrt{r^2 + z^2}} \frac{z}{\sqrt{r^2 + z^2}} \left(1 - \frac{1}{jk\sqrt{r^2 + z^2}}\right). \quad (1.2-12)$$

Now we make the following observations:

(1) For $z \gg \lambda$, the second term in the parentheses is $\ll 1$ and can be ignored.

(2) $z/(r^2 + z^2)^{1/2} \equiv \cos \Phi$, where Φ is the angle between the positive z-axis and the line passing through the points $(\xi, \eta, 0)$ and (x, y, z). Cos Φ is called the obliquity factor and is approximately $\simeq 1$ when the dimensions of the region of interest are small compared to z. Replacing cos Φ by unity is tantamount to the paraxial approximation and is commonly done in Fourier optics.

(3) $(r^2 + z^2)^{1/2}$ is the distance between $(\xi, \eta, 0)$ and (x, y, z). In the paraxial approximation, the quantity in the denominator of Eq. (1.2-12) can be replaced by z.

Hence

$$\tilde{G}(r) \simeq e^{jk(z^2 + r^2)^{1/2}}/j\lambda z \qquad \text{(paraxial approximation)}, \qquad (1.2\text{-}13)$$

and the diffraction integral takes the form

$$U_z(x, y) = \frac{1}{j\lambda z} \int_{-\infty}^{\infty} \int_{-\infty}^{\infty} U_0(\xi, \eta) e^{jk(z^2 + r^2)^{1/2}} \, d\xi \, d\eta. \qquad (1.2\text{-}14)$$

The final approximation is to invoke the so-called Fresnel approximation; i.e.,

$$(z^2 + r^2)^{1/2} = [z^2 + (x - \xi)^2 + (y - \eta)^2]^{1/2}$$

$$\simeq z + (x - \xi)^2/2z + (y - \eta)^2/2z, \qquad (1.2\text{-}15)$$

which the reader will recognize as omitting all but the first two terms of the binomial expansion of $(z^2 + r^2)^{1/2}$. The validity of the Fresnel approximation is discussed by Goodman [1, p. 59]. With the use of the Fresnel approximation in Eq. (1.2-14) we can finally write

$$U_z(x, y) = \frac{e^{jkz}}{j\lambda z} \int_{-\infty}^{\infty} \int_{-\infty}^{\infty} U_0(\xi, \eta) \exp\left\{ j \frac{\pi}{\lambda z} [(x - \xi)^2 + (y - \eta)^2] \right\} d\xi \, d\eta,$$

$$(1.2\text{-}16)$$

which is a central result and is the form of the diffraction integral most often used in Fourier optics. Its repeated, but straightforward, application leads directly to the FTP of a lens. However, the intermediate integrals are rather cumbersome and therefore, for simplicity, we shall do the analysis in one dimension. Then Eq. (1.2-16) will be written as

$$U_z(x) = \frac{e^{jkz/2}}{\sqrt{j\lambda z}} \int_{-\infty}^{\infty} U_0(\xi) \exp\left[j \frac{\pi}{\lambda z} (x - \xi)^2 \right] d\xi. \qquad (1.2\text{-}17)$$

1.3 FOURIER TRANSFORM PROPERTY OF A LENS

Consider the (one-dimensional) configuration shown in Fig. 1.3-1. We shall show that, under certain conditions, the field in the back focal plane of the lens is proportional to $T(u)$, the Fourier transform of $t(x)$, if the latter is illuminated by a plane wave. The technique is to compute the field from one plane to the next in a systematic fashion until the plane of interest is reached. Thus the field immediately after the object $t(x)$ is $U_0(x) = B(x)t(x)$, where, for simplicity, we set $B(x) = 1$, i.e., a unit amplitude plane wave

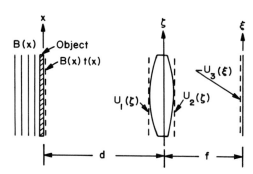

FIG. 1.3-1. Configuration for analyzing the Fourier transform property of a lens. The fields adjacent to the various surfaces are indicated by dashed lines.

traveling parallel to the z-axis. Just before the lens, a distance d away, the field is

$$U_1(\zeta) = K_1(d) \int_{-\infty}^{\infty} t(x) \exp\left[j \frac{\pi}{\lambda d}(\zeta - x)^2 \right] dx, \qquad (1.3\text{-}1)$$

where we used Eq. (1.2-17) with $d \equiv z$ and

$$K_1(d) \equiv \frac{e^{jkz/2}}{\sqrt{j\lambda z}}\Bigg|_{z=d}. \qquad (1.3\text{-}2)$$

The overall lens transmittance is $P(\zeta) \exp[-j(\pi/\lambda f)\zeta^2]$, where $P(\zeta)$ is the pupil function associated with the lens aperture and can be generalized to include aberrations. Typically, for an ideal, nonabsorbing lens of diameter L,

$$P(\zeta) = \text{rect}(\zeta/L), \qquad (1.3\text{-}3)$$

where

$$\text{rect}\left(\frac{\zeta}{L}\right) \equiv \begin{cases} 1, & |\zeta| < L/2 \\ 0, & \text{otherwise.} \end{cases}$$

Right after the lens the field $U_2(\zeta)$ is given by

$$U_2(\zeta) = U_1(\zeta)P(\zeta) \exp\left(-j \frac{\pi}{\lambda f} \zeta^2 \right), \qquad (1.3\text{-}4)$$

and the field in the back focal plane is

$$U_3(\xi) = K_1(f) \int_{-\infty}^{\infty} U_2(\zeta) \exp\left[j \frac{\pi}{\lambda f} (\xi - \zeta)^2 \right] d\zeta. \qquad (1.3\text{-}5)$$

Now using Eq. (1.3-1) in Eq. (1.3-4) and then inserting Eq. (1.3-4) into Eq. (1.3-5) enables us to write, after some algebra,

$$U_3(\xi) = K_1(d)K_1(f)$$

$$\times \exp\left[j\frac{\pi}{\lambda f}\left(1 - \frac{d}{f}\right)\xi^2\right]\right]\int_{-\infty}^{\infty} t(x)\exp\left(-j\frac{2\pi}{\lambda f}x\xi\right)A(x,\xi)\,dx, \quad (1.3\text{-}6)$$

where

$$A(x,\xi) = \int_{-\infty}^{\infty} P(\zeta)\exp\left\{j\frac{\pi}{\lambda d}\left[\zeta - \left(x + \frac{d}{f}\xi\right)\right]^2\right\}d\zeta. \quad (1.3\text{-}7)$$

Because $P(\zeta)$ is a "slowly varying" function, for values of λ that are typical of the optical regime, the integral in Eq. (1.3-7) can be evaluated by the method of stationary phase (see, for example, Papoulis [5, p. 234]). This gives

$$A(x, \xi) = \sqrt{j\lambda d}\, P[x + (d/f)\xi]. \quad (1.3\text{-}8)$$

Inserting this result in Eq. (1.3-6) yields

$$U_3(\xi) = \frac{1}{\sqrt{j\lambda f}}\exp\left[j\frac{\pi}{\lambda f}\left(1 - \frac{d}{f}\right)\xi^2\right]$$

$$\times \int_{-\infty}^{\infty} t(x)P\left(x + \frac{d}{f}\xi\right)\exp\left(-j\frac{2\pi}{\lambda f}x\xi\right)dx, \quad (1.3\text{-}9)$$

where we have discarded the unimportant complex factors $e^{jkd/2}e^{jkf/2}$. Equation (1.3-9) is the final result. Since ξ can be written in terms of the spatial frequency u according to $\xi = u\lambda f$,† we can rewrite Eq. (1.3-9) as

$$U_3(u\lambda f) \equiv \tilde{U}_3(u) = \frac{1}{\sqrt{j\lambda f}}\exp\left[j\pi\left(1 - \frac{d}{f}\right)u^2\lambda f\right]$$

$$\times \int_{-\infty}^{\infty} t(x)P(x + \lambda\,du)e^{-j2\pi ux}\,dx. \quad (1.3\text{-}10)$$

We can now make some interesting observations based on Eq. (1.3-10):

(1) For $d = f$, i.e., the object in the front focal plane of the lens, the phase factor outside the integral vanishes, thereby prompting some writers to suggest (somewhat erroneously) that an exact Fourier transform relation-

---

† This follows from the paraxial theory which allows us to replace tangents by sines and Eq. (1.1-11a). Put another way, with θ denoting the diffraction angle, we have $\tan\theta = \xi/f \simeq \sin\theta = \cos\hat{\alpha} = \alpha = u\lambda$.

ship exists between the front and back focal plane of a lens. As can be seen from Eq. (1.3-10), this is only true if the effect of the pupil is ignored.

(2) For $d = 0$, i.e., the object against the lens, the phase factor does not vanish, but the effect of the pupil vanishes if the physical extent of the object is smaller than the lens aperture. Thus, if D is the maximum dimension of the object and $D < L$, then up to a phase factor and a constant, a Fourier transform relation is indeed observed.

(3) For $d \neq 0$, the fidelity of the Fourier transform of the object will depend on the spatial frequency u. For $D < L$, the lens will act as a low-pass filter. From Eq. (1.3-10) the reader will have no trouble verifying that

for $\qquad |u| < \dfrac{L - D}{2\lambda d}$, $\qquad\qquad$ no attenuation of the spectrum; $\qquad$ (1.3-11a)

for $\quad \dfrac{L - D}{2\lambda d} < |u| < \dfrac{L + D}{2\lambda d}$, $\quad$ partial attenuation of the spectrum; (1.3-11b)

for $\qquad |u| > \dfrac{L + D}{2\lambda d}$, $\qquad\qquad$ total attenuation of the spectrum. $\quad$ (1.3-11c)

The attenuation of high-frequency components in the Fourier spectrum is known as *vignetting*. As can be seen, vignetting is due to the finite lens aperture and can be minimized by making d small. Although a small d does not eliminate the quadratic phase factor in Eq. (1.3-10), the presence of the latter is of no consequence in irradiance spectrum measurements which only involve $|\tilde{U}_3(u)|^2$. Hence it is preferred practice to place the sample as near to the lens as possible when making spectral measurements. This will aid in furnishing a less biased spectrum, especially at high frequencies.

Before turning to the design considerations of optical irradiance spectrum measurements, we conclude this section by giving the two-dimensional form of Eq. (1.3-10), namely,

$$\tilde{U}_3(u, v) = \frac{1}{j\lambda f} \exp\left[j\pi\lambda f \left(1 - \frac{d}{f} \right)(u^2 + v^2) \right]$$

$$\times \int_{-\infty}^{\infty} \int_{-\infty}^{\infty} t(x, y)P(x + \lambda\, du, y + \lambda\, dv)e^{-j2\pi(ux + vy)}\, dx\, dy. \quad (1.3\text{-}12)$$

We have, of course, assumed illumination by a unit amplitude plane wave. If the amplitude of the illumination were C, then this term would appear as a factor in Eq. (1.3-12).

1.4 THE SAMPLE SPECTRUM

We shall use the terms spectrum and irradiance spectrum interchangeably. A coherent-optical spectrum analyzer is shown in Fig. 1.4-1. The sample

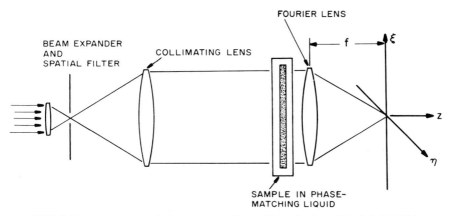

FIG. 1.4-1. A coherent-optical spectrum analyzer. (From Stark and Dimitriadis [10].)

is placed against the lens and is immersed in a phase matching liquid to negate unwanted phase variations. The sample is treated as a two-dimensional real random process, and its transmittance is denoted by $t(x, y)$. The pupil function associated with the Fourier transform lens is given by

$$P_D(x, y) = \begin{cases} 1, & (x^2 + y^2)^{1/2} \le b, \\ 0, & \text{otherwise,} \end{cases} \tag{1.4-1}$$

where b is the radius of the aperture and the subscript D emphasizes the fact that we are dealing with a diffraction-limited system. From Section 1.3 [Eq. (1.3-12)] we see that the field in the back focal plane for this configuration $(d = 0)$ is given by

$$\tilde{U}_3(u, v) \equiv \tilde{F}(u, v) = (1/j\lambda f)e^{j\pi\lambda f(u^2 + v^2)}$$
$$\times \int_{-\infty}^{\infty} \int_{-\infty}^{\infty} t(x, y)P_D(x, y)e^{-j2\pi(ux + vy)}\,dx\,dy, \tag{1.4-2}$$

where

$$u \equiv \xi/\lambda f = \text{horizontal spatial frequency}, \tag{1.4-3}$$
$$v \equiv \eta/\lambda f = \text{vertical spatial frequency}.$$

The mean-squared value of $\tilde{F}(u, v)$ can be written, after some algebra, as

$$\langle|\tilde{F}(u, v)|^2\rangle = \frac{A}{\lambda^2 f^2} \int_{-\infty}^{\infty} \int_{-\infty}^{\infty} R(\alpha, \beta)H_D(\alpha, \beta)e^{-j2\pi(u\alpha + v\beta)}\,d\alpha\,d\beta, \tag{1.4-4}$$

where $A = \pi b^2$, $R(\alpha, \beta)$ is the autocorrelation function of the process $t(x, y)$; i.e.,

$$R(\alpha, \beta) \equiv \langle t(x, y)t(x + \alpha, y + \beta)\rangle \tag{1.4-5}$$

and $H_D(\alpha, \beta)$ is the normalized self-correlation of the diffraction-limited pupil function.† Specifically

$$H_D(\alpha, \beta)$$
$$= \int_{-\infty}^{\infty} \int_{-\infty}^{\infty} P_D(x, y) P_D(x + \alpha, y + \beta) \, dx \, dy \bigg/ \int_{-\infty}^{\infty} \int_{-\infty}^{\infty} P_D^2(x, y) \, dx \, dy.$$

(1.4-6)

The computation of $H_D(\alpha, \beta)$ is given in several places—for example, see Goodman [1, p. 119]. For the case at hand

$$H_D(\alpha, \beta) \equiv H_D(r)$$
$$= \begin{cases} \dfrac{2}{\pi} \left\{ \cos^{-1}\left(\dfrac{r}{2b}\right) - \left(\dfrac{r}{2b}\right) \left[1 - \left(\dfrac{r}{2b}\right)^2\right]^{1/2} \right\}, & r \le 2b, \\ 0, & \text{otherwise,} \end{cases}$$

(1.4-7)

where $r = (\alpha^2 + \beta^2)^{1/2}$. We note that Eq. (1.4-4) is the Fourier transform of a product. Hence it can be written as a convolution:

$$\langle |\tilde{F}(u, v)|^2 \rangle = \frac{A}{\lambda^2 f^2} S(u, v) * 4A \left\{ \frac{J_1[2\pi b(u^2 + v^2)^{1/2}]}{2\pi b(u^2 + v^2)^{1/2}} \right\}^2,$$

(1.4-8)

where $S(u, v) \equiv \mathscr{F}_2[R(\alpha, \beta)]$ is the true spectrum of the process $t(x, y)$, and the term on the right of the asterisk is the Fourier transform of $H_D(r)$. It represents the effect of truncation and tends to *smooth* the true spectrum. However, for large values of b, it tends to behave like a delta function and its smoothing properties are minimal, provided that b is much larger than the reciprocal of the largest correlation interval in the process. Figure 1.4-2 shows the relative insensitivity of the spectrum to different values of b. In practice, the most significant effect of this term is to mak the spectrum at low spatial frequencies. The reason for this is that $t(x, y)$ generally contains a bias transmittance t_0; i.e.,

$$t(x, y) = t_0 + \tau(x, y),$$

(1.4-9)

where $\tau(x, y)$ is the zero-mean process about t_0. Then with $r_c(\alpha, \beta)$ denoting the covariance function of $t(x, y)$, i.e., $r_c(\alpha, \beta) = \langle \tau(x, y)\tau(x + \alpha, y + \beta) \rangle$, we see that

$$R(\alpha, \beta) = t_0^2 + r_c(\alpha, \beta).$$

(1.4-10)

---

† Essentially the optical transfer function.

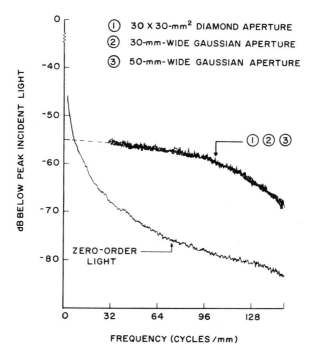

FIG. 1.4-2. Relative insensitivity of the normalized spectrum to aperture size and shape. The spectra shown are those of film grain noise of Royal-X Pan film developed in DK-50 developer.

If we substitute this result in Eq. (1.4-4), we see that $\langle|\tilde{F}(u, v)|^2\rangle$ contains the sum of two terms—one of interest and one proportional to

$$I_0(u, v) = 4At_0^2\{J_1[2\pi b(u^2 + v^2)^{1/2}]/2\pi b(u^2 + v^2)^{1/2}\}^2. \quad (1.4\text{-}11)$$

$I_0(u, v)$ is typically orders of magnitude larger than the term of interest around the origin, resulting in some masking of the spectrum of $t(x, y)$ at very low spatial frequencies. However, these frequencies are frequently not of very great interest, and since methods exist [6] for negating $I_0(u, v)$ around the origin, we shall ignore this effect. Also, the slight smoothing introduced by the finite aperture can be ignored for large b. Thus we shall assume that we have available the sample spectrum

$$\mathscr{S}(u, v) = (\lambda^2 f^2/A)|\tilde{F}(u, v)|^2, \quad (1.4\text{-}12)$$

and the problem is to estimate the true spectrum

$$S(u, v) = \langle\mathscr{S}(u, v)\rangle$$

from $\mathscr{S}(u, v)$. We discuss the nontrivial nature of this problem below.

1.5 STABILITY AND FIDELITY

The sample spectrum given in Eq. (1.4-12) is, for any value of (u, v), a random variable. Thus $\mathscr{S}(u, v)$ is subject to fluctuations and—if used directly to estimate $S(u, v)$—can furnish values which are far from the true value of the spectrum. Figure 1.5-1 shows a typical unsmoothed spectral estimate of film grain noise. How severe are these fluctuations? For many random processes including the Gaussian random process, it can be shown that [6, 7, p. 250]

$$\text{var}[\mathscr{S}(u)]/\langle\mathscr{S}(u)\rangle^2 \simeq 1, \qquad \text{var}[\mathscr{S}(u)] \equiv \text{variance of } \mathscr{S}(u). \quad (1.5\text{-}1)$$

Equation (1.5-1) is true in two dimensions as well. This result implies that $\mathscr{S}(u, v)$ is a very unstable estimate of $S(u, v)$.

The usual procedure for stabilizing the spectrum is to convolve $\mathscr{S}(u, v)$ with a spectral window $W(u, v)$ to generate a smoother estimate of $S(u, v)$ given by

$$\mathscr{S}_W(u, v) = \int_{-\infty}^{\infty} \int_{-\infty}^{\infty} \mathscr{S}(u - \xi, v - \eta) W(\xi, \eta)\, d\xi\, d\eta. \quad (1.5\text{-}2)$$

For large b, the expected value of $\mathscr{S}_W(u, v)$ is simply

$$S_W(u, v) \equiv \langle\mathscr{S}_W(u, v)\rangle = \int_{-\infty}^{\infty} \int_{-\infty}^{\infty} S(u - \xi, v - \eta) W(\xi, \eta)\, d\xi\, d\eta. \quad (1.5\text{-}3)$$

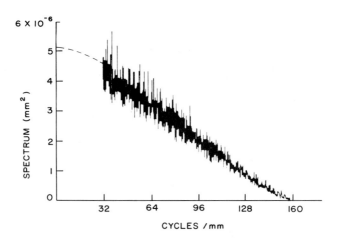

FIG. 1.5-1. Typical unsmoothed spectral estimate of film grain noise. Spectrum obtained with $30 \times 30\text{-mm}^2$ aperture.

How much has this smoothing operation reduced the instability of the spectrum? The answer is furnished by the (one-dimensional) result [7, p. 251]

$$\text{var}[\mathscr{S}_W(u)] = (I/T)S^2(u), \qquad (1.5\text{-}4)$$

where I is the energy in the window; i.e.,

$$I = \int_{-\infty}^{\infty} W^2(u)\, du, \qquad (1.5\text{-}5)$$

T is the length of the record, and $S(u)$ is the true spectrum. The result is easily extended to two dimensions by considering the area of the record instead of its duration, and two-dimensional instead of one-dimensional arguments. Equation (1.5-4) is a significant result which says that the variance of the smoothed estimator can be reduced by decreasing the energy in the window (subject to the normalization constraints that $\int_{-\infty}^{\infty} W(u)du = 1$). However, decreasing the energy in the window tends to increase its width, so that *excessive* smoothing may become a problem. Excessive smoothing affects the fidelity of the estimated spectrum and introduces a bias. The bias, defined by

$$B(u, v) = S_W(u, v) - S(u, v), \qquad (1.5\text{-}6)$$

is generally taken as a measure of fidelity: A large bias implies that the spectrum has been blurred or distorted. High stability by itself is not a suitable criterion for a spectral estimator. For example, we could smooth $\mathscr{S}(u, v)$ to such an extent that there would be no fluctuations in any particular realization nor any variability from sample to sample; i.e., $\mathscr{S}_W(u, v)$ from every sample of $t(x, y)$ would look the same. Unfortunately $\mathscr{S}_W(u, v)$ might very well bear little resemblance to $S(u, v)$. Thus very little information about the process would be conveyed.

The above discussion indicates that stability can be achieved at the price of fidelity, and vice versa. This is indeed the heart of the spectral estimation problem. To give the reader some feel for the problem consider the window

$$W(u) = c[(\sin \pi u c)/\pi u c]^2,$$

where c is a parameter. For this window the variance-to-square mean ratio I/T is $0.667c/T$. It is clear that choosing c small decreases I/T, thereby producing a stable estimate. However, since the main-lobe base width of $W(u)$ is $2/c$, the region of significant smoothing for small c is large, i.e., over many uncorrelated values of $\mathscr{S}(u)$. If there is significant variation in the character of $S(u)$ over frequency intervals less than $2/c$, significant distortion and loss of detail in $\mathscr{S}_W(u)$ will be the case.

1.6 OPTIMUM SMOOTHING OF THE SPECTRUM
WITH A FINITE-SIZE LAG WINDOW

Having demonstrated in the preceding section that smoothing must reflect a compromise between stability and fidelity, we now turn our attention to the problem of actually finding a window $W(\xi, \eta)$ that is optimum in some sense. This problem has been treated in several places under different constraints but always in one dimension. Chapter 6 in Ref. [7] furnishes an excellent review of the various techniques in use.

We note that Eq. (1.5-3)—being a convolution product—can be written as the Fourier transform of a product as

$$S_W(u, v) = \int_{-\infty}^{\infty} \int_{-\infty}^{\infty} R(\alpha, \beta) w(\alpha, \beta) e^{-j2\pi(u\alpha + v\beta)} \, d\alpha \, d\beta, \qquad (1.6-1)$$

where $R(\alpha, \beta)$ has been defined in Eq. (1.4-5), and $w(\alpha, \beta)$ is a spatial lag window and is the inverse Fourier transform of $W(\xi, \eta)$. We shall assume isotropic statistics so that the various quantities depend only on the distance from the origin of the Cartesian plane. Hence

$$R(\alpha, \beta) \equiv R(r), \qquad (1.6\text{-}2a)$$

$$w(\alpha, \beta) \equiv w(r), \qquad (1.6\text{-}2b)$$

$$W(\xi, \eta) \equiv W(\rho), \qquad (1.6\text{-}2c)$$

where $r = (\alpha^2 + \beta^2)^{1/2}$, and $\rho = (\xi^2 + \eta^2)^{1/2}$. The classic constraints on $w(r)$ are [7, p. 243]

$$w(r) = 0, \qquad r > 2c \le 2b, \qquad (1.6\text{-}3)$$

$$w(0) = 2\pi \int_0^{\infty} \rho W(\rho) d\rho = 1. \qquad (1.6\text{-}4)$$

The condition that $w(0) = 1$ is a standard normalization constraint necessary for computing a meaningful bias. The condition that $w(r) = 0, r \le 2b$ follows from the fact that the maximum correlation interval that can be observed through a diffraction-limited system of diameter $2b$ can be no larger than $2b$. However, in general the base width of $w(r)$ is limited to $2c$, where $c < b$. The choice of c exerts control over the degree of smoothing, and its choice is no trivial matter. In general, a small value of c will tend to give $W(\rho)$, a large base width that implies smoothing the spectrum over a large number of frequencies. For large b (i.e., $b \gg c$) this implies averaging over a large number of uncorrelated spectral components. Hence the variance of the estimator will tend to be small, but the bias may be large. When c is large, the base width of $W(\rho)$ will be small. In this case, significant smoothing of the spectrum is confined to a relatively small spectral region so that the bias may be small but the variance will be larger.

We shall now find the spectral window that minimizes the bias subject to a given c. We confine ourselves to the class of windows satisfying $W(\rho) \geq 0$. Only then are we assured that a small bias is indeed associated with a high resolution estimate of $S(u, v)$. To compute the bias

$$B(u, v) = S_W(u, v) - S(u, v) \qquad [\text{Eq. (1.5-6)}] \qquad (1.6\text{-}5)$$

we made the following assumptions:

(i) The first and second partial derivatives of $S(u, v)$ are continuous with respect to u and v.

(ii) $W(\xi, \eta)$ is concentrated about $\xi = \eta = 0$, with its significant values lying within a small circle about the origin.

(iii) The second partial derivatives of $S(\cdot, \cdot)$ are slowing varying throughout the region of significant values of $W(\xi, \eta)$, so that we can replace these derivatives by their values at $\xi = \eta = 0$.

We recall that the smoothed spectrum is related to the true spectrum according to

$$S_W(u, v) = \int_{-\infty}^{\infty} \int_{-\infty}^{\infty} S(u - \xi, v - \eta) W(\xi, \eta) \, d\xi \, d\eta. \qquad (1.6\text{-}6)$$

Applying Taylor's theorem with a remainder enables us to write

$$S(u - \xi, v - \eta) = S(u, v) - \xi \left.\frac{\partial S}{\partial u}\right|_p - \eta \left.\frac{\partial S}{\partial v}\right|_p + \frac{\xi^2}{2} \left.\frac{\partial^2 S}{\partial u^2}\right|_{p^*} + \frac{\eta^2}{2} \left.\frac{\partial^2 S}{\partial v^2}\right|_{p^*}, \qquad (1.6\text{-}7)$$

where the first derivatives are evaluated at $p = (u, v)$ and $p^* = (u^*, v^*)$ is a point on the line segment that joints $(u - \xi, v - \xi)$ and (u, v). The assumption of continuous derivatives is necessary to isolate the effect of $W(\xi, \eta)$. The bias is thus given for this case as

$$B(u, v) = S_W(u, v) - S(u, v)$$

$$= -\frac{\partial S(u, v)}{\partial u} \int_{-\infty}^{\infty} \int_{-\infty}^{\infty} \xi W(\xi, \eta) \, d\xi \, d\eta$$

$$-\frac{\partial S(u, v)}{\partial v} \int_{-\infty}^{\infty} \int_{-\infty}^{\infty} \eta W(\xi, \eta) \, d\xi \, d\eta$$

$$+\frac{1}{2} \int_{-\infty}^{\infty} \int_{-\infty}^{\infty} \left.\frac{\partial^2 S}{\partial u^2}\right|_{p^*} \xi^2 W(\xi, \eta) \, d\xi \, d\eta$$

$$+\frac{1}{2} \int_{-\infty}^{\infty} \int_{-\infty}^{\infty} \left.\frac{\partial^2 S}{\partial v^2}\right|_{p^*} \eta^2 W(\xi, \eta) \, d\xi \, d\eta. \qquad (1.6\text{-}8)$$

For the case of circular symmetry, it is not difficult to show, with $\xi = \rho \cos \theta$, $\eta = \rho \sin \theta$, and $d\xi \, d\eta \to \rho \, d\rho \, d\theta$, that the first two terms on the right side of Eq. (1.6-8) are zero. The bias is thus given for this case by

$$B(u, v) = \frac{1}{2} \int_{-\infty}^{\infty} \int_{-\infty}^{\infty} \left. \frac{\partial^2 S}{\partial u^2} \right|_{p^*} \xi^2 W(\xi, \eta) \, d\xi \, d\eta$$

$$+ \frac{1}{2} \int_{-\infty}^{\infty} \int_{\infty}^{\infty} \left. \frac{\partial^2 S}{\partial v^2} \right|_{p^*} \eta^2 W(\xi, \eta) \, d\xi \, d\eta, \qquad (1.6\text{-}9)$$

where $p^* = (u^*, v^*) = (u - \sigma\xi, v - \sigma\eta)$ and $|\sigma| < 1$. For high-resolution estimates we expect $W(\xi, \eta)$ to be concentrated about $\xi = \eta = 0$, with its significant values lying within a small circle about the origin. If the second partial derivatives of $S(\cdot, \cdot)$ are assumed slowly varying through this region, we can replace the second partial derivatives by their value at $\xi = \eta = 0$. The result of this approximation is to decouple the second partial derivatives from the ξ, η coordinates and thus separate the effect of the window alone on the bias. Thus, with this approximation,

$$B(u, v) = \frac{1}{2} \frac{\partial^2 S(u, v)}{\partial u^2} \int_{-\infty}^{\infty} \int_{-\infty}^{\infty} \xi 2 W(\xi, \eta) \, d\xi \, d\eta$$

$$+ \frac{1}{2} \frac{\partial^2 S(u, v)}{\partial v^2} \int_{-\infty}^{\infty} \int_{-\infty}^{\infty} \eta^2 W(\xi, \eta) \, d\xi \, d\eta, \qquad (1.6\text{-}10)$$

$$B(u, v) = \pi \frac{1}{\omega} \frac{\partial}{\partial \omega} \left[\omega \frac{\partial S(\omega)}{\partial \omega} \right] \int_0^{\infty} \rho^3 W(\rho) \, d\rho. \qquad (1.6\text{-}11)$$

Equation (1.6-11) represents the bias in polar coordinates with

$$\omega = \sqrt{u^2 + v^2}, \qquad \rho = \sqrt{\xi^2 + \eta^2},$$

$$S(u, v) = S(\sqrt{u^2 + v^2}) = S(\omega), \qquad W(\xi, \eta) = W(\sqrt{\xi^2 + \eta^2}) = W(\rho)$$

and

$$\left(\frac{\partial^2}{\partial u^2} + \frac{\partial^2}{\partial v^2} \right) S(u, v) \to \frac{1}{\omega} \frac{\partial}{\partial \omega} \left[\omega \frac{\partial S(\omega)}{\partial \omega} \right].$$

Hence the bias will be minimized when the integral

$$\int_0^{\infty} \rho^3 W(\rho) \, d\rho \qquad (1.6\text{-}12)$$

is minimized subject to the constraint [Eq. (1.6-4)]

$$2\pi \int_0^{\infty} \rho W(\rho) \, d\rho = 1. \qquad (1.6\text{-}13)$$

From Eq. (1.6-12) we can see that only when $W(\rho) \geq 0$ are we assured that a small bias ensures that $W(\rho)$ will be concentrated near the origin.

A. The Minimum-Bias Window Pair

The solution to the bias minimization theorem is obtained with the help of the Féjer–Riesz theorem [8, p. 152]. We note that the one-dimensioned form of this problem has been considered in the literature by Papoulis [9] and that the solution to the two-dimensioned problem has been treated by Stark and Dimitriadis [10]. Hence we shall give the solution here and refer the reader to the listed references for details.

The optimum spectral window that minimizes the bias is given by

$$W_0(\rho) = 4\pi c^2 \zeta_1^2 \{J_0^2(2\pi\rho c)/[\zeta_1^2 - (2\pi\rho c)^2]^2\}, \tag{1.6-14}$$

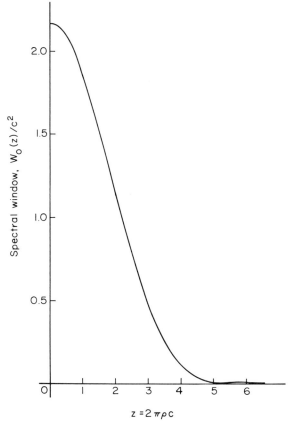

FIG. 1.6-1. The minimum-bias spectral window $W_0(\rho)$ of Eq. (1.6-14). (From Stark and Dimitriadis [10].)

where $J_0(\cdot)$ is the Bessel function of the first kind of order zero, and ζ_1 is the first zero of $J_0(x)$. The function $W_0(\cdot)$ is shown in Fig. 1.6-1. It is proportional to the square of the spread function associated with the optimum apodizer for imaging smooth objects [11, Eq. (39)]. The apodizing properties of this function are discussed in the literature [12]. The window is strongly concentrated about the origin and its envelope for large values of ρ goes as ρ^{-5}. The inverse Hankel transform of $W_0(\rho)$ gives the optimum data window $w_0(r)$. Thus

$$w_0(r) = 2\zeta_1^2 \int_0^\infty \frac{xJ_0^2(x)J_0(xr/c)}{[\zeta_1^2 - x^2]^2} dx, \tag{1.6-15}$$

and

$$w_0(0) = 2\zeta_1^2 \int_0^\infty \frac{xJ_0^2(x)}{[\zeta_1^2 - x^2]^2} dx = 1. \tag{1.6-16}$$

A numerical evaluation of Eq. (1.6-15) gives $w_0(r)$, which is shown in Fig. 1.6-2.

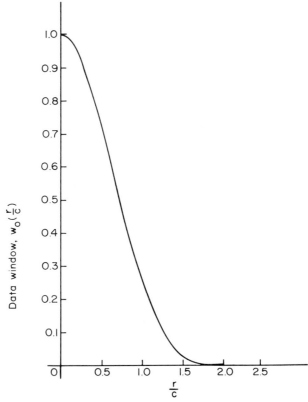

FIG. 1.6-2. The optimum data window $w_0(r)$ of Eq. (1.6-16). (From Stark and Dimitriadis [10].)

B. Comparison with Other Windows for Which $w(r)$ Has Finite Support

Because the variance is proportional to the energy I, given by

$$I = 2\pi \int_0^\infty \rho W^2(\rho) \, d\rho \qquad (1.6\text{-}17)$$

and the bias is proportional to the coefficient D given by

$$D = 2\pi \int_0^\infty \rho^3 W(\rho) \, d\rho, \qquad (1.6\text{-}18)$$

we compute these values for the optimum window $W_0(\rho)$ and compare them with other standard windows.

(i)   Optimum window [Eq. (1.6-14)]:

$$D_0 = 0.144/c^2, \qquad I_0 = 1.148c^2.$$

(ii)   Uniform pupil:

$$W(\rho) = J_1^2(2\pi\rho c)/\pi\rho^2, \qquad D = \frac{1}{2\pi^2 c^2} \int_0^\infty x J_1^2(x) \, dx = \infty, \qquad I = 1.44c^2.$$

We see that for the uniform pupil $I > I_0$. This window, being the Hankel transform of the self-correlation function of a uniform pupil function, is the two-dimensional analog of the Bartlett window. For the latter, the second moment is not a suitable criterion for estimation of the bias, and the bias cannot be simply related to the spectrum $S(\omega)$ [9]. The same considerations apply here as well.

(iii)   Uniform lag window:

$$w(r) = \operatorname{circ}(r/2c), \qquad W(\rho) = (2c/\rho)J_1(4\pi\rho c),$$

$$D = \frac{1}{16\pi^2 c^2} \int_0^\infty x^2 J_1(x) \, dx = \infty, \qquad I = 4\pi c^2.$$

Clearly $I > I_0$. This window is not in the class of nonnegative smoothing windows; it has negative side lobes, and its third moment does not exist. The asymptotic attenuation for large ρ goes as $\rho^{-3/2}$, whereas ours goes as ρ^{-5}.

(iv)   Inverse cosine data window:

$$w(r) = \frac{2}{\pi} \cos^{-1}\left(\frac{r}{2c}\right) \operatorname{circ}\left(\frac{r}{2c}\right), \qquad W(\rho) = 2c \frac{J_0(2\pi\rho c)J_1(2\pi\rho c)}{\rho},$$

$$D = \frac{1}{2\pi^2 c^2} \int_0^\infty x^2 J_0(x)J_1(x) \, dx = \infty, \qquad I = 3.74c^2.$$

As in the previous cases $I > I_0$. The asymptotic attenuation for large ρ goes as ρ^{-2}. Also, as in the previous case, the third moment of $W(\rho)$ does not exist and is not a suitable measure of the bias.

(v)   Modified Parzen window for the circularly symmetric case:

$$W(\rho) = \frac{12}{c^2\pi(3\pi^2 - 16)} \frac{J_1^4(\pi\rho c)}{\rho^4}, \qquad D = \frac{0.179}{c^2},$$

$$I = 0.891c^2 \qquad \text{(by numerical integration)}.$$

We note that $W(\rho)$ is proportional to the fourth power of the Hankel transform of a uniform, circular pupil. The one-dimensional Parzen window is proportional to the fourth power of the Fourier transform of a uniform, truncated pulse. Hence this window can be regarded as the two-dimensional analog of the Parzen window. Its asymptotic attenuation goes as ρ^{-6}. Although the bias is nearly 25% greater than D_0, its variance is less by almost the same amount. When the window size c is readjusted to furnish a variance equal to ours, the bias D is about the same as D_0. These results stand in contrast to the one-dimensional case in which the Parzen window furnishes a 20% greater bias than the optimum window for an equal variance.

C. Smoothing in the Frequency Plane

The smoothing operation given in Eq. (1.5-2) can be attempted with a digital computer or by direct filtering in the focal plane of the Fourier lens.

If a computer is used, the sample spectrum must be converted to an electrical signal, which requires scanning (e.g., a TV camera), A/D conversion, and a numerical evaluation of Eq. (1.5-2) using the window of Eq. (1.6-14). Of course, once the data are in the computer, we have considerable flexibility with respect to realizing a large number of processing operations.

To emphasize that the smoothing operations are being performed in the focal plane, we shall use x_f and y_f to denote Cartesian displacements coordinates in the focal plane. Also, we eliminate the $\sim$ symbol in the back focal plane field $\tilde{F}(u, v)$ [e.g., Eq. (1.4-2)] when the latter is expressed in terms of x_f, y_f; i.e., $\tilde{F}(u, v) = \tilde{F}(x_f/\lambda f, y_f/\lambda f) \equiv F(x_f, y_f)$.

If smoothing is done directly in the focal plane of the lens, a shaded scanning aperture must be used to realize the smoothing associated with the optimum window. Let $g(x_f', y_f')$ denote the shading function (with an adjustable gain parameter included) associated with the scanning aperture, defined with respect to an origin at the center of the aperture. The field behind the aperture is then $g(x_f' - x_f, y_f' - y_f)F(x_f', y_f')$, and the differential output due to the illumination falling on a differential area $d\sigma = dx_f'\, dy_f'$ of the photomultiplier tube is then proportional to

$$dI = |g(x_f' - x_f, y_f' - y_f)F(x_f', y_f')|^2\, dx_f'\, dy_f'.$$

The total output is proportional to the smoothed spectral estimator and is

$$I(x_f, y_f) = \int_{-\infty}^{\infty} \int_{-\infty}^{\infty} |F(x_f', y_f')|^2 |g(x_f' - x_f, y_f' - y_f)|^2 \, dx_f' \, dy_f'. \quad (1.6\text{-}19)$$

The expected value of the smoothed irradiance is

$$\langle I(x_f, y_f) \rangle = A \int_{-\infty}^{\infty} \int_{-\infty}^{\infty} R(\alpha, \beta) Q(\alpha, \beta)$$

$$\times \exp\left[-j2\pi \left(\alpha \frac{x_f}{\lambda f} + \beta \frac{y_f}{\lambda f} \right) \right] d\alpha \, d\beta, \quad (1.6\text{-}20)$$

where

$$Q(\alpha, \beta) \equiv \int_{-\infty}^{\infty} \int_{-\infty}^{\infty} |g(\xi \lambda f, \eta \lambda f)|^2 \exp[-j2\pi(\xi \alpha + \eta \beta)] \, d\xi \, d\eta. \quad (1.6\text{-}21)$$

The smoothed sample spectrum is obtained by normalizing Eq. (1.6-20) and letting $u = x_f/\lambda f$, $v = y_f/\lambda f$. Its expected value is

$$S_W(u, v) = k \int_{-\infty}^{\infty} \int_{-\infty}^{\infty} R(\alpha, \beta) Q(\alpha, \beta) \exp[-j2\pi(u\alpha + v\beta)] \, d\alpha \, d\beta, \quad (1.6\text{-}22)$$

where k is a normalizing constant. To minimize the bias, we require that $Q(\alpha, \beta)$ be a function of circular symmetry $Q(r)$ so that in the optimum case

$$kQ_0(r) = w_0(r). \quad (1.6\text{-}23)$$

Because $w_0(0) = 1$, we require that $k = [Q_0(0)]^{-1}$. If we take the Hankel transform of both sides of Eq. (1.6-23), we obtain

$$|g_0(\lambda f \rho)|^2 = (1/k) W_0(\rho)$$

and

$$|g_0(\lambda f \rho)| = (1/\sqrt{k}) |Y_0(\rho)|, \quad (1.6\text{-}24)$$

where $|Y_0(\rho)| \equiv [W_0(\rho)]^{1/2}$. Thus the shading function is established by Eq. (1.6-24). It is clearly unrealizable, because it requires an infinite pupil. However, $W_0(\rho)$ is nearly zero for values of $2\pi\rho c > 6$, so that we can approximate $|g_0(\lambda f \rho)|$ with a finite pupil function in the frequency plane.

1.7 SMOOTHING WITH A FINITE-BANDWIDTH FREQUENCY WINDOW

In the previous section we found that the optimum amplitude frequency plane function [Eq. (1.6-24)] could only be approximated in practice if we wanted to smooth the sample spectrum directly in the back focal plane of the

Fourier transform lens. The approximation stems from the constraint that $w(r) = 0$ for $r > 2c$, which forces the spectral window to have infinite bandwidth. This constraint, however, is standard in ordinary one-dimensioned waveform processing, since smoothing of the spectrum is generally realized by suitable windowing of the data or its sample autocorrelation function [7, p. 259]. However, in a coherent-optical spectrum analyzer, the sample spectrum is physically available for processing in the BFP of the Fourier transform lens, and it is both practical and desirable to smooth the sample spectrum directly in the frequency plane. For this reason we shall revisit the problem considered in Section 1.6 but apply instead the constraint of finite bandwidth to our spectral window. We shall seek to produce a best estimate, in the sense of fidelity (i.e., minimizing the bias), while holding the stability of the estimator within the bounds specified by the user. This approach has worked well in applications in optical signal processing and pattern recognition (see, for example, Refs. [13–15]) and gives the user precise control over the bandwidth of the window. Reducing the bandwidth of the window ("window closing") enables the user to look for finer detail in the spectrum. Of course, narrowing the window eventually leads to instability, and the question of when to stop the process of narrowing the width of $W(\rho)$ is not governed by rigid rules. However, the user should always recall that the spectrum has to make sense physically, otherwise the results can be of little value.

Smoothing in the frequency plane can be done by analog or digital techniques. Analog smoothing with a uniform window is demonstrated in Figs. 1.7-1 and 1.7-2. Figure 1.7-1 is an estimate of the spectrum of film grain

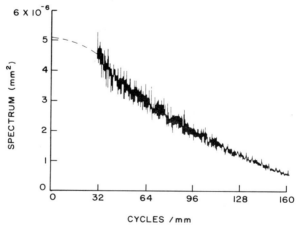

FIG. 1.7-1. Inadequate smoothing of the spectral estimate with a uniform frequency plane window that is too narrow. The window diameter here is 50 μm.

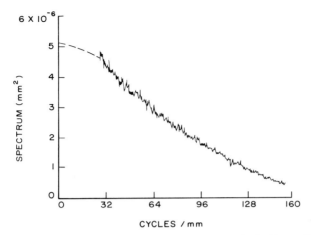

FIG. 1.7-2. The same spectral estimate as in Fig. 1.7-1 smoothed with a uniform 120-μm-diameter window. A smoother spectrum is observed, but the underlying character of the spectrum has not been significantly changed.

noise smoothed with a 50-μm-diameter uniform window. The smoothing is seen to be inadequate, and the smoothed spectrum is seen to be unstable with many noiselike peaks that do not reflect the inherent character of the spectrum. Figure 1.7-2 shows the same sample spectrum smoothed with a 120-μm-diameter uniform window. The spectrum is seen to be smoother, with fewer false peaks. Additional smoothing with still broader windows will

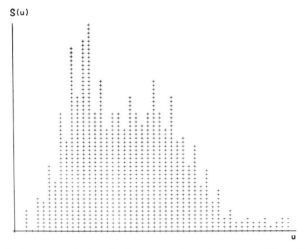

FIG. 1.7-3. An unsmoothed, optically generated, spectrum scanned by a TV camera and digitized. (From Stark and Shao [22].)

$S_w(u)$

u

FIG. 1.7-4. The same spectrum as in Fig. 1.7-4 but smoothed digitally by the optimum filter discussed in Section 1.8. (From Stark and Shao [22].)

eventually increase the bias beyond tolerable bounds and distort the true character of the spectrum.

Smoothing by digital techniques requires additional hardware such as a TV camera, a sampler and A/D converter, and a digital computer. An example of digital smoothing is shown in Figs. 1.7-3 and 1.7-4. The original irradiance spectrum has been digitized and smoothed by digital filtering. Figure 1.7-3 shows the unsmoothed digitized irradiance spectrum of a realization of the overlapping circular grain model of the film grain noise. As before, many noiselike peaks are evident. Figure 1.7-4 shows the smoothed version; smoothing here was done with a nonuniform window (parabolic weighting) of finite bandwidth. This window has optimal properties, and we shall discuss it further in this section. However, before discussing the window optimization problem with a finite-bandwidth constraint, we first consider smoothing by a uniform window. A uniform window is easy to implement and has, as we shall see, certain additional desirable properties.

A. Smoothing with a Uniform Spectral Window of Fixed Bandwidth c

We illustrate how a spectrum-smoothing window reduces the speckle in the sample spectrum by considering a one-dimensional example of an L-truncated random process that consists of N pulses with the center of each uniformly distributed in the interval $[-L/2, L/2]$. We let $y(x)$ denote the pulse shape and let x_i denote its location. The truncated process is described by

$$r(x) = y(x) * \sum_{i=1}^{N} \delta(x - x_i), \qquad (1.7-1)$$

where $\delta(x)$ is the Dirac function and $*$ denotes convolution. We denote the sample spectrum, in this case, by

$$\mathscr{S}(u) \equiv |R(u)|^2/N = |Y(u)|^2[1 + n(u)], \qquad (1.7\text{-}2)$$

where $R(u)$ and $Y(u)$ are the Fourier transforms of $r(x)$ and $y(x)$, respectively. The noise $n(u)$, defined by

$$n(u) \equiv \frac{1}{N} \sum_{\substack{i \neq k}}^{N} \sum^{N} e^{-j2\pi u(x_i - x_k)}, \qquad (1.7\text{-}3)$$

is the cross-term speckle noise; its appearance in Eq. (1.7-2) tends to corrupt our estimate of the true spectrum $S(u) \equiv |Y(u)|^2$ from $\mathscr{S}(u)$. Because x_i and x_k are independent, uniformly distributed random variables, the probability–density function (pdf) of $Z \equiv x_i - x_k$ is

$$f_Z(z) = (1/L)(1 - (|z|/L) \operatorname{rect}(z/2L), \qquad (1.7\text{-}4)$$

and the expected value of $e^{-j2\pi uZ}$ is

$$\langle e^{-j2\pi uZ} \rangle = \operatorname{sinc}^2 uL. \qquad (1.7\text{-}5)$$

From Eqs. (1.7-3) and (1.7-5), we find that

$$\langle n(u) \rangle = N^{-1}(N - 1)N \operatorname{sinc}^2 uL \simeq N \operatorname{sinc}^2 uL$$

for N large. Hence the expected value of $\mathscr{S}(u)$ is given by

$$\langle \mathscr{S}(u) \rangle = S(u)[1 + N \operatorname{sinc}^2 uL].$$

If μ denotes the average number of pulses per unit distance of the truncated process, then

$$N \simeq \mu L, \qquad (1.7\text{-}6)$$

and

$$\langle \mathscr{S}(u) \rangle = S(u)[1 + \mu L \operatorname{sinc}^2 uL] \simeq S(u) \qquad (1.7\text{-}7)$$

for L large and u not near the origin. The variance of $\mathscr{S}(u)$ is

$$\operatorname{var}[\mathscr{S}(u)] = |S(u)|^2 \langle n^2(u) \rangle,$$

where $\langle n^2(u) \rangle$ is given by

$$\begin{aligned}
\langle n^2(u) \rangle = (1/N^2)[&N(N - 1)(N - 2)(N - 3) \operatorname{sinc}^4 uL \\
&+ 2N(N - 1)(N - 2) \operatorname{sinc}^2 uL \\
&+ 2N(N - 1)(N - 2) \operatorname{sinc}^2 uL \operatorname{sinc}^2 2uL \\
&+ N(N - 1) \operatorname{sinc}^2 2uL + N(N - 1)].
\end{aligned} \qquad (1.7\text{-}8)$$

If we substitute Eq. (1.7-6) in Eq. (1.7-8) and let L be large, we obtain for $u \gg L^{-1}$,

$$\langle n^2(u) \rangle \simeq 1,$$

so that

$$\mathrm{var}[\mathscr{S}(u)] / \langle \mathscr{S}(u) \rangle^2 \simeq 1 \qquad (1.7\text{-}9)$$

for L large. Hence we see that, without smoothing, the noise power is as much as the signal power. This is the same result we observed in Eq. (1.5-1). This means that, if we wish to extract a stable estimate of $S(u)$ from $\mathscr{S}(u)$, we must perform a smoothing operation on $\mathscr{S}(u)$. As before our smoothing operation takes the form

$$\mathscr{S}_W(u) = \int_{-\infty}^{\infty} \mathscr{S}(u') W(u - u') \, du', \qquad (1.7\text{-}10)$$

where

$$\mathscr{S}(u) = S(u)[1 + n(u)]. \qquad (1.7\text{-}11)$$

We impose the following conditions on $W(u)$:

$$W(u) = 0, \qquad |u| > c/2, \qquad (1.7\text{-}12\text{a})$$

$$\int_{-\infty}^{\infty} |W(u)|^2 \, du = K, \qquad (1.7\text{-}12\text{b})$$

$$\int_{-\infty}^{\infty} W(u) \, du = 1. \qquad (1.7\text{-}12\text{c})$$

Condition (1.7-12a) reflects our insistence that our filter has finite bandwidth c. Condition (1.7-12b) is a constraint on the energy in the window and also amounts to constraining the variance of the smoothed sample spectrum. Condition (1.7-12c) is a standard normalization constraint that enables us to compare, in a meaningful way, the average values of the smoothed and unsmoothed spectra. For this illustration, we choose the uniform window, for which

$$W(u) = \begin{cases} 1/c, & |u| < c/2, \\ 0, & \text{otherwise.} \end{cases} \qquad (1.7\text{-}13)$$

Condition (1.7-12b) requires that $K = c^{-1}$. The smoothed sample spectrum is given by

$$\mathscr{S}_W(u) = \int_{-\infty}^{\infty} S(u') W(u - u') \, du' + \int_{-\infty}^{\infty} S(u') n(u') W(u - u') \, du'. \qquad (1.7\text{-}14)$$

We ignore the effect of $W(\cdot)$ on $S(\cdot)$ by assuming that $S(u')$ changes insignificantly over the interval $|u - u'| < c/2$. In this case, we replace $S(u')$ inside the integrals by its value at $u' = u$ and take it outside the integral. In this way, we obtain

$$\mathscr{S}_W(u) = S(u)\left[1 + \int_{-\infty}^{\infty} n(u')W(u - u')\,du'\right]. \qquad (1.7\text{-}15)$$

The mean value of $\mathscr{S}_W(u)$ is

$$\langle \mathscr{S}_W(u)\rangle = S(u)\left[1 + \int_{-\infty}^{\infty} \langle n(u')\rangle W(u - u')\,du'\right] \equiv S_W(u)$$

$$\simeq S(u) \qquad \text{for} \quad L \text{ large and } u \gg 1/L. \qquad (1.7\text{-}16)$$

The variance of $\mathscr{S}_W(u)$ is

$$\text{var}[\mathscr{S}_W(u)]$$

$$\simeq |S(u)|^2 \left\langle \left|1 + \int_{-\infty}^{\infty} n(u')W(u - u')\,du'\right|^2\right\rangle - |S_W(u)|^2$$

$$= |S(u)|^2 \int_{-\infty}^{\infty}\int_{-\infty}^{\infty} \langle n(u')n^*(u'')\rangle W(u - u')W(u - u'')\,du'\,du''$$

$$\simeq |S(u)|^2 \left\{\int_{-\infty}^{\infty}\int_{-\infty}^{\infty} \text{sinc}^2[(u' - u'')L]W(u - u')W(u - u'')\,du'\,du''\right.$$

$$\left. + \int_{-\infty}^{\infty}\int_{-\infty}^{\infty} \text{sinc}^2[(u' + u'')L]W(u - u')W(u - u'')\,du'\,du''\right\}. \qquad (1.7\text{-}17)$$

For values of $|u| \gg (2L)^{-1} + (2K)^{-1}$, i.e., for values of u not too near the origin, the second integral can be omitted in relation to the first. In that case

$$\text{var}[\mathscr{S}_W(u)] = |S(u)|^2 K \int_{-1/K}^{1/K} (1 - K|x|)\,\text{sinc}^2\,xL\,dx \simeq \frac{K}{L}|S(u)|^2. \qquad (1.7\text{-}18)$$

Because $S(u) \simeq \langle \mathscr{S}_W(u)\rangle$, we obtain

$$\text{var}[\mathscr{S}_W(u)]/\langle \mathscr{S}_W(u)\rangle^2 \simeq \text{var}[\mathscr{S}_W(u)]/|S(u)|^2 \simeq K/L, \qquad (1.7\text{-}19)$$

which is to be contrasted with Eq. (1.7-9), which holds for the unsmoothed case. The smoothing filter has reduced the variance of the measurement by the factor K/L. As stated earlier, a similar result is given in the literature for the Gaussian process.

In this simple calculation we ignored the effect of $W(u)$ on $S(u)$ for the sake of convenience. In general, however, the effect of $W(u)$ on $S(u)$ leads to a bias in our estimate of the spectrum. In any real situation, the bias

$B \equiv \langle \mathscr{S}_w(u) \rangle - S(u)$ cannot be ignored because it represents the fidelity with which our estimate of $S(u)$ has been extracted from the data. The smoothing action of $W(u)$ on $n(u)$ tends to increase the stability of our results, but high stability coupled with low fidelity is generally undesirable. As always, we would like high fidelity coupled with high stability, i.e., $\langle \mathscr{S}_w(u) \rangle \simeq S(u)$ and no noise. Unfortunately, as we have already seen, these two requirements generally oppose each other, so—as in Section 1.6—we must seek a compromise. We furnish next the band-limited window $W(u)$ that furnishes the best compromise, in the sense that the bias is minimized for a given variance constraint K.

1.8 AN OPTIMUM FINITE-BANDWIDTH FREQUENCY WINDOW

As in Section 1.6, we consider the case of rotational symmetry and assume once more that the bias is given by Eq. (1.6-11), i.e.,

$$B(u, v) = \pi \frac{1}{\omega} \frac{\partial}{\partial \omega} \left[\omega \frac{\partial S(\omega)}{\omega} \right] \int_0^\infty \rho^3 W(\rho) \, d\rho. \tag{1.8-1}$$

With the imposition of the constraint of finite bandwidth, the optimum window function $W(\rho)$ is then the one that minimizes

$$\int_0^\infty \rho^3 W(\rho) \, d\rho \tag{1.8-2}$$

subject to

$$W(\rho) = 0, \quad \rho \geq c \quad \text{(bandwidth constraint)}, \tag{1.8-3a}$$

$$\int_0^\infty \rho W^2(\rho) \, d\rho = K \quad \text{(variance constraint)}, \tag{1.8-3b}$$

$$2\pi \int \rho W(\rho) \, d\rho = 1 \quad \text{(normalization constraint)}. \tag{1.8-3c}$$

The solution to Eq. (1.8-2) subject to the constraints of Eqs. (1.8-3) is obtained with the aid of Lagrange multipliers. The details of the calculation are given in Ref. [13] and will be omitted here. The optimum window is given by

$$W_0(\rho) = \begin{cases} 3K\pi(1 - \frac{3}{2}\pi^2 K\rho^2), & \rho < c_0, \\ 0, & \rho \geq c_0, \end{cases} \tag{1.8-4}$$

where c_0 is the peak-to-null bandwidth of $W(\rho)$ and is related to K according to

$$c_0 = (1/\pi)(2/3K)^{1/2}. \tag{1.8-5}$$

For other windows, the relation between c and K is different from Eq. (1.8-5). A small value of c implies high resolution. If a particular window furnishes a small bias, together with a small bandwidth for a given K, that window can be regarded as an attractive option. We shall see that the optimum window $W_0(\rho)$, when contrasted with other windows, not only furnishes the least bias but also furnishes high resolution (i.e., a small c for a given relative variance K).

Once K is fixed, the parameters that are important for evaluating a particular window are c and the relative bias D, given by

$$D = 2\pi \int_0^\infty \rho^3 W(\rho)\,d\rho. \tag{1.8-6}$$

A. Comparison with Other Windows of Finite Bandwidth

The relative bias D, given in Eq. (1.8-6), and the window size c, which determines the resolution of the smoothed spectrum, are reasonable figures of merit for evaluating the performance of a window under the constraint of fixed relative variance K.

(i)   Optimum window:

$$W_0(\rho) = 3K\pi(1 - \tfrac{3}{2}\pi^2 K\rho^2)\,\mathrm{circ}\!\left(\frac{\rho}{c}\right), \qquad c = \frac{0.26}{\sqrt{K}}, \qquad D = \frac{0.22}{K\pi^2}.$$

(ii)   Pulse window:

$$W(\rho) = 2K\pi\,\mathrm{circ}(\rho/c), \qquad c = \frac{1}{\pi}\sqrt{\frac{1}{2K}} \simeq \frac{0.23}{\sqrt{K}}, \qquad D \simeq \frac{0.25}{K\pi^2}.$$

(iii)   Cone window:

$$W(\rho) = \begin{cases} 4\pi K(1 - 2\pi\sqrt{\tfrac{1}{3}K}\rho), & \rho < c, \\ 0, & \text{otherwise}, \end{cases}$$

$$c = (1/2\pi)\sqrt{3/K} \simeq 0.28/\sqrt{K}, \qquad D \simeq 0.23/K\pi^2.$$

(iv)   Cosine window:

$$W(\rho) = \begin{cases} \dfrac{16K\pi}{\pi + 2}\cos\!\left(4\pi\sqrt{\dfrac{K(\pi - 2)}{\pi + 2}}\,\rho\right), & \rho \le c, \\ 0, & \text{otherwise}, \end{cases}$$

$$c = \frac{1}{8}\sqrt{\frac{\pi + 2}{K(\pi - 2)}} \simeq \frac{0.27}{\sqrt{K}}, \qquad D \simeq \frac{0.22 + \epsilon}{K\pi^2} > D_0.$$

(v)   Sine window:

$$W(\rho) = \begin{cases} \dfrac{4\pi(\pi^2 - 8)}{3\pi^2 - 28} K\left(1 - \sin\left\{\pi(\pi^2 - 8)\sqrt{\dfrac{K}{3\pi^2 - 28}}\,\rho\right\}\right), & \rho < c, \\ 0, & \text{otherwise,} \end{cases}$$

$$c = \frac{1}{2(\pi^2 - 8)}\sqrt{\frac{3\pi^2 - 28}{K}} \simeq \frac{0.34}{\sqrt{K}}, \qquad D \simeq \frac{0.24}{K\pi^2}.$$

The window functions are shown in Fig. 1.8-1. The cosine window is closest to ours in both bias and resolution. With the exception of the uniform window, the other windows have a lower resolution than ours.

In contrast to the problem of optimizing spectral windows, subject to the constraint of a finite lag window size, the differences in performance between these windows are not dramatic. The reason for this is that, in the former case, the side lobe behaviors of the smoothing functions radically influence the size of the bias. In our case, however, the windows have no side lobes and the form of the weighting function over the finite interval ($\rho \leq c$) is not crucial.

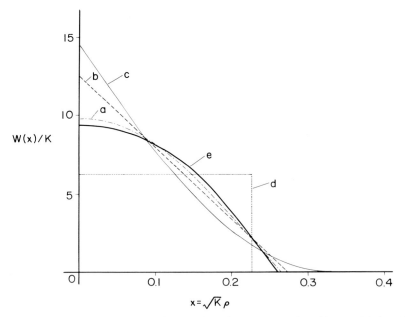

FIG. 1.8-1. Comparison of several smoothing windows: (a) cosine, (b) cone, (c) sine, (d) uniform, (e) optimum. (From Stark et al. [13.])

This is in accordance with what is reported in the literature, namely, that window carpentry, i.e., the form of the weighting function, is much less important than window closing, i.e., the basewidth or support of the window [7, p. 282].

1.9 ESTIMATING THE SPECTRUM AT LOW SPATIAL FREQUENCIES AND AT THE ORIGIN

In Section 1-4, we showed that, when the input sample $t(x, y)$ has an average value t_0, the observed spectrum has a sharp peak at the origin which represents the aperture-diffracted light (sometimes called the zero-order light) and tends to mask the spatial frequency components of the spectrum at and near the origin. Reduction of the zero-order light outside the main lobe can be accomplished, in principle at least, in two ways. The most direct way is to decrease the dimensions of the aperture diffraction pattern by increasing the transverse dimensions of the aperture. However, in practice, the aperture is usually already as large as it can be, being limited by the size of the lens, the uniformity of the collimated beam, and the off-axis capability of the optical system. The second approach is to use appropriate pupil functions that suppress the zero-order light by apodization.

The pupil function can be generated by an aperture with uniform (unity) transmission, together with a suitable contour, or by an absorbing filter of appropriate transmission with a "normal" contour. The latter technique offers the advantage of not confining the desired effect to any one or, at most, a few particular directions in the frequency plane as is characteristic of apodizing by contour shaping. However, the construction of accurate absorbing filters is in itself a difficult task, since great care must be taken to avoid filter-generated noise. The other possibility—apodization by aperture shaping—avoids this problem, providing the contour is reasonably constructed, as is not difficult in practice.

When contoured apertures are used, the power spectrum should be measured along a direction of apodization. Typical aperture contours that apodize effectively along selected directions are the diamond and Gaussian contours. If the random process has isotropic statistics, the measurement of the spectrum in any direction is sufficient to obtain second-moment information about the process; if the process is not isotropic, rotating the sample with respect to the aperture will effectively allow the measurement of the power spectrum along different directions. However, such contoured apertures are not adequate for measuring the power spectrum at zero spatial frequency because their diffraction patterns do not have nulls at the origin. Perhaps the simplest aperture function which has the property of furnishing

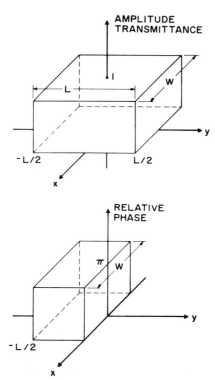

FIG. 1.9-1. The π-phase step: amplitude and phase characteristics. (From Stark *et al.* [15].)

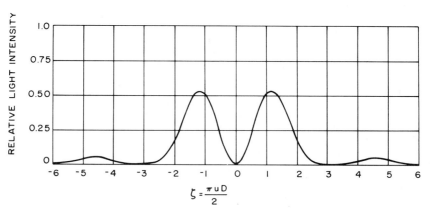

FIG. 1.9-2. Diffraction pattern of a π-phase step. In practice, a π-phase step can be made from two flat glass plates of equal width. The 180-deg phase shift is obtained by rotating one plate with respect to the other, thereby controlling the optical path difference between the plates. Here D is the aperture width, u the spatial frequency, and ζ a dimensionless variable. (From Stark *et al.* [15].)

a null at dc is the π-phase step shown in Fig. 1.9-1. The π-phase step is a one-cycle phase grating that produces a relative diffraction intensity given by

$$H_\pi(u, v) = LW \operatorname{sinc}^2 \tfrac{1}{2}uL \operatorname{sinc}^2 vW \sin^2 \tfrac{1}{2}\pi uL, \qquad (1.9\text{-}1)$$

where L and W are the dimensions of the step. The side lobe response of a π-phase step is shown in Fig. 1.9-2. When the π-phase step is used in spectral density measurements, there is a loss of resolution in the measurement of the spectrum along the frequency axis parallel to the direction of the phase grating. However, the loss of resolution will generally not be serious enough to affect the fidelity of the spectrum, provided that the dimensions of the aperture are much larger than the largest significant correlation interval of interest. When viewed in the frequency domain, the loss of resolution in the expected value of the spectrum is the result of smoothing the true spectrum $S(u, v)$ with a spectral window function such as that given in Eq. (1.9-1). Thus

$$S_\pi(u, v) = \int_{-\infty}^{\infty} \int_{-\infty}^{\infty} S(\xi, \eta) H_\pi(u - \xi, v - \eta) \, d\xi \, d\eta. \qquad (1.9\text{-}2)$$

The weight given to $S(\xi, \eta)$ at $\xi = u$ is zero, which is quite the opposite of what a typical aperture—which would normally give a maximum weight at $\xi = u$—would do.

The significant averaging over frequency extends over the two lobes on either side of $\xi = u$, which furnish a null-to-null width in frequency of $4L^{-1}$. For an ordinary rectangular aperture of the same width, the significant

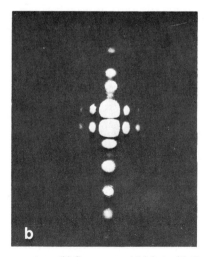

FIG. 1.9-3. (a) Diffraction pattern of a square aperture. (b) Same as part (a) but with the π-phase step in place. (From Stark *et al.* [15].)

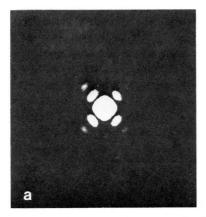 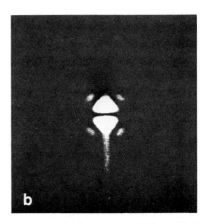

FIG. 1.9-4. (a) Diffraction pattern of a diamond aperture. (b) Same as part (a) but with the π-phase step in place. (From Stark *et al.* [15].)

averaging over frequency is restricted to the main lobe of the diffraction pattern, which has a null-to-null width of $2L^{-1}$.

Figures 1.9-3a and 1.9-4a show the diffraction patterns for square and diamond apertures, respectively; Figs. 1.9-3b and 1.9-4b show the nulls generated at the origin when a π-phase step is used with the same apertures. A π-phase step can be built from two equally thick glass plates. Reference 15 discusses a particular design.

1.10 EXTRAPOLATING THE FOURIER SPECTRUM OF SPATIALLY BOUNDED OBJECTS

A significant and historic problem in optics is extrapolating the Fourier spectrum of a spatially bounded object from a segment of the spectrum. The problem arises because of diffraction effects and the fact that lenses are not infinitely large. This means that optical systems have limited passbands and act as (two-dimensional) low-pass filters. Hence the image is reconstructed from only the low-pass portion of the spectrum and is therefore devoid of high frequency information. Often, obtaining the spectrum is an end in itself, and it is desirable to extrapolate the spectrum beyond the passband of the optical system.

In theory, it is possible to obtain the entire spectrum of the object from only a known surface segment, provided that the object is spatially bounded. The dual of this result is often used in signal processing, whereby a band-limited signal is extrapolated from a known segment of the signal [16].

There are a number of two-dimensional restoration schemes [1, p. 133]

[16], and they are based on the following two well-known fundamental theorems regarding analytic functions.

Theorem 1. The two-dimensional Fourier transform of a spatially bounded function $f(x, y)$ is an analytic function in the spatial frequency uv-plane.

Theorem 2. If any analytic function in the uv-plane is known exactly over a finite region of that plane, then the function is uniquely determined over the entire region of analyticity.

Although spectral extrapolation algorithms are not new, there have been some recent developments in recursive, relatively efficient, extrapolation algorithms especially suitable for computer applications. One of the best known of the recursive procedures is the one developed by Gerchberg [17] and Papoulis [18]. We refer to it as the G-P algorithm and extend it to two dimensions. We shall first describe how the algorithm is applied to a band-limited space domain object when only a two-dimensional segment of the object is given. We shall then consider the more interesting—at least from the point of view of applications to optics—dual problem of extrapolating the spectrum of a spatially bounded object when a surface segment of the spectrum is obtained.

The problem can be stated as follows. Let it be required to determine the band-limited function $f(x, y)$ for all (x, y) if $f(x, y)$ is given over a region $\mathscr{A} \equiv \{(x, y): |x| < L/2, |y| < W/2\}$ and $F(u, v) \equiv \mathscr{F}_2[f(x, y)] = 0$ for $|u| > B_x$, $|v| > B_y$. We form the function

$$g(x, y) = \begin{cases} f(x, y), & (x, y) \in \mathscr{A}, \\ 0, & \text{elsewhere}, \end{cases} \tag{1.10-1}$$

and use it as a starting point in the recursive algorithm described below. Thus let $f_1(x, y) \equiv g(x, y)$. Now we compute its Fourier transform (FT)

$$F_1(u, v) = \mathscr{F}_2[f_1(x, y)]. \tag{1.10-2}$$

Since we know that $F(u, v) = 0$ outside the rectangle $|u| > B_x$, $|v| > B_y$, we form the "corrected" function

$$G_1(u, v) = F_1(u, v) \, \text{rect}(u/2B_x) \, \text{rect}(v/2B_y). \tag{1.10-3}$$

The inverse FT of $G_1(u, v)$ yields $g_1(x, y)$. Now we correct our estimate by incorporating our a priori knowledge of $f(x, y)$ over $\mathscr{A}$. Hence we form the correction

$$f_2(x, y) = \begin{cases} f(x, y), & (x, y) \in \mathscr{A}, \\ g_1(x, y), & \text{elsewhere}. \end{cases} \tag{1.10-4}$$

These steps represent the initial cycle. The nth cycle $(n > 1)$ is then

$$F_{n-1}(u, v) = \mathscr{F}_2[f_{n-1}(x, y)], \tag{1.10-5a}$$

$$G_{n-1}(u, v) = F_{n-1}(u, v)\,\mathrm{rect}(u/2B_x)\,\mathrm{rect}(v/2B_y), \tag{1.10-5b}$$

$$g_{n-1}(x, y) = \mathscr{F}_2^{-1}[G_{n-1}(u, v)], \tag{1.10-5c}$$

$$f_n(x, y) = \begin{cases} f(x, y), & (x, y) \in \mathscr{A}, \\ g_{n-1}(x, y), & \text{elsewhere.} \end{cases} \tag{1.10-5d}$$

Papoulis then shows that

$$\int_{-\infty}^{\infty} \int_{-\infty}^{\infty} |F(u, v) - F_n(u, v)|^2 \, du\, dv$$

$$< \int_{-\infty}^{\infty} \int_{-\infty}^{\infty} |F(u, v) - F_{n-1}(u, v)|^2 \, du\, dv, \tag{1.10-6}$$

and—when the data are continuous (a.e.)—that

$$\lim_{n \to \infty} \int_{-\infty}^{\infty} \int_{-\infty}^{\infty} |F(u, v) - F_n(u, v)|^2 \, du\, dv = 0. \tag{1.10-7}$$

Thus not only is the mean square error reduced at every step but, by Eq. (1.10-7), convergence is ensured. In the case of discrete data, the rms error may still decrease but the convergence of $f_n(x, y)$ to $f(x, y)$ generally does not occur, since the discrete signal does not have the analyticity property that the continuous signal has.

The dual problem has the same structure. Here we are given that $f(x, y)$ is spatially bounded to a rectangle $|x| < L/2, |y| < W/2$, and we observe its spectrum $F(u, v)$ over a region $\mathscr{B} = \{(u, v): |u| < B_x, |v| < B_y\}$. The algorithm for extrapolating the spectrum for all (u, v) then takes the following form. We form the function

$$G(u, v) = \begin{cases} F(u, v), & (u, v) \in \mathscr{B}, \\ 0, & \text{elsewhere.} \end{cases} \tag{1.10-8}$$

and use it as our first estimate of the spectrum. Thus let $F_1(u, v) \equiv G(u, v)$. Its inverse FT gives $f_1(x, y)$. Since we know that the function is spatially bounded, we form the correction

$$g_1(x, y) = f_1(x, y)\,\mathrm{rect}(x/L)\,\mathrm{rect}(y/W). \tag{1.10-9}$$

With $G_1(x, y) \equiv \mathscr{F}_2[g_1(x, y)]$, we form the function

$$F_2(u, v) = \begin{cases} F(u, v), & (u, v) \in \mathscr{B}, \\ G_1(u, v), & \text{elsewhere.} \end{cases} \tag{1.10-10}$$

This correction incorporates the a priori knowledge we have from observing the spectrum over the passband of the system. The nth cycle ($n > 1$) then involves the following calculations:

$$f_{n-1}(x, y) = \mathscr{F}_2^{-1}[F_{n-1}(u, v)], \tag{1.10-11a}$$

$$g_{n-1}(x, y) = f_{n-1}(x, y)\, \text{rect}(x/L)\, \text{rect}(y/W), \tag{1.10-11b}$$

$$G_{n-1}(u, v) = \mathscr{F}_2[g_{n-1}(x, y)], \tag{1.10-11c}$$

$$F_n(u, v) = \begin{cases} F(u, v), & (u, v) \in \mathscr{B}, \\ G_{n-1}(u, v), & \text{elsewhere.} \end{cases} \tag{1.10-11d}$$

By duality, it can be shown that

$$\int_{-\infty}^{\infty} \int_{-\infty}^{\infty} |f(x, y) - f_n(x, y)|^2 \, dx \, dy$$

$$< \int_{-\infty}^{\infty} \int_{-\infty}^{\infty} |f(x, y) - f_{n-1}(x, y)|^2 \, dx \, dy, \tag{1.10-12}$$

and, analogously to Eq. (1.10-7), that convergence is ensured.

Other extrapolation algorithms are discussed by Cadzow [19] and Sabri and Steenaart [20]. The vital question of ill-posedness, e.g., how well these algorithms work when noise is present, is discussed in an excellent and noteworthy paper by Youla [21].

1.11 CONCLUSION

In this chapter we have explored the basic principles associated with obtaining and measuring Fourier spectra using optical systems. We reviewed some basic principles of diffraction theory and showed under what circumstances an ideal lens would produce a good Fourier transform of an object. We considered the tradeoff between stability and fidelity in measuring the power spectrum and derived some smoothing windows that had certain optimal properties. We briefly reviewed the problem of measuring the spectrum at low spatial frequencies and at the origin. Finally we considered the problem of space-limited or band-limited extrapolation. Space and time constraints forced our discussion in this area to be necessarily brief, but we felt that the interesting work going on in this area should be mentioned. In the subsequent chapters of this book, numerous applications of the Fourier properties of optical systems will be presented.

REFERENCES

[1] J. W. Goodman (1968). "Introduction to Fourier Optics." McGraw-Hill, New York.
[2] W. T. Cathey (1974). "Optical Information Processing and Holography." Wiley (Interscience), New York.
[3] R. J. Collier, C. B. Burckhardt, and L. H. Lin (1971). "Optical Holography." Academic Press.
[4] A. Erdélyi (1953). "Higher Transcendental Functions," Vol. 2. McGraw-Hill, New York.
[5] A. Papoulis (1968). "Systems and Transforms with Applications in Optics." McGraw-Hill, New York.
[6] H. Stark (1968). Tech. Rep. T-1/006-1-00. Riverside Res. Inst., New York.
[7] M. G. Jenkins and D. G. Watts (1968). "Spectral Analysis and its Applications." Holden-Day, San Francisco, California.
[8] N. I. Achieser (1956). "Theory of Approximation." Ungar, New York.
[9] A. Papoulis (1973). *IEEE Trans. Inf. Theory* **IT-19**, 9–12.
[10] H. Stark and B. Dimitriadis (1975). *J. Opt. Soc. Am.* **65**, 425–431.
[11] A. Papoulis (1972). *J. Opt. Soc. Am.* **62**, 1423–1429.
[12] P. Jacquinot and B. Roizen-Dossier (1964). *In* "Progress in Optics" (E. Wolf, ed.), Vol. III, p. 31. North-Holland, Amsterdam.
[13] H. Stark, D. Lee, and B. Dimitriadis (1975). *J. Opt. Soc. Am.* **65**, 1436–1442.
[14] H. Stark, D. Lee, and B. W. Koo (1976). *Appl. Opt.* **15**, 2246–2249.
[15] H. Stark, W. R. Bennett, and M. Arm (1969). *Appl. Opt.* **8**, 2165–2172.
[16] C. W. Barnes (1966). *J. Opt. Soc. Am.* **56**, 575–578.
[17] R. W. Gerchberg (1974). *Opt. Acta* **21**, 709–720.
[18] A. Papoulis (1975). *IEEE Trans. Circuits Syst.* **CAS-22**, 735–742.
[19] J. A. Cadzow (1979). *IEEE Trans. Accoust., Speech, Signal Process.* **ASS-27**, 4–11.
[20] M. S. Sabri and W. Steenaart (1978). *IEEE Trans. Circuits Syst.* **CAS-25**, 74–78.
[21] D. C. Youla (1978). *IEEE Trans. Circuits Syst.* **CAS-25**, 694–701.
[22] H. Stark and G. Shao (1977). *Appl. Opt.* **10**, 1670–1674.

Chapter 2

Pattern Recognition via Complex Spatial Filtering

SILVERIO P. ALMEIDA and GUY INDEBETOUW

DEPARTMENT OF PHYSICS
VIRGINIA POLYTECHNIC INSTITUTE
   AND STATE UNIVERSITY
BLACKSBURG, VIRGINIA

INTRODUCTION

The concept of spatial filtering, that is, the manipulation of spatial frequencies in order to alter the properties of an image, has been known for over 100 years as a result of the work of Abbe. During the late fifties and the sixties the use of Fourier techniques, the systematic exploitation of the analogy between optical systems and electrical networks, and the application of holographic techniques for generating complex spatial filters, contributed to the rapid development of coherent optics.

A brief account of the historical development of optical processing is given in Section 2.1. The basic properties of coherent spatial filtering systems are reviewed in Section 2.2, while Section 2.3 deals mostly with practical matters such as recording techniques, film response, and multiplexing. The critical attributes of optical correlators such as sensitivity to input scale and orientation are also discussed.

While no pretense to completeness is made, the aim of Section 4 is to give through some examples a general idea of the span of applications of optical spatial filtering. The discussion will, however, be restricted to spatial spectrum analysis and matched filtering. As an example of the latter, the application of matched filtering to the monitoring of water pollution by measuring diatom population is treated in more detail.

41

2.1 HISTORICAL OVERVIEW

The work of Ernst Abbe in the 1870s on the theory of image formation [1–3] had a profound impact on the discipline which was later called *Fourier optics*: the branch of optics which resulted from the interaction between optics and the communication and information sciences. In his quest not only to improve the quality of microscopic imaging (he was then working for Carl Zeiss) but also to explain scientifically the basis of these improvements, Abbe laid the cornerstone upon which much of today's optical processing rests.

Abbe's rediscovery and interpretation of the Smith–Helmholtz invariance principle in centered optical imaging systems [4–6] led him to formulate the well-known *Abbe* or *sine condition*: The condition for stigmatic imaging of off-axis object points is that the quantity $n \times \sin u$ takes the same value in image space as in object space. Here X is the off-axis coordinate of either an object point P or its image, n is the index of refraction in object or image space, and u is the angle of an imaging ray through P with the optical axis. Of most relevance to the remainder of this chapter is Abbe's observation that the imaging of fine details in an object was directly affected by the numerical aperture (or angular acceptance) of the objective lens. This eventually led to formulation of the image formation process as a diffraction phenomenon and to the description of the imaging instrument as a *filter* for the spatial frequencies of the object.

Prior to Abbe's work, the development of the wave theory of light had provided a new understanding of light phenomena such as diffraction and interference and had set up the mathematical framework for later analysis of his theory. Among the early contributors, one might mention Grimaldi's early work on diffraction phenomena, which was published in 1665. Later, Hooke [7], in an attempt to explain certain interference phenomena, described light as a propagating vibration, and in 1678 Huygens [8] proposed his wave theory of light based on the "wavelets envelope principle." This concept is still widely used in many textbooks to describe diffraction phenomena. For more than a century, however, Huygen's ideas were overshadowed by the corpuscular theory of light published by Newton about 1700. For the time, Newton's model was philosophically more acceptable and could indeed explain simply many geometrical or ray optical phenomena.

Experimental evidence of the wave nature of light was obtained in 1801 with the qualitative but important work of Thomas Young on interference phenomena [9]. The synthesis by Augustin Fresnel in 1816 of Huygen's theory and Young's observations led to the first great success of the wave theory of light: Fresnel's spectacular interpretation of diffraction phenomena [10]. Some of Fresnel's calculations were later experimentally confirmed by

Arago. After the formidable contribution of Maxwell to the theory of light, a scalar wave diffraction theory with a more solid mathematical basis was set forth by Gustav Kirchhoff. This theory, modified and further refined, is known today as the Rayleigh–Sommerfeld scalar diffraction theory [6, 11].

Following Abbe's work, Porter [12], whose experiments published in 1906 confirmed Abbe's theory, further developed the theory of image formation in the microscope, while Airy and Lord Rayleigh [13] introduced the important notion of limit of resolution and perfect or *diffraction-limited* instruments. Microscopic imaging was also the motivation of Frits Zernike [14, 15], who in 1935 introduced his Nobel prize winning phase-contrast microscope. In this instrument, small phase objects are made visible by introducing a phase difference between direct and diffracted light. This task is accomplished by a pupil mask which, in modern technology, would be called a spatial filter. Other early examples of spatial filtering were the Schlieren and Foucault knife-edge techniques which essentially use masking of portions of the spatial frequency spectrum of a wave front in order to transform phase variations into a visible intensity pattern.

A key turning point in optical image processing occurred in the middle of the century with the work of Duffieux [16] who introduced Fourier analysis as a powerful analytical tool in image formation theory. Treating a general imaging instrument as a linear filter, he established the Fourier transform relationship between the energy distributions in the image or object plane and in the pupil plane and described the image intensity distribution as a convolution product between the object intensity distribution and an instrumental point spread function (i.e., the impulse response). Duffieux's ideas were to bear their first fruit in the fifties. During this decade, an abundant number of papers concerning the Fourier treatment of optical imaging were published.

The paper by Marechal and Croce [17] was perhaps the first to point clearly to the analogy between image formation in an optical instrument and a communication network. Both are vehicles for carrying, transferring, or transforming information (spatial in one case and temporal in the other) and, within certain restrictions discussed in Chapter 1 of this book, can be described as linear-invariant processes. Sharing the same mathematical description has understandably strengthened the ties between the two disciplines. In particular, the optical implementation of some concepts which were well known in system theory has been extremely useful. The analogy between image enhancement and transfer function equalization has been one of the first examples to be successfully exploited. Presently, topics such as optical feedback systems and nonlinear and space-variant systems are the subject of intense research. Some of these topics will be reviewed in this book.

The analogy between optics and communications science, or between optical filters and network filters, was further exploited by authors such as Blanc-Lapierre [18], Cheatham and Kohlenberg [19], Elias [20], and Rhodes [21]. At first, image formation theory greatly benefited from this insight. Among the most important, one might mention the work of Hopkins [22, 23], Jacquinot *et al.* [24], Linfoot [25, 26], Marechal [27], Selwyn [28], and Steel [29]. It was recognized quite early that the similarity between optics and communications could be used successfully not only in systems analysis but also in system synthesis. The essential idea behind optical system synthesis is that a specific transformation between the object and image distribution can be accomplished by controlling the light distribution in the pupil of the apparatus with masks or filters. The *double-diffraction* setup introduced by Marechal and Croce [17, 30] has become the classic experimental arrangement for optical filtering work. The "4-F" system is of particular interest because of its simple mathematical description. It consists of a two-lens afocal imaging system with the object plane in the front focal plane of lens 1, the image plane in the back focal plane of lens 2, and the pupil plane in the coincident back focal plane of lens 1 and front focal plane of lens 2. In this system, not only are the input, output, and pupil planes easily accessible but, for an aberration-free system, an exact Fourier transform holds between the input (object) and pupil (filter) plane as well as between the pupil and output (image) plane.

With the advent of the laser, which furnished an intense light with a high degree of spatial and temporal coherence, the field of coherent-optical processing flourished rapidly. During the late fifties and early sixties a sizable amount of theoretical and experimental work was published on the subject. A great number of practical applications of coherent-optical techniques were also proposed during this period of time. Interestingly, contributions to the new discipline of optical processing came both from the field of optics, with, for example, the work of Marechal and Croce [31, 32], Marquet [33], Tsujiuchi [34, 35], and Jacquinot and Dossier [36], and from the field of information science, with the work of O'Neil [37], Turin [38], and Cutrona [39], just to mention a few. This multidisciplinary character has no doubt contributed to the rapid growth of the new discipline.

One should also mention that, during the same period of time, holography, invented by Gabor [40, 41] in 1948, and which for many reasons is usually associated with coherent optics, was being developed into a spectacular technique by researchers such as Leith, Upatniek, and numerous followers [42–44].

The early application of coherent-optical image processing made use of rather simple spatial filters such as binary masks, simple amplitude (no-phase) distribution [36], or phase masks of elementary geometry such as rings and disks [35]. The method of production of these filters was either photographic or vacuum disposition. An attempt to use polarization as a way to control the phase of the spatial filter has also been made [45–47]. Clearly, as the filter structure becomes more complex, the technological difficulty rapidly reaches unreasonable limits. A very important advance was made in 1963 by Vander Lugt who proposed a holographic-like technique for constructing spatial filters with continuously controllable amplitude and phase transmittance. With this new technique, which will be discussed in Section 2.2, it became possible to construct arbitrarily complex matched filters, hence to address the important problem of detecting a signal in a noisy background by optical means [48, 49]. As is well known from information theory, this is essentially a problem of pattern recognition and requires, at least in the case of stationary noise, a filter with a transfer function proportional to the complex conjugate of the signal spectrum, i.e., a *matched filter*. In order to generate such a filter optically, Vander Lugt proposed to record what is essentially a Fourier hologram of the signal. One sideband of this hologram would then reconstruct a complex amplitude distribution proportional to the desired transfer function. Vander Lugt's technique opened up a whole new realm of optical processing in general and optical pattern recognition in particular.

The properties and characteristics of holographic matched filters have been studied in great detail. Ways to control their performance systematically have also been sought. One limitation of the holographically recorded matched filter is that the searched signal must exist physically (most often in form of a transparency). As shown by A. Lohmann [50–52] and others [53, 54], this constraint can be avoided if the matched filters are generated by a computer. Another important point concerning synthetic filters is that their properties (for example, choice of a particular bandpass in the spectrum or enhancement of the filter sensitivity to some particular input features) can, in principle, be controlled with considerable accuracy. One of the most important characteristics of the coherent matched filtering technique is its high sensitivity to the size and orientation of the input. For many applications, this is a serious drawback. Optical correlators capable, in principle, of recognizing an input independently of either its size or its orientation have been applied recently by Casasent and Psaltis [55, 56].

A matched filtering technique using incoherent light has also been proposed. In theory, incoherent-optical processors have much better signal-to-noise characteristics than similar coherent systems. Different approaches to incoherent processing have been studied. In addition to template matching techniques [57–60], Lohmann has proposed a system in which holographic spatial filters are used with incoherent light [61, 62]. Real-time incoherent-to-coherent transducers [63] such as the PROM [64–66] and the liquid crystal light valve [67] have also been used as input for coherent correlators, for example, in the work of Gara [68]. Another incoherent technique, which makes use of either a spatial [69, 70] or temporal [71] carrier frequency to generate bipolar point spread functions with an incoherent source, can also be applied to optical correlation [72]. These techniques, as well as others proposed in earlier work [73–76] can, in principle, overcome the most serious drawbacks of classic incoherent-optical correlators: the large dc bias and the restriction to real, positive point spread functions. Systems using partially coherent light have also been studied [77–79]. The simplest way to obtain the correlation of two images is to translate a diffusely illuminated transparency of the first image in front of a transparency of the second one. The amount of transmitted light as a function of shift is a measure of the correlation. Recently, a new kind of correlator has been studied which performs the same operation without any motion [80–82]. Instead, a dispersive element is used to produce different lateral shifts of the input for different wavelengths of the source. A *wavelength-coded correlation* can therefore be obtained in real time.

2.2 COMPLEX SPATIAL FILTERING

A. Coherent-Optical Processor

As discussed in Chapter 1, the ease with which a lens can perform the Fourier transformation of a two-dimensional distribution is certainly one of the most important assets of coherent-optical processing. Perhaps even more important is the fact that the spectrum of the input is physically accessible and therefore can be manipulated simply by placing masks or optical filters in the Fourier transform plane. The optical processor of Fig. 2.2-1 is sometimes called, for obvious reason, a 4-F system. It is one of several possible optical configurations that permit signal processing.

An input transparency of complex amplitude transmittance $g_1(x_1, y_1)$ is placed in the front focal plane of lens L_1 and illuminated by a plane-parallel beam of uniform intensity and zero phase. The amplitude distribution in the back focal plane P_f of the lens is, within certain conditions which are discussed in Section 1.3 of Chapter 1 (see also Goodman [83]), represented by the Fourier transform $G_1(u, v)$ of the input. Neglecting some proportionality factors,

$$G_1(u, v) = \mathscr{F}_2[g_1(x_1, y_1)]$$

$$= \int_{-\infty}^{\infty} \int_{-\infty}^{\infty} g_1(x_1, y_1)e^{-j2\pi(ux_1 + vy_1)} \, dx_1 \, dy_1, \qquad (2.2\text{-}1)$$

where λ is the wavelength and (u, v) are the rectangular spatial frequencies

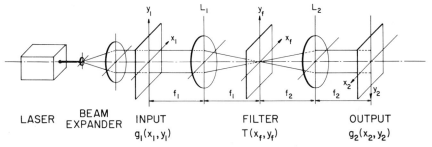

FIG. 2.2-1. Coherent-optical processor.

of the input. They are related to the coordinate (x_f, y_f) in the Fourier plane by

$$u = x_f/\lambda f_1 \quad \text{and} \quad v = y_f/\lambda f_1, \tag{2.2-2}$$

where f_1 is the focal length of lens L_1. If a filter transparency of amplitude transmittance $T(x_f, y_f)$ is placed in the plane P_f, the amplitude distribution just after the transparency becomes

$$U_f(x_f, y_f) = G_1(x_f, y_f) \cdot T(x_f, y_f). \tag{2.2-3}$$

The second lens L_2 of the processor performs a second Fourier transform of $U_f(x_f, y_f)$, leading to an amplitude distribution in its back focal plane P_2 given by

$$g_2(x, y) = \mathscr{F}_2[G_2(u', v')], \tag{2.2-4}$$

where $u' = x_f/\lambda f_2 = u/M$, $v' = y_f/\lambda f_2 = v/M$, and $M = f_2/f_1$ is the lateral magnification of the imaging system. Use of Eqs. (2.2-3) and (2.2-4), leads to the well-known relationship between input and output of a linear-invariant coherent system. The output spectrum then becomes the product of the input spectrum with the transfer function $H(u, v)$ which is proportional to the amplitude transmittance of the pupil mask $T(x_f, y_f)$, i.e.,

$$G_2(u, v) = G_1(u, v) \cdot H(u, v), \quad H(u, v) = T(\lambda f u, \lambda f v). \tag{2.2-5}$$

Equivalently, the output can be represented as the convolution of the input— scaled by a magnification factor M—with a point spread function (impulse response) $h(x_2, y_2)$

$$g_2(x_2, y_2) = \int_{-\infty}^{\infty} \int_{-\infty}^{+\infty} g(Mx_1, My_1) h(x_2 - Mx_1, y_2 - My_1) \, dx_1 \, dy_1,$$

$$\tag{2.2-6}$$

where

$$h(x_2, y_2) = \mathscr{F}_2[H(u, v)].$$

In short notation, the convolution product is represented by the symbol $*$, and Eq. (2.2-6) takes the form

$$g_2(x, y) = g_1(x, y) * h(x, y). \qquad (2.2\text{-}7)$$

B. Complex Spatial Filter

The coherent-optical processor of Fig. 2.2-1 is capable of performing a general linear-invariant transformation as expressed by Eqs. (2.2-5) and (2.2-6). This is possible, of course, as long as the complex-valued filter transmittance $T(x_f, y_f)$ can be constructed. Complex spatial filters have been found useful in many applications. Various names have been given to these filters according to the context in which they are used: The terms *holographic filters* and *Fourier holograms* refer to the technique usually involved in filter production. In image enhancement, they might be called inverse filters or deblurring filters, while in pattern recognition, their most common names are Vander Lugt filters (after their originator) or matched filters.

As stated in the beginning of this chapter, an interferometric technique widely used to record any complex filter for which the point spread function is known was introduced in 1963 by Vander Lugt [49]. It consists of recording the interference pattern between the desired filter transfer function and a mutually coherent reference beam. This is essentially a Fourier hologram of the point spread function. The process is similar to the modulation technique in information theory where a complex-valued function (signal) can be recorded as a real-valued function on a carrier frequency as long as the sampling theorem is satisfied; i.e., the carrier frequency must be at least twice as large as the signal cutoff frequency.

C. Holographic Recording

The most straightforward technique for recording a complex-valued holographic filter is shown in Fig. 2.2-2. A transparency of amplitude transmittance proportional to the desired point spread function $h(x, y)$ is placed in the front focal plane of lens L_1 and illuminated by a plane wave. The distribution in the back focal plane is made to interfere with a plane-parallel reference beam U_R tilted by an angle θ with respect to the x_f-axis. The total complex amplitude U_T in plane P_f is then

$$U_T(x_f, y_f) = U_R(x_f, y_f) + H(x_f, y_f), \qquad (2.2\text{-}8)$$

where

$$U_R(x_f, y_f) = R \exp(-j2\pi u_0 x_f). \qquad (2.2\text{-}9)$$

R^2 is a measure of the reference-to-object beam energy ratio, and u_0 is the

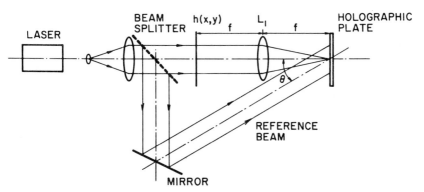

FIG. 2.2-2. Interferometric arrangement for recording a holographic filter.

carrier spatial frequency given by

$$u_0 = \sin\theta/\lambda. \qquad (2.2\text{-}10)$$

Th plane reference wave can also be regarded as the uniform spectrum of a point source $\delta(x_1 - x'_R, y_1)$ located at the coordinate $(x'_R, 0)$ in the object plane. In the paraxial approximation, one has

$$x'_R = \lambda f u_0 = f \sin\theta. \qquad (2.2\text{-}11)$$

From Eqs. (2.2-8) and (2.2-9), we see that the total irradiance in plane P_f is given by

$$I(x_f, y_f) = |U_T(x_f, y_f)|^2$$

$$= R^2 + |H(x_f, y_f)|^2 + RH(x_f, y_f)\exp(+j2\pi u_0 x_f)$$

$$+ RH^*(x_f, y_f)\exp(-j2\pi u_0 x_f). \qquad (2.2\text{-}12)$$

If we write the complex function $H(x_f, y_f)$ as

$$H(x_f, y_f) = |H(x_f, y_f)|\exp[j\phi(x_f, y_f)], \qquad (2.2\text{-}13)$$

the expression for the irradiance $I(x_f, y_f)$ can be rewritten as

$$I(x_f, y_f) = R^2 + |H(x_f, y_f)|^2$$

$$+ 2RH(x_f, y_f)\cos[2\pi u_0 x_f + \phi(x_f, y_f)]. \qquad (2.2\text{-}14)$$

This expresssion shows explicitly how the phase $\phi(x_f, y_f)$ is encoded as a modulation of the spatial carrier.

At this point, it is customary to assume that this irradiance is recorded linearly on some suitable medium. High-resolution photographic emulsions are often used for their high information capacity and relative low cost. Linear recording implies that the amplitude transmittance of the developed plate or film is proportional to the irradiance, i.e.,

$$T(x_f, y_f) \propto I(x_f, y_f). \tag{2.2-15}$$

Linear recording over an extended dynamic range is extremely difficult to achieve by photographic means. It is, however, useful to carry the analysis of this ideal case; departure from it will be discussed later.

D. Coherent-Optical Correlation

If a filter of amplitude transmittance $T(x_f, y_f)$ as described in Eq. (2.2-15) is placed in the filter plane of a double-diffraction setup such as that shown in Fig. 2.2-1, the resulting system is a processor with a transfer function proportional to Eq. (2.2-12) or (2.2-14). With an amplitude distribution $g_1(x_1, y_1)$ in the input plane, and using Eqs. (2.2-6), (2.2-15), and (2.2-12), we can write for the amplitude distribution in the output plane

$$U_2(x_2, y_2) = \mathscr{F}_2[G_1(u, v) \cdot T(u, v)]$$

$$= \int_{-\infty}^{+\infty} \int_{-\infty}^{+\infty} R^2 G_1(u, v) e^{-j2\pi(ux_2 + vy_2)} \, du \, dv$$

$$+ \int_{-\infty}^{+\infty} \int_{-\infty}^{+\infty} |H(u, v)|^2 G_1(u, v) e^{-j2\pi(ux_2 + vy_2)} \, du \, dv$$

$$+ \int_{-\infty}^{+\infty} \int_{-\infty}^{+\infty} R H(u, v) G_1(u, v) e^{j2\pi ux_R} e^{-j2\pi(ux_2 + vy_2)} \, du \, dv$$

$$+ \int_{-\infty}^{+\infty} \int_{-\infty}^{+\infty} R H^*(u, v) G_1(u, v) e^{-j2\pi ux_R} e^{-j2\pi(ux_2 + vy_2)} \, du \, dv. \tag{2.2-16}$$

By using elementary properties of the Fourier transform we can also write Eq. (2.2-16) as

$$U_2(x_2, y_2) = R^2 g_1(x_2, y_2)$$

$$+ \int_{-\infty}^{+\infty} \int_{-\infty}^{+\infty} \int_{-\infty}^{+\infty} \int_{-\infty}^{+\infty} h(\xi, \eta) h^*(\xi + \alpha - x_2, \eta + \beta - y_2)$$

$$\times g(\alpha, \beta) \, d\xi \, d\eta \, d\alpha \, d\beta$$

$$+ R \int_{-\infty}^{+\infty} \int_{-\infty}^{+\infty} h(x_2 + x_R - \xi, y_2 - \xi) g(\xi, \eta) \, d\xi \, d\eta$$

$$+ R \int_{-\infty}^{+\infty} \int_{-\infty}^{+\infty} (\xi - x_2 + x_R, \eta - y_2) g(\xi, \eta) \, d\xi \, d\eta \tag{2.2-17}$$

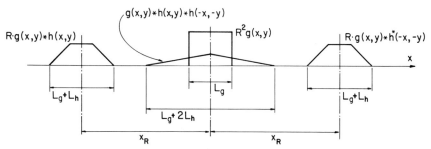

FIG. 2.2-3. Output of an optical processor with a holographic filter.

or, in shorthand notation, as

$$U_2(x_2, y_2) = R^2 g_1(x_2, y_2)$$
$$+ g_1(x_2, y_2) * h(x_2, y_2) * h^*(-x_2, -y_2)$$
$$+ R g_1(x_2, y_2) * h(x_2, y_2) * \delta(x_2 + x_R, y_2)$$
$$+ R g_1(x_2, y_2) * h^*(-x_2, -y_2) * \delta(x_2 - x_R, y_2). \quad (2.2\text{-}18)$$

In these expressions, $x_R = M x_R'$ is the abscissa of the point in the output plane where the reference beam used to record the holographic filter would come to focus. The different output terms are shown in Fig. 2.2-3. The first two terms of Eq. (2.2-18) are components centered at the origin of the output plane. The third term is the convolution product of the input $g_1(x_2, y_2)$ with the desired point spread function $h(x_2, y_2)$. The convolution with the delta function shifts the term along the x_2-axis and centers it at $(-x_R, 0)$. The last term is the correlation of the input and the point spread function centered at $(x_R, 0)$ in the output plane. Clearly, if x_R is large enough, the different terms can be separated in the output plane.

E. Other Geometrical Arrangements

The setup described so far, which can be used in a coherent processor or to record a holographic filter, is certainly not the only possible one. Many different arrangements have been proposed [84]; they all are two-beam interferometric configurations of either the Mach–Zehnder or the Rayleigh type. Each has its advantages and drawbacks. The setup of Fig. 2.2-4, for example, was proposed by Vander Lugt. It allows the size of the spectrum to be varied and might be used to perform a scale search of the input. The scaling of the spatial frequencies is in this case given by

$$u = x_f / \lambda d \quad \text{and} \quad v = y_f / \lambda d, \quad (2.2\text{-}19)$$

where d is the adjustable distance between input plane and Fourier plane.

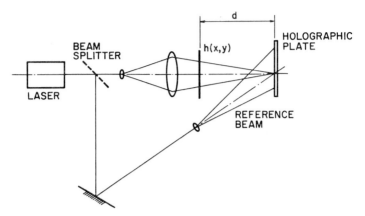

FIG. 2.2-4. Recording arrangement allowing scale changes. (After Vander Lugt [84].)

The arrangement of Fig. 2.2-5 might be of interest if the size of the input is small or if the recording medium has limited resolution. In this case, the reference offset angle must be as small as possible. The setup in Fig. 2.2-6 serves the same purpose and allows an even smaller reference angle. The additional beam splitter might, however, introduce undesirable distortions and aberrations.

Very often, holographic filters are recorded on high resolution photographic emulsions (1000 lines/mm or more). In this case, it is customary to use a large offset angle for the reference beam in order to separate the different output terms. Usually, only one of the sidebands is of interest. In pattern recognition, for example, the term of interest is usually the correlation product. With a large offset angle, this term is easily isolated with an arrangement such as that shown in Fig. 2.2-7.

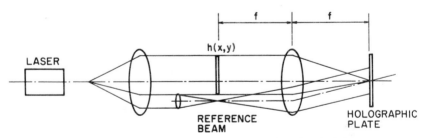

FIG. 2.2-5. Recording setup using a minimum number of optical elements.

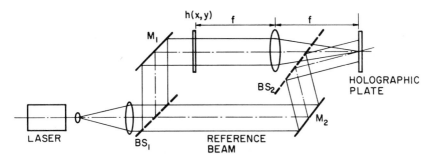

FIG. 2.2-6. Recording arrangement of the Mach–Zehnder type allowing small reference angles. BS, Beam splitter; M, mirrors.

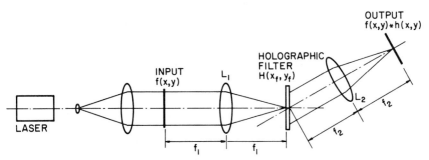

FIG. 2.2-7. Optical correlator using a holographic filter.

2.3 PATTERN RECOGNITION VIA MATCHED FILTERING

A. Matched Filters

As is well known in communication theory, the optimum filter for extracting a known signal $s(t)$ from stationary noise $n(t)$, is a matched filter with a transfer function

$$H(v) = kS^*(v)/|N(v)|^2, \qquad (2.3\text{-}1)$$

where $S(v)$ is the signal spectrum, $|N(v)|^2$ is the noise spectral density, and k is a constant. For the optical counterpart of this matched filter, one can simply replace the temporal variables t and v by their spatial equivalents (x, y) and (u, v), respectively. An optical filter matched to the signal $g(x, y)$ should have a transfer function proportional to the complex conjugate of the signal spectrum:

$$H(u, v) = kG^*(u, v)/|N(u, v)|^2, \qquad (2.3\text{-}2)$$

where

$$H(u, v) = \mathscr{F}[h(x, y)] \quad \text{and} \quad G^*(u, v) = \mathscr{F}[g^*(-x, -y)].$$

In the case where the noise spectral density can be assumed constant, the filter transfer function and impulse response become

$$H(u, v) = k'G^*(u, v) \quad \text{and} \quad h(x, y) = k'g^*(-x, -y). \quad (2.3\text{-}3)$$

Clearly, such a filter can be constructed by the holographic technique described earlier. A linearly recorded Fourier hologram of the signal $g(x, y)$ will reconstruct, in one of the sidebands, a distribution of amplitude proportional to the point spread function $g^*(-x, -y)$.

If an input transparency $f(x, y)$ contains a feature $g'(x - x_0, y - y_0)$ similar to the signal $g(x, y)$ but centered at (x_0, y_0), as well as other features $f'(x, y)$, the output of the processor in the "correlation channel" can be written as

$$U_2(x_2, y_2) = g'(x_2, y_2) * g^*(-x_2, -y_2) * \delta(x - x_R - x_0, y - y_0)$$
$$+ f'(x_2, y_2) * g^*(-x_2, -y_2) * \delta(x - x_R, y - y_R). \quad (2.3\text{-}4)$$

The point $(x_R, 0)$, which is the location of the image of the reference point source, is often chosen as the origin of coordinates for the correlation distribution. This is assumed in the discussion that follows. The first term of Eq. (2.3-4) is characterized by an autocorrelation peak centered at (x_0, y_0), indicating the location of the match. The other features lead to cross-correlation terms in which the energy is spread over a larger area. The correlation peak intensity is the quantity usually taken as a measure of the similitude between the input and the signal.

B. Bandwidth Consideration

The transfer function of a linearly recorded holographic filter was found to contain four important terms: a dc term, a low-frequency component, and two sidebands centered at the frequency of the spatial carrier. When addressed by an input $f(x, y)$, the response of this filter is the convolution of $f(x, y)$ with the impulse response $h(x, y)$. This results in an amplitude distribution in the output plane given by Eq. (2.2-18).

If L_g and L_h measure, respectively, the maximum extent of the input and of the impulse response, the different output terms, shown in Fig. 2.2-3, have the following dimensions: The two terms centered at the origin have extent L_g and $L_g + 2L_h$, respectively, and the extent of both the convolution and the correlation is $L_g + L_h$. From Fig. 2.2-3, the condition for nonoverlapping of the different output terms is found to be

$$x_R = L_g + \tfrac{3}{2}L_h \quad (2.3\text{-}5)$$

The minimum spatial carrier frequency is then given by $u_c = x_R/\lambda f$. Figure 2.3-1 shows the response of a spatial filter matched to the contour of a geometrical pattern when addressed by the same pattern as input. The central dc term is similar to the contour of the input, the two sidebands clearly show the difference between the convolution and the correlation term. Most of the energy of the autocorrelation term is concentrated in a sharp peak. This signifies that the most extended dc term, which is very weak, will not interfere much with detection of the sharp correlation peak even if they overlap each other. It is therefore often possible to relax the bandwidth requirement given in Eq. (2.3-5) to

$$x_R > L_g/2. \tag{2.3-6}$$

This might be of interest if the recording medium has a limited information storage capacity or space–bandwidth product. Film nonlinearity, however, will usually introduce higher harmonics centered at multiples of the carrier frequency. Although weaker, these terms can eventually interfere with the correlation spot if their dimensions are larger than $2x_R$.

C. Filter Optimization

The optimum filter for a given application depends strongly on the constraints that the intended application might dictate. The matched filter defined in Eq. (2.3-1) was found to maximize the detection signal-to-noise ratio when the noise was signal-independent, stationary, and additive [85]. For a temporal signal, the noise spectral density can often be considered uniform (white noise). For many problems, especially those that occur often in Fourier optics, this assumption is not valid. In pattern recognition, for example, the problem might be to recognize or extract one element from a possible class. The other members of the class are considered noise. Clearly, this noise can, in general, be highly correlated to the signal; in this case, the matched filter described by Eq. (2.3-3) is not necessarily optimum.

Other criteria might be used to define the optimum filter. In image enhancement, for example, the most widely used criterion is to minimize the least mean square error between the original and the restored image. It was found by Helstrom [86] that, with this definition, the optimum filter in the presence of random, zero-mean, additive noise had the form

$$H = \frac{1}{G} \frac{\phi_0/\phi_n}{(\phi_0/\phi_n) + 1/|G|^2}, \tag{2.3-7}$$

where ϕ_0 and ϕ_n are the power spectra of the image and noise distribution, respectively. These quantities are often assumed to be constant over the signal spectrum. $H(u, v)$ and $G(u, v)$ are, as before, the filter transfer function and the signal spectrum, respectively. Such filters were successfully used by

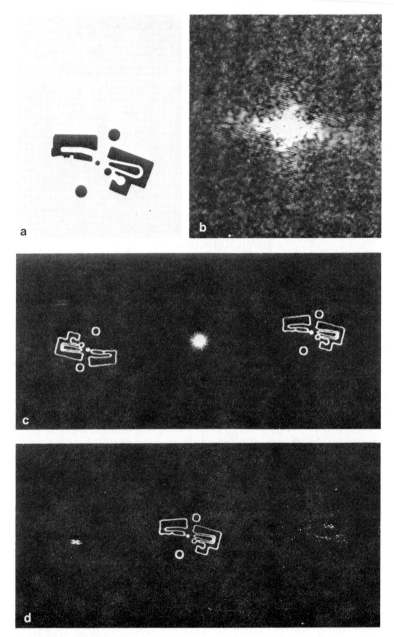

FIG. 2.3-1. (a) Optical correlator input, (b) holographic matched filter, (c) impulse response of (2.3-6), and (d) output of the correlator (autocorrelation spot on the left side).

Horner [87, 88] to restore images blurred by motion and turbulence. Filter optimization, with the constraints that they be passive and that their transfer function be no greater than unity, has been studied by Gallagher and Liu [89, 90], while Yatagai [91] considered the important case of multiplicative noise. Although most filter optimization studies deal with image restoration, some are directly concerned with pattern recognition problems, such as those treated by Cathey [92], Ullman [93], and Watrasiewicz [94].

It is interesting to remark that the filter described by Eq. (2.3-7) leads, in the noise-free case, to the inverse filter with a transfer function

$$H(u, v) = 1/G(u, v). \tag{2.3-8}$$

This filter is mostly used in image restoration, where its function is to compensate for a blurring transfer function $G(u, v)$ by equalizing the overall transfer function. By noting that $1/G(u, v) = G^*(u, v)/|G(u, v)|^2$, this filter can be produced by the same holographic techniques as those used for the matched filter. The additional amplitude division filter can, with some care, be produced photographically [95–99]. The inverse filter can also be used for pattern recognition. In this case, $G(u, v)$ is the spectrum of the searched signal. When addressed by an input $f(x, y)$, the response of the filter can be written as the Fourier transform of the spectral plane distribution:

$$U_f(u, v) = F(u, v)G^*(u, v)/|G(u, v)|^2. \tag{2.3-9}$$

The autocorrelation case, $f(x, y) \equiv g(x, y)$ is characterized, in theory, by a uniform amplitude distribution in the Fourier plane (just after the filter). In practice this distribution is limited by the size of the filter. The autocorrelation spot produced by this filter is therefore a duplicate of the point spread function of the optical system with a clear pupil (no filter). This spot is a sharper distribution than the signal autocorrelation produced by the matched filter described earlier. In general, it leads to a much higher selectivity (ability of the system to distinguish small signal variations).

Inverse filtering, however, can only be approximated. Since the transmittance of the film is constrained to the domain 0 to 1, the zeroes of the signal spectrum cannot be represented accurately. This, however, is not necessary nor even desirable. If noise is taken into account, the optimum filter is obtained by attenuating the filter's transmittance proportionally to the noise spectral density [99]. As the noise is often unknown or too complicated to be adequately represented, one can simply assume a uniformly distributed noise power spectrum and reduce the filter's transmittance in the regions where the signal-to-noise ratio (S/N) is most likely to be low, i.e., near the zeroes of the signal's spectrum. Whether this inverse filter, the matched filter previously described, or any other variant is best suited for a given application strongly depends on the signal-to-noise ratio and criteria involved. High selectivity and sensitivity to noise usually go together and require some type of design compromise. An entire range of filters with varying selectivity can, for example, be described by an amplitude transmittance

$$H(u, v) = \begin{cases} KG^{-1}(u, v)P(u, v) & \text{for } |G(u, v)| > K, \\ \exp[\,j \arg G^*(u, v)]P(u, v) & \text{for } |G(u, v)| < K, \end{cases} \quad (2.3\text{-}10)$$

where $P(u, v)$ is the pupil function defined as

$$P(u, v) = \begin{cases} 1 & \text{within signal bandwidth,} \\ 0 & \text{outside,} \end{cases}$$

and the adjustable parameter K is used to vary the degree of clipping of the filter. Figure 2.3-2 shows the evolution of this filter as the S/N decreases, for an input $\text{rect}(x/a)$. In situations where the S/N is high, K may be made very small. The filter is practically an inverse filter over the entire bandwidth. Only small areas around the zeroes of $H(u, v)$ are clipped. As the S/N decreases, harder clipping is necessary in order to reduce the relative contribution of the spectral bands near the zeroes of the spectrum. A higher value of K is necessary. For $K = \max[H(u, v)]$, for example, the filter is clipped over the entire bandwidth. The phase of the signal spectrum, however, is still accurately recorded. For even an smaller S/N, further attentuation is necessary in the region of small signal. This leads naturally to the matched filter whose transmittance is proportional to the complex conjugate of the signal spectrum.

D. Film Nonlinearity

In the previous analysis, it was assumed that the holographic filter was linearly recorded or that its amplitude transmittance could be precisely controlled. If the filters are produced photographically, linear recording is only possible in a very limited range of exposure where the amplitude transmittance-versus-exposure curve can be approximated to a straight line. Power spectra of inputs usually used in pattern recognition (such as aerial photographs, high contrast recordings of inputs, shadowgraphs, or contours of opaque objects) have an envelope decreasing roughly as $(u^2 + v^2)^{-1}$, where u and v are the spatial frequencies. To record such a spectrum linearly over a range of frequency from 1 to 50 mm^{-1} would require a dynamic range greater than three decades.

One must stress, however, that the lack of linearity displayed by most recording materials, and by photographic emulsions in particular, is not necessarily a drawback but, on the contrary, has often been exploited as a desirable feature. The complex transfer function of a holographically re-

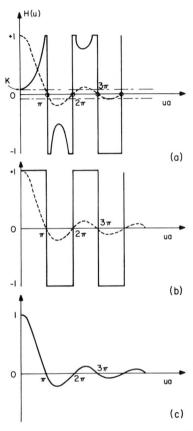

FIG. 2.3-2. Evolution of a filter transmittance from an inverse filter (a) to a matched filter (c) as the signal-to-noise ratio increases.

corded filter has been described as one of the sidebands (first-order diffraction) reconstructed by this hologram. Two recording parameters—the exposure value and the ratio of reference to object beam irradiance—are available to control the holographic diffraction efficiency. In principle, one could also control the contrast and dynamic range of the recorded hologram through the properties of the emulsion and the developer, although in most cases a recording medium with very high contrast is chosen in order to obtain maximum diffraction efficiency. By varying the reference-to-object beam ratio the user can choose the band of spatial frequencies in the spectrum which will exhibit the highest holographic fringe contrast and therefore the highest diffraction efficiency. The exposure value can also be used to control the spectral zones which are underexposed or saturated and therefore will not diffract at all. In general, filters tuned to higher frequencies (corresponding to finer input structure) are more selective.

Systematic enhancement of higher spatial frequencies using additional amplitude filters have been studied by Lowenthal and Belvaux [100–102]. If an additional filter of amplitude transmittance given by

$$T(u, v) = (2j\pi\sqrt{u^2 + v^2})^{2n} \qquad (2.3\text{-}11)$$

is used together with a matched filter with a transfer function $G^*(u, v)$, the amplitude distribution in the Fourier plane that results from an input $f(x, y)$ is

$$U_f(u, v) = T(u, v) \cdot G^*(u, v). \qquad (2.3\text{-}12)$$

For $n = 1$, for example, one has [103]

$$\mathscr{F}_2[T(u, v)] = t(x, y) = \left(\frac{d^2}{dx^2} + \frac{d^2}{dy^2}\right)\delta(x, y). \qquad (2.3\text{-}13)$$

This leads to an output amplitude distribution

$$U_2(x, y) = \left(\frac{d^2}{dx^2} + \frac{d^2}{dx^2}\right)\delta(x, y) * f(x, y) * g^*(-x, -y). \qquad (2.3\text{-}14)$$

Since the convolution operation commutes with addition and differentiation, the output can be written as

$$U_2(x, y) = \mathbf{V}f(x, y)\mathbf{V}g^*(-x, -y), \qquad (2.3\text{-}15)$$

where $\mathbf{V}$ is the gradient operator. Similarly, with $n = 2$ in Eq. (2.3-11), the output becomes

$$U_2(x, y) = \Delta f(x, y) * \Delta g^*(-x, -y), \qquad (2.3\text{-}16)$$

where Δ is the Laplacian operator. The gradient or Laplacian filters allow a correlation of the derivatives of the signal and the input to be performed. Such correlations have been found to increase spectacularly the selectivity and signal-to-noise ratio of correlators used for pattern recognition [102].

By saturating the dc and low frequency content of the input spectrum, film nonlinearities can perform an operation quite similar to that of a derivative filter, although less systematically. The influence of the photographic process on the selectivity of the filters used for pattern recognition has been extensively studied [104–113]. The optimum amount of saturation, as well as the optimum frequency band to be used for a particular problem, is obviously a function of the type of input, and the system requirements and must be defined (often empirically) for each application.

Besides low frequency saturation, other techniques have been proposed to improve the signal-to-noise ratio of optical correlators. Vander Lugt and Rotz, for example, proposed interchanging the role of signal and input [114] (i.e., introducing the input as a Fourier transform hologram instead of the signal) with some gain in the S/N and a relaxation of the positioning requirements. Such a system, however, would only be practical if each input must be correlated with many signals (it is easier to change the input than to change the filter) or if a real-time device is used to introduce the Fourier-transformed input into the system. For binary inputs, gain in the S/N can also be obtained by correlating complementary patterns [115].

E. Computer-Generated Spatial Filter

The best known technique for generating a matched filter is to record a Fourier transform hologram of the desired signal. Although the photographic process can be controlled, to some extent, it leaves few degrees of freedom for optimization of the filter. In spite of these limitations, photographic recording is the easiest and cheapest technique so far. To achieve more precise control of the information contained in the filter, the most versatile technique (at least if the signal has a relatively simple form) is to generate the filters by a computer, using known techniques of computer-generated holograms.

Whether the spatial filter is intended to recognize a particular signal or to measure some deviation from an ideal signal, it is often imperative to select carefully the information to which the filter is matched. Diverse techniques for producing computer-generated matched filters have been proposed and realized by Lohmann, Huang, Lee, and others [51–54, 116–119]. Synthetic filters have also been found very useful in producing exact derivative or gradient correlation filters [50, 120] and in shaping the response curve of matched filters used for quality control [121].

F. Multiplexing

The analogy between optical matched filtering and similar techniques in communication theory suggests the possibility of optical multiplexing in order to correlate an input with a set of different signals. Originally, Vander Lugt [84, 122] proposed superimposing the filters on the same holographic plate but with each filter on a different carrier. The filters can then be addressed simultaneously by the input, and the different correlation peaks are measured by an array of detectors or by a TV tube scanning the output plane. Similar schemes in which coded reference beams are used to unscramble the outputs were proposed by Gabor [123] and La Machia and White [124]. Leith et al. [125] coded the superimposed holograms by rotating the plate for each exposure, a technique quite similar to θ-modulation [126]. The superposition, either coherent (all filters recorded together) or incoherent (filters sequentially recorded) of matched filters is limited by the dynamic range of the recording medium and the drastic decreases in diffraction efficiency (proportional to the square of the number of superimposed filters) of superimposed holograms. Other multiplexing techniques in which the different filters are spatially separated in the filter plane avoid this difficulty. Groh, for example [127], used point holograms to produce multiple images of the input spectrum. The spatially separated filters are recorded with different carriers, and an array of detectors picks the outputs of the simultaneously addressed filters. Koch and Rabe [128] used a similar technique with a special phase plate multiplying the images [129]. They compared their results with a time sharing method where the filters are sequentially addressed. This last technique imposes the least stringent space–bandwidth (SBW) product requirement for the recording medium and the optical system but, being essentially serial, is slower. Color multiplexing using a multiple-wavelength source has also been proposed recently [130, 131].

Multiplexing techniques have been used to process an input either simultaneously or sequentially with a set of filters. Some applications have been, for example, the simultaneous correlation of the shadow of a back-illuminated three-dimensional input with different shapes corresponding to different possible settings of the object [132–134]. Sequentially addressed multiple filters have also been used to overcome the size and/or orientation sensitivity problem of most optical correlators by storing side by side a number of filters matched to different sizes and orientations of the input.

G. Orientation and Size Mismatch

Three important points have restricted the practical application of coherent-optical pattern recognition. They concern the selection of information and the sensitivity of the technique to input size and orientation. The success of many practical systems based on optical correlation very much depends on how well these problems can be solved. We discuss the sensitivity to size and orientation first.

The degradation of the correlation peak with input orientation and size mismatch has been studied for at least three different types of inputs: relatively simple geometrical forms [135], more complicated but still regular patterns such as diatoms [136] and alphanumerics [122], and aerial imagery with irregular structures and very short correlation lengths [137].

The influence of size variations on the correlation peak clearly depends on the space–bandwidth product of the input and on the band of spatial frequencies to which the filter is tuned. The influence of orientation mismatch follows the same general pattern, with more severe degradation for an input with a larger space–bandwidth product. Typical values of S/N drops for given size and orientation mismatches are given in Table 2.3-1.

Some qualitative arguments can be put forward to explain the effect of an angular mismatch between input and stored signal. If the filter is tuned to a bandpass around the spatial frequency u_c, which corresponds to a coordinate $x_c = \lambda f u_c$ in the filter plane, a simple criterion for finding the amount of tolerable rotation between filter and input is that the linear displacement at a radial distance x_c in the filter plane be smaller than the size of a resolution cell. A resolution cell in the filter plane has a dimension $\lambda f / L_0$, where L_0 is the input size and f is the focal length of the lens. The maximum rotation is found to be

$$\Delta\theta = 1/L_0 u_c.$$

For a 30-mm input format and a filter tuned to 5 lines/mm, the value of $\Delta\theta$ is about 0.5 deg, which is close to the rotation experimentally found to give a S/N drop of 3 dB for highly structured inputs.

Input size and orientation were already of concern to Vander Lugt [2-84] who in his early papers proposed using a converging-beam geometry in order to scan and search for the scale of the input spectrum by moving the input along the z-axis. This technique allows size ranges up to about 20 or 30% to be analyzed. Moving parts, however, are not very appealing in such a system. Not only does it exclude real-time work, but mechanical motion is difficult to control within the required accuracy. The converging-beam geometry may also introduce new problems and aberrations.

Techniques for the orientation search have also been proposed. Since rotation of the input and its spectrum are identical, the orientation search can be performed by rotation of either the input or the filter. Input rotation, however, leads to a slow system if the entire output field must be searched (to pick and measure correlation peaks) after each rotation step of a few degrees

TABLE 2.3-1

Influence of Size Variation and Orientation Mismatch on the Output S/N for Two Different Types of Input[a]

Type of pattern	Size variation (%)	Drop in S/N (dB)	Orientation mismatch (deg)	Drop in S/N (dB)
Geometrical (particles, micromechanics, diatoms, contours, etc.)	5	2–5	5	~3
Structured (aerial imagery, etc.)	1	~10	<0.5	~3

[a] Input format ~35 mm, filter tuned to a special frequency between 5 and 10 lines/mm.

or less. A better solution (proposed by Vander Lugt) is to rotate the filter and to compensate for the change in the direction of the diffracted beam, which is caused by rotation of the holographic fringes, by using a wedge which rotates together with the filter and brings the first diffracted order back on-axis. The output correlation field is then stationary on-axis. Time sharing techniques have the disadvantage of greatly lengthening the time required to analyze a single-input scene. As mentioned previously, multiplex techniques have also been proposed to solve this problem by searching simultaneously for a set of signals with different orientations or sizes [133, 134]. When the input is such that only a few different signals must be multiplexed, filter superposition (space sharing) might be a viable solution. If it is, however, necessary to superpose the spectra of more than approximately 10 inputs, the drop in diffraction efficiency and signal-to-noise ratio becomes intolerable.

A totally different approach has been proposed by Casasent and Psaltis [55, 56]. Their technique rests on the observation that a Mellin transform is scale-invariant and can be made equivalent to a combination of a coordinate transformation followed by a Fourier transform (this last being easily performed by optical means). The Mellin transform, however, is not invariant to input shift (recall that a Fourier transform is shift-invariant but is not scale-invariant). It is, in principle, possible to combine the two techniques to perform a scale and shift-invariant correlation-like operation. It has also been demonstrated that rotation invariance can be included in this scheme. The technique requires an initial step of input preparation in which the input $f(x, y)$ is first Fourier-transformed (optically) and the modulus of its spectrum $|F(u, v)|$ is extracted (by a quadratic detector).

Ablation of the phase spectrum is necessary to make the system shift-invariant. The result is that the system operates on the input autocorrelation rather than on the input itself. Next, a change to polar coordinates,

$$\rho = (u^2 + v^2)^{1/2}, \qquad \theta = \tan^{-1}(v/u), \tag{2.3-17}$$

leads to the new real function $F_p(\rho, \theta)$. This geometrical transformation is made to separate the effects of size variation (which affects ρ only) from the orientation mismatch (which affects θ only). In the polar coordinates (ρ, θ), a rotation is expressed by a shift of part of the input along the θ-axis. This is invariant under Fourier transformation. The next operation to be performed on $F_p(\rho, \theta)$ is a Mellin transform (size-invariant) in ρ. This can be done by first introducing a coordinate stretching $\rho = e^{\alpha}$ followed by a Fourier transform in α:

$$\mathcal{M}_\rho[F(\rho, \theta)] = \int_0^\infty F(\rho, \theta)\rho^{-j\omega-1} \, d\rho$$

$$= \int_{-\infty}^{+\infty} F(e^\alpha, \theta)e^{-j\alpha\omega} \, d\alpha = \mathcal{F}_\alpha[F(e^\alpha, \theta)]. \tag{2.3-18}$$

The log scaling can be performed electronically or with computer-generated masks [138] such as those used by Bryngdahl [139] to perform generalized geometrical mapping. The scaled function $F_p(e^\alpha, \theta)$ must be recorded (at best on a real-time reusable spatial light modulation) and Fourier-transformed (eventually optically). The correlation operation based on the described hybrid transformation is, although not simple to implement, scale-, position-, and orientation-invariant.

H. Information Selection

The problem of information selection was mentioned in the preceding section. It concerns the question of which features should be stored in the holographic filter to optimize a particular recognition system. Each application will of course dictate its own requirements. Many pattern recognition problems in which some signals must be detected in an input cannot be solved by the straightforward technique of correlating the input with each signal directly. A better approach might be to correlate the input with a set of pattern features or characteristics in order to increase the number of discrimination criteria.

When applied to quality control [135, 140], matched filtering usually requires filters tuned to the high frequency content of the input in order to detect small defects or small changes in the input. In the diatom recognition problem (which will be described in detail in the next section), the filter should be able to discriminate sharply between two different species but should be insensitive to size and orientation within each species. Unless the information content of the filter is carefully selected, these two requirements are not compatible. For this particular application, the use of averaged filters, matched to the average shape of an input, have been found very useful [141]. Fujii and Almeida have shown, for example, that with such filters the tolerance in size variation and input orientation for a particular input could be increased to 50% and 45 deg, respectively. For an identical performance, a classic matched filter has tolerances of a few percent for the size and about 3 deg for the input orientation.

I. Practical Considerations

Although some special purpose optical correlators have been built and are commercially available, most of them are still in the laboratory stage with little resemblance to what might be called an engineered prototype. As a result, one seldom finds material describing all the little tricks and know-how necessary to make things work." Some very good studies, however, have analyzed some of the practical aspects of optical correlation. These results might serve as guidelines as to what to look for and what to expect in optical matched filtering.

First of all, one needs a laser and some beam expanders, mirrors, and beam splitters. These are standard elements which need no further description. The lenses are another matter, particularly the first lens which is used to perform the Fourier transform, or more exactly to bring the far-field diffraction pattern of the input into its back focal plane. Many lenses specially designed to perform a Fourier transformation have been described in the literature and have appeared on the market [142–145]. For most applications, however, it is sometimes difficult to justify their price unless inputs of large space–bandwidth products (10^8 or higher) are processed or unless space is an important factor in the system (some of these lenses have an effective focal length on the order of 600 mm with a distance between input and output plane of about 400 mm).

By far the most commonly used Fourier transform lenses are simple achromat doublets or triplets. These objectives are corrected to focus a parallel bundle of light into a point on-axis and perform quite well for inputs of reasonable space–bandwidth product. For example, a 25-mm input with a resolution of 40 lines/mm (SBW product 10^6) can be satisfactorily handled by an F/8 or F/5 doublet of focal length 500 mm. (It will only be used at F/15 where its aberrations are not critically severe.)

The role of the quality of the lens must, however, not be underestimated. In one of the early applications of matched filtering—fingerprint identification—the improvement of performance and S/N when using a specially designed lens was found to be quite noticeable [102]. The effects of lens aberration on the correlation peak was studied by Vander Lugt [2-84] and more recently by Casasent and Luu [146]. The influence of wave front distortion has also received attention [147, 148]. Lens aberrations are important if, as is often the case, a small signal is searched for in an extended input. For example, the search for a runway in an aerial photograph or the search for a particular diatom species in a large microscope field of view. Moreover, it is always assumed that the optical system is not only linear but also space-invariant. This is only true, of course, if the aberrations and distortions of the optics are negligible or are themselves space-invariant, which is rarely the case. Another factor which limits the space invariance of an optical matched filtering system and can be easily underestimated is the filter's substrates. Holographic filters are usually recorded on high-resolution emulsions which have a typical thickness of 10 μm or more. These filters must be treated as thick diffraction gratings for which Bragg effects are not negligible. This effect has been studied by Brousseau and Arsenault [149], Winzer and Kachel [150], and Douklias and Shamir [151] who, in a typical experiment, reported a drop of 50% in the autocorrelation peak for a 25-mm lateral displacement of the input. An easily observable consequence of this effect is that, if a filter is recorded, processed, and exactly repositioned, the autocorrelation peak will not be located at the point in the out-plane where the reference beam would have come to focus. To relocate the peak at this point the input must be shifted by, typically, a few milimeters. This small angular mismatch is due to emulsion shrinkage and the corresponding shift of the Bragg angle.

Another important factor in the degradation of the correlation which has been studied is the effect of filter displacement. This is particularly important if many spatially separated filters must be sequentially addressed using either a beam deflector or a mechanical translation of the filters. An estimate of the required accuracy for the filter positioning is that it be smaller than a resolution cell in the filter plane; i.e.,

$$\Delta x_f = \lambda f / L_0, \tag{2.3.19}$$

where L_0 is the input size and f the focal length. For typical values of $\lambda \approx 0.5 \, \mu m$, $f \approx 500 \, mm$, $L_0 \approx 25 \, mm$, one finds $\Delta x_f \approx 10 \, \mu m$. Accuracy on the order of a few micrometers is usually recommended and is not too difficult to achieve with precise mechanical translation or beam deflectors.

Other reasons for the possible degradation of performance have also been studied. For example, Pernick [152] and Herman [153] have analyzed the optimum lens configuration and the effect of vignetting, while Tischer [154] has published an extensive report on the effects of deformations of the filter's substrate (plates or films). The effects of misalignments [155], filter vibrations [156], and dirt, speckle, and spurious reflections on optical elements [157], as well as the practical design of beam splitters, *in situ* filter processing, and automated output analysis [158] have also received some attention.

2.4 APPLICATIONS OF SPATIAL FILTERING

Following the paper of Vander Lugt in 1963, which described an efficient way of producing matched filters for signals of arbitrary complexity, a great number of applications are proposed, studied, tested and, in a few cases, engineered into working instruments. The optical approach to pattern recognition is by no means the only or best one. A number of comparative studies on optical and digital techniques have been published [159–161]. In the last few years, optical digital processing, which typically combines an optical system with digital technology, has received considerable attention. These topics, however, will not be covered here. After a brief review of the application of spatial filtering to image processing, we shall discuss in this section the application of two classic analog optical techniques: spectral analysis or spectral sampling, and holographic matched filtering. A review of optical pattern recognition techniques including these two has been published recently by Ullman [93]. Early applications of matched filtering are described in a paper by Vander Lugt [162].

Whether optical matched filtering is a viable solution for particular problem must be decided individually for each application. There are no general rules except that, apparently, the coherent-optical systems which have been most successful so far have dealt with situations in which a relatively simple (and easily implemented optically) operation has to be performed repetitively on a large amount of data. In these cases, full advantage can be taken of the optical processor's most attractive feature which is its ability to parallel-process information at very high data rates.

The applications of spatial filtering have touched so many different disciplines that it would be a formidable task to try to draw up a complete list. The examples mentioned here were chosen to illustrate the diversity of applications where spatial filtering has been shown to be a useful tool.

A. Image Processing

From the work of Marechal and Croce [17], the use of spatial filters in coherent-optical imaging systems has been recognized as a powerful technique. Since, in such systems, the amplitude transmittance of a pupil mask in the Fourier plane is the transfer function of the system, it is very simple to manipulate the spatial frequency content of an input with masks or filters. Operations such as contrast enhancement (low spatial frequency suppression) [17, 37], extraction of periodic signals from random noise [163], and elimination of periodic noise such as raster lines or halftone data [37] are easily implemented.

Simple amplitude [36, 164] or amplitude and phase filters [35, 165, 166] have also been very successful for image enhancement and aberration compensation. Their construction by vacuum film deposition, however, imposes severe technological limits on their complexity. Holographic filters have opened up new possibilities: in particular, in transfer function equalization problems such as deblurring [167] and aberration compensation by inverse filtering. Similar techniques have been applied to the restoration of radio and thermographic images [168–170].

B. Spectrum Analysis

On the border line between image processing and pattern recognition, a simple and useful technique for analyzing pictorial data has been the extraction of characteristic features by masking the spectrum with binary (opaque or transparent) filters. This technique has been applied in processing seismic [171, 172] and other geophysical data [173–175] and in analyzing bubble chamber photographs [176, 177]. Its implementation in incoherent light [178] has been studied.

Pattern recognition via spectral analysis, partly because of its simplicity, has been a very successful tool. The technique takes full advantage of the simplicity and ease with which a lens can furnish the Fourier transform, hence the irradiance spectrum of a two-dimensional input. Because the irradiance or power spectrum is invariant with respect to the position of the input, it is possible to extract from the input a characteristic spectral signature which can be used to identify or classify the input. This signature can be obtained by sampling the spectrum with a set of fixed detectors or by scanning the spectrum with an aperture (often a rotating vertical slit or wedge). This technique has been applied to numerous problems such as the recognition and classification of small particles with different shapes [179, 180] or the automatic pattern recognition in photographic imagery [181–183]. A re-fined variant of the technique has also been successfully used by Stark *et al.* to measure the distribution of particles of different sizes [184–186]. Particle sizing will be discussed in much greater detail in Chapters 3 and 4. Power spectrum masking or intensity spatial filtering has also been proposed for the inspection of integrated circuit photomasks [187, 188]. In the area of quality control, samples of the power spectrum were used to identify objects on an automated assembly line [189] and to measure, with remarkable accuracy, the diameter of precision bores [190]. Another application of optical spectrum analysis has been in the control of surface roughness using one-dimensional Fourier transforms [191].

High speed analysis of power spectra [192] (up to more than 1000 images/sec) has been performed with a specially designed detector con-sisting of 32 annular ring elements, each selecting a narrow frequency band on one half of the sensor, and 32 wedged elements on the other half. The applications of such a device range from high speed image analysis, as illustrated in the classification of ground features and are correlation [193, 194], to small particle or biological cell analysis [195]. Backed by a mini-computer for analyzing the data, the technique was used to screen lung x rays and was able to diagnose pneumoconiosis (black lung) with a remarkable 90% accuracy [196, 197].

Another important application of spatial filtering in regard to the automated screening of biomedical samples can be found in the work of Hultzler [198] and in the recently developed and remarkably successful automatic screening of cervical cytological samples [199–201].

C. Applications of Matched Filtering

One of the first proposed applications of optical matched filtering was to the problem of automatic character recognition [102, 162]. Much work has been done in trying to apply Vander Lugt filters to automatic character reading [94, 106, 202, 203]. This has, however, not been an easy application [204]. One of the reasons is that the very selective matched filters which can be constructed to obtain a high signal-to-noise ratio are very intolerant of any character variation such as size, orientation, or defects. Even if the problem of real-time conversion of the input into a transparency suitable for a coherent system is assumed solved, character reading remains problematic. In spite of the difficulties, the subject is still actively researched [205–208]. Multiplexing techniques [209–211], as well as detailed analysis of the cross-correlation distribution [211], have been proposed to increase the number of discrimination criteria.

Another well-known area of application for optical matched filtering has been fingerprint identification [35, 102, 212, 213] and personalized ID card checking [214]. These systems, proposed in the late sixties have not yet come to fruition, most probably because the vast improvement in digital pattern recognition techniques has made these early optical systems, often relying on so irreproducible a process as the photographic one, awkward and out of date.

Some interesting work has been done by Vienot and his group in the area of handwriting analysis [215–217], where the objective is to recognize and extract the similarities or differences in handwritten documents, hopefully pointing out a forgery. This technique has also been used to draw an "average portrait" by extracting the common features from a set of facial descriptions [215].

D. Radar data processing

The application of optical processing to radar data has received considerable attention [218–222]. The optical processing of synthetic aperture radar information was one of the earliest and foremost successes of optical computing. A review on the subject has been recently published by Leith [223]. In the same volume, Casasent reviews some recent work [224] on optical processing of phased array data [225, 226]. The application of optical processing to chirp and pulse-Doppler radar has also been discussed [227–230]. Much work has also been published on the optical display and simulation of the radar ambiguity function [231–236].

The joint transform correlation, although not new, has recently received some attention in radar signal processing because of its real-time capability. In an early system described by Weaver and Goodman [237], both input and signal are placed side by side, their centers separated by a distance x_0, in the input plane; and their spectra are recorded simultaneously. The linearly processed (in intensity) record, containing interference fringes of average frequency $x_0/\lambda f$, is analyzed in a second Fourier transform optical system. The resulting off-axis first diffraction orders lead to output amplitude distributions proportional to the correlation of input and signal. Weaver used such a system to detect radar signals with low signal-to-noise ratio [238]. A more recent joint transform correlator was used by Guilfoyle to process range Doppler radar ambiguity functions [239], while Lee et al. proposed a real-time dual-axis version using a thermoplastic spatial light modulator, thereby eliminating the photographic recording step [240].

Matched filtering has found another important domain of application in bandwidth compression and data reduction. It was recognized early by Leith [241] and Vander Lugt and Rotz [242] that the transmission of processed data instead of raw data would result, in many cases, in an enormous saving of the transmitter bandwidth—an invaluable gain in an area such as space communication. More recent aspects of the subject have been studied by Casasent and Klimas [243, 244].

E. Space applications

On-board optical data processing has been used in earth reconnaissance and planetary spacecraft. The main reason for its use was to compress the transmitted signal bandwidth. It led, however, to some other interesting research. System design, for example, became a primary concern. Compact, foldable optical processors using parabolic mirrors instead of lenses and a laser diode as a coherent source were built (see, for example, Hussain-Abidi [245, 246]). More recently, Fienup et al. [247, 248] have studied the performance of compact optical matched filter processors using holographic optics.

Optical matched filtering techniques have also been proposed for space navigation. Automated tracking using planetary or lunar features has been studied by Holeman and Welch [249], while Gorstein et al. [250] used matched filtering techniques to search and track unique star features and determine space vehicle attitude. The real-time comparison of an image captured by sensors with a bank of reference maps has also been proposed [251, 252] for automated missile guidance.

F. Aerial imagery

The screening of aerial imagery usually involves very large amounts of data; a great deal of time and effort could conceivably be saved by efficient preprocessing. The search for characteristic features or forms in aerial reconnaissance imagery using spatial filtering has been studied by Rotz and Greer [253] and Lieb *et al.* [254, 255]. A recent article by Balasubramanian [256] describes the application of image processing to photogrammetry.

Similar screening and image analysis techniques have been applied to other types of imagery. For example, Vander Lugt [257] described a hybrid optical–digital system using an optical processor coupled to a PDP8 computer for the special task of analyzing cloud motion in ATS III satellite imagery. Gee *et al.* have used matched filtering for the detection of meteor trails [258], and Vander Lugt and Mitchel have proposed a dual-channel optical processor for signature recognition of nuclear detonation [259].

G. Industrial quality control

Aside from the many applications of spectral analysis to industrial processes mentioned earlier, matched filtering has also been found useful in quality control. The early detection of fatigue, for example, is a very serious problem in the production of complex mechanical structures. Matched filtering was used by Marom [140, 260] to detect minute surface deformations and microcracks which are an early sign of mechanical fatigue. Similar techniques were used by Bond *et al.* [261] and Jenkins and McIlwain [262] to spot early failure of solder joints. Automated inspection of mass-produced micromechanical elements using optical correlation techniques has been studied by Indebetouw *et al.* [121, 135], while Gara [68, 263] has used real-time optical correlation to trace and track moving objects. In the two last examples, an attempt is made to provide, through matched filtering techniques, a "smart eye" for the robots used in automated assembly lines.

In another area of quality control, a matched filtering technique has been developed by Grumet to evaluate the modulation transfer function (MTF) of optical systems [264].

Particle sizing is yet another area where optical correlation has been found useful. The size of potentially harmful particles is often a very important factor in determining the degree of danger they represent. Tschudi *et al.*, for example, have used computer-generated holograms in a multichannel system which simultaneously sorts particles in the micrometer range [265].

H. Biomedical application

A few of the numerous applications of coherent-optical techniques to biomedical sciences have already been mentioned. Recent reviews of the subject have appeared in the literature [266, 267]. The application of image enhancement and aberration compensation techniques to radiographic and thermographic data has been particularly significant [168–170]. Power spectrum analysis has also become a very powerful tool in automated analysis of biomedical samples [195–201]. Holographic spatial filtering plays an important role in image deblurring. It has also been proposed that matched filters be used for the screening of biomedical data [268]. Conceivably, computer-generated holographic matched filters could be used, together with certain scattering models of biological cells [269] for such screening.

Another area of application of matched spatial filtering has been in water pollution monitoring by diatom population measurements [136], a subject treated in more detail in the remainder of this chapter.

I. Diatom identification

Early interest in the identification of diatoms (algae) by optical means [270–274] was generated because of the important roles that diatoms play in water pollution monitoring. By obtaining distributions of diatoms as a function of species and sampling the water at given intervals one can monitor the degree of water quality. Human identification of diatoms is a tedious and time-consuming task requiring a trained diatom taxonomist; therefore, this method is not practical. One would like to develop a rapid optical method for both identifying and counting diatoms. The method we chose is that of complex matched spatial filtering which has been discussed in the preceding sections.

The diatoms considered are about 40 μm in length and are composed of a silicon dioxide skeleton. The shape of a diatom is usually symmetric about two axis, and it has a very detailed interior made of stria or a gratinglike structure. Within a given species diatoms are nearly identical. Typical features can be seen in the 25 diatoms shown in Fig. 2.4-1. Figure 2.4-2 shows 19 cross-correlation signals obtained through 4 of the 19 diatoms on the photograph to the left. Averaging was necessary to account for the slightly rotated interiors as well as depth of focus [274].

Since the diatoms are phase objects, the first step in their identification was to convert them into a suitable input format for the optical processor. This was done by photographing the specimens through a phase-contrast microscope. The resulting 35-mm negatives were then made into positive transparencies in order to reduce the background noise. A matched spatial filter was prepared for each or the 25 diatoms shown in Fig. 2.4-1. This set of diatoms was carefully selected to test the effectiveness of matched spatial filtering in identifying diatoms. The filters were recorded such that their spatial frequencies gave a good signal-to-noise autocorrelation. We purposely avoided aiming for the maximum signal-to-noise ratio in order to avoid excessive performance sensitivity to size and orientation of the input. Shown in Fig. 2.4-3 is the 25 × 25 matrix output composed of 625 correlation signals. Each of the 25 matched filters was compared against the entire set (25)

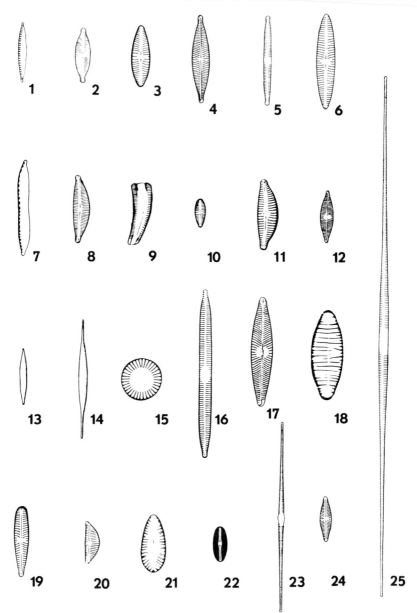

FIG. 2.4-1. Twenty-five diatoms each photographed through a phase-contrast microscope and used as input to a coherent-optical processor.

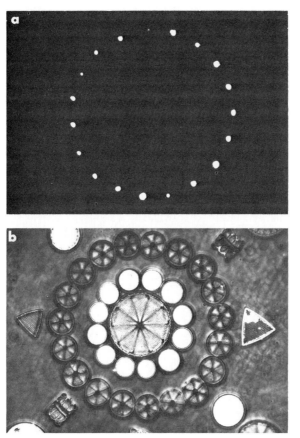

FIG. 2.4-2. (a) Diatoms arranged in a circle by a taxonomist. (b) Correlation signals obtained with averaged matched spatial filter using 4 of the 19 diatoms shown in (a). (After Almeida *et al.* [274].)

used as the input. It is seen that the diagonal terms, which have been normalized, are the autocorrelation signals, while the off-diagonal signals are the cross-correlations. The 100% success in being able to distinguish all 25 diatoms from each other demonstrates the high degree of pattern recognition possible with matched spatial filters [275, 283].

To avoid excessive sensitivity to size and rotation of the input, the filters for the 25 diatoms were tuned for the lower spatial frequencies. Thus, in effect, little of the detailed structure of the diatom's interior skeleton played a role in the recognition process. Going one step further Fujii and Almeida [141] showed that matched spatial filters made of diatom sketches using only an outline could identify certain diatoms. In a particular case, as shown in

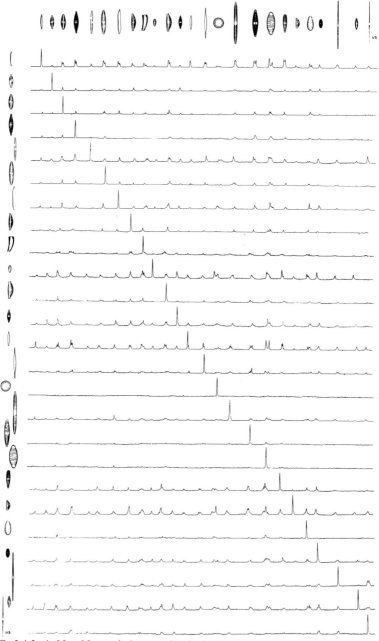

FIG. 2.4-3. A 25 × 25 correlation matrix obtained with matched spatial filters of each diatom compared against the entire 25. The diagonal autocorrelation signals have been normalized. (After Almeida *et al.* [275].)

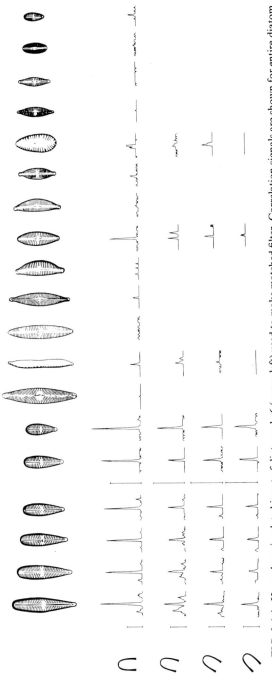

FIG. 2.4.4. Horseshoe simulated input of diatoms 1–6 (upper left) used to make matched filter. Correlation signals are shown for entire diatom set (top row); also as a function rotation to about 45 deg. (After Fujii and Almeida [141].)

Fig. 2.4-4, the input filter was rotationally invariant to about ± 45 deg and size-invariant to about a 50% size variation.

Some understanding of simulated filters can be obtained by studying the Fourier transforms of individual and averaged diatoms, as was done by Almeida and Fujii [276]. While simulated filters may not work for all

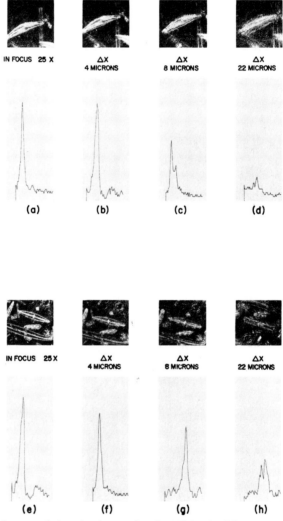

FIG. 2.4-5. Autocorrelation signals as a function of depth of focus using a microscope coherent processor. (After Partin *et al.* [277].)

specimens, they are easier to realize on film than the previously discussed multiplexed filters [133].

The use of phase-contrast transparencies in diatom identification has been seen to be quite successful. However, input preparation of the film is time-consuming, and it would be better to use the glass microscope slide containing diatoms directly as the input. This has recently been done by Partin *et al.* [277], using an interference phase-contrast microscope. The study also shows how the autocorrelation signals vary as function of depth of focus in

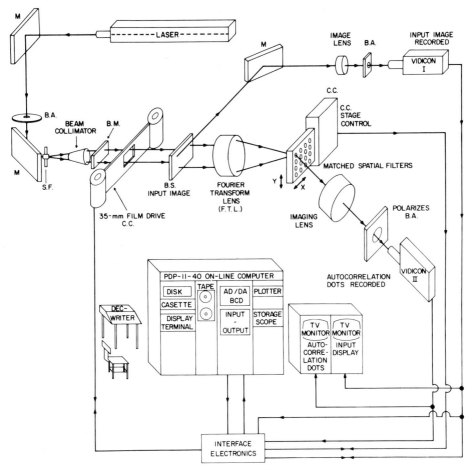

FIG. 2.4-6. Hybrid coherent-optical processor. S.F., spatial filter; M, mirror; B.S. beam splitter; F.T.L., Fourier transform lens; B.M., beam modulator; R.B., reference beam; B.A., beam attenuator; C.C., computer controlled. (After Almeida *et al.* [272].)

the microscope; see Fig. 2.4-5. The coherent microscope processor could be further improved by using simulated matched spatial filters with it.

The use of coherent microscope processors [277–281] is a step toward real-time diatom identification. One could also consider a liquid crystal light value [67] or an incoherent-to-coherent transducer such as the PROM [64–66] to perform a similar input function. A main consideration in going to real-time input will be the resolution and contrast requirements of the transducers. Cost and reproducibility of results are also important factors to bear in mind.

The hybrid processor used to perform diatom recognition from input transparencies is shown in Fig. 2.4-6. The processor's optics and functions have already been described in earlier sections. In addition, the processor is interfaced on a on-line basis to a PDP11-40 computer. The computer controls the filter plate holder to an x, y-position accuracy of 1 μm. It also has a slow scan 8-bit digitizer interfaced to read out the correlation signals for later analysis [273]. The input transparency is located in a 360-deg rotatable liquid gate to remove film phase variations. The spatial filters can also be placed in a rotating liquid gate [2-282]. Rotation of the filter has the advantage of a fast scan for randomly oriented input signals. The layout of interfaces to the hybrid processor is shown in the block diagram in Fig. 2.4-7. More details are given by Almeida et al. [273].

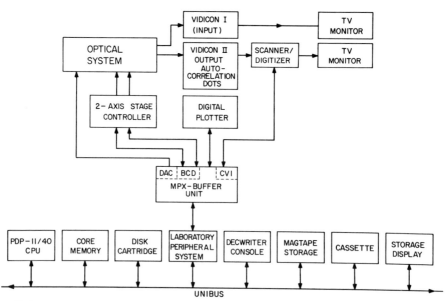

FIG. 2.4-7. Optical–electronic interface of the hybrid processor. (After Almeida et al. [273].)

2.5 CONCLUSION

Coherent-optical processing has had an interesting history of application. In particular, the technology of pattern recognition by complex matched spatial filters has greatly improved, and the technique itself is being used in novel applications. Despite some significant existing problems with the development of coherent-optical processors, pattern recognition via complex spatial filtering holds great promise for new and exciting applications.

REFERENCES

[1] E. Abbe (1873). *Arch. Microsc. Anat.* **9**, 413.
[2] E. Abbe (1873). *Jena. Ges. Med. Naturwiss.* **129.**
[3] E. Abbe (1880). *Carlsberg Rep. Phys.* **16**, 303.
[4] R. Clausius (1864). *Pogg. Ann.* **1**, 121.
[5] M. Helmholtz (1874). *Pogg. Ann. Jubelband* **557.**
[6] M. Born and E. Wolf (1970). "Principles of Optics." Pergamon, New York.
[7] R. Hooke (1665). *Micrographia* **47.**
[8] C. Huygens (1690). "Traité de la Lumière." Leyden, Netherlands.
[9] T. Young (1802). *Philos. Trans. R. Soc. London* **12**, 387.
[10] A. J. Fresnel (1816). *Ann. Chim. Phys.* **2**, 239.
[11] A. Sommerfeld (1896). *Math. Ann.* **47**, 317.
[12] A. B. Porter (1906). *Philos. Mag.* **1**, 121.
[13] Lord Rayleigh (1896). *Philos. Mag.* **42**, 167.
[14] F. Zernike (1935). *Z. Tech. Phys.* **16**, 454.
[15] F. Zernike (1964). *In* "Nobel Lectures." Elsevier, Amsterdam.
[16] P. M. Duffieux (1970). "L'Integrale de Fourier et ses Applications a l'Optique." Masson, Paris. (Orig. publ., 1946.)
[17] A. Marechal and P. Croce (1953). *C. R. Acad. Sci. Paris.* **237**, 607–609.
[18] A. Blanc-Lapierre (1953). *Symp. Microwaves Opt., McGill Univ., Montreal.*
[19] T. P. Cheatham and A. Kohlenberg (1954). *IRE* **4**, 6.
[20] P. Elias (1953). *J. Opt. Soc. Am.* **43**, 229–232.
[21] J. E. Rhodes (1953). *Am. J. Phys.* **21**, 337.
[22] H. H. Hopkins (1951). *Proc. R. Soc. London, Ser. A* **208**, 263.
[23] H. H. Hopkins (1953). *Proc. R. Soc. London, Ser. A* **217**, 408.
[24] P. Jacquinot, P. Boughton, and B. Dossier (1949). "La Theorie des Images Optiques." Rev. Opt., Paris.
[25] E. H. Linfoot (1964). "Optical Image Evaluation." Focal Press, New York.
[26] E. H. Linfoot (1955). *J. Opt. Soc. Am.* **45**, 808.
[27] A. Marechal (1951). *Natl. Bur. Stand. (U.S.), Circ. No.* 526.
[28] E. W. H. Selwyn (1951). *Natl. Bur. Stand. (U.S.), Circ. No.* 526.
[29] W. H. Steel (1953). *Rev. Opt.* **32**, 4.
[30] A. Marechal and M. Francon (1970). "Diffraction Structure des Images." Masson, Paris.
[31] A. Marechal, P. Croce, and K. Dietzel (1958). *Opt. Acta* **5**, Hors Ser. 256.
[32] P. Croce (1956). *Rev. Opt.* **35**, 569.
[33] M. Marquet (1959). *Opt. Acta* **6**, 404.
[34] J. Tsujiuchi (1960). *Opt. Acta* **7**, 243–261.
[35] J. Tsujiuchi (1963). *Prog. Opt.* **2**, 133–180.

[36] P. Jacquinot and B. Dossier (1964). *Prog. Opt.* **3**, 31–186.

[37] E. L. O'Neil (1956). *IRE Trans. Inf. Theory* **IT-2**, 56–65.

[38] G. L. Turin (1960). *IRE Trans. Inf. Theory* **IT-6**, 311–329.

[39] L. J. Cutrona (1964). *Proc. IEEE* **1**, 101–108.

[40] D. Gabor (1949). *Proc. R. Soc. London, Ser. A* **197**, 454–487.

[41] D. Gabor (1951). *Proc. R. Soc. London, Ser. B* **378**, 449–469.

[42] E. Leith and J. Upatniek (1962). *J. Opt. Soc. Am.* **52**, 1123–1130.

[43] E. Leith and J. Upatniek (1963). *J. Opt. Soc. Am.* **53**, 1377–1381.

[44] E. Leith and J. Upatniek (1964). *J. Opt. Soc. Am.* **54**, 1295–1301.

[45] T. M. Holladay and J. D. Gallatin (1966). *J. Opt. Soc. Am.* **56**, 869–872.

[46] M. Osterberg (1947). *J. Opt. Soc. Am.* **37**, 726.

[47] A. S. Marathay (1969). *J. Opt. Soc. Am.* **59**, 748–753.

[48] A. B. Vander Lugt (1968). Rep. No. 459-22-T. Radar Lab., Inst. Sci. Technol., Univ. of Michigan, Ann Arbor.

[49] A. B. Vander Lugt (1964). *IEEE Trans. Inf. Theory* **IT-10**, 139–145.

[50] A. W. Lohmann and D. P. Paris (1968). *Appl. Opt.* **7**, 651–655.

[51] B. R. Brown and A. W. Lohmann (1966). *Appl. Opt.* **5**, 967–969.

[52] A. W. Lohmann, D. P. Paris, and M. W. Werlich (1967). *Appl. Opt.* **6**, 1139–1140.

[53] W. H. Lee and M. O. Greer (1974). *Appl. Opt.* **13**, 925–930.

[54] K. Campbell, G. W. Wecksung, and C. R. Mansfield (1974). *Opt. Eng.* **13**, 175–188.

[55] D. Casasent and D. Psaltis (1976). *Opt. Commun.* **17**, 59–63.

[56] D. Casasent and D. Psaltis (1976). *Appl. Opt.* **15**, 1795–1799.

[57] L. S. G. Kovasnay and A. Arman (1975). *Rev. Sci. Instrum.* **28**, 793.

[58] L. P. Horwitz and G. L. Shelton (1961). *Proc. IRE* **49**, 1975.

[59] D. McLachlan (1962). *J. Opt. Soc. Am.* **52**, 454–459.

[60] G. L. Roger (1975). *Opt. Laser Technol.* **8**, 153–162.

[61] J. D. Armitage and A. W. Lohmann (1965). *Appl. Opt.* **4**, 461–467.

[62] A. W. Lohmann and H. W. Werlich (1971). *Appl. Opt.* **12**, 670–672

[63] D. Casasent (1978). *In* "Applied Optics and Engineering" (R. Kingslake, ed.). Academic Press, New York.

[64] P. Nisenson and S. Iwasa (1972). *Appl. Opt.* **11**, 2760–2767.

[65] P. Nisenson and R. A. Sprague (1975). *Appl. Opt.* **4**, 2602–2606.

[66] B. A. Horwitz and F. J. Corbett (1978). *Opt. Eng.* **17**, 353–364.

[67] J. Grinberg, A. Jacobson, W. Bleha, L. Miller, L. Fraas, D. Boswell, and G. Myer (1975). *Opt. Eng.* **43**, 217–225.

[68] A. D. Gara (1977). *Appl. Opt.* **16**, 149–153.

[69] A. W. Lohmann (1977). *Appl. Opt.* **16**, 261–263.

[70] W. Stoner (1977). *Appl. Opt.* **16**, 1451–1453.

[71] W. T. Rhodes (1977). *Appl. Opt.* **16**, 265–267.

[72] A. Furman and D. Casasent (1979). *Appl. Opt.* **18**, 660–665.

[73] A. W. Lohmann (1959). *Opt. Acta* **6**, 319.

[74] D. H. Kelly (1961). *J. Opt. Soc. Am.* **51**, 1095–1101.

[75] E. A. Trabka and P. G. Roetling (1964). *J. Opt. Soc. Am.* **54**, 1242–1252.

[76] K. Bromley, M. A. Monahan, J. F. Bryant, and B. J. Thompson (1969). *Appl. Phys. Lett.* **14**, 67.

[77] R. E. Williams (1965). *IEEE Trans. Inf. Theory* **IT-4**, 499–507.

[78] B. M. Watrasiewicz (1969). *Opt. Acta* **16**, 321–326.

[79] S. V. Gupta and D. Sen (1972). *Opt. Acta* **19**, 137–154.

[80] M. O. Bartlett (1979). *Opt. Commun.* **29**, 37.

[81] A. Lacourt (1978). *Opt. Commun.* **27**, 47–52.

[82] J. P. Goedgebuer and R. Gazeu (1979). *Opt. Commun.* **27,** 53–56.

[83] J. W. Goodman (1968). "Introduction to Fourier Optics." McGraw-Hill, New York.

[84] A. Vander Lugt (1966). *Appl. Opt.* **5,** 1760–1765.

[85] J. B. Thomas (1969). "An Introduction to Statistical Communication," Chap. 5. Wiley, New York.

[86] C. W. Helstrom (1967). *J. Opt. Soc. Am.* **57,** 297–303.

[87] J. L. Horner (1969). *J. Opt. Soc. Am.* **59,** 553–558.

[88] J. L. Horner (1970). *Appl. Opt.* **9,** 167–171.

[89] B. Liu and N. C. Gallagher (1974). *J. Opt. Soc. Am.* **64,** 1227–1236.

[90] N. C. Gallagher and B. Liu (1975). *J. Opt. Soc. Am.* **65,** 182–187.

[91] T. Yatagai (1976). *Opt. Commun.* **19,** 236–239.

[92] W. T. Cathey (1971). *J. Opt. Soc. Am.* **61,** 478–482.

[93] J. R. Ullman (1974). *Opto-electronics (London)* **6,** 319–332.

[94] B. M. Watrasiewicz (1972). *Opt. Laser Technol.* **4,** 288.

[95] G. W. Stroke and R. G. Zech (1967). *Phys. Lett. A* **25A,** 89–90.

[96] G. W. Stroke, G. Indebetouw, and C. Puech (1967). *Phys. Lett. A* **26A,** 443–444.

[97] S. I. Ragnarson (1970). *Phys. Scr.* **2,** 145–153.

[98] J. W. Goodman and H. B. Strübin (1973). *J. Opt. Soc. Am.* **63,** 50–58.

[99] J. W. Goodman (1977). *In* "Coherent Optical Engineering" (F. T. Arecchi *et al.*, eds.), pp. 263–280. North-Holland Publ., Amsterdam.

[100] S. Lowenthal and Y. Belvaux (1967). *Opt. Acta* **14,** 245–258.

[101] S. Lowenthal and J. Belvaux (1966). *C. R. Hebd. Seances Acad. Sci., Ser. B* **262,** 413.

[102] S. Lowenthal and J. Belvaux (1967). *Rev. Opt.* **1,** 1–64.

[103] R. M. Bracewell (1965). "The Fourier Transform and its Applications." McGraw-Hill, New York.

[104] R. A. Binns, A. Dickinson, and B. M. Watrasiewicz (1968). *Appl. Opt.* **7,** 1047–1051.

[105] K. J. Petrosky and S. H. Lee (1971). *Appl. Opt.* **10,** 1968–1969.

[106] M. J. Caufield and W. T. Maloney (1969). *Appl. Opt.* **8,** 2354–2356.

[107] J. Bulabois, A. Caron, and J. C. Vienot (1969). *Opt. Laser Technol.* **8,** 191–195.

[108] D. A. Ansley, P. J. Peters, and J. C. Cassidy (1969). *In AGARD Conf. Proc.* p. 50.

[109] R. O. Webb and F. M. Shofner (1971). *NASA [Spec. Publ.] SP* **NASA SP-299,** 153–166.

[110] R. A. Gonsalvez and R. M. Dumais (1973). *Opt. Eng.* **12,** 43–46.

[111] D. Casasent and A. Furman (1977). *Appl. Opt.* **16,** 1662–1669.

[112] W. J. Dallas, R. Linde, and M. Weiss (1978). *Opt. Lett.* **3,** 247–249.

[113] F. T. S. Yu and G. C. Kung (1972). *J. Opt. Soc. Am.* **62,** 147–149.

[114] A. Vander Lugt and F. B. Rotz (1970). *Appl. Opt.* **9,** 215–221.

[115] U. Wagner (1971). *Opt. Commun.* **3,** 130–132.

[116] B. R. Brown and A. W. Lohmann (1969). *IBM J. Res. Dev.* **13,** 160–168.

[117] T. S. Huang *et al.* (1971). *Proc. IEEE* **59,** 1586–1609.

[118] S. H. Lee (1973). *Pattern Recognition* **5,** 21–35.

[119] A. W. Lohmann and D. P. Paris (1967). *Appl. Opt.* **6,** 1739–1742.

[120] S. H. Lee (1974). *Opt. Eng.* **18,** 196–207.

[121] G. Indebetouw (1977). *Appl. Opt.* **16,** 1944–1950.

[122] A. Vander Lugt, F. B. Rotz, and A. Klooster (1965). *In* "Optical and Electro-Optical Information Processing" (J. T. Tippet *et al.*, eds.), pp. 125–141. MIT Press, Cambridge, Massachusetts.

[123] D. Gabor (1965). *Nature (London)* **208,** 422.

[124] J. T. La Machia and D. L. White (1968). *Appl. Opt.* **7,** 91–94.

[125] E. N. Leith, A. Kozma, J. Upatniek, J. Marks, and N. Massey (1966). *Appl. Opt.* **5,** 1303–1311.

[126] J. D. Armitage and A. W. Lohmann (1965). *Appl. Opt.* **4,** 399–403.
[127] G. Groh (1970). *Opt. Commun.* **1,** 454–456.
[128] M. Koch and G. Rabe (1972). *Opt. Commun.* **5,** 73–77.
[129] H. Dammann and K. Gortler (1971). *Opt. Commun.* **3,** 312–315.
[130] N. K. Shi (1978). *Opt. Lett.* **3,** 85–87.
[131] S. K. Case (1979). *Appl. Opt.* **18,** 1890–1894.
[132] J. C. Vienot, J. Bulabois, and L. R. Guy (1971). *Opt. Commun.* **1,** 431–434.
[133] S. P. Almeida and J. K. T. Eu (1978). *Appl. Opt.* **17,** 163.
[134] G. Indebetouw (1976). Rep. No. TS-13, IAP. Univ. of Berne, Berne.
[135] G. Indebetouw, T. Tschudi, and G. Herziger (1976). *Appl. Opt.* **15,** 510.
[136] S. P. Almeida and J. K. T. Eu (1976). *Appl. Opt.* **15,** 510–515.
[137] D. Casasent and A. Furman (1977). *Appl. Opt.* **16,** 1652–1661.
[138] D. Casasent and C. Szcutowski (1976). *Opt. Commun.* **19,** 217–222.
[139] O. Bryngdahl (1974). *J. Opt. Soc. Am.* **64,** 1092–1099.
[140] E. Marom (1970). *Appl. Opt.* **9,** 1385–1391.
[141] H. Fujii and S. P. Almeida (1979). *Appl. Opt.* **18,** 1654–1662.
[142] K. Von Bièren (1971). *Appl. Opt.* **10,** 2739–2743.
[143] R. E. Williams, K. Von Bieren, and M. Morales (1975). *Appl. Opt.* **14,** 2944–2954.
[144] C. G. Wynne (1974). *Opt. Commun.* **12,** 266–269.
[145] C. G. Wynne (1974). *Opt. Commun.* **12,** 270–274.
[146] D. Casasent and T. K. Luu (1978). *Appl. Opt.* **17,** 1701–1708.
[147] T. K. Luu and D. Casasent (1979). *Appl. Opt.* **18,** 791–795.
[148] H. W. Rose, T. L. Williamson, and F. T. S. Yu (1971). *Appl. Opt.* **10,** 515–518.
[149] N. Brousseau and M. H. Arsenault (1975). *Appl. Opt.* **14,** 1679–1682.
[150] G. Winzer and I. Kachel (1973). *Opt. Acta* **20,** 359–364.
[151] N. Douklias and J. Shamir (1973). *Appl. Opt.* **12,** 364–367.
[152] B. J. Pernick (1971). *Am. J. Phys.* **39,** 959–960.
[153] S. Herman (1969). *Proc. IEEE* **57,** 346–347.
[154] F. J. Tischer (1968). *NASA Rep.* **NGR-34-002-038 /51.**
[155] C. A. Pipan (1968). *Opt. Spectra* **11,** 55–56.
[156] A. Vander Lugt (1967). *Appl. Opt.* **6,** 1221–1229.
[157] A. Hamori and G. L. Bencze (1977). *Opt. Commun.* **21,** 347–350.
[158] G. Winzer and N. Donklias (1971). *Isr. J. Technol.* **9,** 275–279.
[159] K. Preston (1972). *Proc. IEEE* **60,** 1216–1231.
[160] D. J. Granrath and B. R. Hunt (1977). *SPIE Semin. Proc.* **118.**
[161] D. C. Kovalski (1960). *Bendix Tech. J.* **2,** 78–80.
[162] A. B. Vander Lugt (1968). *Opt. Acta* **15,** 1–33.
[163] H. Damman and M. Koch (1971). *Opt. Commun.* **3,** 251–253.
[164] G. Lansraux (1965). *In* "Optical and Electro-Optical Information Processing" (J. T. Tippett *et al.*, eds.), pp. 59–68. MIT Press, Cambridge, Massachusetts.
[165] P. Considine and R. Profio (1968). *SPIE Semin. Proc.* **16.**
[166] B. Thompson (1974). *SPIE Semin. Proc.* **52.**
[167] J. Tsujiuchi and G. W. Stroke (1971). *In* "Applications of Holography" (E. S. Barrekette *et al.*, eds.), pp. 259–308. Plenum, New York.
[168] H. C. Becker, P. H. Meyers, and C. M. Nice (1968). *IEEE Trans. Bio. Eng.* **BME-15,** 186–195.
[169] J. B. Minkoff, S. K. Hilal, W. F. Konig, M. Arm, and L. B. Lambert (1968). *Appl. Opt.* **7,** 633–639.
[170] G. A. Krusos (1974). *Opt. Eng.* **13,** 208–218.
[171] M. B. Dobrin, A. L. Ingalls, and T. A. Long (1965). *Geophysics* **30,** 1144–1178.

[172] M. B. Dobrin (1968). *IEEE Spectrum* **5**, 59–66.
[173] P. L. Jackson (1965). *Geophysics* **30**, 5.
[174] P. L. Jackson (1965). *Appl. Opt.* **4**, 419–427.
[175] M. J. Pincus and M. B. Dobrin (1966). *J. Geophys. Res.* **71**, 4861.
[176] D. G. Falconer (1966). *Appl. Opt.* **5**, 1365–1369.
[177] M. S. White and D. E. Hall (1969). *SPIE Semin. Proc.* **18**.
[178] M. Lasserre and R. W. Smith (1974). *Opt. Commun.* **12**, 260–265.
[179] H. J. Tiziani, B. M. Beyeler, and W. Witz (1970). *Opt. Laser Technol.* **5**, 75–79.
[180] B. H. Kaye and A. G. Naylor (1972). *Pattern Recognition* **4**, 195–199.
[181] G. C. Lendaris and G. L. Stanley (1970). *Proc. IEEE* **58**, 198–216.
[182] K. Tanaka and K. Osawa (1972). *Pattern Recognition* **4**, 251–262.
[183] H. L. Hasden (1978). *SPIE Semin. Proc.* **201**.
[184] H. Stark, G. Lee, and B. Dimitriadis (1975). *J. Opt. Soc. Am.* **65**, 1436–1442.
[185] H. Stark, D. Lee, B. W. Kao (1978). *Appl. Opt.* **15**, 2246–2249.
[186] H. Stark and G. Shao (1977). *Appl. Opt.* **16**, 1670–1674.
[187] L. S. Watkins (1969). *Proc. IEEE* **4**, 1634–1639.
[188] N. N. Axelrod (1972). *Proc. IEEE* **4**, 447–448.
[189] G. Indebetouw (1977). *Rev. Sci. Instrum.* **48**, 547–549.
[190] A. L. Flamholz and H. A. Frost (1973). *IBM J. Res. Dev.* **17**, 509–518
[191] B. J. Pernik (1979). *Appl. Opt.* **18**, 796–801.
[192] N. Jensen, H. K. Kasdan, D. C. Mead, and J. Thomasson (1974). *IEEE Opt. Comput. '74* pp. 14–17.
[193] S. Nybert *et al.* (1971). *Photogram. Eng.* **6**, 547.
[194] N. Jensen (1973). *Photogram. Eng.* **12**, 1321.
[195] P. F. Mullaney and P. N. Dean (1970). *Biophys. J.* **10**, 764.
[196] H. Stark and D. Lee (1976). *IEEE Trans. Syst., Man Cybern.* **6**, 788–793.
[197] R. P. Kruger, W. B. Thompson, and A. F. Turner (1974). *IEEE Trans. Syst., Man Cybern.* **4**, 40–49.
[198] P. J. S. Hultzler (1977). *Appl. Opt.* **16**, 2264–2272.
[199] B. Pernik, R. E. Kopp, J. Lisa, J. Mendelsohn, M. Stove, and R. Wohler (1978). *Appl. Opt.* **17**, 21–34.
[200] R. Wohler, J. Mendelsohn, R. E. Kopp, and B. J. Pernik, (1978). *Appl. Opt.* **17**, 35–42.
[201] B. Pernik, S. Jost, R. Herold, R. E. Kopp, J. Mendelsohn, and R. Wohler (1978). *Appl. Opt.* **17**, 43–51.
[202] S. Hard and T. Eeuk (1973). *Pattern Recognition* **5**, 75–82.
[203] J. M. Fournier and J. C. Vienot (1971). *Isr. J. Technol.* **9**, 281.
[204] G. Winzer (1974). *IEEE Opt. Comput. '74* pp. 9–13.
[205] D. Casasent and D. Furman (1976). *Appl. Opt.* **15**, 1690–1691.
[206] J. M. Fournier (1977). *In* "Application of Holography" (E. Marom, ed.), pp. 533–539, 549–554. Pergamon, New York.
[207] D. Casasent and D. Furman (1977). *Appl. Opt.* **16**, 1663–1669.
[208] J. E. Hinds, D. Casasent, and E. Caimi (1979). *SPIE Semin. Proc.* **177**.
[209] G. Winzer and N. Donklias (1972). *Opt. Laser Technol.* **10**, 222.
[210] M. Nakajima, T. Morikawa, and K. Sakurai (1972). *Appl. Opt.* **11**, 362–372.
[211] R. A. Gonsalves, P. S. Considine, and A. A. Shea (1977). *SPIE Semin. Proc.* **118**.
[212] V. V. Horvath, J. M. Holeman, and C. Q. Lemmond (1967). *Laser Focus* **6**, 13.
[213] J. T. Thomasson (1969). *SPIE Semin. Proc.* **18**.
[214] K. K. Sutherlin and E. M. Christy (1969). *SPIE Semin. Proc.* **18**.
[215] J. C. Vienot, J. Duvèrnoy, G. Tribillon, and J. C. Tribillon (1973). *Appl. Opt.* **12**, 950–960.
[216] J. C. Vienot and J. Duvernoy (1976). *Appl. Opt.* **15**, 523–529.

[217] J. Duvernoy (1972). *Opt. Commun.* **7,** 142–145.

[218] R. O. Harger (1970). "Synthetic Aperture Radar Systems." Academic Press, New York.

[219] R. A. Shulman (1970). "Optical Data Processing." Wiley, New York.

[220] L. J. Cutrona, E. N. Leith, C. J. Palermo, and L. J. Porcello (1960). *IRE Trans. Inf. Theory* **6,** 386–400.

[221] L. J. Cutrona (1965). *In* "Optical and Electro-Optical Information Processing" (J. T. Tippett *et al.,* eds.), pp. 83–124. MIT Press, Cambridge, Massachusetts.

[222] K. Preston (1972). "Coherent Optical Computers." McGraw-Hill, New York.

[223] E. N. Leith (1978). *In* "Optical Data Processing" (D. Casasent, ed.), Vol. 23, pp. 89–116. Springer-Verlag, Berlin and New York.

[224] D. Casasent (1978). *In* "Optical Data Processing" (D. Casasent, ed.), Vol. 23, pp. 241–281. Springer-Verlag, Berlin and New York.

[225] L. Lambert, M. Arm, and A. Aimette (1965). *In* "Optical and Electro-Optical Information Processing" (J. T. Tippet *et al.,* eds.), pp. 715–748. MIT Press, Cambridge, Massachusetts.

[226] D. Casasent and F. Casasayas (1975). *IEEE Trans. Aerosp. Electron. Syst.* **AES-11,** 65.

[227] D. Casasent and F. Casasayas (1975). *Appl. Opt.* **14,** 1364–1372.

[228] D. Casasent and D. Psaltis (1977). *Proc. IEEE* **65,** 77–84.

[229] D. Casasent and D. Psaltis (1976). *Appl. Opt.* **15,** 2015.

[230] H. Stark, F. B. Tuteur, and M. Sayar (1971). *Appl. Opt.* **10,** 2728–2733.

[231] P. W. Woodward (1963). "Probability and Information Theory with Applications to Radar." Pergamon, Oxford.

[232] R. J. Mark, II, J. F. Walkup, and T. F. Krile (1977). *Appl. Opt.* **16,** 746–750.

[233] R. A. K. Said and D. C. Cooper (1973). *Proc. Int. Electr. Eng.* **120,** 423.

[234] E. N. Leith *et al.* (1970). *IEEE Trans. Aerosp. Electron. Syst.* **AES-6,** 832–840.

[235] B. A. Horwitz (1977). *SPIE Semin. Proc.* **118.**

[236] G. W. Deley (1970). *In* "Radar Handbook" (M. I. Skolnik, ed.), Chap. 3. McGraw-Hill, New York.

[237] C. S. Weaver and J. W. Goodman (1966). *Appl. Opt.* **5,** 1248–1249.

[238] C. S. Weaver, S. D. Ramsey, J. W. Goodman, and A. M. Rosie (1970). *Appl. Opt.* **9,** 1672–1682.

[239] P. Guilfoyle (1977). *SPIE Proc.* **118.**

[240] T. C. Lee, J. Rebholtz, and P. Tamura (1979). *Opt. Lett.* **4,** 121–123.

[241] E. N. Leith (1968). *IEEE Trans. Aerosp. Electron. Syst.* **AES-4,** 879–885.

[242] A. Vander Lugt and F. B. Rotz (1974). Rep. No. 4594-40R. Radar Lab., Inst. Sci. Technol., Univ. of Michigan, Ann Arbor.

[243] D. Casasent and E. Klimas (1978). *Appl. Opt.* **17,** 2058–2063.

[244] D. Casasent (1977). *SPIE Semin. Proc.* **118.**

[245] A. S. Hussain-Abidi (1973). *NASA* [*Spec. Publ.*] *SP* **NASA SP-299,** 163–172.

[246] A. S. Hussain-Abidi (1973). *Pattern Recognition* **5,** 3–11.

[247] J. R. Fienup and C. D. Leonard (1979). *Appl. Opt.* **18,** 631–640.

[248] J. R. Fienup, W. S. Colburn, B. J. Chang, and C. D. Leonard (1977). *SPIE Semin. Proc.* **118.**

[249] J. M. Holeman and T. D. Welch (1967). *Space-Aeronaut.* **6,** 104.

[250] M. Gorstein, J. N. Hallock, and J. Volge (1970). *Appl. Opt.* **9,** 351–358.

[251] D. Casasent and M. Saverino (1977). *SPIE Semin. Proc.* **118.**

[252] R. W. Paulson, E. Price, J. Hodor, and J. Barney (1977). *SPIE Semin. Proc.* **118.**

[253] F. B. Rotz and M. O. Greer (1974). *SPIE Semin. Proc.* **45.**

[254] K. G. Leib, R. A. Bondurant, S. Hsiao, R. Wohlers, and R. H. Herold (1978). *Appl. Opt.* **17,** 2892–2899.

[255] K. G. Leib, R. A. Bondurant, and S. Hsiao (1976). Final Rep. Night Vision Lab., Fort Belvoir, Virginia (Contract DA AG 53-75-C0199).

[256] N. Balasubramanian (1978). In "Optical Data Processing" (D. Casasent, ed.), Vol. 23, pp. 119–148. Springer-Verlag, Berlin and New York.

[257] A. Vander Lugt (1971). NASA [Spec. Publ.] SP NASA SP-299, 131–137.

[258] T. M. Gee, C. W. Allen, and K. S. Cliffton (1971). NASA [Spec. Publ.] SP NASA SP-299, 145–148.

[259] A. Vander Lugt and R. H. Mitchel (1969). Opt. Acta 16, 453–461.

[260] E. Marom (1969). Laser Focus 19, 43–45.

[261] R. L. Bond, R. E. Biessner, J. Lankford, and W. W. Bradshaw (1971). NASA [Spec. Publ.] SP NASA SP-299, 177–182.

[262] R. B. Jenkins and H. C. McIlwain (1971). NASA [Spec. Publ.] SP NASA SP-299, 183–192.

[263] A. D. Gara (1979). Appl. Opt. 18, 172–174.

[264] A. Grumet (1977). Appl. Opt. 16, 154–159.

[265] T. Tschudi, G. Herziger, and A. Engle (1974). Appl. Opt. 13, 245–248.

[266] G. Reynolds, J. Zuckerman, D. Micler, and W. Dye (1973). Opt. Eng. 12, 23–35.

[267] J. Caufield (1978). In "Optical Data Processing" (D. Casasent, ed.), Vol. 23, pp. 199–238. Springer-Verlag, Berlin and New York.

[268] R. L. Bond, M. K. Mazunder, M. K. Testerman, and D. Hsich (1973). Science 149, 571.

[269] A. Brunsting and P. F. Mullaney (1972). Appl. Opt. 11, 675–680.

[270] S. P. Almeida, D. R. Del Balzo, J. Cairns, Jr., K. L. Dickson, and G. R. Lanza (1972). Trans. Kans. Acad. Sci. 74, 257.

[271] J. Cairn, Jr., K. L. Dickson, G. R. Lanza, S. P. Almeida, and D. Del Balzo (1972). Arch. Mikrobiol. 83, 141.

[272] S. P. Almeida, J. K. T. Eu, P. F. Lai, J. Cairns, Jr., and K. L. Dickson (1976). Proc. Int. Opt. Comput. Conf., Capri, Italy, pp. 113–115.

[273] S. P. Almeida, J. K. T. Eu, and P. F. Lai (1977). IEEE Trans. Meas. Instrum. IM-26, 312–316.

[274] S. P. Almeida, J. K. T. Eu, P. F. Lai, J. Cairns, Jr., and K. L. Dickson (1977). In "Application of Holography and Optical Data Processing" (E. Marom et al., eds.), pp. 573–579. Pergamon, New York.

[275] S. P. Almeida, S. K. Case, J. M. Fournier, H. Fujii, J. Cairns, Jr., K. L. Dickson, and P. Pryfogle (1978). In "Optica Hoy y Mañana" (Proc. I.C.O.-11, Madrid), pp. 351–354.

[276] S. P. Almeida and H. Fujii (1979). Appl. Opt. 18, 1663–1669.

[277] J. Partin, S. P. Almeida, and H. Fujii (1979). In "Holography in Medicine and Biology" (G. von Bally, ed.), Springer Series in Optical Sciences, Vol. 18, pp. 73–76. Springer-Verlag, Berlin and New York.

[278] J. Partin, S. P. Almeida, and H. Fujii (1979). SPIE Semin. Proc. 177.

[279] S. K. Case, S. P. Almeida, W. J. Dallas, and J. M. Fournier (1978). Appl. Phys. 17, 287–293.

[280] J. Partin, S. P. Almeida, and H. Fujii (1979). SPIE Semin. Proc. 185.

[281] J. Partin, S. P. Almeida, G. Indebetouw, and H. Fujii (1979). SPIE Semin. Proc. 201.

[282] H. Fujii, S. P. Almeida, and J. E. Dowling (1980). Appl. Opt. 19, 1190–1195.

[283] J. Cairns, Jr., K. L. Dickson, P. Pryfogle, S. P. Almeida, S. K. Case, J. M. Fournier, and H. Fujii (1979). Water Res. Bull. 15, 148–163.

Chapter 3

Particle Identification and Counting by Fourier-Optical Pattern Recognition

WALLACE L. ANDERSON

ELECTRICAL ENGINEERING DEPARTMENT
UNIVERSITY OF HOUSTON
HOUSTON, TEXAS

INTRODUCTION

It is a remarkable fact that small particle scattering, if considered on a very fundamental scale, accounts for most of the optical phenomena we observe in daily life and in the laboratory. By *fundamental*, we refer to quite small particles indeed, namely, electrons, for which a lower limit of diameter may not, in fact, exist. Since the electron's charge-to-mass ratio is high, it responds enthusiastically to an oscillatory electromagnetic field, absorbing energy and reradiating it throughout a large range of angles. This interaction is *elementary* scattering; it is the basis for virtually all changes we observe in the direction of propagation of electromagnetic waves. When electrons are bound to atoms, the details of the binding give rise to a great variety of wavelength sensitivity and so create spectral effects and absorption. All such effects are strengthened or modified in a great many different ways when atoms are collected together in various configurations.

Reflection and refraction are collective electron scattering phenomena of a particular kind, where the driven oscillations are very much in unison. The combination of the scattered with the unscattered portion of the radiation occurs in a very orderly manner, with reinforcement along specific ray directions and destructive interference in transverse directions.

In the context of light scattering, the term *particle* is generally understood to be a small object composed of many atoms in solid or liquid form. When

89

such objects are small compared with a wavelength, the electrons of their atoms move almost simultaneously in the incident light field. The scattering by particles of different sizes differs only in magnitude, being greater where more electrons participate in the action and lesser for fewer electrons. Their scattering patterns are like those of an elementary electric dipole [1], with no features allowing a distinction between different sizes or shapes.

Scattering of this kind is the well-known *Rayleigh scattering*, so named in honor of its having been analyzed by Lord Rayleigh at about the end of the last century [2]. Its strength varies inversely as the fourth power of the wavelength, thus accounting for the blue of the overhead sky and the red of the rising or setting sun.

When particle dimensions are some appreciable fraction of a wavelength or more, relative temporal and spatial phase lags in the induced currents on and within the particles can give rise to highly complicated interactions with the incident radiation, with drastic effects on scattering patterns. The problem of calculating such patterns with full mathematical rigor is remarkably intractable, in general, even for relatively simple shapes and structural details. The first person to solve the boundary value problem for a homogeneous isotropic sphere in a plane wave field was Mie [3], using a series of spherical harmonics and spherical Bessel functions.

For relatively large particle sizes (compared with wavelength) the concepts of refraction and reflection, which can be characterized to a considerable degree by geometrical or ray optics, are sometimes useful in accounting for different aspects of particle scattering behavior. An elementary but partially adequate explanation of a rainbow can, in fact, be based on a ray optics treatment, as demonstrated (laboriously) by Descartes in 1637. These concepts are of no real help to us here. Our interest is in the scattering patterns themselves, regardless of whether different contributions to them can or cannot be described by refraction instead of diffraction, or whatever.

At any rate, combinations of the effects of shape, size, composition, and inhomogeneity give rise to an enormous variety of scattering behavior over a wide range of angle. It is this variety, of course, on which the success of our pattern recognition efforts depends. As noted above, the lack of variety in Rayleigh scattering precludes our being able to distinguish between different particles in that regime; however, the opportunity is available, if needed, to distinguish Rayleigh scatterers considered *as a single class* from others having more complex patterns.

We should note additionally that, although the theoretical treatment in the following sections presumes the validity of the Fresnel diffraction approximation, confined to a quite limited range of angle, we do not interpret it as being a barrier to the use of a wider angular range. Ideally we should like to make use of as much optical information as possible, which is to say, the widest possible extent of the spatial frequency spectrum. The optical micro-scope, for example, when being operated at the limit of its resolving power, utilizes scattering angles greater than 90 deg. The propriety of extending the angular range for pattern recognition purposes in our case rests principally on the requirement that particles of the same type but at different locations in the object space contribute almost equally to light power at the *same* locations in the scattering pattern. One condition for meeting this require-ment is that the size of the region containing the particles be small compared with the distance to the pattern; another is that the particle distribution not be so dense that many of the particles intercept and rescatter energy already scattered from others.

Particles of interest to us often, if not usually, occur in shapes and compositions not at all lending themselves to the rigorous computation of their scattering properties. For this reason we perceive little to be gained by concerning ourselves with scattering theory. The reader who wishes to pursue such theory will find no dearth of excellent texts and other publica-tions in which lengthy bibliographies of past work are given [4, 5]. But usually the Fourier-optical approach to particle analysis will have to rely on measured patterns or on patterns inferred indirectly from experiment.

3.1 THE FOURIER-OPTICAL APPROACH

An outstanding merit of the use of far-field scattering from small regions of illuminated particles for inference of their numbers and identities becomes evident when viewed in the framework of Fourier optics. It consists in the fact that, regardless of where an object may be located in the area of illumination, its spectrum is always centered on the optic axis [6, 7]. Information as to its position occurs strictly as a phase factor, in accord with the "shifting theorem" of the Fourier integral. In the power spectrum, consisting of the magnitude of the Fourier spectrum squared, this information then disappears. When multiple, randomly assorted objects are present, however, their power spectrum contains contributions from cross-products involving the phase factors for the different object locations [8]. Positional information is then, strictly speaking, retained, albeit in a rather intractable form. As a conse-quence, at any given point in the Fourier pattern the amplitude for N objects may assume a value ranging from zero to N times the incident amplitude, according to whether the interference is such as to yield total cancelation or maximum possible reinforcement, respectively. The power, correspondingly, may range from zero to N^2 times the incident power. The pattern as it would be observed for a single object is present in an average sense but displays a random modulation, fluctuating violently as a function of the transverse spatial coordinates [9].

Since the positional information giving rise to this phenomenon is, for our purposes, of no interest, we are compelled to regard it as noise. Its effects must be mitigated if we are to attend usefully to the portion of the information that we need for estimation of particle numbers and types. Fortunately, as will be discussed at greater length in the following sections, this noise can be dealt with very effectively to the point where it becomes of but minor concern. Under these circumstances we may conclude that, by taking the power spectrum, we serve a primary aim of what pattern information analysts call *feature extraction*, which is to say, we eliminate an inconsequential part of the data. As will be seen, moreover, the data retained are in a form that proves especially advantageous for subsequent operations.

The obliteration of phase information related to position that occurs— on the average—in the power spectrum should not be assumed to apply to phase information in the sense in which this term would be understood by a microscopist. The extent to which phase information of the latter type is preserved can perhaps be best illustrated by example. Consider an amplitude transmittance given by

$$t(r, \theta) = \begin{cases} \exp(j\pi/2), & 0 \le r \le r_0, \\ 1, & r_0 < r < \infty. \end{cases} \tag{3.1-1}$$

This is a "pure phase" object, disk-shaped, with radius r_0, centered at the origin and surrounded by open space. In the transform plane we obtain a delta function at the origin due to the concentration of all the unscattered light. In addition we have the quantity

$$F(\rho) = -(1 - j)r_0 J_1(2\pi r_0 \rho)/\rho. \tag{3.1-2}$$

The irradiance will be proportional to the magnitude squared, or

$$|F(\rho)|^2 = 2r_0^2 J_1^2(2\pi r_0 \rho)/\rho^2, \tag{3.1-3}$$

which is the *Airy pattern*, of long familiarity in optics. Except for the factor 2 (a result of its being transparent) it is precisely the same as the pattern we would obtain for an opaque disk of the same radius. Obviously, then, it does not *uniquely* characterize the object. But the phase information is certainly retained and, as mentioned above, is preserved in an average sense for a random ensemble of such objects. The fact that—as in this instance—certain types of phase information may be indistinguishable from certain types of amplitude information is of relatively little importance in practice. Particle classification tasks generally require consideration of only a very limited number of types, among which the possibilities for confusion due to this phenomenon have a low probability of occurrence.

In the case of angular variations—either in amplitude or phase—in the object transmittance, it may again be instructive to consider a specific instance. First let us imagine a prototype object with real transmittance:

$$t(r, \theta) = \begin{cases} \frac{1}{2} + \frac{1}{2}\cos n\theta, & 0 \le r \le r_0, \\ 0, & r_0 < r < \infty, \end{cases} \tag{3.1-4}$$

where n is an integer. Its transform can easily be shown to consist of an Airy pattern arising from the first term, and another pattern for which a somewhat more cumbersome expression is necessary. We shall concern ourselves only with the general form of the latter. It depends on the product of a factor $\cos n\phi$ or $\sin n\phi$ and a factor which is an infinite sum of Bessel functions of the first kind, order $n + 2k + 1$, $k = 0, 1, 2, 3, \ldots$, and argument $2\pi r_0\rho$ [10].

Two general observations can be made. The first is cautionary: One should not fall into the trap of supposing that an ensemble average of the spectrum leads to a pattern that merely looks like the transform of the angularly constant part of $t(r, \theta)$ [i.e., $t_0(r, \theta) = \frac{1}{2}, 0 \le r \le r_0; t_0(r, \theta) = 0, r_0 < r$]. Ensemble averaging does indeed remove the dependence on $\cos n\phi$ or $\sin n\phi$ but leaves intact the radial factor, which because of the angular variation in transmittance contains high order Bessel functions of appreciable magnitude.

The second observation is with respect to the details of the radial variations in spectrum due to the angularly varying transmittance. The lowest order Bessel function in the sum mentioned above is $n + 1$. For $k \ge 2$, $J_k(x)$ is relatively small in the range (roughly) $0 \ge x \ge k - 1$. Between $k - 1$ and k it increases rapidly, reaching its first (and greatest) value for x approximately equal to k. Thereafter it goes through a series of alternately negative and positive excursions, diminishing in magnitude roughly as $x^{-1/2}$. The more rapid the angular variations in transmittance as a function of θ, the greater the value of ρ at which the first peak of the spectrum occurs. Power is scattered into increasingly large angles from the optic axis.

The effects of angular variations in phase as compared with those in amplitude are again somewhat difficult to discuss in complete generality, but it may be helpful once more to resort to a simple example. Let us consider the transmittance

$$t(r, \theta) = \begin{cases} e^{jm\theta}, & 0 \le r \le r_0, \\ 0, & r_0 < r < \infty, \end{cases} \tag{3.1-5}$$

with m integer. Its transform is

$$2\pi e^{jm(\phi - \pi/2)} \int_0^{r_0} J_m(2\pi\rho r)r \, dr, \tag{3.1-6}$$

for which the radial portion is again given, for $m > 0$, by an infinite sum of Bessel functions of the first kind and order $2k + m + 1$, $k = 0, 1, 2, 3, \ldots, \infty$.

The same conclusions apply as previously; larger values of m, hence higher frequency angular variations, lead to larger amounts of diffraction into higher radial frequencies. Again, of course, the relationship of power spectrum to transmittance is not unique, but again the practical importance of this point will usually be of small consequence.

Summarizing these remarks, then, we can point to two significant conclusions concerning the use of the Fourier power spectrum for small particle counting and identification. One is that it produces a pattern from which particle positional information is essentially deleted. This, together with the fact that all unscattered light is brought to a focus in a small region on the optic axis permits the entire pattern recognition effort to be addressed to the information of importance. The second conclusion is that phase information relevant to particle identification, even when occurring as angular variations in phase, is not badly degraded in the ensemble average power spectrum. This fact can be of great importance, especially in dealing with the frequently transparent—or nearly so—particles of physiological origin [11].

In the sections that follow, most of our discussion will revolve around what the writer regards as the most fundamental principles of the Fourier-optical approach to particle classification and counting. Familiarity and personal bias may of course play a significant part in our judgment as to what is fundamental and what is not. We freely acknowledge, moreover, the likelihood that specific tasks and specific purposes, even though falling within the domain we have defined for ourselves in this chapter, may not be especially well accommodated or enlightened by the principles and conclusions presented here. For instance, it may be simpler and yet perfectly adequate, in many cases, to invert data by specialized *ad hoc* procedures rather than by least mean squares (LMS) analysis, despite the theoretical optimality of the latter. The enormous variety of possibilities for such procedures clearly precludes, however, any hope we might have of doing justice to them here.

We will be interpreting the term *Fourier-optical* according to what we feel is the typical contemporary understanding of this phrase. It appears to us that it does not include volume scattering of radiation, as is so frequent a topic of the literature dealing with atmospheric transmission of light or of transmission by colloidal particle suspensions in liquids. One expects that Fourier optics generally involves some appreciable dependence on plane-to-plane relationships of light amplitudes, wherein spatial frequency concepts and the two-dimensional Fourier integral play an important role. It can, of course, be argued that Fourier optics is merely a viewpoint, and one which does not alter the underlying physical processes of light diffraction in any way. But viewpoint or not, we shall adopt this general attitude as our guide.

Before entirely dismissing volume scattering, however, we should point out that inversion of the measured data associated with this subject frequently makes use of least mean squares analysis, a topic of central importance in later sections of this chapter. We therefore cannot afford to ignore the valuable lessons available to us in the literature in this field. Its development has occured over a long period of time and has led to many results directly relevant to our concerns.

One of the first practical applications of holography (and, therefore, eminently Fourier-optical) was to record, holographically, a volume containing small aerosols, illuminated by pulsed laser [12]. The reconstructed image was then searched in depth, as well as transversely, and the particle Fraunhofer patterns were used to infer their sizes. Since this evidently does not provide a Fourier-optical means of counting as well as sizing, we do not include it in our subsequent discussion, except to note a recent marriage of this concept to a counting method based on *composite* Fraunhofer diffraction patterns [13]. We shall return to this briefly in Section 3.8.

3.2 THE FOURIER SPECTRUM OF RANDOMLY ASSORTED SCATTERERS

For a function $f(r, \theta)$ prescribed in terms of polar coordinates, the Fourier transform is given by [14]

$$F(\rho, \phi) = \int_0^\infty r\, dr \int_0^{2\pi} f(r, \theta) \exp[-j2\pi r\rho \cos(\theta - \phi)]\, d\theta. \quad (3.2\text{-}1)$$

Since $f(r, \theta)$ is a periodic function of θ, it may be expanded in the Fourier series

$$f(r, \theta) = \sum_{m=-\infty}^{\infty} f_m(r) \exp(jm\theta). \quad (3.2\text{-}2)$$

Substitution into Eq. (3.2-1) and interchange of the order of summation and integration yield the expression

$$F(\rho, \phi) = \sum_{m=-\infty}^{\infty} \exp\left[jm\left(\phi - \frac{\pi}{2}\right)\right] \int_0^\infty f_m(r) J_m(2\pi r\rho)r\, dr, \quad (3.2\text{-}3)$$

where J_m is the Bessel function of the first kind and mth order. This is of essentially the same form as Eq. (3.2-2), as can be made more explicit by defining

$$F_m(\rho) = \int_0^\infty f_m(r) J_m(2\pi r\rho)r\, dr \quad (3.2\text{-}4)$$

and rewriting as

$$F(\rho, \phi) = \sum_{m=-\infty}^{\infty} (-j)^m F_m(\rho) \exp(jm\phi). \tag{3.2-5}$$

Equation (3.2-4) is (aside from a factor 2π) the Hankel transform of order m [15].

We shall use $f(r, \theta)$ as a prototype expression for the complex amplitude transmittance of one of a number N of identical objects, and $F(\rho, \phi)$ as the corresponding prototype transform. Since our observations are to be concerned strictly with relative values of irradiance in the Fourier plane, there is no need for us to take account of the magnitude and phase factors by which the Fourier transform differs from its optical realization.

First considering rotation through an angle α, we observe that the transform is rotated by exactly the same amount; i.e., for an object function $f(r, \theta - \alpha)$ the transform is $F(\rho, \phi - \alpha)$. We next consider a translation in the object plane in the direction θ_0 and through a distance r_0. The transform is modified only with respect to a phase factor, given by $\exp[-j2\pi\rho r_0 \times \cos(\theta_0 - \phi)]$. The resultant transform expression when the effects of rotation and translation are both included is given, in relation to the prototype, by

$$F(\rho, \phi - \alpha) \exp[-j2\pi\rho r_0 \cos(\theta_0 - \phi)]. \tag{3.2-6}$$

We assign random variables α_k, r_k, and θ_k to the kth of our N object functions. The transform for the entire assembly can then be written as

$$F_N(\rho, \phi) = \sum_{k=1}^{N} F(\rho, \phi - \alpha) \exp[-j2\pi\rho r_k \cos(\theta_k - \phi)], \tag{3.2-7}$$

or in terms of the coefficients $F_m(\rho)$ as

$$F_N(\rho, \phi)$$

$$= \sum_{m=-\infty}^{\infty} F_m(\rho)$$

$$\times \sum_{k=1}^{N} \exp\left\{j\left[m\left(\phi - \frac{\pi}{2} - \alpha_k\right) - 2\pi\rho r_k \cos(\theta_k - \phi)\right]\right\}. \tag{3.2-8}$$

We shall discuss the statistics associated with this expression in Section 3.3; for now, let us merely remark that, if we assume that any value of phase can occur with equal probability in the sum over k, the expected value of this sum is known from probability theory (i.e., the random walk problem) to be $N^{1/2}$. [A proof of this result is nicely elucidated in Sommerfeld's classic text, "Optics" (1954, pp. 192–193).]

The expected value of the irradiance is then given by

$$\langle I_N(\rho) \rangle = \langle |F_N(\rho, \phi)|^2 \rangle = N \sum_{m=-\infty}^{\infty} |F_N(\rho)|^2, \qquad (3.2\text{-}9)$$

with the angle brackets denoting the ensemble average. From Parseval's formula this must also be

$$\langle I_N(\rho) \rangle = \frac{N}{2\pi} \int_0^{2\pi} |F(\rho, \phi)|^2 \, d\phi, \qquad (3.2\text{-}10)$$

which is clearly N times the spatial average, over the angle ϕ, of the power spectrum for an individual particle [16].

The treatment of polydisperse mixtures is a straightforward extension of the above. Let us imagine that there are M classes of particles in the mixture and that each of these has a sufficiently unique averaged power spectrum [Eq. (3.2-9) or (3.2-10)] distinguishable from those of the others. (The question as to what constitutes distinguishability—and the sufficiency of such distinguishability—will be deferred to Section 3.5.) It will be convenient now to regard the total number of particles as being characterized by a column matrix $\mathbf{N}$ having elements $N_1, N_2, N_3, \ldots, N_M$, representing the numbers in each of the M categories (and that we intend, by our operations upon the measurements, to infer).

The expression for the complex amplitude in the polydisperse case is an extension of Eq. (3.2-8), replacing N by the vector $\mathbf{N}$ with elements corresponding to the number of each particle type:

$$F_{\mathbf{N}}(\rho, \phi)$$

$$= \sum_{n=1}^{M} \sum_{m=-\infty}^{\infty} F_{nm}(\rho)$$

$$\times \sum_{k=1}^{N_n} \exp\left\{ j\left[m\left(\phi - \alpha_{kn} - \frac{\pi}{2} \right) - 2\pi\rho r_{kn} \cos(\theta_{kr} - \phi) \right] \right\}, \qquad (3.2\text{-}11)$$

where the index n denotes the nth type. As one might expect from the monodisperse case,

$$\langle |F_{\mathbf{N}}(\rho, \phi)|^2 \rangle = \sum_{n=1}^{M} N_n \sum_{m=-\infty}^{\infty} |F_{nm}(\rho)|^2. \qquad (3.2\text{-}12)$$

We observe that this is a linear equation which, upon abbreviating the left-hand side as $\langle I(\rho) \rangle$, can be written

$$\langle I(\rho) \rangle = \sum_{n=1}^{M} G_n(\rho) N_n, \qquad (3.2\text{-}13)$$

where

$$G_n(\rho) = \sum_{m=-\infty}^{\infty} |F_{nm}(\rho)|^2 = \int_0^{2\pi} |F_n(\rho, \phi)|^2 \, d\phi. \qquad (3.2\text{-}14)$$

The function G_n is the expected value of the power spectrum for a particle of the nth type, taking into account all its possible angular orientations [16].

The procedure for arriving at an optimum estimate of N will be discussed in Section 3.4, after we have had an opportunity to examine the equations above more closely and with attention to statistical detail.

3.3 STATISTICAL CHARACTERISTICS OF IRRADIANCE PATTERNS

We shall begin with Eq. (3.2-7) restated in Cartesian coordinates:

$$F_N(u, v) = \sum_{k=1}^{N} F_k(u, v) \exp[-j2\pi(ux_k + vy_k)]. \qquad (3.3\text{-}1)$$

Let $\chi_k(u, v)$ be the phase of $F_k(u, v)$ together with any phase effect contributed by the optical system; then our diffraction pattern amplitude is proportional to [9, 17]

$$F_N(u, v) = \sum_{k=1}^{N} |F_k(u, v)| \exp\{-j2\pi[ux_k + vy_k + \chi_k(u, v)]\}. \qquad (3.3\text{-}2)$$

For the sake of brevity we shall put $F_k = |F_k(u, v)|$ and $\beta_k = 2\pi[ux_k + vy_k + \chi_k(u, v)]$, whereby

$$F_N(u, v) = \sum_{k=1}^{N} F_k \exp(-j\beta_k). \qquad (3.3\text{-}3)$$

The irradiance is given by

$$I_N(u, v) = |F_N(u, v)|^2 = \sum_{k=1}^{N} \sum_{j=1}^{N} F_k F_j \cos(\beta_k - \beta_j). \qquad (3.3\text{-}4)$$

We define a and b:

$$a = \sum_{k=1}^{N} F_k \cos \beta_k, \qquad (3.3\text{-}5)$$

$$b = \sum_{k=1}^{N} F_k \sin \beta_k. \qquad (3.3\text{-}6)$$

This gives us

$$F_N = a - jb \qquad (3.3\text{-}7)$$

and

$$I_N = a^2 + b^2. \tag{3.3-8}$$

Any light which passes through the particles without scattering, together with light through the open area between particles, comes to a focus in a small region around the origin. For a diffraction-limited optical system the extent of this region (as measured by u and v) is of the order $1/X$ in the x-direction and $1/Y$ in the y-direction, where X and Y are half-widths of the object plane.

For large N, the central limit theorem permits the conclusion that a and b are normally distributed. Expected values relevant to us are

$$\langle a \rangle = \sum_{k=1}^{N} \langle F_k \rangle \langle \cos \beta_k \rangle = 0,$$

$$\langle b \rangle = \sum_{k=1}^{N} \langle F_k \rangle \langle \sin \beta_k \rangle = 0,$$

$$\langle ab \rangle = \sum_{k=1}^{N} \sum_{j=1}^{N} \langle F_k F_j \rangle \langle \cos \beta_k \sin \beta_k \rangle = 0, \tag{3.3-9}$$

$$\langle a^2 \rangle = \sum_{k=1}^{N} \sum_{j=1}^{N} \langle F_k F_j \rangle \langle \cos \beta_k \cos \beta_j \rangle = \frac{1}{2} \sum_{k=1}^{N} \langle F_k^2 \rangle,$$

$$\langle b^2 \rangle = \sum_{k=1}^{N} \sum_{j=1}^{N} \langle F_k F_j \rangle \langle \sin \beta_k \sin \beta_j \rangle = \frac{1}{2} \sum_{k=1}^{N} \langle F_k^2 \rangle.$$

The joint probability density function of a and b is

$$p(a, b) = (1/2\pi\sigma^2) \exp[-(a^2 + b^2)/2\sigma^2], \tag{3.3-10}$$

where $2\sigma^2 = \sum_{k=1}^{N} \langle F_k^2 \rangle$. The probability density function for I_N is appropriately written as the conditional density function or the likelihood function

$$p(I/N) = \begin{cases} (1/2\sigma^2) \exp(-I/2\sigma^2), & I \geq 0, \\ 0, & I < 0. \end{cases} \tag{3.3-11}$$

The expected value of I_N is

$$\langle I_N \rangle = \int_0^\infty I p(I|N) \, dI = \sigma^2. \tag{3.3-12}$$

If the χ_k are uniformly distributed in ϕ,

$$\langle F_k^2 \rangle = \frac{1}{2\pi} \int_0^{2\pi} |F(\rho, \phi)|^2 \, d\phi. \tag{3.3-13}$$

(This requirement on χ_k is met if the α_k are uniformly distributed and if any phase error introduced by the optical system is a function only of ρ.) Then, from Eqs. (3.2-9) and (3.2-10),

$$\langle I_N \rangle = \langle |F_N(\rho, \phi)|^2 \rangle = N \sum_{m=-\infty}^{\infty} |F_m(\rho)|^2$$

$$= \frac{N}{2\pi} \int_0^{2\pi} |F(\rho, \phi)|^2 \, d\phi = N \langle F_k^2 \rangle. \tag{3.3-14}$$

Since evidently $\langle F_k^2 \rangle$ is independent of k, the distinction conveyed by this index can be dropped. It will be convenient to identify a function $G(\rho) \equiv \langle F_k^2 \rangle$, so that

$$\langle I_N \rangle = GN. \tag{3.3-15}$$

The variance of I_N is easily determined to be

$$\langle (I_N - \langle I_N \rangle)^2 \rangle = \text{var } I_N = G^2 N^2. \tag{3.3-16}$$

For a maximum likelihood estimate (MLE) of N, the usual procedure is to determine the value $\hat{N}$ of N for which the maximum of the logarithm of the likelihood function, Eq. (3.3-11), occurs with respect to variations in N. One readily finds that in this case $\hat{N} = I/G$. The variance is given by

$$\text{var } [\hat{N}(I) - N] = \langle [\hat{N}(I) - N]^2 \rangle = N^2. \tag{3.3-17}$$

This satisfies the Cramer–Rao lower bound with an equality; i.e.,

$$N^2 = \left[-\left\langle \frac{\partial^2 \ln p(I|N)}{\partial N^2} \right\rangle \right]^{-1}, \tag{3.3-18}$$

whereby the estimate $\hat{N}$ must be efficient. It is also unbiased.

The standard deviation expressed as a percentage with respect to N is therefore 100%. Obviously this is too large for the estimate to have very much utility. One means of reducing it is to take a number of independent trials and average the results. The outcome of the MLE may be more or less self-evident, but the exercise is brief, so we shall provide it here in the interest of thoroughness. For L independent trials the likelihood function will be

$$p_L(\mathbf{I}|N) = \prod_{l=1}^{L} \frac{1}{G_l N} \exp\left(\frac{-I_l}{G_l N}\right), \tag{3.3-19}$$

where I_l is the irradiance observed in the lth trial, G_l is G at the radius at which I_l is observed, and I is the set of values of I_l. Differentiating the logarithm of this function and setting the result equal to zero gives, for the estimate of N,

$$\hat{N}(I) = \frac{1}{L} \sum_{l=1}^{L} \frac{I_l}{G_l}. \tag{3.3-20}$$

The ensemble average of $\hat{N}$ is evidently N. The estimate is unbiased and efficient, with variance

$$\text{var}(\hat{N} - N) = N^2/L. \tag{3.3-21}$$

As will be discussed later, it would be desirable from an optical processing viewpoint to improve the estimate not by processing numerous samples but rather by simple summation over an area encompassing a sufficient number of statistically independent irradiance values to effect the desired result. Leaving aside, for the moment, the question as to how many such values can be anticipated per unit area, we again consider a MLE based on the sum

$$I_s = \sum_{l=1}^{L} I_{Nl}, \tag{3.3-22}$$

where I_{Nl} is $I_N(\rho_l, \phi_l)$ (or $I(\rho_l, \phi_l)/N$). Its probability density function is obtained as a convolution of the density functions of I_{Nl}:

$$p(I_s|N) = p(I_1|N) * p(I_2|N) * \cdots * p(I_L|N), \tag{3.3-23}$$

where the asterisk represents convolution and the designation $(I_l|N)$ is equivalent to I_{Nl}. It will be advantageous to assume that I_s is the sum of irradiances at the same ρ but different ϕ; then G will have the same value throughout. For a given l,

$$p(I_l|N) = \begin{cases} (1/GN) \exp(-I_l/GN), & I_l \geq 0, \\ 0, & I_l < 0, \end{cases} \tag{3.3-24}$$

and we find the likelihood function to be

$$p(I_s|N) = \frac{I_s^{L-1}}{(GN)^L(L-1)!} \exp\left(-\frac{I_s}{GN}\right). \tag{3.3-25}$$

This is known as a *Pearson III* distribution [10] with mean LGN and variance LG^2N^2. The MLE of N is found to be

$$\hat{N} = I_s/LG \tag{3.3-26}$$

with variance

$$\text{var}(\hat{N} - N) = N^2/L, \tag{3.3-27}$$

as in the case of the L separate observations.

Now addressing ourselves to the polydisperse case, we begin with Eq. (3.2-11), written in somewhat different form as

$$F_{\mathbf{N}}(\rho, \phi) = \sum_{n=1}^{M} \sum_{k=1}^{N_n} |F_n(\rho, \phi - \alpha_k)|$$

$$\times \exp\{-j[\rho r_{kn} \cos(\theta_{kn} - \phi) + \psi_{kn}]\}, \tag{3.3-28}$$

with ψ_{kn} representing the sum of the phase factor of $F_n(\rho, \phi - \alpha_k)$ and that introduced by the optical system, if any. Let the total phase be represented by β_{knl}; then we can write

$$F_{\mathbf{N}}(\rho_l, \phi) = \sum_{n=1}^{M} \sum_{k=1}^{N_n} F_{knl} \cos \beta_{knl} - j \sum_{n=1}^{M} \sum_{k=1}^{N_n} F_{knl} \sin \beta_{knl}. \quad (3.3\text{-}29)$$

As we did in Eqs. (3.3-3) through (3.3-8) we write the abbreviated equivalent:

$$F_{\mathbf{N}}(\rho_l, \phi) = a_l - jb_l \quad (3.3\text{-}30)$$

of which the magnitude squared is

$$I_{\mathbf{N}}(\rho_l, \phi) = a_l^2 + b_l^2. \quad (3.3\text{-}31)$$

Using the same reasoning as before we find that the likelihood function (probability density function) is

$$p(I_{\mathbf{N}}|\mathbf{N}) = [1/\langle I_{\mathbf{N}}(\rho_l)\rangle \exp[- I_{\mathbf{N}}(\rho_l)/\langle I_{\mathbf{N}}(\rho_l)\rangle], \quad (3.3\text{-}32)$$

where

$$\langle I_{\mathbf{N}}(\rho_l)\rangle = \sum_{n=1}^{M} G_n(\rho_l)N_n \quad (3.3\text{-}33)$$

and

$$G_n(\rho_l) = \frac{1}{2\pi} \int_0^{2\pi} |F_n(\rho_l, \phi)|^2 \, d\phi. \quad (3.3\text{-}34)$$

Evidently $G_n(\rho_l)$ is the expected value of the irradiance at ρ_l scattered by a particle of the nth type.

The likelihood equation can be constructed as follows. We form the $1 \times P$ matrix

$$\ln p(I_{\mathbf{N}}|\mathbf{N}) = \ln p(I_1|\mathbf{N}) \ln p(I_2|\mathbf{N}) \ldots \ln p(I_P|\mathbf{N}), \quad (3.3\text{-}35)$$

where $I_l = I_{\mathbf{N}}(\rho_l)$. The likelihood equation is then

$$\mathbf{V}_{\mathbf{N}} \ln p(I_{\mathbf{N}}|\mathbf{N})|_{\mathbf{N}=\mathbf{N}_{ML}} = 0. \quad (3.3\text{-}36)$$

The gradient operator $\mathbf{V}_{\mathbf{N}}$ is defined to be an $M \times 1$ column matrix with elements $\partial/\partial N_n$, $n = 1, 2, 3, \ldots, M$. The operation gives as a result an $M \times P$ matrix with elements

$$L_{nl} = [G_n(\rho_l)/\langle I_l\rangle^2](I_l - \langle I_l\rangle). \quad (3.3\text{-}37)$$

The likelihood equation is therefore satisfied by

$$I_l = \langle I_l\rangle = \sum_{n=1}^{M} G_n(\rho_l)\hat{N}_n \quad (3.3\text{-}38)$$

for $l = 1, 2, 3, \ldots, P$. In matrix form we have

$$I = G\hat{N}_{ML} \qquad (3.3\text{-}39)$$

as the prescription for the maximum likelihood estimate of N, $\hat{N}_{ML}$ [9].

In general, however, this equation has no solution. The best that can be done with it is to estimate $\hat{N}_{ML}$ on the basis of some criterion or another as to what constitutes the best estimate; however, the result then will not generally be a maximum likelihood estimate. Before we can decide how best to resolve this problem we first need to follow up the implications of the result, Eq. (3.3-27), as considered in the context of optical processing.

Therefore, as our concluding effort in this section, we shall take up the question as to how many statistically independent information elements are contained per unit area in the diffraction pattern. Equation (3.3-27) indicates the importance of this question for the monodisperse case; obviously it must be equally important in the analysis of polydisperse particle mixtures. [In the latter case, the number of information elements governs the extent to which the model equation matrix is redundant, or "overdetermined." If this number is n_0 and the rank of the matrix is R, the factor $(n_0 - R)^{-1}$ acts similarily to L^{-1}, in Eq. (3.3-27), to reduce the variance of the estimates.]

We shall approach this problem by calculating the spatial autocorrelation R_I of the irradiance for the monodisperse case. The lateral extent of the region within which R_I assumes a nonnegligible value can be taken as the average spacing between independent irradiance values.

We assume that the system is spatially invariant, and therefore

$$R_I = \langle I_N(u_1, v_1)I_N(u_2, v_2)\rangle = R_I(\Delta u, \Delta v), \qquad (3.3\text{-}40)$$

where $\Delta u = u_2 - u_1$, $\Delta v = v_2 - v_1$. Each of the I_N is represented as a double sum, as in Eq. (3.3-4). Since the details are slightly tedious, we go on to the result:

$$R_I(\Delta u, \Delta v) = (N^2 - N)G_1 G_2 + NB^2$$
$$+ (N^2 - N)A^2 \operatorname{sinc}^2\!\left(\frac{\Delta u X}{\pi}\right)\operatorname{sinc}^2\!\left(\frac{\Delta u Y}{\pi}\right), \qquad (3.3\text{-}41)$$

where X and Y are the dimensions of the illuminated rectangle in the object plane and

$$G_1 \equiv \frac{1}{2\pi}\int_0^{2\pi} |F(\rho_1, \phi_1)|^2 \, d\phi_1,$$

$$G_2 \equiv \frac{1}{2\pi}\int_0^{2\pi} |F(\rho_2, \phi_2)|^2 \, d\phi_2,$$

$$A \equiv \frac{1}{2\pi} \int_0^{2\pi} |F(\rho_1, \phi_1)| \|F(\rho_2, \phi_2)| \, d\phi_1, \tag{3.3-42}$$

$$B^2 \equiv \frac{1}{2\pi} \int_0^{2\pi} |F(\rho_1, \phi_1)|^2 |F(\rho_2, \phi_2)|^2 \, d\phi_1,$$

where the subscripts 1 and 2 of u and v have been carried over directly to the corresponding polar coordinates. In the case of small Δu and Δv (or $\Delta \rho$ and $\rho \Delta \phi$) we can make the approximations

$$G_1 G_2 \approx A^2 \approx B^2 \approx G_1^2 \approx G_2^2. \tag{3.3-43}$$

We define the autocorrelation coefficient as

$$\gamma = (\langle I_{N1} I_{N2} \rangle - \langle I_{N1} \rangle \langle I_{N2} \rangle)/\sigma_1 \sigma_2, \tag{3.3-44}$$

where $I_{N1} = I_N(u_1, v_1)$, $I_{N2} = I_N(u_2, v_2)$, and σ_1, σ_2 are their respective standard deviations. With the approximations indicated, this reduces to

$$\gamma \approx \mathrm{sinc}^2(2\Delta u X) \, \mathrm{sinc}^2(2\Delta v Y). \tag{3.3-45}$$

Similarly, for a circular aperture of radius R defining the area of illumination of the object plane, we obtain

$$\gamma = \mathrm{sinc}^2 2\rho \, \Delta \phi R \tag{3.3-46}$$

for ρ constant, and

$$\gamma = \mathrm{sinc}^2 2\Delta \rho R \tag{3.3-47}$$

for ϕ constant. In terms of actual distances in the diffraction plane, Eq. (3.3-45) becomes

$$\gamma \approx \mathrm{sinc}^2(2X \, \Delta x_f / \lambda f) \, \mathrm{sinc}^2(2Y \, \Delta y_f / \lambda f), \tag{3.3-48}$$

where λ is the wavelength and f is the effective focal distance (distance from object plane to observation plane when particles are illuminated with a converging beam). The first zeros of the sinc functions, on either side of the maximum, occur for the arguments ± 1. Evidently, then, the distance over which the correlation coefficient has appreciable value and outside of which it very rapidly diminishes is of the order

$$\delta_c = \lambda f / R. \tag{3.3-49}$$

The area comprising such a region has been termed a *correlation cell* [18]. The inverse of δ_c is a spatial frequency equal to the cutoff frequency for a diffraction-limited coherent imaging system. In accord with the variance value N^2 [Eq. (3.3-17)], violent fluctuations in irradiance will occur in traversing a path through some number of such cells; these constitute the phenomenon called *speckle*. Its magnitude depends on the number of scatterers, but its size (as measured, for instance, by δ_c) does not, as long as N is fairly large.

When the optical system parameters are specified, it is evident that an approximate idea as to the number of statistically independent irradiance values in a given area is readily obtainable using the result above. If, for example, $\lambda = 633$ nm and the ratio of focal length to diameter is about 10 or 20, a few square millimeters of area are sufficient to reduce the standard deviation from 100 to 1%.

As will be seen in the following sections, these results are of considerable importance in pointing the direction to take in implementing the estimation.

3.4 DATA INVERSION: GENERAL CONSIDERATIONS

The terms *feature selection* and *feature extraction* reflect a tremendous variety of possible techniques of data reduction—some based upon rational considerations, and some less so. *Feature* represents simply any characteristic or specific combination of characteristics present either in the raw data or in the results of any transformations performed on it. One sees that the vagueness of the word is a quite suitable match to its usage in this case.

Typically, feature selection and extraction play an important if not critical role in pattern information processing by digital computer. It is not infrequently the case that the data comprising the information are so voluminous as to obviate their examination in any great detail. The art and science of feature selection consists in the astuteness with which a choice is made of some manageable fraction of the information. At the very least, the choice must not be such as to relinquish the minimal aspects essential for correct interpretation. In addition it should of course furnish the required degree of confidence in the decisions or estimates.

As noted in a previous section of this chapter, the Fourier-optical approach as applied to particle identification and counting inherently furnishes two examples of feature extraction. One is by way of the fact that particle position information is automatically suppressed in the irradiance spectrum. In addition, the unscattered—and therefore uninformative—portion of the transmitted illumination is brought to a focus, permitting its easy removal from observation.

Beyond those two instances, however, optical processing includes no further feature extraction, except in the rather incidental sense of the choice that is made of the range of scattering angle within which observations are taken.

More conventional examples of feature extraction are contained in some procedures whereby the diffraction pattern is not processed optically, but rather is sampled at selected points, and sample values then handled numerically. In the instance of relatively simple particle distributions distinctive features may occur so obtrusively as to make their choice obvious. An example is the diameters of the Airy rings in the Fraunhofer diffraction pattern generated by a collection of monodisperse particles with circular cross section. From the Airy ring diameters the particle diameters can be deduced immediately from a table of zeroes of the Bessel function $J_1(x)$. If the particles are disperse in diameter and/or shape, but not too widely so, the diffraction rings may still be visible with sufficient contrast to permit a good estimate of average particle diameter. (A device designed, in fact, for this very purpose is Young's eriometer, which presumably dates from the early nineteenth century [19].)

When only a relatively few sample values of scattered power are measured, it becomes feasible to explore a variety of linear and nonlinear operations upon them for the purpose of finding and enhancing any tendency they may have to fall into distinct groupings or clusters corresponding uniquely to the classifications required. This advantage of flexibility as compared with the far more restricted options that are practicable when processing a very large number of light values is, however, gained at the expense of greater statistical uncertainty in the measurements, forfeiting the noise smoothing effects of the redundancy that is intrinsically available in the diffraction pattern. The question whether, despite such drawbacks, a small number of samples may be quite sufficient for the task is one that can be answered only with respect to specific cases.

In an experiment employing a "one-at-a-time" flow of unstained human leukocytes through a focused beam, success was reported in distinguishing among lymphocytes, monocytes, and granulocytes using scattered light measurements at only two locations [20]. Although this example departs considerably from the Fourier-optics mode discussed here, it illustrates the point that a very limited number of measurements may be adequate for certain discriminatory tasks. In such cases feature selection may be based either on previously determined characteristics of the expected patterns or on trial and error. The greater the number of features, the less feasible it becomes to use random trial and error.

The outcome usually desired as a result of a pattern recognition effort is the estimate of some set of values of parameters that occur as random variables underlying the generation of the pattern. The mathematical model is exemplified in a fairly general sense by a linear integral equation (Fredholm equation of the first kind):

$$y(\rho) = \int_{\sigma_1}^{\sigma_2} G(\rho, \sigma)x(\sigma)\, d\sigma, \qquad (3.4\text{-}1)$$

where $x(\sigma)$ is the unknown parameter function, $G(\rho, \sigma)$ is a known connecting link, or mapping kernel, from the parameter space to the observation space, and the functional $y(\rho)$ represents a measurable result [21]. The discrete version of this is the equation

$$\mathbf{Y} = \mathbf{GX}, \tag{3.4-2}$$

where now $\mathbf{X}$ is an $M \times 1$ matrix with elements $X_j, j = 1, 2, 3, \ldots, M$; $\mathbf{Y}$ is a $P \times 1$ matrix with elements $Y_i, i = 1, 2, 3, \ldots, P$; and $\mathbf{G}$ is a $P \times M$ matrix with elements G_{ij}.

For our purposes, Eq. (3.4-2) is incomplete, since it fails to take into account the inevitable error involved in making the measurements. If such error is additive, we can characterize it by the quantity

$$\boldsymbol{\varepsilon} = \mathbf{W} - \mathbf{Y}, \tag{3.4-3}$$

where $\mathbf{W}$ represents actual measured values. The model equation is then

$$\mathbf{W} = \mathbf{GX} + \boldsymbol{\varepsilon}, \tag{3.4-4}$$

and the problem to which we address ourselves is that of inferring the best estimate of $\mathbf{X}$ that is possible under the circumstances.

If "best" is defined in a certain way and if certain conditions apply to $\boldsymbol{\varepsilon}$, it becomes possible to specify precisely the linear operation on $\mathbf{W}$ required for an optimal estimate of $\mathbf{X}$. We may therefore note that, regardless of whatever feature selection and extraction may have entered into determination of the $\mathbf{W}$-values, any further exercise of this nature is no longer at the discretion of the observer. The inversion algorithm takes the matter out of our hands entirely. It automatically provides optimal selection of the features of the $\mathbf{W}$ pattern.

We now proceed to the problem of inverting Eq. (3.4-4). If the expected value of $\boldsymbol{\varepsilon}$, denoted by $\langle \boldsymbol{\varepsilon} \rangle$, is zero and if the expected value of $\boldsymbol{\varepsilon}\boldsymbol{\varepsilon}^T$ is $\mathbf{V}_{\varepsilon\varepsilon}$, a known covariance matrix, the minimum variance unbiased estimator of $\mathbf{X}$ is given by

$$\hat{\mathbf{X}} = [(\mathbf{G}^T\mathbf{V}_{\varepsilon\varepsilon}^{-1}\mathbf{G})^{-1}\mathbf{G}^T\mathbf{V}_{\varepsilon\varepsilon}^{-1}]\mathbf{W}. \tag{3.4-5}$$

This result is known as the *Gauss–Markov theorem* [22]. Since it minimizes the quadratic form

$$Q_1 = (\mathbf{W} - \mathbf{GX})^T\mathbf{V}_{\varepsilon\varepsilon}^{-1}(\mathbf{W} - \mathbf{GX}) = \boldsymbol{\varepsilon}^T\mathbf{V}_{\varepsilon\varepsilon}^{-1}\boldsymbol{\varepsilon}, \tag{3.4-6}$$

it is sometimes referred to as a *weighted least squares* estimate.

The corresponding equation if $\mathbf{V}_{\varepsilon\varepsilon} = \sigma^2\mathbf{U}_1$, where $\mathbf{U}_1$ is the identity matrix, is

$$\hat{\mathbf{X}} = [(\mathbf{G}^T\mathbf{G})^{-1}\mathbf{G}^T]\mathbf{W}. \tag{3.4-7}$$

which minimizes the quadratic form

$$Q_2 = (\mathbf{W} - \mathbf{GX})^T(\mathbf{W} - \mathbf{GX}) = \varepsilon^T\varepsilon, \tag{3.4-8}$$

which is just the sum of the squared error terms. This is sometimes called the *unweighted least squares* estimate. The unqualified designation *least squares* estimate can then be considered to encompass both possibilities: weighted and unweighted. The reader should be cautioned, however, that this usage is not universal in the literature.

As mentioned previously, the estimates of Eqs. (3.4-5) and (3.4-7) will not, in general, be maximum likelihood. It is known, however, that in the event that the statistics of the random process are normal, these estimates *are* maximum likelihood as well as linear minimum variance unbiased.

3.5 APPLICABILITY OF THE MODEL; STABILITY OF ESTIMATES

We have defined ε to represent the differences between measurements $\mathbf{W}$ and those predicted by the model equation $\mathbf{Y} = \mathbf{GX}$. As we have noted, in order for the estimate of Eq. (3.4-5) to be minimum variance unbiased, it is necessary that $\langle\varepsilon\rangle = 0$ and that $\langle\varepsilon\varepsilon^T\rangle$ be a known covariance matrix. We now need to examine the questions as to whether the particle analysis problem can be fitted to this model and whether these conditions on ε are met.

We shall assume that ε can be written as [9]

$$\varepsilon = \varepsilon_s + \varepsilon_m, \tag{3.5-1}$$

where ε_s is the error due to the speckle or intrinsic light irradiance variations, and ε_m is the error introduced by the measuring instrumentation. Equations (3.3-32) and (3.3-33) tell us what we can anticipate for the ε_s portion of the error. The mean value of ε_s is zero, but $\langle\varepsilon_s\varepsilon_s^T\rangle$ depends on the number of particles. Since these numbers are *a priori* unknown, so is $\langle\varepsilon_s\varepsilon_s^T\rangle$. Equation (3.4-5) cannot be used, and Eq. (3.4-7) will not generally be minimum variance.

Fortunately the high density per unit area of statistically independent information cells provides a straightforward way to deal with this problem. If each measurement includes a sufficient area, the ε_s contribution to error can be made extremely small. Then ε_m will dominate. Without this provision for suppressing ε_s we would have the model equation

$$\mathbf{J} = \mathbf{I} + \varepsilon_m = \langle\mathbf{I}\rangle + \varepsilon_s + \varepsilon_m = \mathbf{GN} + \varepsilon_s + \varepsilon_m, \tag{3.5-2}$$

with ε_s a function of $\mathbf{N}$. Here $\mathbf{J}$ is the measured irradiance, $\mathbf{I}$ is the "true" irradiance, $\langle\mathbf{I}\rangle$ is the expected value of $\mathbf{I}$ in view of the irradiance statistics, and $\mathbf{N}$ is the set of particle numbers in the various categories. By suppressing

ε_s we have

$$\mathbf{J} = \langle \mathbf{I} \rangle + \varepsilon_m = \mathbf{GN} + \varepsilon_m, \qquad (3.5\text{-}3)$$

for which we can quite reasonably entertain the presumption that the matrix $\langle \varepsilon_m \varepsilon_m^T \rangle$ is independent of $\mathbf{N}$ and can be determined or estimated in most circumstances. It is therefore possible to assume with some degree of equanimity that the least mean squares analysis is not only usable but essentially optimum as well.

Our next point of concern lies in a more detailed examination of the LMS inversion process. It will be instructive, for our purposes, to think of each column of the matrix $\mathbf{G}$ as a vector, let us say $\mathbf{g}_i$, in P dimensions. The distinction between any two particle types i and j is represented, obviously, by the distinction between $\mathbf{g}_i$ and $\mathbf{g}_j$. Evidently $\mathbf{g}_i$ and $\mathbf{g}_j$ must be linearly independent for all i and j; otherwise the inverse of the matrix $\mathbf{G}^T\mathbf{G}$ or $\mathbf{G}^T\mathbf{V}_{\varepsilon\varepsilon}^{-1}\mathbf{G}$ will not exist.

In practice it is perhaps unlikely that linear dependence between any two columns of $\mathbf{G}$ would go unnoticed. A more probable occurrence would be the case where, for instance, an attempt is being made to distinguish between particles with similar but not identical scattering patterns. An example might be where type i is defined to be "red blood cells with 7-μm diameter" and type j is "red blood cells with 7.05-μm diameter." Two such types are likely to have very similar optical characteristics, and then $\mathbf{g}_i$ and $\mathbf{g}_j$ may be regarded as "almost" linearly dependent. This is a rather imprecise description, however. The critical point is the extent to which noise (i.e., error) in the measurements can add enough uncertainty to make the angular separation between $\mathbf{g}_i$ and $\mathbf{g}_j$ indiscernible, on the average.

A tighter characterization of the principles involved here can be based on a geometric development. In the simplest case the LMS analysis minimizes $\varepsilon^T\varepsilon = \sum \varepsilon_i^2$. We shall define, therefore, a figure of merit denoted by $\Phi(\mathbf{N})$, with

$$\Phi(\mathbf{N}) = \varepsilon^T\varepsilon. \qquad (3.5\text{-}4)$$

The elements $N_1, N_2, \ldots, N_n$ of $\mathbf{N}$ are always presumed to be large numbers in our case, so no essential error will result from treating them as continuous in the parameter space. We assume also that Φ is continuous and that $\Phi(\mathbf{N})$ is its unique unconstrained minimum. We then pose a question that can be stated in several ways, of which one is: How sensitive is Φ, in the region of its minimum, to changes in the elements of $\mathbf{N}$? Or we could as well ask: Are there any directions in the parameter space along which the minimum of Φ is much more poorly defined than it is in other directions?

Mathematically this question can be pursued as follows. The region of $\mathbf{N}$-space in which Φ differs from its minimum by the amount Δ or less is defined by

$$\Phi(\mathbf{N}) - \Phi(\hat{\mathbf{N}}) \leq \Delta. \tag{3.5-5}$$

The surface bounding this region is evidently an $(M-1)$-dimensional surface for which the equation is

$$\Phi(\mathbf{N}) = \Phi(\hat{\mathbf{N}}) + \Delta. \tag{3.5-6}$$

We can expand $\Phi(\mathbf{N})$ in a Taylor series around the point $\mathbf{N} = \mathbf{N}$ and, if Δ is sufficiently small, Φ will be well approximated by

$$\Phi(\mathbf{N}) = \Phi(\hat{\mathbf{N}}) + [\nabla\Phi(\hat{\mathbf{N}})]^T \, \delta\mathbf{N} + \tfrac{1}{2}\delta\mathbf{N}^T\mathbf{H}(\hat{\mathbf{N}}) \, \delta\mathbf{N}, \tag{3.5-7}$$

where we define

$$\nabla\Phi(\hat{\mathbf{N}}) = \nabla\Phi|_{\mathbf{N}=\hat{\mathbf{N}}} \tag{3.5-8}$$

and $\mathbf{H}(\hat{\mathbf{N}})$ is the matrix with elements

$$H_{mn}(\hat{\mathbf{N}}) = \left.\frac{\partial^2\Phi}{\partial N_m \, \partial N_n}\right|_{\mathbf{N}=\hat{\mathbf{N}}}. \tag{3.5-9}$$

Since we have specified an unconstrained minimum, the term that is linear in $\delta\mathbf{N}$ must vanish. [As noted in the previous section, it is this condition that yields the formula (3.4-7)]. It follows that the region of interest is specified by

$$\delta\mathbf{N}^T\mathbf{H}(\hat{\mathbf{N}}) \, \delta\mathbf{N} \leq 2\Delta. \tag{3.5-10}$$

For the equality, this is an ellipsoid in the parameter space.

We now need to calculate $\mathbf{H}$. We have

$$\Phi(\mathbf{N}) = (\mathbf{J} - \mathbf{GN})^T(\mathbf{J} - \mathbf{GN})$$
$$= \mathbf{J}^T\mathbf{J} - \mathbf{J}^T\mathbf{GN} - (\mathbf{GN})^T\mathbf{J} + (\mathbf{GN})^T(\mathbf{GN})$$
$$= \mathbf{J}^T\mathbf{J} - 2\mathbf{J}^T\mathbf{GN} + \mathbf{N}^T\mathbf{G}^T\mathbf{GN}. \tag{3.5-11}$$

Then

$$\frac{\partial\Phi}{\partial N_m} = \frac{\partial}{\partial N_m}\left\{-2\sum_j\sum_k J_j G_{jk}N_k + \sum_i\sum_j\sum_k N_i G_{ji}G_{jk}N_k\right\}$$
$$= -2\sum_j J_j G_{jm} + \sum_i\sum_j N_i G_{ji}G_{jm} + \sum_j\sum_k G_{jm}G_{jk}N_k$$

and

$$\frac{\partial^2\Phi}{\partial N_m \, \partial N_n} = \sum_j G_{jn}G_{jm} + \sum_j G_{jm}G_{jn},$$

with the result

$$H_{mn} = 2 \sum_j G_{jn} G_{jm}$$

or

$$\mathbf{H} = 2\mathbf{G}^T\mathbf{G}. \tag{3.5-12}$$

From Eq. (3.5-10),

$$\delta\mathbf{N}^T(\mathbf{G}^T\mathbf{G})\,\delta\mathbf{N} \le \Delta. \tag{3.5-13}$$

To find $\delta\mathbf{N}$ of minimum length satisfying this equation we employ the method of Lagrange multipliers; i.e., we form the function

$$f(\delta\mathbf{N}, \gamma) = \delta\mathbf{N}^T\,\delta\mathbf{N} - \gamma[\delta\mathbf{N}^T(\mathbf{G}^T\mathbf{G})\,\delta\mathbf{N} - \Delta]$$

and set

$$\frac{\partial f}{\partial(\delta\mathbf{N})} = 0 = 2\delta\mathbf{N} - 2\gamma(\mathbf{G}^T\mathbf{G})\,\delta\mathbf{N},$$

whereby

$$(\mathbf{G}^T\mathbf{G})\,\delta\mathbf{N} = \gamma^{-1}\,\delta\mathbf{N}. \tag{3.5-14}$$

Let us identify $\lambda = \gamma^{-1}$; then this equation states that $\delta\mathbf{N}$ is an eigenvector of $\mathbf{G}^T\mathbf{G}$ associated with the eigenvalue λ.

With this result, it follows immediately that

$$\delta\mathbf{N}^T(\mathbf{G}^T\mathbf{G})\,\delta\mathbf{N} = \lambda\delta\mathbf{N}^T\,\delta\mathbf{N}. \tag{3.5-15}$$

The hypersurface specified by Eq. (3.5-6) is also specified by setting the left-hand side of this equation equal to Δ. Thus,

$$\delta\mathbf{N}^T\,\delta\mathbf{N} = \Delta/\lambda, \tag{3.5-16}$$

which tells us that the length of the eigenvector $\delta\mathbf{N}$ is $\sqrt{\Delta/\lambda}$.

Since $\mathbf{G}^T\mathbf{G}$ is a real symmetric matrix, there exists an orthogonal matrix $\mathbf{U}$ such that

$$\mathbf{G}^T\mathbf{G} = \mathbf{U}\Lambda\mathbf{U}^T = \mathbf{U}\Lambda\mathbf{U}^{-1}, \tag{3.5-17}$$

where Λ is diagonal with values $\lambda_1, \lambda_2, \ldots, \lambda_M$. Thus,

$$\delta\mathbf{N}^T(\mathbf{G}^T\mathbf{G})\,\delta\mathbf{N} = \delta\mathbf{N}^T(\mathbf{U}\Lambda\mathbf{U}^T)\,\delta\mathbf{N}. \tag{3.5-18}$$

We define a change in variables such that

$$\delta\mathbf{X} = \mathbf{U}^T\,\delta\mathbf{N}, \tag{3.5-19}$$

which results in the canonical form

$$\delta X^T \Lambda \, \delta X = \sum_{i=1}^{M} \lambda_i x_i^2, \qquad (3.5\text{-}20)$$

representing a rotation of coordinate axes in the parameter space. Equating to Δ gives us

$$\sum_{i=1}^{M} \frac{x_i^2}{\sqrt{\Delta/\lambda_i}} = 1 \qquad (3.5\text{-}21)$$

from which we see that $\sqrt{\Delta/\lambda_i}$ is the semiaxis of the ellipsoid in the direction x_i of our new coordinate system.

We now see that the uncertainty in the parameter estimates is relatively small along coordinate axes associated with large eigenvalues, and relatively large along those associated with small eigenvalues. A similar indication of this fact is given by the covariance of N:

$$V_{NN} = \langle (\hat{N} - N)(\hat{N} - N)^T \rangle, \qquad (3.5\text{-}22)$$

which can be shown to be

$$V_{NN} = (G^T V_{\varepsilon\varepsilon}^{-1} G)^{-1} \qquad (3.5\text{-}23)$$

for the weighted LMS case. For the unweighted case, $V_{\varepsilon\varepsilon} = \sigma^2 U_I$, where U_I is the identity matrix, and so then

$$V_{NN} = \sigma^2 (G^T G)^{-1}. \qquad (3.5\text{-}24)$$

The transformation (3.5-17) can be interpreted in the form

$$U^{-1}(G^T G)^{-1} U = \Lambda^{-1}, \qquad (3.5\text{-}25)$$

which shows that, in the rotated coordinate system, a covariance matrix

$$V_{XX} = U^{-1} V_{NN} U = \Lambda^{-1} \qquad (3.5\text{-}26)$$

can be found which is diagonal, with values $1/\lambda_1$, $1/\lambda_2$, etc. The variance of x_i is thus $1/\lambda_i$.

It is not difficult to extend the analysis of Eqs. (3.5-4) through (3.5-21) to the case of weighted least squares. The critical role played by $G^T G$ for the unweighted case is then played by $G^T V_{\varepsilon\varepsilon}^{-1} G$ for the weighted case, as can be seen also from the form of V_{NN}. It is important in this connection to recognize the influence of the noise, which, although not appearing explicitly in Eq. (3.5-21), is evident in the two forms of V_{NN} above. If a principal axis transformation is applied to $G^T V_{\varepsilon\varepsilon}^{-1} G$ in the same way as that which led to Eq. (3.5-21), the noise effects will be directly incorporated into the uncertainty region ellipsoid, modifying its axes by individually different amounts. The parallel effect in the unweighted case requires division of λ_i by the constant factor σ^2, thus changing all axes in the same proportion.

It should be noted here that minimization of the quadratic forms Q_1 and Q_2 [Eqs. 3.4-6) and (3.4-8)] results also in a "minimization" of $\mathbf{V_{NN}}$ in the sense of minimizing some reasonable measure of $\mathbf{V_{NN}}$. The determinant is sometimes used for this purpose, but other measures are possible [23].

The picture conveyed by either Eq. (3.5-21) or Eq. (3.5-26), of course, requires that a transformation be made back to the original parameter space coordinates in order to obtain an interpretation in terms of $N_1, N_2, N_3, \ldots, N_M$. The parameter covariance matrix $\mathbf{V_{NN}}$ communicates this interpretation more directly, albeit somewhat less elegantly. It should be understood that values of $\mathbf{N}$ will *not invariably* be badly determined as the result of large indeterminacies in the canonical variables. For instance, suppose that for a two-parameter space the ellipse

$$\delta N_1^2 - 1.99\, \delta N_1\, \delta N_2 + \delta N_2^2 = 1$$

gives the boundary of an uncertainty region with $\Delta = 1$. The semimajor axes are along the line $\delta N_1 = \delta N_2$, and the semiminor axes are along $\delta N_1 = -\delta N_2$. The length of the semimajor axis is approximately 14, while that of the semiminor axis is about 0.25, indicating that estimates of $N_1 + N_2$ will be about 7.5 times more uncertain than estimates of $N_1 - N_2$. Nonetheless, the individual estimates of N_1 and N_2 are fairly well determined, falling within an interval of magnitude one in uncertainty. The result shows that, if for any reason we might wish to choose $N_1 \neq \hat{N}_1$ and $N_2 \neq \hat{N}_2$, we have some latitude available to do so but should keep $N_1 - N_2 = \hat{N}_1 - \hat{N}_2$ for best reliability.

The ratio of the largest to the smallest eigenvalue of the matrix $\mathbf{G^T G}$ or $\mathbf{G^T V_{\varepsilon\varepsilon}^{-1} G}$ is sometimes referred to as a *condition number*. Its square root gives a quick measure of the relative extent to which noise will be "magnified" by the matrix instability, for the worst case [24]. Matrices having large condition numbers are said to be ill-conditioned, or poorly conditioned. The condition number depends on the scale factors of the parameters as well as the directions of the vectors $\mathbf{g}_i$ representing columns of $\mathbf{G}$. Consider, for instance, a two-parameter case with elements $1, 1, 0, 0$ for $\mathbf{g}_1$ and $0, 0, 1, 1$ for $\mathbf{g}_2$. Evidently $\mathbf{g}_1$ and $\mathbf{g}_2$ are orthogonal, and $\mathbf{G^T G}$ has eigenvalues $\lambda_1 = \lambda_2 = 2$. If the noise is isotropic, the signal-to-noise ratio will be the same for $\hat{N}_1$ and $\hat{N}_2$. Let us compare this with a second example with elements $1, 1, 0, 0$ for $\mathbf{g}_1$ and elements $0, 0, 0.1, 0.1$ for $\mathbf{g}_2$. The two vectors are still orthogonal, but now the eigenvalues of $\mathbf{G^T G}$ are $\lambda_1 = 2$, $\lambda_2 = 0.02$. The signal-to-noise ratio applicable to the estimate $\hat{N}_2$ is only one-tenth that of the ratio for $\hat{N}_1$. Note, however, that if the scale by which we measure N_2 is changed so that it becomes N_2', where $N_2' = N_2/10$, the condition number goes back to 1. This of course does not change the accuracy of the estimate of N_2, since the uncertainty in N_2' must be multiplied by 10 to recover its original scale. It does, however, illustrate that scale factors enter the picture in an important way. The condition number in itself is not completely informative unless it is based on the eigenvalues of $\mathbf{G}^T \mathbf{V}_{\varepsilon\varepsilon}^{-1} \mathbf{G}$; in other words, the nature and level of the noise must be included to obtain the total picture.

Other perspectives of the stability problem can be gained in various ways, for instance, by expressing the integral Eq. (3.4-1) in terms of the Fourier spectrum of its kernel $G(\rho, \sigma)$ [25]. We should mention also that certain procedures can sometimes be used to improve the conditioning or mitigate the effects of ill-conditioning [26]. But the best approach to a problem with poor conditioning is to avoid it. This does not necessarily mean abandoning it altogether; it may necessitate only a consideration of the physical nature of the problem and elimination of the particular linear representations that are the inherent cause of the difficulty. In our case, for instance, there is no point in attempting to distinguish between particles with virtually identical scattering patterns; one would be foolish to expect reliable estimates under these conditions.

3.6 IMPLEMENTING THE INVERSION: HYBRID METHODS

The irradiance values in the diffraction pattern obtained with a Fourier-optical system are obviously at the disposal of the experimenter, for whatever means he may wish to apply in using them. One possible choice is measuring the power levels in selected increments of area and using numerical calculations to infer the estimates. Since such calculations may be lengthy, they are most frequently approached through some form of digital computation. Systems employing this concept are appropriately referred to as *hybrid*, signifying the combination of optical and digital processes.

Stark and colleagues [27–30] employed the hybrid approach in a number of experiments estimating counts of both circular and noncircular particles. An important aspect of their work was their consideration of the speckle noise which, for a relatively limited light power sampling area, may contribute significantly to errors in the data.

Their treatment of this problem is presented in the first two of the papers [27, 28] cited above. It is based on the derivation of an optimum (minimum bias, and nominally specified variance) smoothing filter for a circular sampling aperture. (The term apodization is frequently used in the optics literature to describe such use of a weighting function to vary the transmission within the area of an aperture.) With $W(\rho)$ representing the irradiance smoothing function, and ρ as the radial frequency, they derive the optimum filter for two different circumstances: the first with a constraint on the spatial extent of the data lag window (in the object, i.e., particle plane); the second with a constraint on the extent of the window in the spatial frequency or observation plane. As they point out, the first of these is not strictly realizable as a spatial frequency filter, because the data truncation necessarily generates an infinite frequency spread. In the second case, they show that, for a circular window of radius C_0, a normalizing condition

$$2\pi \int_0^{C_0} \rho W(\rho)\, d\rho = 1, \qquad (3.6\text{-}1)$$

and a (relative) variance K given by

$$\int_0^{C_0} \rho W^2(\rho)\, d\rho = K, \qquad (3.6\text{-}2)$$

the optimum characteristic is

$$W_0(\rho) = \begin{cases} 3K\pi(1 - 3\pi^2 K\rho^2/2), & \rho < (1/\pi)(\sqrt{2}/3K), \\ (2/\pi C_0^2)(1 - \rho^2/C_0^2), & \rho < C_0, \\ 0, & \rho \geq C_0. \end{cases} \qquad (3.6\text{-}3)$$

They give a number of comparisons with other possible choices of $W(\rho)$ and also show experimentally that estimates based on smoothed spectral samples are considerably superior to those obtained when unsmoothed readings are employed. (See also their other two papers for additional experimental evidence involving both circular and noncircular particles.)

Another aspect of the work by Stark and associates is contained in the Stark–Shao paper [30], where they demonstrate the use of a training procedure, required when the individual patterns characteristic of separate classes of particles are not experimentally or theoretically available. Then the scattering matrix **G** must itself be statistically inferred from the patterns of particle mixtures. A generalized matrix inverse based on minimization of the mean square error is as applicable here as it is for estimating **N** when **G** is known. The most obvious procedure is to form a matrix $\tilde{\mathbf{N}}$ with elements N_{ik}, where i represents the ith training sample and k the kth value of radius at which an observation is made. The set of observed patterns is also arranged as a matrix, say $\tilde{\mathbf{W}}$, with elements W_{ik}. Then

$$\tilde{\mathbf{W}} = \tilde{\mathbf{N}}\mathbf{G}, \qquad (3.6\text{-}4)$$

and the estimate of **G** is given by

$$\mathbf{G} = (\tilde{\mathbf{N}}^T\tilde{\mathbf{N}})^{-1}\tilde{\mathbf{N}}^T\tilde{\mathbf{W}}. \qquad (3.6\text{-}5)$$

The values of **N** for the different training samples must of course differ sufficiently from each other to ensure stability in the inversion.

Training of a hybrid system for cell recognition and counting is also treated by Carlson and Lee [31], who emphasize the adaptive possibilities of such a system, to accommodate changes in particle categories, for example. Their data analysis was not based on least mean squares inversion but rather on *ad hoc* selection and the use of features in both linear and nonlinear combinations.

Commercial versions of hybrid systems manufactured by Recognition Systems, Inc., have been employed in a variety of applications for many years. Frequency plane sampling is accomplished by them with a single silicon chip on which there are 64 separate photodetection areas, half of them as sectors of about 8 deg each and half as concentric annular (actually, half-annular) regions. The 64 measurements are processed digitally to extract a relatively small number of features that can then be used as signatures applicable to the particular problem concerned. These systems appear to be enjoying a widening domain of use in industrial, military, and other applications. The use of this detector has been reported by some investigators mentioned elsewhere in this chapter [32]. It is of course a much more convenient device than mechanically scanning apparatus and furnishes a reasonable number of discrete features of the diffraction pattern. Sampling apertures are sufficient to provide appreciable smoothing of spatial noise; in addition, the radial weighting inherent in polar coordinate geometry is of benefit to the signal-to-noise ratio as a function of distance from the optic axis. It thus carries with it some of the same advantages as optical processing, to be discussed in the next section.

Results of experiments by investigators mentioned above and others will be discussed in Section 3.8.

3.7 IMPLEMENTING THE INVERSION OPTICALLY

The use of optical inversion to obtain the least mean squares estimates directly from the diffraction pattern is conceptually simple but requires a few adjustments, which we shall now discuss.

We have represented the data available for measurement by the quantity **J**, understood to consist of elements $J(\rho_i)$, $i = 1, 2, \ldots, P$. A more logical representation from the analog point of view is simply its functional form $J(\rho)$. This may seem foolhardy by virtue of the fact that, because of the granular nature of the irradiance pattern, J will display violent fluctuations in magnitude if observed on a fine scale. However, the power observed over some appreciable angular width $\Delta\phi$, within a small increment of radius $\Delta\rho$, will vary fairly smoothly, in accordance with our conclusions in Section 3.3.

Since the optical processing step is to be strictly passive, it cannot accommodate weighting functions with values less than zero or greater than one. It is of course not difficult to scale the operation to any range of magnitude we choose. To provide for negative values of weighting functions we must also introduce a bias which, as we shall see, will necessitate a correction to the light power measurements before interpreting them in terms of particle numbers.

When viewed from the vantage point of optical processing, particle diffraction patterns are easily perceived to offer the opportunity of carrying out an appreciable number of operations in parallel. Circular symmetry allows us to consider any sector of angular width $\Delta\phi$ as being statistically equivalent to any other. The inversion matrix consists of M rows, one for each particle category, and this suggests that the diffraction pattern should be divided into M sectors, one for each of the M operations. We need one additional sector to give the information we need for bias correction, as we shall explain below. If we choose to make all the sectors of equal width and assume negligible spacing between adjacent edges, the angular width of each will be $2\pi/(M + 1)$.

Let us take the total radial extent over which measurements are to be made as $\rho_0 \leq \rho \leq \rho_1$, where ρ_0 is selected so as to exclude the small region of unscattered light on and near the optic axis. We define $J_j(\rho)$ as the average irradiance within $\Delta\phi_1$. Then the power within the angle $\Delta\phi_j$ and incremental radial distance $d\rho$ is given by

$$L_j(\rho)\, d\rho = \frac{2\pi}{M + 1}\, \rho J_j(\rho)\, d\rho = \frac{2\pi}{M + 1}\, \rho[\langle I(\rho)\rangle + \varepsilon_j(\rho)]\, d\rho. \quad (3.7\text{-}1)$$

We shall assume that, because of area averaging, $\varepsilon_j(\rho)$ is principally measurement error, with negligible speckle noise. We presume that $\langle \varepsilon_j \rangle = 0$ for all j; thus the expected values of J_j are all $\langle I(\rho)\rangle$.

We define $\mathbf{G}(\rho)$ to be a $1 \times M$ matrix with elements consisting of the scalar functions $G_k(\rho)$. Next, we introduce $\mathbf{Q}(\rho)$ such that

$$\mathbf{Q}(\rho) = \frac{2\pi}{M + 1}\, \rho \mathbf{G}(\rho). \quad (3.7\text{-}2)$$

Then

$$\frac{2\pi}{M + 1}\, \rho\langle I(\rho)\rangle = \frac{2\pi}{M + 1}\, \rho\mathbf{G}(\rho)\mathbf{N} = \mathbf{Q}(\rho)\mathbf{N}. \quad (3.7\text{-}3)$$

The unweighted LMS estimation is based on minimizing the expression

$$\int_{\rho_0}^{\rho_1} [L_j(\rho) - \mathbf{Q}(\rho)\mathbf{N}]^T [L_j(\rho) - \mathbf{Q}(\rho)\mathbf{N}]\, d\rho \quad (3.7\text{-}4)$$

and leads to the result [16]

$$\tilde{\mathbf{N}} = \left[\int_{\rho_0}^{\rho_1} \mathbf{Q}^T(\rho)\mathbf{Q}(\rho)\, d\rho \right]^{-1} \left[\int_{\rho_0}^{\rho_1} \mathbf{Q}^T(\rho)L_j(\rho)\, d\rho \right]. \qquad (3.7\text{-}5)$$

This specifies an estimation of all elements of $\mathbf{N}$ from the light power in the jth sector, contrary to our earlier stated intent to multiplex the estimation process. The necessary adjustment as well as the scaling required can be made by defining a weight function

$$W_j(\rho) = \alpha_j \sum_{k=1}^{M} \left[\int_{\rho_0}^{\rho_1} \mathbf{Q}^T(\rho)\mathbf{Q}(\rho)\, d\rho \right]^{-1}_{jk} Q_k(\rho), \qquad (3.7\text{-}6)$$

where α_j is the factor needed for scaling. The estimator for N_j is then

$$\hat{N}_j = \alpha_j^{-1} \int_{\rho_0}^{\rho_1} W_j(\rho)J_j(\rho)\, d\rho, \qquad (3.7\text{-}7)$$

which makes use strictly of light in the jth sector.

To extend these results to the weighted LMS analysis involves no essential difficulty insofar as its formal expression is concerned. We replace the covariance matrix $\mathbf{V}_{\varepsilon\varepsilon}$, applicable to discrete values of ρ, by the covariance function

$$V_\varepsilon(u, v) = \langle \varepsilon(u)\varepsilon(v) \rangle, \qquad (3.7\text{-}8)$$

where $\varepsilon(x) = L(x) - \mathbf{Q}(x)\mathbf{N}$. Its inverse is needed and is defined by the requirement that, for any function $g(x)$ [33],

$$\int_{\rho_0}^{\rho_1} \int_{\rho_0}^{\rho_1} V_\varepsilon^{-1}(u, x)V_\varepsilon(\sigma, u)g(\sigma)\, du\, d\sigma = g(x). \qquad (3.7\text{-}9)$$

Alternatively V_ε^{-1} can be specified by use of the Dirac delta function [34]

$$\int_{\rho_0}^{\rho_1} V_\varepsilon^{-1}(u, x)V_\varepsilon(\sigma, u)\, du = \delta(x - \sigma). \qquad (3.7\text{-}10)$$

We now minimize the expression

$$\int_{\rho_0}^{\rho_1} \int_{\rho_0}^{\rho_1} [L(u) - \mathbf{Q}(u)\mathbf{N}]^T V_\varepsilon^{-1}(u, v)[L(v) - \mathbf{Q}(v)\mathbf{N}]\, du\, dv \qquad (3.7\text{-}11)$$

with respect to variations in $\mathbf{N}$. The result is

$$\hat{\mathbf{N}} = \Delta^{-1} \int_{\rho_0}^{\rho_1} \int_{\rho_0}^{\rho_1} \mathbf{Q}^T(u)V_\varepsilon^{-1}(u, v)L(v)\, du\, dv, \qquad (3.7\text{-}12)$$

where

$$\Delta = \int_{\rho_0}^{\rho_1} \int_{\rho_0}^{\rho_1} \mathbf{Q}^T(u) V_\varepsilon^{-1}(u, v) \mathbf{Q}(v) \, du \, dv.$$

In the event that V_ε^{-1} peaks sharply in the vicinity of $u = v$ and rapidly becomes negligible as u differs increasingly from v, it may suffice to take

$$V_\varepsilon^{-1}(u, v) = \sigma^{-2}(u) \, \delta(u - v). \tag{3.7-13}$$

Then

$$\hat{\mathbf{N}} = \left[\int_{\rho_0}^{\rho_1} \mathbf{Q}^T(\rho)\sigma^{-2}(\rho)\mathbf{Q}(\rho) \, d\rho \right]^{-1} \left[\int_{\rho_0}^{\rho_1} \mathbf{Q}^T(\rho)\sigma^{-2}(\rho)L(\rho) \, d\rho \right], \tag{3.7-14}$$

which is a relatively simple modification of Eq. (3.7-5). The corresponding modification of $W_j(\rho)$ should be apparent.

As mentioned previously, we cannot perform the operation of Eq. (3.7-7) by direct optical processing, inasmuch as $W_j(\rho)$ may be negative for some values of ρ. A quantity we *can* obtain optically, however, is

$$R_j = \int_{\rho_0}^{\rho_1} [W_j(\rho) + \rho f_j(\rho)] J_j(\rho) \, d\rho, \tag{3.7-15}$$

where $\rho f_j(\rho)$ is a biasing term selected so that its sum with $W_j(\rho)$ is equal to or greater than zero. [The reader should bear in mind that integration over angle $\Delta \phi_j$ is implicit in this expression, as is also the factor ρ in $W_j(\rho)$. It will be convenient to let ρ be explicit in connection with integration of the bias term.]

In most circumstances the simplest choice of $f_j(\rho)$ may be the best, namely, $f_j(\rho) = b$, where b is a constant. Then

$$R_j = \int_{\rho_0}^{\rho_1} [W_j(\rho) + \rho b] J_j(\rho) \, d\rho \tag{3.7-16}$$

and

$$\hat{N}_j = \alpha_j^{-1} \left[R_j - b \int_{\rho_0}^{\rho_1} J_j(\rho)\rho \, d\rho \right] = \alpha_j^{-1}(R_j - B_j). \tag{3.7-17}$$

The second term on the right-hand side of Eq. (3.7-17) is of course not available to us. However, we should be able to obtain a good estimate of it from a separate "clear" sector of the optical mask, with transmission one everywhere. Let us arbitrarily assign the $(M + 1)$th sector to this purpose and let

$$B = b \int_{\rho_0}^{\rho_1} J_{M+1}(\rho)\rho \, d\rho = \hat{B}_j. \qquad (3.7\text{-}18)$$

Then, for all N_j,

$$\hat{N}_j = \alpha_j^{-1}(R_j - B). \qquad (3.7\text{-}19)$$

The subtraction required here can be done in a variety of ways, ranging from manual to analog or digital electronic. There seems to be no need for us to pursue this point further.

It remains to consider how best to synthesize the transmittance $W_j(\rho) + \rho b$. It is not a simple task to do so by use of variable-density photographic procedures. A preferable approach is to let the jth sector consist of complementary regions having only the two values of transmittance zero and one. In other words, the angular width of the transparent region is designed to vary as a function of ρ in exact concordance with the sum $W_j(\rho) + \rho b$.

We can express this more formally by defining

$$\phi_{k0} = k2\pi/(M + 1), \qquad k = 1, 2, 3, \ldots, M. \qquad (3.7\text{-}20)$$

Then

$$R^j = \int_{\rho_0}^{\rho_1} \int_{\phi_{j0}}^{\phi_j(\rho)} J_j(\rho) \, d\rho \, d\phi, \qquad (3.7\text{-}21)$$

where $\phi_j(\rho) = \phi_{j0} + W_j(\rho) + \rho b$. This describes directly the actual physical process of the jth measurement: It is simply a reading of total light power through the jth sector. All the R_j are obtainable simultaneously if a separate photodetector–amplifier channel is furnished for each. The R_j can also be read sequentially using a single channel and an aperture scanning through all the sector locations, one by one.

Generally the $W_j(\rho)$ will differ appreciably—and in some cases drastically—when different wavelengths are used to illuminate the particles. In such instances the provision of polychromatic illumination "tuned" to outstanding differences in color response can greatly enhance the system capabilities for particle discrimination. Infrared and ultraviolet as well as visible light illumination are perfectly feasible. One way to make use of the information gained in this way is to combine it additively. For example, if two wavelengths λ_1 and λ_2 are used, we form the expression

$$\hat{N}_j = \sum_{m=1}^{2} \alpha_j^{-1}(\lambda_m) \int_{\rho_0}^{\rho_1} W_j(\rho, \lambda_m) J_j(\rho, \lambda_m) \, d\rho. \qquad (3.7\text{-}22)$$

Of course it is not necessary that the interval of integration be identical for the two λ's. But in any case, readings corresponding to Eq. (3.7-21) must be taken separately, whether or not within the jth sector location, biases subtracted, scaling applied, and results added.

Another kind of optical inversion is involved in the optical Wiener filter [35], the design of which is based on the Wiener filter (sometimes termed *optimum smoothing filter*, and sometimes called the *Wiener–Kolmogoroff filter*) developed in the context of one-dimensional signal processing [36, 37]. Wiener filtering as originally conceived was hampered with respect to its implementation by the causality constraint that applies to any real-time operations on time domain signals. This constraint does not apply to optical spatial filtering and permits the Wiener concept to be realized in totality, without compromise.

Wiener filtering and the generalized matrix inversions discussed in preceding sections have one feature in common, and that is the fact that both are based on a minimization of the squared error. However, Wiener filtering presumes that the task at hand is one of recovering a minimally noisy, minimally distorted signal from some background of interference. There are only two dependent variables: one is signal, the other, noise. The latter may be a composite of effects due to a variety of separate causes; in any case, its power spectrum must be known, as must also that of the signal, in order to design the filter. The formula for the filter transfer function is given in the case of uncorrelated signal and noise by

$$H(\omega) = G_{ss}(\omega)/[G_{ss}(\omega) + G_{nn}(\omega)], \qquad (3.7\text{-}23)$$

where G_{ss} and G_{nn} are the power spectral densities of signal and noise, respectively [38]. In the optical realization, these quantities are expressed as functions of the radial spatial frequency.

Ward *et al.*, [39–41] constructed optical Wiener filters based on this formula for estimating the concentrations of specific cell types in whole blood smears. A photographic procedure was used as follows (in one of two realizations employed): A plate was exposed to the power spectrum of the signal (i.e., the cell type to be detected); then a positive was made, each step being controlled to produce a final gamma of -2 and, therefore, an intensity transmittance proportional to the square of G_{ss}. A second plate was exposed to the power spectrum of signal and noise combined (whole blood with normal cell concentrations) and developed to a gamma of $+2$. With the plates sandwiched, the required amplitude transmittance (the square root of the intensity transmittance) was obtained in accordance with Eq. (3.7-23).

3.8 EXPERIMENTAL INVESTIGATIONS AND RESULTS

Most of the experimental data and particle analyses using Fourier-optical methods have been obtained with hybrid systems. Ward [42] sampled the diffraction pattern of a collection of spherical particles ranging from 7.5 to 57.5 μm in diameter using a series of annular apertures. Individual apertures were rotated successively in front of a light-collecting lens and focused onto a single photodetector. A least mean squares matrix inversion was used to calculate an 11-element histogram of the particle size distribution. Some details have been given by Thompson [43].

A particle size analyzer based on composite pattern analysis in the Fraunhofer plane was investigated by Cornillault [44] in application to particles from about 2.5 to 80 μm in diameter. Ten measurements were performed, each through a "window" centered at a different radius. The photocell was positioned behind each window in succession to obtain the light readings. Illumination was with a helium–neon laser. Reproducibility of calculated size distributions for different sets of measurements was reported to be within about 2%, and results were described as being in good agreement with other particle sizing methods.

Optical filters arranged in the form of seven annular rings concentric with the optic axis were used by McSweeney and Rivers [45] for measuring size distributions of particles suspended in water. The fibers in each annulus were brought together in a bundle, each bundle conveying its light to a separate photodetector. Tests were conducted using opaque circular spots recorded on photographic film as the "particles." Calculations were made using two measurements for a mixture of two sizes of particles; results were reported as encouraging, but no details were given.

Stark *et al.* [28] presented data for five runs with three classes of circular particles, termed small, medium, and large. Populations were constant: 50 small, 25 medium, 15 large. A LMS analysis of the raw (unsmoothed) data samples gave particle number estimates with standard deviations of 37% (small), 34% (medium), and 18% (large). Corresponding average error magnitudes were 36, 34, and 17%, respectively. With filtered spectral data the estimates were much improved, yielding standard deviations of 3, 6.4 and 9.4%, in the same order. Error magnitudes were again similar: 2.8, 5.6, and 8.0%. The tendency for the smoothed data to give proportionately higher error for the large particles in contrast to the raw data, which gave proportionately higher error for small particles, suggests the possibility that the smoothing filter was suppressing some of the vital low spatial frequency information in readings closest to the optic axis.

Stark *et al.*, [29] gave results of an experiment using noncircular particles, again consisting of small, medium, and large classes. Particle numbers were 35, 30, and 20, respectively, in each of these categories. Again, five runs were made; results were given only for smoothed data, and had standard deviations of 3.1, 6.7, and 9.7% in the order mentioned. Error magnitudes were 2.3, 5.3, and 9%. The tendency for greater error in the large particle estimates is again evident; indeed, the authors made specific note of the experimental problem of sampling at the lowest spatial frequencies, while maintaining blockage of the uninformative central portion, and described to some extent the inaccuracies experienced when suppressing such frequencies.

An investigation by Stark and Shao [30] addressed the more demanding task of estimating particle class signatures before engaging in particle count estimation. Thirteen mixtures of 300-, 500-, and 700-μm circular particles were used as training samples, and patterns analyzed as discussed in Section 3.6. Particle count estimates were then made for the mixtures based on the learned signatures. For unsmoothed data, the overall percentage of error (the sum of all error magnitudes divided by the sum of all particles) was given as 17.5%. For smoothed data the corresponding figure was 9.2%, with 11.9, 8.8, and 7.1% representing the individual small, medium, and large classes, respectively. The authors again pointed out the necessity for care in dealing with the lowest range of spatial frequencies. Sources of error cited by them as being principally responsible for estimation inaccuracies were (1) inadequate light power at high spatial frequencies, (2) laser output power variation, and (3) problems with the near-zero spatial frequencies.

Carlson and Lee [31] also used training samples to determine the elements of an inversion algorithm for counting reticulocyte concentrations in whole blood. Reticulocytes are red blood cells newly released from the bone marrow, which have not yet disposed of the mitochondria and ribosomes they possess while in the infant and adolescent stages of their existence. The fine structure ($\sim 1 \mu$m) characteristic of these remnants scatters appreciably more light into the high range of the spatial frequency spectrum than do mature cells. Carlson and Lee sampled the diffraction pattern at six locations, selected so as to emphasize prominent distinctions between spectra of reticulocytes and normal cells. Twenty-five training samples were employed. The potential of such a system for adaptive processing of diffraction data was pointed out, and considerable attention given to procedures for selection of the most significant samples and testing for goodness of fit of the estimator model. An estimated standard deviation of 2.05% reticulocyte count per 1000 cells resulted from their experiment and was characterized as representing an improvement over results of other methods.

Size distributions of kerosene droplets in a fuel spray have been estimated by Fourier-optical methods by Swithenbank and colleagues [46]. Droplets ranging from 5 to 500 μm in diameter were classified into 31 size groups using 31 semicircular concentric photosensitive rings located in the Fourier plane. A collimated helium–neon laser beam was used for illumination of the particles, which were generated in the form of a narrow jet. Particles were modeled as circular disks, and each detector annulus was located at the position of maximum energy for the average-diameter particle in its range;

i.e., at the maximum of $J_1^2(2\pi r_k \rho)/\rho$, where r_k is the radius of an average particle in group k. It was noted by the authors that the measured size distribution would differ from the correct one if, as expected, some droplets moved more rapidly than others; however, a correction could be made if the velocity distribution were known. The data were arranged in square matrix form corresponding to the 31 measurements and 31 size groups. Inversion was performed by minicomputer. The accuracy of the technique was estimated to be about 2%.

Ewan [13] has investigated application of the system used by Swithenbank *et al.* [46] to analysis of particle images reconstructed from holograms. Earlier (Section 3.1) we mentioned the work of Thompson and colleagues [12], who were the first to use holograms to store aerosol particle images, permitting their reconstruction and detailed analysis throughout the volume originally illuminated. The holograms studied by Ewan were made with collimated light from a pulsed ruby laser. Diffraction pattern measurements were made with a 31-element semicircular photodetector array, as described in the preceding paragraph. During a 2-sec period, 100 sets of data were averaged. The light energy profile was curve-fitted to a variable parameter size distribution equation. Results using the hologram reconstructions were compared with on-line particle size estimates and showed good agreement. It appears that the effect of the out-of-focus particles was not great, as indeed it should not be for quite small cross sections of particle flow.

Apparently the only experimental arrangements for particle identification and counting by analog optical processing have been those of Ward *et al.* [39–41] using Wiener filters, and Dodge and Anderson [47] using a least mean squares matrix inversion (as described in Section 3.7). Wertheimer and Wilcock [48] have also used Fourier plane filtering to monitor flow parameters in a stream of spherical particles; however, no classification was encompassed by their method, which was designed to furnish light flux amounts in proportion to the second, third, and fourth powers of the particle radii. The masks they used were constructed by suitable shaping of three separate open areas in an otherwise opaque disk and so are related in principle to the masks described by Anderson and Beissner [16] and used by Dodge and Anderson [47]. From the three signals they obtained, they could estimate on a continuous basis the average volume (and mass) flow, the mean circular radius (i.e., the radius of the cross-sectional area), the mean spherical radius, and the area standard deviation.

A portion [39, 41] of the Wiener filtering effort reported by Ward *et al.* was directed to the estimation of reticulocyte concentrations in whole blood, a problem later addressed by Carlson and Lee [31], as mentioned earlier in this section. Their filter construction method has been described in the previous section. Estimation accuracy as indicated by the standard deviation of results was found to be comparable to that of technician counting, using a microscope.

Ward *et al.* [40] also constructed and used Wiener filters for estimation of the concentrations of two kinds of white cells: lymphocytes and granulocytes. Together, these two classes comprise about 95% of all leukocytes, the only remaining component being monocytes. Leukocytes are outnumbered by red cells, however, in a ratio of roughly 500 or 1000 to 1 and so represent a rather small "signal" in a rather large background of "noise." Fortunately (from the viewpoint of optics), their diameters are appreciably larger, on the average, than those of red cells, and in addition their nuclei, granules, and often irregular morphology give rise to appreciable spectral differences, both with respect to one another and with respect to the red cells. Thrombocytes, or platelets, are the other major formed constituent of the blood, but their diameters average less than half those of red cells. Even with such distinctions in optical characteristics, however, blood cell classification and counting remains a challenging task. These investigators reported encouraging results in this instance, with estimation errors comparing favorably with those of a trained technician.

In the experiments by Dodge and Anderson [47] the diffraction patterns of mixtures of small latex spheres suspended in a water–glycerin mixture were analyzed by the unweighted least mean squares method. Both optical and digital inversions were used and compared with each other as well as with particle concentrations calculated from the mixing ratios with respect to the original samples (for which the concentrations were provided by the supplier, Dow Chemical). A hemocytometer was also used to check the numbers deduced from the mixing ratios. One set of spheres was of 2.02 μm mean diameter with standard deviation 0.0135; the other contained a range of diameters from about 4 to 7 μm, with standard deviation 1.51. Each was treated as a *single* type; in other words, all preparations were regarded as two-component mixtures. Scattering patterns were carefully measured for each of the two types using reference concentrations that gave area densities of about $2.35 \times 10^6/\text{cm}^2$ for the 2-μm particles and about $2.0 \times 10^5/\text{cm}^2$ for the others. Correlation index between the two patterns was 0.69. The polar angle included in the patterns was about 8.5 deg.

Construction of the masks used in the optical inversion was based on the principles outlined in Section 3.7. The weighting functions as converted to angular width were calculated from the experimentally measured patterns and plotted in a large format on white poster board. Transmission sections were blackened, and the resultant patterns photographed with exactly the demagnification required. To avoid any possibility of nonuniform transmission in the "clear" areas, these were carefully excised.

Mixtures of the reference concentrations were prepared in the ratios

5/95, 10/90, 25/75, 50/50, 75/25, 90/10, and 95/5 (percent). Using these as test samples, light power measurements were made both with the masks, and sampling at closely spaced intervals (50 samples) without the masks. There was close agreement between optical and digital inversions, with a mean difference (magnitude) of about 0.4% and a standard deviation of 0.6%. These were of the same order as the differences between estimated and nominal percentages: For the optical inversions, the mean magnitude difference was 0.9% and the standard deviation 1.1%. Comparable figures for digital inversions were mean 1.2% and standard deviation 1.3%. Deviations from nominal tended to be greatest at the 5/95 and 95/5 ratios, as might be expected.

In terms of actual particle numbers rather than percentages, the deviation of estimated from nominal values can, of course, be greater than the figures given above; for instance, a 1% error with reference to a 10/90 concentration represents a 10% number error for the minority constituent and a 1.1% number error for the majority constituent. For the optical inversions the mean deviation of particle number estimates from nominal was about 5%, with standard deviation about 9%, while for the digital inversions the corresponding figures were about 6 and 11%, respectively. If the troublesome 5/95 and 95/5 results are omitted, these figures all reduce to the vicinity of 2–5%.

A considerable number of possible improvements in apparatus and procedures were suggested by these experiments, and many have since been incorporated. Work is continuing in directions to be reported on another occasion.

A few other Fourier-optical methods should perhaps be mentioned because of their intrinsic interest in the area of particle analysis, despite their not meeting the criterion prescribed earlier for this chapter that we would discuss only classification and counting on a multiple basis, i.e., not one by one. One such method has already been mentioned. This is the particle size analysis system developed by Thompson and colleagues [12] using pulsed laser holograms for the storage of instantaneous droplet distributions throughout some volume, followed by a scanning readout procedure.

Another Fourier-optical method has been investigated in application to the screening of Pap smears for evidence of cervical cell abnormalities [32]. Cell images were stored photographically or electro-optically, and their Fourier power spectra generated individually, one by one. A 64-element silicon detector sampled the average spectrum incident on each of 32 wedge-shaped regions, or sectors, and on each of 32 ring-shaped or annular regions. Six features available from these readings were selected and used as a basis for discrimination between normal and abnormal cell types. The experiments indicated a good potential for effective use of the technique.

Another instance of one-at-a-time cell identification has been cited earlier in this chapter [20] and was based on the use of a liquid flow system for carrying the cells through the region of illumination. A more recent example involving some of the same investigators [49] has utilized the 64-element silicon detector array described in the preceding paragraph. Evidently the tendency in these efforts has been in the direction of obtaining more scattering information to aid in the analysis.

Optical matched filters [49] have been investigated by Fujii *et al.*, [50] for the purpose of classifying diatoms. A portion of their work has been concerned with reduction of the matched filter sensitivity to size and orientation. Simplified simulations of the specimens were used in place of the specimens themselves to generate filters having lower specificity for these and other optical details not relevant to the identifications. Rotatable matched filters were also used in order to accommodate all orientation possibilities. Image plane correlation peaks exceeding suitable threshold values were counted by scanning. Almeida and Fujii made some excursions into the study of frequency plane patterns of the diatoms and appeared to recognize the possibility of classification by power spectral analysis.

Tuerke *et al.* [51] reported the use of a Fourier-optical method for determining diameters of nucleated spherical biological cells. It is based on the detailed shape of their power spectra, examined cell by cell. The authors pointed out that diffraction by the nucleus could be brought into prominence relative to that by the cytoplasm by immersing the cells in a medium with an index of refraction matched to the latter. Their experiments included uniform gold and glass spheres as well as different types of cells as test objects.

3.9   SUMMARY AND CONCLUSIONS

We have presented in this chapter a treatment of Fourier-optical methods applied to problems of identifying and counting small particles in polydisperse mixtures. Essentially all the emphasis has been on analysis of the composite scattering pattern representing the contributions from a large number of particles illuminated simultaneously.

In Section 3.1 we looked at the question of what type of object information is retained and what is discarded in the power spectrum, inasmuch as this is the principal entity with which our exposition is concerned. Since it is sometimes thought that the power spectrum is devoid of phase information, we took some pains to clarify this term and point out that such an assumption is incorrect.

We derived the expression for the power spectrum of a system of randomly located and oriented scatterers in Section 3.2, and in Section 3.3 we examined its statistical characteristics. The consequences of these characteristics as they affect the use of a generalized matrix inversion procedure for the estimation of particle numbers and types are discussed in Sections 3.4 and 3.5. We considered also the possibility that instability may arise in the inversion and showed how this possibility could be analyzed and related to parameter estimates.

The use of a hybrid system implementation of the inversion algorithm was considered in Section 3.6, where we emphasized the necessity for the exercise of care in performing spectral sampling. Characteristics of a smoothing filter for application to this task were mentioned. An example was cited of training a hybrid system in case the scattering patterns of individual constituents of a mixture were initially unknown.

Methods of optical data inversion were described in Section 3.7, with most attention given to implementation of the generalized matrix inversion.

In Section 3.8 the results of several experimental investigations involving a variety of scattering objects were described. Brief mention was made also of some instances of Fourier-optical analysis based on individual examination of small objects rather than of their collective scattering patterns.

In summarizing the progress made in the development of this application of Fourier optics, we must conclude that, as measured by rate of transition from theory to practice, it has been rather slow. In the admittedly somewhat biased view of the writer, the approach holds forth a high degree of untapped potential, and so perhaps a more vigorous pursuit could have been hoped to occur.

The need to analyze collections of small particles can hardly be called widespread, of course. Examples mentioned in the preceding section illustrate some diversity in application areas, suggesting the possibility that investigators may not always be thoroughly cognizant of the optics literature. This might account for a tendency for some of the investigations to have been pursued, apparently, *ab origine*, with little or no building upon previous work by others.

Perhaps this is to be expected in an embryonic field and may correct itself with a quickening pace of activity or by a natural process of information diffusion. One suspects that the obscurity which seems to surround this area of Fourier optics might be lifted by a clear-cut laboratory success, particularly if achieved in an instance such as blood cell typing and counting, where the response of lay and professional observers alike may be sufficiently visceral to command high and sustained interest. We shall conclude by expressing the hope that within some reasonable time some such success may be attained.

REFERENCES

[1] P. M. Morse and H. Feshbach (1953). "Methods of Theoretical Physics." McGraw-Hill, New York.

[2] Lord Rayleigh (J. W. Strutt) (1899). "Scientific Papers," Vol. 1, pp. 87, 104.

[3] G. Mie (1908). *Ann. Phys. (Leipzig)* **25**, 377.

[4] H. C. van de Hulst (1957). "Light Scattering by Small Particles." Wiley, New York.

[5] M. Kerker (1969). "The Scattering of Light and other Electromagnetic Radiation." Academic Press, New York.

[6] K. S. Shifrin and A. Y. Perelman (1967). *In* "Electromagnetic Scattering" (R. L. Rowell and R. S. Stein, eds.), pp. 131–167. Gordon & Breach, New York.

[7] W. L. Anderson (1968). *J. Opt. Soc. Am.* **58**, 1566–1567A.

[8] J. W. Goodman (1976). *J. Opt. Soc. Am.* **66**, 1145–1149.

[9] W. L. Anderson and S. Y. Shen (1979). *J. Opt. Soc. Am.* **69**, 1684–1690.

[10] M. Abramowitz and I. A. Stegun (1965). "Handbook of Mathematical Functions." Dover, New York.

[11] W. L. Anderson (1980). *IEEE Trans. Pattern Anal. Mach. Intell.* **PAMI-2**, 458–463.

[12] B. J. Thompson (1963). *J. Soc. Photo-Opt. Instrum. Eng.* **2**, 43–46; G. B. Parrent, Jr. and B. J. Thompson (1964). *Opt. Acta* **11**, 183–193; B. A. Silverman, B. J. Thompson, and J. H. Ward (1964). *J. Appl. Meteorol.* **3**, 792; B. J. Thompson (1965). *Jpn. J. Appl. Phys.* **4**, Suppl. 1, 302–307.

[13] B. C. R. Ewan (1980). *Appl. Opt.* **19**, 1368–1372.

[14] J. W. Goodman (1968). "Introduction to Fourier Optics." McGraw-Hill, New York.

[15] G. N. Watson (1966). "A Treatise on the Theory of Bessel Functions." Cambridge Univ. Press, London and New York.

[16] W. L. Anderson and R. E. Beissner (1971). *Appl. Opt.* **10**, 1503–1508.

[17] S. Y. Shen and W. L. Anderson (1974). *J. Opt. Soc. Am.* **64**, 1399A.

[18] J. W. Goodman (1965). *Proc. IEEE* **53**, 1688–1700.

[19] R. W. Ditchburn (1963). "Light," Vol. 1, pp. 206–207. Wiley (Interscience), New York.

[20] G. C. Salzman, J. M. Cromwell, J. C. Martin, T. T. Trujillo, A. Romero, P. F. Mullaney, and P. M. La Bauve (1975). *Acta Cytol.* **19**, 374–377.

[21] S. Twomey (1965). *J. Franklin Inst.* **279**, 95–109.

[22] T. D. Lewis and P. L. Odell (1971). "Estimation in Linear Models." Prentice-Hall, Englewood Cliffs, New Jersey.

[23] Y. Bard (1974). "Nonlinear Parameter Estimation." Academic Press, New York.

[24] C. Lanczos (1956). "Applied Analysis." Prentice-Hall, Englewood Cliffs, New Jersey.

[25] S. Twomey and H. B. Howell (1967). *Appl. Opt.* **6**, 2125–2131; S. Twomey (1977). "Introduction to the Mathematics of Inversion in Remote Sensing and Indirect Measurements," Development in Geomathematics, Vol. 3. Elsevier, Amsterdam.

[26] A. S. Householder (1964). "Theory of Matrices in Numerical Analysis." Dover, New York; G. H. Golub (1965). *Numer. Math.* **7**, 206–216; B. Noble and J. W. Daniel (1977). "Applied Linear Algebra." Prentice-Hall, Englewood Cliffs, New Jersey.

[27] H. Stark and B. Dimitriadis (1975). *J. Opt. Soc. Am.* **65**, 425–431.

[28] H. Stark, D. Lee, and B. Dimitriadis (1975). *J. Opt. Soc. Am.* **65**, 1436–1442.

[29] H. Stark, D. Lee, and B. W. Koo (1976). *Appl. Opt.* **15**, 2246–2249.

[30] H. Stark and G. Shao (1977). *Appl. Opt.* **16**, 1670–1674.

[31] F. P. Carlson and C. K. Lee (1978). *IEEE Trans. Biomed. Eng.* **BME-25**, 361–367.

[32] B. Pernick, R. E. Kopp, J. Lisa, J. Mendelsohn, H. Stone, and R. Wohlers (1978). *Appl. Opt.* **17**, 21–34; R. Wohlers, J. Mendelsohn, R. E. Kopp, and B. J. Pernick (1978). *Appl. Opt.* **17**, 35–42; B. Pernick, S. Jost, R. E. Kopp, J. Mendelsohn, and R. Wohlers (1978). *Appl. Opt.* **17**, 43–51.

[33] J. M. Crowell, R. D. Hiebert, G. C. Salzman, B. J. Price, L. S. Cram, and P. F. Mullaney (1978). *IEEE Trans. Biomed. Eng.* **BME-25**, 519–526.

[34] J. H. Laning, Jr. and R. H. Battin (1956). "Random Processes in Automatic Control." McGraw-Hill, New York.

[35] W. L. Anderson and L. S. Berger (1969). *Phys. Lett. A* **29A,** 619–620.

[36] N. Wiener (1949). "The Extrapolation, Interpretation, and Smoothing of Stationary Time Series with Engineering Applications." Wiley, New York.

[37] A. Kolmogorov (1941). *Bull. Acad. Sci. USSR, Ser. Math.* **5,** 3–14.

[38] H. W. Bode and C. E. Shannon (1950). *Proc. IRE* **38,** 417–425.

[39] J. E. Ward, F. P. Carlson, and J. D. Heywood (1972). *J. Opt. Soc. Am.* **62,** 722A.

[40] J. E. Ward, F. P. Carlson, and J. D. Heywood (1973). *J. Opt. Soc. Am.* **63,** 1307A.

[41] J. E. Ward, F. P. Carlson, and J. D. Heywood (1974). *IEEE Trans. Biomed. Eng.* **BME-21,** 12–20.

[42] J. H. Ward (1968). *J. Opt. Soc. Am.* **58,** 1566A.

[43] B. J. Thompson (1977). *Proc. IEEE* **65,** 62–76.

[44] J. Cornillault (1972). *Appl. Opt.* **11,** 265–268.

[45] A. McSweeney and W. Rivers (1972). *Appl. Opt.* **11,** 2101–2102.

[46] J. Swithenbank, J. M. Beer, D. S. Taylor, D. Abbot, and G. C. McCreath (1977). *Prog. Astronaut. Aeronaut.* **53,** 421–447.

[47] C. Dodge and W. L. Anderson (1974). *J. Opt. Soc. Am.* **64,** 544A.

[48] A. L. Wertheimer and W. L. Wilcock (1976). *Appl. Opt.* **15,** 1616–1620.

[49] A. Vander Lugt (1964). *IEEE Trans. Inf. Theory* **IT-10,** 139–145.

[50] H. Fujii and S. P. Almeida (1979). *Appl. Opt.* **18,** 1659–1662; S. P. Almeida and H. Fujii (1979). *Appl. Opt.* **18,** 1663–1667; H. Fujii, S. P. Almeida, and J. E. Dowling (1980). *Appl. Opt.* **19,** 1190–1195.

[51] B. Tuerke, G. Seger, M. Achatz, and W. von Seelen (1978). *Appl. Opt.* **17,** 2754–2761.

Chapter 4

Signal Processing Using Hybrid Systems

JAMES R. LEGER† AND SING H. LEE

DEPARTMENT OF ELECTRICAL ENGINEERING AND COMPUTER SCIENCES
UNIVERSITY OF CALIFORNIA, SAN DIEGO
LA JOLLA, CALIFORNIA

INTRODUCTION

In Chapter 3, W. Anderson showed how optical Fourier transforms can be used directly to solve an important estimation problem. In Chapter 2, S. P. Almeida and G. Indebetouw discussed how a number of problems in pattern recognition can be dealt with by Fourier plane filtering. The examples given in these earlier chapters show that optical systems using the Fourier transform property of a lens offer some unique advantages over electronic systems. The inherent two-dimensional nature of an optical system permits it to accept large two-dimensional arrays of data very easily. As a result, image processing, phased array radar processing, solutions to two-dimensional partial differential equations, synthetic aperture radar, and many other problems of a two-dimensional nature have made extensive use of optical systems. A second advantage optical systems enjoy is parallel computation. By this we mean that all the data points of an image pass through the optical system together, or in parallel, rather than one point at a time. Besides offering a tremendous speed advantage, since many operations can be performed simultaneously, this parallelism eliminates the conversion from two-dimensional spatial data to one-dimensional temporal data and back

---

† Present address: **3M Company** St. Paul, Minnesota.

again. A third advantage of an optical system is its ability to operate at tremendous data rates. Since the signals traveling through a passive optical system propagate at the speed of light, the theoretical throughput of optical systems is enormous (though we will see later that this rate is not always achievable). Finally, a fourth advantage of Fourier optics is that the Fourier transform is simple to perform. With the Fourier transform as the basic building block, it is straightforward to design systems for performing correlation, convolution, differentiation, matched filtering, etc.

It is also apparent, however, that at the current state of the art a purely optical system has some drawbacks which make certain tasks difficult or impossible to perform. The first is that systems based on Fourier optics are inherently analog. As with other analog systems, both electrical and mechanical, high accuracy is difficult to achieve. A second problem is that optical systems by themselves cannot be used to make decisions. The simplest type of decision task might be based on a comparison of the output of the optical system with a stored value. This type of operation cannot be performed presently without help from electronics. A third problem which arises is that optical systems are difficult to "program," in the sense that we are used to programming general purpose digital electronic computers. Purely optical systems can be designed to perform specific tasks (analogous to a "hardwired" electronic computer) but cannot be used where a great deal of flexibility is required. Finally, to apply optical systems to problems other than images (for example, radar data), spatial light modulators have to be built to convert electrical signals into two-dimensional optical ones. Coherent-optical processors have the further restriction that the input to the optical system must be coherent light.

The deficiencies of optical systems happen to be the strong points of some electronic systems. For instance, accuracy, control, and programming flexibility are all traits of a digital electronic computer. The idea of combining electronic technology with optical systems thus follows quite naturally as a means of applying the fast processing ability and parallelism of optics to a wider range of problems. These systems are called *hybrid*, since they are composed of both optical and electronic subsystems.

In this chapter, we hope to introduce several of the design criteria involved in configuring a hybrid processor. We will then discuss several experimental systems to illustrate how each of these criteria has been applied to practical situations. Special emphasis has been placed on hybrid systems which clearly demonstrate an extension of optical processing to new areas through the hybrid approach. These include radar signal processing, generalized linear transformations, and incoherent systems for image and signal processing.

4.1 A GENERALIZED HYBRID SYSTEM AND ITS DESIGN CONSIDERATIONS

Hybrid systems have been designed to perform many different tasks. Because of this diversity, they often bear little resemblance to one another. Figure 4.1-1 is a block diagram of a generalized hybrid system which includes most of the elements of practical systems now being built.

Since a hybrid optical electronic system is, by definition, a combination of a purely optical system and an electronic one, interfacing between the two systems becomes one of the primary design problems. Three interface subsystems are shown in Fig. 4.1-1. The input interface will be discussed first in Section 4.1,A. Its job is to convert the raw input signal (electrical, incoherent-optical, or coherent-optical) into a two-dimensional optical signal suitable as input to the optical processor. The output interface will be covered next in Section 4.1,B. It simply converts the two-dimensional optical output of the optical processor into the desired output format. This could be as simple as displaying the optical output on a screen, or could require converting the optical output into an electrical signal. Finally, an interface is provided for controlling the optical processor. This makes it possible to "program" the processor automatically. Since the control interface depends heavily on the specific processor used in the hybrid system, we will postpone discussing it until Section 4.3 on practical systems.

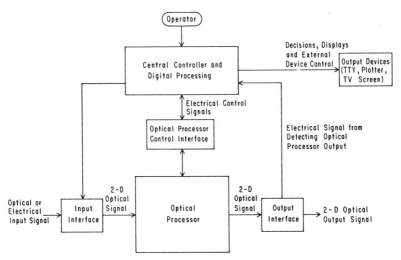

FIG. 4.1-1. Block diagram of a generalized hybrid system. The optical processor is interfaced to other subsystems in three places. The operator controls the entire system through the central controller.

The three interfaces are frequently connected to a central controller, which is usually a digital electronic computer because of its flexibility and ease of programming. It is then possible for the operator to control the input, analyze the output, and change the operation of the optical processor simply by programming the digital computer.

The final block in Fig. 4.1-1 is the optical processor itself. Since this section is devoted to a generalized system, we again defer detailed descriptions of specific optical processors until later. In Section 4.1,C, we shall limit ourselves to taking a look at optical processors in general by exploring the capabilities and pitfalls of their common root—the Fourier transform.

A. Interfacing the Input Signal to the Optical Processor

The input interface's function is to convert an electrical signal or an incoherent-optical signal into a two-dimensional spatially modulated light field. This is one of the most difficult areas in the design of a hybrid system because of the stringent requirements of this interface. The first requirement is high speed. It was stated in the introduction that the throughput of an optical system is tremendous—essentially given by the speed of light through the optical elements. However, this rate is never achieved in a realistic system which is limited by the speed of the input and output interfaces. A second characteristic of the input interface which must be matched to the optical processor is high space–bandwidth product (SBP). The SBP of optical systems is often as high as 10^6–10^7 resolvable points. Thus, the interface must be capable of high resolution over a fairly large area. Finally, if the optical processor requires coherent light for its operation, the interface must be able to preserve the optical quality of the wave front while controlling the amplitude and/or phase of a coherent light source spatially.

Incoherent-optical systems which will accept incoherent light at the input can make use of television technology to provide both electrical-to-optical and optical-to-optical interfaces. Television monitors are capable of generating a two-dimensional incoherent image from an electrical signal. By including a TV camera in the system, any image can be scaled, amplified, and coupled into the optical system. High resolution TV systems are capable of space bandwidth products of 10^6. Even at the modest speed of one frame every $\frac{1}{30}$ sec, the throughput rate is already 3×10^7 points/sec.

Light emitting diodes (LEDs) have been used in some cases as a substitute for the television monitor to interface electrical signals to the incoherent-optical processor. An array of LEDs can handle many electrical signals simultaneously, converting them to optical signals in parallel. Because of the parallel addressing and fast response of a LED, these arrays can be run at a faster rate than a TV system. The SBP is generally lower than for

TV, however. For instance, a typical array may consist of a matrix of 32×32 LEDs. To reduce the amount of drive circuitry, the LEDs can be addressed one column at a time with 32 parallel channels. The rise time of an LED limits the minimum pulse width to 100 nsec. Thus, the entire array can be addressed in 3.2 μsec. The throughput rate is then 3×10^8 points/sec.

Interfacing to the input of a coherent-optical system is generally more difficult than in the case of an incoherent system. There is no easy way of generating a coherent image as we generated an incoherent image using a TV monitor or LED array. Instead one usually starts with an expanded laser beam and spatially modulates the beam by changing its amplitude or phase in specific areas.

The easiest way to modulate the beam spatially is with a photographic transparency of the input information, usually placed in a liquid gate to remove any phase variations due to nonuniform film thickness. If the information happens to be stored on film, this is a perfectly good way of introducing the data into the coherent-optical system.

Another method of spatial modulation which is applicable to a few specific problems in pattern recognition is to place the object itself in the laser beam. Even though the object is opaque, in certain cases, the information contained in the diffraction pattern of the input object itself is enough for pattern recognition purposes. Two examples of this method which work quite nicely are the inspection of cloth (where the expanded laser beam is passed right through a piece of cloth) and the inspection of hypodermic syringes [1].

The vast majority of applications for coherent processors, however, cannot make use of the previous two methods of spatial light modulation. For this reason, a large amount of research and development has gone into perfecting real-time spatial light modulators. A comprehensive review of this field is far beyond the scope of this chapter. (A more thorough review is contained in Ref. [2].) Instead, we will briefly describe three representative devices to illustrate some of the characteristics of spatial light modulators which must be considered in designing a hybrid system. The first device to be described is the Pockels readout optical modulator (PROM). It is representative of optically addressed modulators with memory. The second device is the liquid crystal light valve (LCLV). Both the PROM and the LCLV are optically addressable, although the incoherent image must be continuously present during the operation of the LCLV. The third device is the electron beam-addressed deuterated potassium dihydrogen phosphate (DKDP) crystal or TITUS tube, which is an example of an electrical-to-optical converter. How some of these devices are used in coherent-optical radar signal processing is discussed in Chapter 5.

The Pockels readout optical modulator [3] acts somewhat like a real-time photographic film (Fig. 4.1-2). It is first exposed by a write image, then read

FIG. 4.1-2. Photograph of the PROM device. A fiber-optic cable for providing an erase
flash is visible in back of the bisox crystal. For amplitude modulation, an analyzer is added after
the crystal to convert a polarization change to an amplitude change.

out by coherent light, and finally erased when a new image is desired. The
heart of the PROM device is a thin slice of bismuth silicon oxide (bisox). This
crystal has two very desirable properties: (i) It exhibits a linear electro-optic
effect (Pockels effect), and (ii) it is a photoconductor. The linear electro-optic
effect, as the name implies, is one where an applied electric field causes a
physical change in the crystal lattice, which in turn changes the index of
refraction of the crystal proportional to the applied field. This effect can be
used to modulate either the phase or the amplitude of the coherent readout light.

The thin slice of bisox is coated on both sides with a dielectric layer
(Parylene) and transparent electrodes. Initially, a voltage approximately
equal to the half-wave voltage of the crystal (3.9 kV) is applied to the electrodes across the bisox. During the write cycle, the crystal is exposed to a
white or blue write image. The blue light is readily absorbed by the crystal,
and in the process produces hole–electron pairs as a result of the photo-
conductivity of the bisox. Thus, in regions of the crystal exposed to this blue
light (corresponding to bright areas of the write image), the crystal is dis-
charged. Then, the write light is removed and a spatially varying potential

proportional to the write image intensity remains across the crystal. During the read cycle, red coherent light is passed through the crystal. The bisox crystal absorbs very little light in the red region of the optical spectrum, so the read beam does not discharge the crystal. Upon passing through the crystal, the linearly polarized read beam becomes elliptically polarized via the linear electro-optic effect, where the degree of ellipticity depends upon the potential. By passing this elliptically polarized light through an analyzer, the read light becomes spatially intensity-modulated. This modulation corresponds to the original write image. The erase cycle consists of a brief flash of white light from a xenon flash lamp.

After an image is stored on the PROM, it is possible to apply an additional electric field to the crystal by applying voltage to the electrodes during readout. The effect of this constant field is to shift the background level of the image. This effect, known as baseline subtraction, is useful for performing contrast enhancement, contrast inversion, level slicing, and image subtraction. It also can provide a π phase shift so that negative as well as positive real values of a transmittance function can be recorded.

Table 4.1-1 provides a list of various optical and electrical characteristics of the PROM. These characteristics are for a transmissive readout as described in the previous paragraph. The device has also been fabricated for use in the reflection mode by including a dielectric reflector on one surface of the bisox crystal. The main advantages of a reflection device are reduced voltage requirements (by one-half) and cancellation of optical activity effects.

The liquid crystal light valve [4], as the name implies, performs the same task on a two-dimensional light field as a transistor does on a one-dimensional electric current: It controls the spatial intensity of one image by means

TABLE 4.1-1

Optical and Electrical Characteristics of a Typical PROM Device

Cycle time	33 msec
Phase distortion	$\frac{1}{10}$ wavelength rms
Contrast ratio	$10^4 : 1$
Resolution	15 lp/mm (50% MTF),
(2400 ergs/cm^2 write-in power)	60 lp/mm (10% MTF)
Write-in energy density (404 nm)	50 ergs/cm^2 at $1/e$ point,
	1000 ergs/cm^2 at 1% point
Wavelength restrictions	400–500 nm write-in,
	> 550 readout
Optical aperture	18–38 mm
External accessories required	2-kV-dc power supply
	reversible at 30 Hz, xenon
	flash lamp

of another. Hence, when the controlling image is removed, the controlled image vanishes. Although this device has no memory, it is very well suited for real-time conversion of incoherent signals (typically from a television monitor) to coherent ones. It also has obvious applications as a spatial light amplifier.

A picture of the light valve is shown in Fig. 4.1-3, and a diagram of its construction in Fig. 4.1-4. The electro-optic material in this device is a biphenyl nematic liquid crystal. It is sandwiched between a transparent electrode-coated glass substrate on one side and a dielectric mirror coating on the other. A light blocking layer (CdTe) and a photoconductive layer (CdS) are placed after the dielectric coating, with a transparent electrode-coated glass or fiber-optic faceplate completing the structure. A small (5–15 V rms) ac voltage is applied across the transparent electrodes. The amount of this voltage which drops across the liquid crystal is a function of

FIG. 4.1-3. Photograph of the liquid crystal light valve. Coherent light is reflected from the front of the light valve. The incoherent image is applied from the back. A polarizing beam splitter, visible at the lower left, provides amplitude modulation.

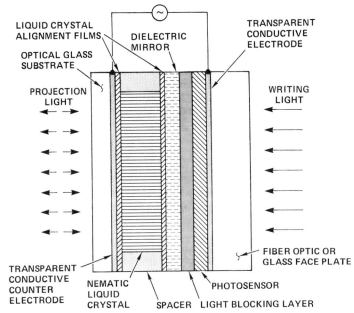

FIG. 4.1-4. Block diagram of the liquid crystal light valve. An incoherent image is projected onto the valve from the right side. An expanded laser beam, projected onto the valve from the left side, picks up a horizontal polarization which is spatially proportional to the intensity of the incoherent image. The coherent light reflects off the dielectric mirror in the center of the device and exits from the left. An external analyzer converts the polarization modulation into intensity modulation in the coherent beam. (From Bleha *et al.* [4].)

the impedance of the photoconductor, which in turn is a function of the amount of write light falling on a specific region of the photoconductor. Thus, by applying a write-in image from the right side (Fig. 4.1-4), a spatially varying electric field is applied across the liquid crystal material proportional to the intensity of the write-in image. When a readout beam is applied to the left side of the device, it passes through the liquid crystal material, reflects off the dielectric mirror coating, passes through the liquid crystal in the reverse direction, and exits the device from the left. The liquid crystal uses both the twisted nematic effect and optical birefringence (both of which are functions of the electric field) to control spatially the state of polarization of the readout beam. When this spatially modulated beam is passed through an analyzer, the polarization modulation is converted into intensity modulation. A list of the performance characteristics of the liquid crystal light valve is provided in Table 4.1-2.

The last spatial light modulator we shall describe is an electron beam-addressed potassium dideuterium phosphate crystal or TITUS tube [5]. In

TABLE 4.1-2

*Optical and Electrical Characteristics
of the Liquid Crystal Light Valve*

Cycle time	50 msec
Phase distortion	$\frac{3}{4}$ wavelength across entire face
Contrast ratio	100:1 at one wavelength
Resolution	30 lp/mm (50% MTF),
	40 lp/mm (10% MTF)
Write-in power (white light)	0.1 μW/cm^2 (toe of
	sensitometry curve),
	100 μW/cm^2 (shoulder of
	sensitometry curve)
Wavelength restrictions	None (write-in obeys CdS
	sensitivity)
Optical aperture	46 mm (maximum)
External accessories required	10-V rms audio oscillator

many ways, the DKDP modulator is similar to the PROM device. It has memory, write, read, and erase cycles, and the modulation of the light is produced by the linear electro-optic effect. Unlike the PROM, however, DKDP is not a photoconductor, so the spatially varying potential must be applied to the crystal in a different way. One way which has been quite successful is to deposit the charge on the crystal using an electron gun. By scanning a beam of electrons across the face of the crystal and modulating the electron current by the input signal, a (one-dimensional) temporal signal is converted into a (two-dimensional) spatial charge distribution. This in turn produces a spatially varying electric field which causes the crystal to modulate the readout light via the linear electro-optic effect. Erasure is provided by a second electron gun with a much lower acceleration potential and defocused spot.

A picture of an experimental version of the electron-beam-addressed DKDP device is shown in Fig. 4.1-5, along with a schematic. Because electron guns are used, the crystal must be contained in a vacuum. Also, for improved resolution and memory characteristics, the crystal must be operated at its ferroelectric transition temperature ($-50°C$), necessitating the use of a two-stage Peltier cell cooling system. Table 4.1-3 summarizes the performance of this device.

B. Interfacing the Output of an Optical Processor to an Electronic System

The interface between a two-dimensional optical field and a one-dimensional electronic signal always involves some sort of sampling. A television

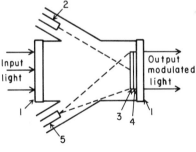

(b)

FIG. 4.1-5. The electron beam-addressed DKDP device. (a) Photograph of device. (b) Schematic. Components are (1) optical windows, (2) write electron gun, (3) DKDP target crystal, (4) transparent electrode, and (5) erase electron gun. (From Casasent [5].)

camera, for instance, samples the optical pattern in the vertical direction because of its raster scan format, while preserving the continuous nature of the horizontal direction. A charge-coupled device on the other hand is composed of an array of discrete sensors and therefore performs sampling in both the horizontal and vertical directions. If the optical system is interfaced

TABLE 4.1-3

*Optical and Electrical Characteristics of the Electron
Beam-Addressed DKDP Spatial Light Modulator*

Cycle time	33 msec
Phase distortion	$< \lambda/4$
Contrast ratio	10,000 : 1
Resolution	20 lp/mm (50% MTF),
	40 lp/mm (limiting)
Wavelength restrictions	None
Optical aperture	50×50 to 75×75 mm

with a digital computer, the input to the computer must be discrete, again requiring sampling. It is therefore of utmost importance that the designer be well acquainted with the theory of sampling and able to apply it properly. We start this section with a brief review of the sampling theorem and then proceed to discuss two problems frequently encountered in sampled systems. These are: (i) aliasing caused by a violation of the sampling theorem, and (ii) the effects of a sample aperture of finite size.

1. The Two-Dimensional Sampling Theorem and Aliasing†

A continuous function $g(x, y)$, which is band-limited, has a Fourier transform $G(f_x, f_y)$ which is zero for frequencies $|f_x| > B_x$ and $|f_y| > B_y$. Making use of the one-dimensional sampling function $\text{comb}(x)$ which is defined as

$$\text{comb}(x) = \sum_{n=-\infty}^{\infty} \delta(x - n), \qquad (4.1\text{-}1)$$

we can write the two-dimensional rectangularly sampled version of the continuous function $g(x, y)$ as

$$g_s(x, y) = g(x, y)\,\text{comb}(x/a)\,\text{comb}(y/b), \qquad (4.1\text{-}2)$$

where the sample spacings are a and b in the x- and y-directions, respectively. Using the Fourier transform relationship

$$\mathscr{F}\{\text{comb}(x)\} = \text{comb}(f_x), \qquad (4.1\text{-}3)$$

we can write the Fourier transform of Eq. (4.1-2) as

$$G_s(f_x, f_y) = G(f_x, f_y) * ab\,\text{comb}(af_x)\,\text{comb}(bf_y), \qquad (4.1\text{-}4)$$

where $*$ indicates a convolution operation, and $G_s(f_x, f_y)$ and $G(f_x, f_y)$ are

† References [6, pp. 21–25] and [7, Chapter 4].

the Fourier transforms of $g_s(x, y)$ and $g(x, y)$, respectively. $G_s(f_x, f_y)$ is plotted in Fig. 4.1-6. It shows that the effect of the comb function sampling is to replicate the original Fourier distribution $G(f_x, f_y)$ around points spaced at intervals $1/a$ and $1/b$ in the f_x- and f_y-directions, respectively. These replicated versions of the original distribution are called aliases. It is obvious from Fig. 4.1-6b that the necessary criterion for preventing overlap of the aliases is

$$a \leq 1/2B_x \quad \text{and} \quad b \leq 1/2B_y. \tag{4.1-5}$$

Thus, the sample spacing must be fine enough to sample the highest frequency sine wave at least twice. Equation (4.1-5) is called the Nyquist criterion. If this criterion is satisfied, then the original continuous function $g(x, y)$ can be recovered simply by applying a low-pass filter to the sampled function $g_s(x, y)$, where the cutoff frequencies of the filter are B_x and B_y. The spectrum of the continuous function is then recovered:

$$G(f_x, f_y) = G_s(f_x, f_y) \, \text{rect}(f_x/2B_x) \, \text{rect}(f_y/2B_y), \tag{4.1-6}$$

where

$$\text{rect}(f_x/2B_x) = \begin{cases} 1, & |f_x/2B_x| \leq \tfrac{1}{2}, \\ 0, & \text{otherwise.} \end{cases}$$

The mathematical operation on the sampled function $g_s(x, y)$ then consists of a convolution:

$$g(x, y) = g_s(x, y) * \left[4B_x B_y \, \text{sinc}(2B_x x) \, \text{sinc}(2B_y y) \right], \tag{4.1-7}$$

where

$$\text{sinc}(x) = \left[\sin(\pi x) \right]/\pi x.$$

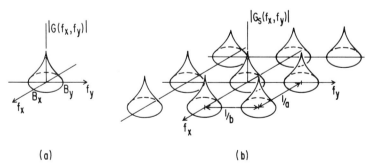

(a)   (b)

FIG. 4.1-6. The Fourier spectra of (a) a continuous band-limited function, and (b) the same function sampled at intervals a and b in the x- and y-directions. Overlap of replicas (aliases) in (b) is prevented when $a \leq 1/2B_x$ and $b \leq 1/2B_y$.

Equation (4.1-7) is called the Whittaker–Shannon sampling theorem and is commonly used to recover continuous functions from sampled ones.

What if for some reason we were unable to satisfy the Nyquist criterion in Eq. (4.1-5)? The aliases in Fig. 4.1-6b would then start to overlap each other. It is obvious that there is no low-pass filter we can apply to this spectrum to recover only the spectrum of the unsampled function in Fig. 4.1-6a, since by passing all the components of one alias, we cannot avoid passing the overlapped components of the adjacent aliases. Thus, we are unable to recover the original function perfectly. This condition is called aliasing and is a potential source of noise or distortion in sampled systems.

Aliasing results in spurious spatial frequency components in both the sampled and reconstructed (low-passed) images. Thus, by simply sampling an image and looking at the sampled version, we may have injected an aliasing noise which in general is impossible to remove completely (some restoration techniques are given, however, in Ref. [8, Chapter 7]). An example of this effect is shown in Fig. 4.1-7. Part (a) shows the continuous image of a spoke target which is chosen because of the wide range of spatial frequencies present. Part (b) is an undersampled version of the spoke target (128×80 samples) causing aliasing to occur in both the f_x and f_y spatial frequencies. It is apparent that, as the spatial frequency in the x-direction increases (i.e., as we approach the center of the spoke target along a centered vertical line), the information is at first accurately preserved. Then, at one point, as we continue to increase the spatial frequency of the original, the sampled version decreases in frequency, which can be attributed to the aliasing or overlapping of an adjacent replica whose negative frequency components are inversely ordered. Aliasing noise can be avoided obviously by increasing the sample rate (decreasing the sampling intervals a and b of Eq. (4.1-2)). Figure 4.1-7c shows the result of sampling the spoke target at twice the rate used in part (b) in the horizontal direction and three times the rate in the vertical direction. This is equivalent to separating the aliases in Fig. 4.1-6 by two and three times the original distances in the f_x- and f_y- directions, respectively. If the sampling rate cannot be increased to cover completely the bandwidth of the picture [as is the case in image (c)], then the picture should be low-passed *before* the sampling takes place to reduce the bandwidth of the image. Image bandwidth can be reduced optically by applying an appropriate pupil function in the imaging system of the camera. Figure 4.1-7d shows the case where the lens was misfocused slightly, providing a low-pass filtering of the image, and the spoke target was sampled as in image (c).

Low-passing the original image before sampling is especially important when noise is present, because the noise bandwidth is frequently much larger

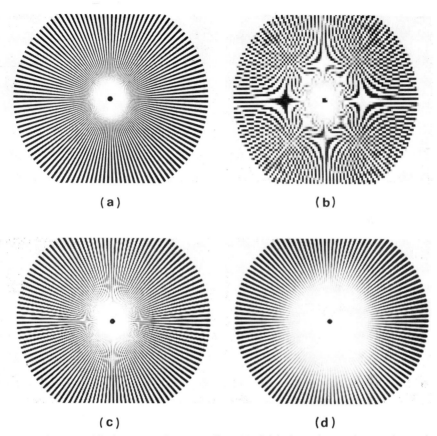

FIG. 4.1-7. Aliasing errors from sampling. (a) Original continuous image of a spoke target. (b) Sampled spoke target. Sampling rate is the same for horizontal and vertical directions. (c) Sampling rate is increased by two horizontally and three vertically. (d) Same sample rate as in (c), but camera lens is slightly misfocused.

than the signal bandwidth. Figure 4.1-8 shows the result of sampling a one-dimensional signal plus noise at a rate which satisfies the Nyquist criterion for the signal but not the noise. Although the signal does not alias (c), the noise does (d). The effect of this is to "fold" the higher frequency noise back into the lower frequencies, hence on top of the signal, leading to additional noise errors. It is therefore advisable to filter the signal and noise as much as possible (without seriously degrading the signal), so that as much noise power can be eliminated as possible *before* sampling.

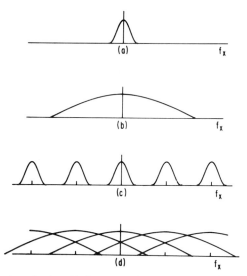

FIG. 4.1-8. Aliasing of noise. (a) Spectrum of a continuous signal. (b) Spectrum of continuous noise. (c) Spectrum of a sampled signal showing no aliasing errors. (d) Spectrum of sampled noise showing aliasing errors.

2. The Effects of a Sample Aperture of Finite Size†

When we applied the sampling theorem in the previous section, a series of delta functions was used to sample a continuous function. In practice, the aperture of the detector which samples the image must be finite in size. Often, its size is not small compared with the sample spacing, and approximating it by a delta function is inappropriate. We would like to determine the effect of a finite-size aperture on the resulting sampled image.

If the original intensity distribution is $g(x, y)$, the output of the detector consists of a spatial integration of this distribution over its aperture:

$$g_p(x', y') = \int_{-\infty}^{\infty} \int_{-\infty}^{\infty} g(x, y)p(x - x', y - y') \, dx \, dy, \qquad (4.1\text{-}8)$$

where $p(x, y)$ is the aperture function of the detector, and $g_p(x', y')$ is the measured value at point (x', y'). We recognize Eq. (4.1-8) as a convolution of the original function $g(x, y)$ with the aperture function with reversed coordinates $p(-x, -y)$. Thus, the aperture acts like a low-pass filter with a transfer function $\mathscr{F}\{p(-x, -y)\}$. If the aperture is circular, as is often the case, the Fourier transform of the original function will be "rolled off" by the $J_1(\rho)/\rho$ function.

† Reference [7, Chapter 4].

When the detector is used to sample an optical field, we can describe the detector output as

$$g_p(x, y)\, \text{comb}(x)\, \text{comb}(y) = [g(x, y) * p(-x, -y)]\, \text{comb}(x)\, \text{comb}(y). \quad (4.1\text{-}9)$$

Equation (4.1-9) shows that sampling an image with a finite aperture is equivalent to sampling the filtered image with a series of delta functions, where the impulse response of the filter is given by the aperture function.

All detectors are of finite size. Often for practical reasons, we must use large detectors. For instance, we may be measuring a very weak signal and need a large collecting area to obtain a reading above noise level. Can we do anything to correct for the roll-off effect in Eqs. (4.1-8) and (4.1-9)? Since we know the effect of the detector aperture is simply to multiply the Fourier transform of the original data by a known filter function, we could in principle multiply the Fourier transform of the measured values by the inverse filter function to correct them. This method, called inverse filtering [9], works quite well when there is very little or no noise. However, in the presence of noise, other methods such as Wiener filtering [10] and maximum entropy [11] have proven more effective. It should also be pointed out that the low-pass effect on a finite aperture is not always undesirable. We see from the previous section that this effect helps to band-limit images which otherwise may exceed the Nyquist frequency, thus helping to prevent aliasing.

C. Design Considerations Related to the Fourier Transform

So far we have been discussing hybrid system design in terms of inter-facing between optical and electronic systems. We now turn our attention to the optical processor and consider those tasks to which the optical part of a hybrid computer is best suited. We would like to provide a deep under-standing of the properties of the Fourier transform and the correlation integral, which form the basis of many optical processors. This understanding will guide us in later sections in using these powerful properties most effec-tively and avoiding their limitations.

We first discuss the invariance properties of linear transforms, with emphasis on shift and scale invariance. Second, the statistical instability of the Fourier transform is briefly reviewed. Third, the data compression property of the Fourier transform is studied. Finally, the Fourier transform is used to compute the correlation integral as a means of performing pattern recognition. The effects of spatial shift, rotation, and scale change of the input are examined. These studies are aimed at aiding hybrid system design.

1. *Invariance Properties of Linear Transforms*

In many applications one wants to measure certain characteristics of an image which do not change when the image is shifted. This would arise, for example, in pattern recognition, where the pattern might be misregistered or its location unknown. Let us represent the shift operation by a linear operator $\mathscr{L}$ defined such that

$$\mathscr{L}\{g(x, y)\} = g(x - a, y - b), \tag{4.1-10}$$

where a and b are real numbers. This operator is shown schematically in Fig. 4.1-9a. We would like to study the characteristics of $g(x, y)$ which do not change when operated on by $\mathscr{L}$. Thus, we would like to decompose $g(x, y)$ into a set of functions, where each function has the property

$$\mathscr{L}\{\psi(x, y)\} = \lambda\psi(x, y). \tag{4.1-11}$$

To within a complex constant λ, then, $\psi(x, y)$ would remain unchanged by the operator $\mathscr{L}$. These functions are the eigenfunctions of $\mathscr{L}$, as shown schematically in Fig. 4.1-9b.

It can be verified by direct substitution that the eigenfunctions of $\mathscr{L}$ are given by

$$\psi(x, y) = e^{rx + sy}, \tag{4.1-12}$$

where r and s are complex numbers. Substituting Eq. (4.1-12) into Eq. (4.1-11), we obtain

$$\mathscr{L}\{e^{rx + sy}\} = e^{r(x - a) + s(y - b)} = (e^{-ra - sb})e^{rx + sy}. \tag{4.1-13}$$

If now we impose the additional constraint that all eigenvalues λ must have unity magnitude, we see that r and s must be completely imaginary. [The reasons for requiring λ to have unity magnitude will become clear when the Fourier power spectrum is introduced in Eq. (4.1-22).] Defining

$$i2\pi f_x = r, \qquad i2\pi f_y = s, \tag{4.1-14}$$

where f_x and f_y are real numbers, the eigenfunctions become

$$\psi(x, y) = e^{i2\pi(f_x x + f_y y)}. \tag{4.1-15}$$

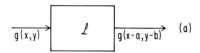

FIG. 4.1-9. A "black box" representation of the linear shift operator. (a) The function $g(x, y)$ is shifted by an amount a and b in the x- and y-directions, respectively. (b) The effect of shifting the eigenfunction $\psi_i(x, y)$ is simply to multiply it by a complex constant λ_i.

It can be shown that these eigenfunctions are orthonormal and complete. We can therefore decompose any original function $g(x, y)$ into a linear combination of these eigenfunctions:

$$g(x, y) = \sum_{f_x = -\infty}^{\infty} \sum_{f_y = -\infty}^{\infty} G(f_x, f_y)e^{i2\pi(f_x x + f_y y)}. \qquad (4.1\text{-}16)$$

Equation (4.1-16) has very special significance with respect to the shift operator $\mathscr{L}$. It asserts that, although shifting $g(x, y)$ will, in general, result in a completely different function of x and y, shifting each one of the terms of the summation on the right-hand side of Eq. (4.1-16) will change that term by at most a complex constant with unity magnitude.

Since f_x and f_y take on all real values, the summation signs can be replaced by integral signs:

$$g(x, y) = \int_{-\infty}^{\infty} \int_{-\infty}^{\infty} G(f_x, f_y)e^{i2\pi(f_x x + f_y y)} \, df_x \, df_y. \qquad (4.1\text{-}17)$$

Using the orthogonality relation

$$\int_{-\infty}^{\infty} \int_{-\infty}^{\infty} e^{i2\pi(f_x x + f_y y)} e^{-i2\pi(f_x' x + f_y' y)} \, dx \, dy = \delta(f_x - f_x', f_y - f_y'), \qquad (4.1\text{-}18)$$

we can solve for the coefficients of the eigenfunctions:

$$G(f_x, f_y) = \int_{-\infty}^{\infty} \int_{-\infty}^{\infty} g(x, y)e^{-i2\pi(f_x x + f_y y)} \, dx \, dy. \qquad (4.1\text{-}19)$$

We recognize this as the two-dimensional Fourier transform of the original function $g(x, y)$. We now see one of the reasons why the Fourier transform is so useful: It provides us with the coefficients of an expansion in terms of the eigenfunctions of the linear shift operator. The shift property of Fourier transforms is also evident. Applying the shift operator to Eq. (4.1-17),

$$\mathscr{L}\{g(x, y)\} = g(x - a, y - b)$$

$$= \int_{-\infty}^{\infty} \int_{-\infty}^{\infty} G(f_x, f_y)\mathscr{L}\{e^{i2\pi(f_x x + f_y y)}\} \, dx \, dy$$

$$= \int_{-\infty}^{\infty} \int_{-\infty}^{\infty} \lambda(f_x, f_y)G(f_x, f_y)e^{i2\pi(f_x x + f_y y)} \, dx \, dy, \qquad (4.1\text{-}20)$$

where $\lambda(f_x, f_y) = e^{-i2\pi(f_x a + f_y b)}$ from Eq. (4.1-13). Equation (4.1-19) then becomes

$$\lambda(f_x, f_y)G(f_x, f_y) = \int_{-\infty}^{\infty} \int_{-\infty}^{\infty} g(x - a, y - b)e^{-i2\pi(f_x x + f_y y)} \, dx \, dy. \qquad (4.1\text{-}21)$$

Thus, the Fourier transform of a shifted function is equal to the Fourier transform of the original function multiplied by a complex function with unity magnitude. We have in effect designed this property into the linear transform by choosing the eigenfunctions of the shift operator to perform the expansion.

The Fourier power spectrum is given by $|G(f_x, f_y)|^2$. The usefulness of this quantity is apparent when applied to the shifted function $g(x - a, y - b)$. The Fourier power spectrum of Eq. (4.1-21) is

$$|\lambda(f_x, f_y)G(f_x, f_y)|^2 = |G(f_x, f_y)|^2, \qquad (4.1\text{-}22)$$

since we restricted $\lambda(f_x, f_y)$ to always have the property $|\lambda(f_x, f_y)| = 1$. Equation (4.1-22) indicates that the power spectrum of the shifted function is identical to that of the unshifted function. Thus, the Fourier power spectrum provides characteristics of a function which are completely independent of shift. This is illustrated schematically in Fig. 4.1-10.

It is interesting to extend the analysis utilizing eigenfunctions of operators to operations other than shift. For example, we may want to study the characteristics of an object which are independent of scale change. This is the case in pattern recognition when the distance from the pattern to the observer is unknown. We want to determine the eigenfunctions of an operator $\mathcal{M}$ which is defined by

$$\mathcal{M}\{g(x, y)\} = g(ax, by). \qquad (4.1\text{-}23)$$

The eigenfunctions must be solutions to the equation

$$\mathcal{M}\{\psi(x, y)\} = \lambda\psi(x, y). \qquad (4.1\text{-}24)$$

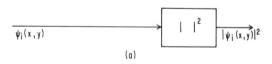

(a)

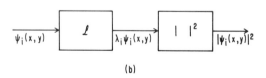

(b)

FIG. 4.1-10. (a) The effect of squaring the magnitude of the eigenfunction $\psi_i(x, y)$. (b) The linear operator $\mathcal{L}$ is followed by the nonlinear operation of squaring the magnitude. If eigenfunctions of $\mathcal{L}$ are chosen with unity magnitude eigenvalues, the resulting squared magnitude functions are identical to the result in (a) and so are completely unaffected by the operator $\mathcal{L}$.

It is apparent that the set of polynomials

$$\psi(x, y) = x^r y^s, \tag{4.1-25}$$

where r and s are complex numbers, satisfies Eq. (4.1-24), since by direct substitution,

$$\mathcal{M}\{x^r y^s\} = (ax)^r (by)^s = [a^r b^s] x^r y^s. \tag{4.1-26}$$

Again, we would like the eigenvalues to be of unity magnitude. This requirement is met if r and s are pure imaginary, since by defining new variables for r and s as in Eq. (4.1-14), the magnitude of the eigenvalues is

$$\left| a^{i2\pi f_x} b^{i2\pi f_y} \right| = \left[(a^{i2\pi f_x} b^{i2\pi f_y})(a^{-i2\pi f_x} b^{-i2\pi f_y}) \right]^{1/2} = a^0 b^0 = 1. \tag{4.1-27}$$

Equation (4.1-25) then becomes

$$\psi(x, y) = x^{i2\pi f_x} y^{i2\pi f_y}, \tag{4.1-28}$$

where f_x and f_y are real. We can decompose an arbitrary function $g(x, y)$ in terms of the eigenfunctions given in Eq. (4.1-28):

$$g(x, y) = \int_{-\infty}^{\infty} \int_{-\infty}^{\infty} G(f_x, f_y) x^{i2\pi f_x} y^{i2\pi f_y} \, df_x \, df_y. \tag{4.1-29}$$

The orthogonality relationship for these new eigenfunctions is given by

$$\int_0^\infty \int_0^\infty \frac{1}{xy} x^{i2\pi f_x} y^{i2\pi f_y} x^{-i2\pi f'_x} y^{-i2\pi f'_y} \, dx \, dy = \delta(f_x - f'_x, f_y - f'_y). \tag{4.1-30}$$

Solving Eq. (4.1-29) for $G(f_x, f_y)$ by applying the relation in Eq. (4.1-30), we have

$$G(f_x, f_y) = \int_0^\infty \int_0^\infty g(x, y) x^{-i2\pi f_x - 1} y^{-i2\pi f_y - 1} \, dx \, dy. \tag{4.1-31}$$

Equation (4.1-31) is known as the two-dimensional Mellin transform, where the complex frequency is chosen to be purely imaginary.

The relationship between the Mellin transform and the Fourier transform can be seen by a change in variables:

$$x = e^\xi, \qquad dx = e^\xi \, d\xi, \qquad y = e^\eta, \qquad dy = e^\eta \, d\eta. \tag{4.1-32}$$

Equation (4.1-31) then becomes

$$G(f_x, f_y) = \int_{-\infty}^{\infty} \int_{-\infty}^{\infty} g(e^\xi, e^\eta) e^{-i2\pi(f_x \xi + f_y \eta)} \, d\xi \, d\eta. \tag{4.1-33}$$

The Mellin transform can be thought of as a Fourier transform of a geometrically distorted function $g(e^\xi, e^\eta)$. We will see in Section 4.2,B that this fact enables a hybrid computer to perform a two-dimensional Mellin transform.

It is evident from Eq. (4.1-29) that applying the scaling operator $\mathcal{M}$ to the function $g(x, y)$ merely multiplies the Mellin transform $G(f_x, f_y)$ by a complex function with unity magnitude. In a way analogous to the Fourier power spectrum, we can define a Mellin power spectrum as the squared magnitude of the Mellin transform. Thus, this power spectrum provides characteristics of a function which are completely independent of the size of the original function.

2. Statistical Instability of the Fourier Power Spectrum

Up to this point, we have been treating images as though they were deterministic signals—that is, ones which can be described by explicit mathematical functions. Often, however, images from real physical objects are not completely deterministic, and therefore the nondeterministic portion must be modeled as a random process.

Measuring the Fourier power spectrum of a random process involves some unique difficulties not encountered in measuring deterministic signals. In this section, we briefly describe the statistical instability of the Fourier power spectrum and a method for increasing the stability. Instability in a power spectral estimate implies that the variance of the estimate does not approach zero as the number of data points increases. A much more thorough description of the problem and various optimal solutions can be found in Chapter 1 of this book (see Sections 1.4–1.8).

Consider the problem of trying to measure the power spectrum of a certain random process. We start by obtaining a specific realization of the process, taking the Fourier transform, and squaring the magnitude of the result. There is no question that we have measured the correct power spectrum of that one realization, but what does that indicate about the power spectrum of the random process it was taken from? Specifically, if we were to take another realization, would we obtain the same result? The answer to this question is well known to those who have shined a laser beam through a diffuser and observed the far-field diffraction pattern (equivalent to the Fourier power spectrum of the diffuser). The section of the diffuser illuminated by the laser can be considered to be a realization of a random process. In this case, the random process represented by the diffuser can only be described by its autocorrelation function. Since a diffuser is an array of independent phase transmittances, its autocorrelation function can be approximated by a delta function. Its power spectrum (which is simply the Fourier transform of the autocorrelation function) should therefore be uniform. We know from practical experience, however, that the far-field diffraction pattern of a diffuser is far from uniform, but instead has a *speckle* appearance. It can be shown [12, pp. 543–548] that the standard deviation of this variation is about as large as its mean value. Furthermore, increasing the illuminated area of the diffuser does not decrease the standard deviation. Thus, an estimate of the power spectrum based on a measurement of this pattern is said to be unstable.

We now decide to illuminate a different portion of the diffuser with the laser beam. This represents another realization of the same random process. Although the far-field diffraction pattern is roughly the same, we notice on closer inspection that the dark and bright spots in the speckle pattern have moved around in a seemingly random way. It seems reasonable therefore that we could obtain a better estimate of the power spectrum of the underlying random process by averaging the power spectra of several realizations of the process. In our case, the average of many far-field diffraction patterns produced by different regions of the diffuser would eventually result in a uniform distribution as originally expected.

It can be shown [13, pp. 239–248] that an equivalent averaging effect can be obtained by passing the sample spectrum through a spectral window. This corresponds to convolving the measured spectral estimate with a window function which is designed to smooth the sample spectrum. Many examples of different spectral windows and their effects can be found in Chapter 1 of this book.

3. *The Data Compression Property of the Fourier Transform*†

An optical system generally has a tremendous capacity for processing large amounts of data in parallel. However, when this processor is incorporated into a hybrid system, the output of the processor must interface with an electronic system of much smaller processing capacity. It is evident therefore that a properly designed optical system must be able to provide a large data compression. It must somehow be able to sift through all the redundant and extraneous input data and provide the electronic system with a highly efficient set of important features. Many systems to be studied later in this chapter use the Fourier transform to perform the task of dimensionality reduction. It is the purpose of this section to provide an understanding of the sense in which the Fourier transform has this property.

Intuitively, we can see that studying a picture point by point in image space produces data which describe a local region. On the other hand, a single data point from the Fourier transform is a global descriptor of the image (i.e., all points in the image have some influence on a single data point in the Fourier transform). Further, the information contained in a set of data points of the Fourier transform of an image in some sense describe the

---

† Reference [14, Chapter 9].

texture of the image, since the measurement of spatial frequencies describes whether the image is rough or smooth.

These intuitive feelings can be put on firm ground by introducing the formalism of random vectors and general linear transformations. We start by expressing the two-dimensional image $f(x, y)$ in a one-dimensional discrete format. This can be done, for instance, by sampling the image and scanning it row by row to form a one-dimensional lexicographically ordered column vector (see Andrews and Hunt [15, p. 40]). We can therefore define the image as a vector $\mathbf{x}$ in n space. We also assume that this image is a member of some class of images, and that we know something about the first- and second-order statistics of this class (i.e., the mean vector and autocovariance matrix). We would like to transform this image so that the resultant image $\mathbf{y}$ is in a lower dimensional space (dimensionality reduction) and contains the important features of the original image.

A linear orthonormal transformation can be expressed as

$$\mathbf{y} = \mathbf{\Psi x}, \tag{4.1-34}$$

where $\mathbf{\Psi}^+$ is a matrix composed of m orthonormal column vectors $[\mathbf{\psi}_1, \mathbf{\psi}_2, \ldots, \mathbf{\psi}_m]$. Thus,

$$\mathbf{\psi}_i^+ \mathbf{\psi}_j = \begin{cases} 1, & i = j, \\ 0, & i \neq j. \end{cases}$$

If the transformation $\mathbf{\Psi}$ completely spans the original image space, the number of column vectors m must equal the dimension of the input vector n. In this case, we can solve for $\mathbf{x}$ exactly by

$$\mathbf{x} = \mathbf{\Psi}_{(n \times n)}^+ \mathbf{y} = [\mathbf{\psi}_1, \mathbf{\psi}_2, \ldots, \mathbf{\psi}_n]\mathbf{y} = \sum_{i=1}^{n} \mathbf{\psi}_i y_i. \tag{4.1-35}$$

If we want the transformation $\mathbf{\Psi}$ to perform some data compression however, we require the dimension of $\mathbf{y}$ to be less than the dimension of $\mathbf{x}$. Thus, we would like to keep only m components of the $\mathbf{y}$ vector and fill the remaining $(n - m)$ components with constants b_i. Then $\mathbf{x}$ can no longer be expressed exactly, and an error vector $\mathbf{\epsilon}$ results:

$$\mathbf{x} = \sum_{i=1}^{m} \mathbf{\psi}_i y_i + \sum_{i=m+1}^{n} \mathbf{\psi}_i b_i + \mathbf{\epsilon}. \tag{4.1-36}$$

Solving for this error term and expressing $\mathbf{x}$ as in Eq. (4.1-35) we have

$$\mathbf{\epsilon} = \sum_{i=1}^{n} \mathbf{\psi}_i y_i - \sum_{i=1}^{m} \mathbf{\psi}_i y_i - \sum_{i=m+1}^{n} \mathbf{\psi}_i b_i = \sum_{i=m+1}^{n} \mathbf{\psi}_i (y_i - b_i). \tag{4.1-37}$$

A data reduction transform must not only reduce the dimensionality of the output but also preserve the important features of the input. Our criterion for a good transformation will be to minimize the mean square error ε in Eq. (4.1-37). The mean square error is given by

$$E\{\|\varepsilon\|^2\} = E\{ \sum_{i=m+1}^{n} \sum_{j=m+1}^{n} (y_i - b_i)(y_j - b_j)\boldsymbol{\psi}_i^+ \boldsymbol{\psi}_j\}$$

$$= E\{ \sum_{i=m+1}^{n} (y_i - b_i)^2\} \qquad (4.1\text{-}38)$$

from the orthonormality of $\boldsymbol{\psi}_i$ defined in Eq. (4.1-34), where $E\{\ \}$ is the expected value of the quantity in braces. The best choices for the constants b_i are given by

$$\frac{\partial}{\partial b_i} E\{(y_i - b_i)^2\} = -2[E\{y_i\} - b_i] = 0 \qquad (4.1\text{-}39)$$

or

$$b_i = E\{y_i\}.$$

Making this substitution for b_i in Eq. (4.1-38) and using $y_i = \boldsymbol{\psi}_i^+ \mathbf{x}$,

$$E\{\|\varepsilon\|^2\} = \sum_{i=m+1}^{n} E\{(y_i - E\{y_i\})(y_i - E\{y_i\})\}$$

$$= \sum_{i=m+1}^{n} \boldsymbol{\psi}_i^+ E\{(\mathbf{x} - E\{\mathbf{x}\})(\mathbf{x} - E\{\mathbf{x}\})^+\}\boldsymbol{\psi}_i$$

$$= \sum_{i=m+1}^{n} \boldsymbol{\psi}_i^+ \mathbf{R}_x \boldsymbol{\psi}_i, \qquad (4.1\text{-}40)$$

where

$$\mathbf{R}_x = E\{(\mathbf{x} - E\{\mathbf{x}\})(\mathbf{x} - E\{\mathbf{x}\})^+\}$$

or the covariance matrix of $\mathbf{x}$.

It is shown in the Appendix that the orthonormal vectors $\boldsymbol{\psi}_i$ which minimize the mean square error in Eq. (4.1-40) are the eigenvectors of $\mathbf{R}_x$. Thus, they are given by solving

$$\mathbf{R}_x \boldsymbol{\psi}_i = \lambda_i \boldsymbol{\psi}_i \qquad (4.1\text{-}41)$$

for its set of eigenvectors $\boldsymbol{\psi}_i$ and corresponding eigenvalues λ_i. This is in general a difficult problem to solve analytically. However, in certain special cases it becomes tractable. We first assume that the dimensionality of the matrix is very large ($n \to \infty$). This allows us to write Eq. (4.1-41) in a continuous form with limits extending to $\pm \infty$. Thus, Eq. (4.1-41) becomes

$$\int_{-\infty}^{\infty} R(x, x')\psi(x', \eta)\, dx' = \lambda(\eta)\psi(x, \eta). \tag{4.1-42}$$

If we furthermore assume that the statistics are wide-sense stationary (i.e., the mean and covariance are not functions of absolute position), we can express the covariance $R(x, x')$ as a function of $(x - x')$ only. Thus, Eq. (4.1-42) becomes

$$\int_{-\infty}^{\infty} R(x - x')\psi(x', \eta)\, dx' = \lambda(\eta)\psi(x, \eta). \tag{4.1-43}$$

This equation is easily solved by Fourier transform techniques. Taking a one-dimensional Fourier transform of both sides with respect to x, we have

$$S(f_x)\hat{\psi}(f_x, \eta) = \lambda(\eta)\hat{\psi}(f_x, \eta), \tag{4.1-44}$$

where

$$S(f_x) = \mathscr{F}\{R(x)\}$$

and

$$\hat{\psi}(f_x, \eta) = \mathscr{F}\{\psi(x, \eta)\}.$$

The only set of functions $\hat{\psi}(f_x, \eta)$ which will satisfy Eq. (4.1-44) for all $S(f_x)$ are delta functions of the form

$$\hat{\psi}(f_x, \eta) = \delta(f_x - \eta), \tag{4.1-45}$$

where η can take on any value from $\pm \infty$. This implies that the original basis functions $\psi(x, \eta)$ are of the form

$$\psi(x, \eta) = e^{i2\pi\eta x}. \tag{4.1-46}$$

Thus we see that, under the conditions of large dimensionality and stationarity, the linear transformation which provides the best data compression in a least mean square error sense is the Fourier transform.

4. *Applying the Fourier Transform to Correlation Functions*

One of the close relatives of the Fourier transform is the convolution integral. A convolution of two functions is easily performed by Fourier-transforming each function, multiplying the two transforms, and inverse-transforming the result. Correlation integrals are equally easy to perform, if the complex conjugate of one of the transforms is used in the product. Pattern recognition has made extensive use of the correlation integral for detecting the presence of a particular pattern. We will show in this section in what sense correlation is suited to this task and then examine some of the limitations of the technique when applied to two-dimensional pattern recognition.

Initially, the statistical theory of pattern recognition and signal detection was developed for processing one-dimensional radar data. One of the most basic problems was to determine whether a specific signal was present in a measurement which includes both signal and noise. This same theory is directly applicable to two-dimensional pattern recognition. It can be shown that the optimum linear statistic G for determining whether a test pattern $h(x, y)$ is present when the input image $g(x, y)$ contains linearly additive white noise is given by the inner product relationship

$$G = \int \int g(x', y')h^*(x', y') \, dx' \, dy'. \tag{4.1-47}$$

In the simple case where the functions $g(x, y)$ and $h(x, y)$ have values of only 1 and 0, this is equivalent to a template matching, where templates of $g(x, y)$ and $h^*(x, y)$ are superimposed and all their common regions added up.

We now compare the desired statistic given in Eq. (4.1-47) with the correlation between $g(x, y)$ and $h(x, y)$ given by

$$C(x, y) = g(x, y) \star h^*(x, y) = \int \int g(x', y')h^*(x' - x, y' - y) \, dx' \, dy', \tag{4.1-48}$$

where $\star$ is the symbol for correlation. It is clear that the statistic G is given by the correlation integral evaluated at its origin:

$$G = C(0, 0) = \int \int g(x', y')h^*(x', y') \, dx' \, dy'. \tag{4.1-49}$$

Thus, the correlation integral can be used to generate the desired inner product.

It is natural to ask what the rest of the correlation function $C(x, y)$ signifies. Inspection of Eq. (4.1-48) shows that the correlation integral measured at a point (a, b) off the origin is simply generating the inner product statistic G between an input function $g(x, y)$ and a shifted pattern $h(x - a, y - b)$. Thus, $C(a, b)$ can be used to determine whether the shifted version of the pattern $h(x - a, y - b)$ is present in the input.

The correlation operation can easily be performed by optical means using Vander Lugt filters. Hence pattern recognition systems can be constructed which can detect the presence of a particular pattern independent of its location. In fact, by measuring the coordinates where the correlation is maximum (and exceeds the decision threshold), the location of the pattern in the input can be determined.

The correlator's ability to detect patterns independent of spatial position is a direct consequence of the shift-invariance property of Fourier transforms previously discussed. It should come as no surprise then that other variations of the input pattern may have a severe effect on the detectability. Specifically, size changes and rotations have been shown to degrade the quality of the correlation [16, pp. 132–134]. We can determine the severity of the degradation caused by a change in size of a square pattern by evaluating the correlation between a pattern function $h(x, y)$ of size $2L \times 2L$ and a scaled version of this pattern $h(ax, ay)$ (since a size change corresponds to equal scaling in x and y):

$$h(x, y) \star h^*(ax, ay)$$

$$= \int_{-\infty}^{\infty} \int_{-\infty}^{\infty} h(x', y') \operatorname{rect}\left(\frac{x'}{2L}\right) \operatorname{rect}\left(\frac{y'}{2L}\right) h^*(ax' - x, ay' - y)$$

$$\times \operatorname{rect}\left(\frac{ax'}{2L} - x\right) \operatorname{rect}\left(\frac{ay'}{2L} - y\right) dx' \, dy'. \qquad (4.1\text{-}50)$$

Consider first the effect of a single two-dimensional complex spatial frequency of amplitude $A(f_x, f_y)$ and size $2L \times 2L$:

$$h(x, y) = A(f_x, f_y) \operatorname{rect}\left(\frac{x}{2L}\right) \operatorname{rect}\left(\frac{y}{2L}\right) \exp(-i2\pi f_x x) \exp(-i2\pi f_y y). \quad (4.1\text{-}51)$$

The scaled version of Eq. (4.1-51) is given by

$$h(ax, ay) = A(f_x, f_y) \operatorname{rect}\left(\frac{ax}{2L}\right) \operatorname{rect}\left(\frac{ay}{2L}\right)$$

$$\times \exp(-i2\pi f_x ax) \exp(-i2\pi f_y ay), \qquad (4.1\text{-}52)$$

where a is the scaling factor. We assume without loss of generality that $a < 1$. (For cases where $a > 1$, the roles of $h(x, y)$ and $h(ax, ay)$ can be switched.) Then the resulting detection statistic given by the cross-correlation evaluated at the origin is

$$G(f_x, f_y) = \{h(x, y) \star h^*(ax, ay)\}_{x=0}$$

$$= \int_{-\infty}^{\infty} \int_{-\infty}^{\infty} |A(f_x, f_y)|^2 \operatorname{rect}\left(\frac{x'}{2L}\right) \operatorname{rect}\left(\frac{y'}{2L}\right)$$

$$\times \exp[-i2\pi f_x(1 - a)x'] \exp[-i2\pi f_y(1 - a)y'] dx' \, dy'. \quad (4.1\text{-}53)$$

We recognize Eq. (4.1-53) as the Fourier transform of the two-dimensional rectangle function with transform variables $f_x(1 - a)$ and $f_y(1 - a)$. Hence

$$G(f_x, f_y) = |A(f_x, f_y)|^2 4L^2 \operatorname{sinc}[2f_x(1 - a)L] \operatorname{sinc}[2f_y(1 - a)L]. \quad (4.1\text{-}54)$$

We now consider the case of a real image composed of many spatial frequency components. Assuming that the various spatial frequency components are uncorrelated, as scale changes occur and cause one spatial frequency component to impinge on a neighboring one, no contributions to the correlation are obtained. In this case, the total detection statistic is given by

$$G = \int_{-\infty}^{\infty} \int_{-\infty}^{\infty} G(f_x, f_y) \, df_x \, df_y = 4L^2 \int_{-\infty}^{\infty} \int_{-\infty}^{\infty} |A(f_x, f_y)|^2$$
$$\times \operatorname{sinc}[2f_x(1-a)L] \operatorname{sinc}[2f_y(1-a)L] \, df_x \, df_y. \qquad (4.1\text{-}55)$$

Here $|A(f_x, f_y)|^2$ is recognized as the power spectrum of the original function $h(x, y)$. If no scaling error were present ($a = 1$), this whole spectrum would contribute to the detection statistic, since the sinc function would become unity. If, however, there is a scaling error, the original spectrum is rolled off by a sinc function, resulting in a lower value of G. The severity of this degradation is determined by the image power spectrum $|A(f_x, f_y)|^2$, the amount of scale change $(1 - a)$, and the image size L. We can make several predictions based on these results. First, high spatial frequencies are more affected than low ones by a scale change. It follows that an image with a spectrum rich in high frequencies is very sensitive to scaling errors. Second, the amount of the power spectrum $|A(f_x, f_y)|^2$ useful for detection [under the main lobe of the sinc function in Eq. (4.1-55)] decreases as the difference in scale $(1 - a)$ becomes larger. Finally, this amount decreases as the picture size (L) increases. This last statement implies that a given scaling error sets a limit on the useful or effective space–bandwidth product of $h(x, y)$. (Using an input with a larger space–bandwidth product does not enhance the detection of the object.) The effective bandwidth of $h(x, y)$ can be defined as the area under the main lobe of the sinc function [where the actual bandwidth of $h(x, y)$ is assumed no smaller than this width]. Thus

$$BW_{\text{eff}} = \left| \frac{1}{L(1-a)} \right|^2. \qquad (4.1\text{-}56)$$

Since the area of the pattern $h(x, y)$ is $(2L)^2$, we have an expression for the effective space–bandwidth product of $h(x, y)$:

$$SBP_{\text{eff}} = \left| \frac{1}{L(1-a)} \right|^2 (2L)^2 = \left| \frac{2}{(1-a)} \right|^2, \qquad (4.1\text{-}57)$$

which is a function only of scaling error.

An upper bound on the detection statistic G in Eq. (4.1-55) can be calculated by replacing $|A(f_x, f_y)|^2$ with the largest value obtained by $|A(f_x, f_y)|^2$.

Then Eq. (4.1-55) becomes

$$G \leq 4L^2 K \int_{-\infty}^{\infty} \int_{-\infty}^{\infty} \operatorname{sinc}[2Lf_x(1-a)] \operatorname{sinc}[2Lf_y(1-a)] \, df_x \, df_y$$

$$= 4L^2 K/4L^2(1-a)^2 = K/(1-a)^2, \tag{4.1-58}$$

where

$$K = \max\{|A(f_x, f_y)|^2\}.$$

Thus, if the spectrum of the image $|A(f_x, f_y)|^2$ is flat and we consider a scale difference $(1-a)$ which is large enough so that the effective SBP is less than the SBP of the image, the detection statistic G is proportional to $(1-a)^{-2}$. The intensity of the optical correlator $I_p = |G|^2$ is the quantity which is actually measured. Hence I_p is proportional to $(1-a)^{-4}$.

This theory has been compared with experimental values in Ref. [17]. Figure 4.1-11 shows a graph of I_p versus scale change for a particular aerial photograph. Curve FF corresponds to a 35-mm aperture, and curve AF to an 8-mm aperture of the photograph. Two things are evident from this graph. First, both curves vary as $(1-a)^{-4}$ for scale changes greater than 1%. This substantiates the above prediction. Second, the difference between a large SBP picture (curve FF) and a smaller SBP one (curve AF) is small for scale changes greater than 1%. This is evidently because the effective SBP defined in Eq. (4.1-57) is smaller than the SBP of both pictures. Thus, using

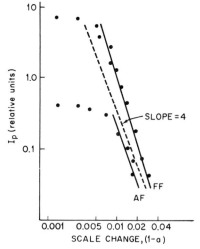

FIG. 4.1-11. The intensity of a correlation peak (I_p) is plotted against change in scale $(1-a)$. FF corresponds to a full 35-mm picture, and AF to an 8-mm apertured version of this picture. (From Casasent and Furman [17].)

a picture with a larger SBP than SBP_{eff} does not enhance the detection. For scale changes less than 1%, however, SBP_{eff} becomes larger than the SBP of the small picture. Hence the larger picture (curve FF) results in a considerably larger value of I_p than the smaller picture (curve AF) in this region. The lack of the $(1 - a)^{-4}$ dependence for small $(1 - a)$ is also expected, as stated after Eq. (4.1-58).

Rotational errors occur when the input pattern is oriented differently than the original pattern. The effect of a rotational error is very similar to that of a scaling error. Specifically, an image with a high SBP is more susceptible to a given rotation error than one with a low SBP. The effect of this error on the square of the detection statistic I_p is shown in Fig. 4.1-12. A complete aerial image (FULL) and three apertured regions of that image (A-C) were used. The apertured images were $\frac{1}{16}$ the area of the full image. Image A was of a rural scene, hence had a small bandwidth. The bandwidth of image C was somewhat higher (structured image), and the bandwidth of B was the highest (corresponding to an urban scene). The effects of both increased bandwidth and increased aperture are evident from this figure. The entire image (FULL) was much more sensitive to rotational error than the apertured ones (A, B, and C). Also, the apertured image of highest bandwidth (B) was more sensitive to rotation error than the low bandwidth image

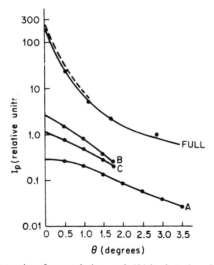

FIG. 4.1-12. The intensity of a correlation peak (I_p) is plotted against rotation of the input (θ). The areas of the images used in curves (A–C) were $\frac{1}{16}$ the area of the image used in the curve labeled (FULL). The image used in curve (A) was a rural scene (small bandwidth), curve (B) used an urban scene (high bandwidth), and curve (C) used a structured scene (medium bandwidth). (From Casasent and Furman [17].)

(A). Both these results imply that, the higher the SBP, the more severe the effects of rotation error on detectability.

It should be apparent from this discussion that rotation and scale effects must be considered when designing an optical pattern recognition system based on correlation. Often, hybrid systems can be designed to compensate for some of these effects. We shall see in Sections 4.2 and 4.3 various methods for accomplishing this.

4.2 HYBRID SYSTEMS BASED ON OPTICAL POWER SPECTRUM MEASUREMENTS

We have seen in Section 4.1, C that the Fourier power spectrum has some unique properties which make it suitable for various signal processing tasks. Specifically, the data compression property can be exploited to perform pattern recognition and classification on a reduced number of data points. As was shown in Chapter 3, size measurements can be made on objects whose locations are unknown or random. Finally, estimates of the power of a specific spatial frequency component can be made to determine such characteristics as roughness and texture (as in Chapter 11).

A hybrid optical electronic system for measuring the power spectrum is diagrammed in Fig. 4.2-1. A spatial light modulator, described in Section 4.1,A, is used to couple the input signal to the coherent-optical system. A lens produces the two-dimensional Fourier transform of the input data in its back focal plane. The intensity of this light (corresponding to the Fourier power spectrum) is sampled and detected (often by the same device), digitized, and fed into a digital computer for analysis. In this section we first describe

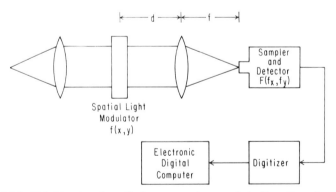

FIG. 4.2-1. Hybrid system for optical power spectrum analysis. Two-dimensional information (such as an image) modulates a coherent plane wave by using a spatial light modulator. The Fourier transform of this light distribution occurs in the back focal plane of the lens. The intensity in this plane is sampled, detected, digitized, and entered into a digital computer.

two of the most common methods of detecting the light intensity in the focal plane of the Fourier transform lens. Sampling problems and finite aperture effects of these devices will be explored. The second half of the section will describe several applications of power spectrum analysis. Particular attention will be paid to the solutions each hybrid system offers to the problems of rotational sensitivity, optimal data compression, statistical stability, and scale sensitivity.

A. Intensity-Measuring Devices

Several devices are currently available for measuring a two-dimensional light intensity pattern. These include close-circuit television vidicon tubes, charge-coupled device (CCD) detectors, and two-dimensional arrays of photodetectors. An example of a photodetector array which is very useful in image power spectrum analysis is shown in Fig. 4.2-2. It consists of 32 semicircular ring detectors on one half, and 32 wedge-shaped detectors on the other half. This design takes advantage of the symmetry in the power spectrum of a real function, allowing radial and angular sampling to be performed

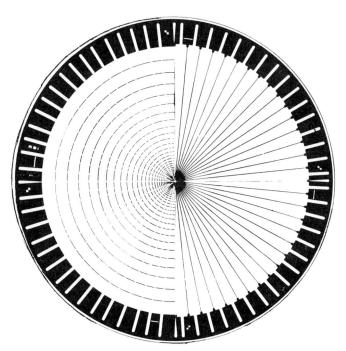

FIG. 4.2-2. The wedge–ring detector, consisting of 32 semicircular ring detectors on one half and 32 wedge-shaped detectors on the other half. (From George and Kasdan [19].)

simultaneously. The radial sampling provides information about the distribution of spatial frequences in the image. The wedge sampling on the other hand provides information on the orientation of the spatial frequencies. Although the power spectrum itself is sensitive to rotation and scale change, the ring detectors provide information independent of rotation and the wedge detectors provide information independent of scale.

We recall from Section 4.1,C and also from Chapter 1 that the statistical stability of the power spectral estimate is important to consider and that increased stability can be achieved by averaging over a range of spatial frequencies. Because the resolution of frequency components by the Fourier transform lens is far finer than the width of the detectors in the wedge–ring device, the output from each detector is the average of many frequency components. This averaging smooths the power spectrum, and we would expect to obtain stable spectral estimates.

It is interesting to compute the amount of data compression which results from using a wedge–ring detector to sample the Fourier power spectrum of an image. If the original image was 2×2 cm and had a limiting resolution of 50 lines/mm, the SBP of the image will be 10^6. If the signal-to-noise ratio of the film limits the gray scale resolution to 64 shades of gray, a 6-bit word can be used to digitize each picture element (where the number of picture elements is equal to the space bandwidth product). Thus, without the help of the Fourier transform, we would require the analysis and storage of 6×10^6 bits. If on the other hand the signal-to-noise of the wedge–ring detector is high enough to require 12 bits to digitize each detector output, and all 64 outputs are used, the total amount of data in this case will be $12 \times 64 = 768$ bits. Thus, a data compression of almost 10^4 results from using the wedge–ring detector to sample the Fourier power spectrum. We also conclude from Section 4.1,C that, assuming certain statistical properties hold, the data compression which results is "good" in a least squares error sense.

The data compression offered by the wedge–ring detector is at times too severe and the spatial frequency resolution too coarse. For these applications another useful detector may be the vidicon tube in a conventional closed circuit television camera. The spatial resolution of a TV vidicon is much higher than that offered by most detector arrays. Both vertical and horizontal resolutions of up to 1000 lines can be obtained. One pays the price for this increased spatial resolution in two ways. First, a complete picture takes $\frac{1}{30}$ sec to generate (compared to 10 μsec for the wedge–ring detector). Second, the gray scale resolution is worse than the wedge–ring detector because the signal-to-noise ratio (SNR) is lower. We can improve the SNR and thus increase the gray scale resolution at the expense of speed by averaging several successive pictures of the same image to reduce the time-dependent noise. The results of applying this method are shown in Fig. 4.2-3. The right side of this figure is a single frame of an image recorded at low light level and high camera gain. The high noise level produced by the TV camera is very evident. The left side shows the result of averaging 64 successive frames of the noisy image. If the noise is time-dependent and uncorrelated, it can be shown that the SNR improves as the square root of the number of frames in the average. Hence we expect an improvement of eight times in the SNR of the right side of the image.

In the process of converting the two-dimensional light intensity into a one-dimensional function of time, the TV camera must sample the intensity pattern in one direction. When a horizontal raster scan is used, the sampling takes place in the vertical direction. Whenever a quantity is sampled, the question of aliasing error must be addressed. In the present context, the TV camera is used to sample a power spectrum of an image. There are two

FIG. 4.2-3. Reduction of time-dependent noise by time integration. The right side is a single frame of a noisy TV camera image. The left side is after averaging 64 frames.

effects which help to band-limit this function and prevent aliasing. The first is the finite resolution of the Fourier transform lens. The second is the finite size of the electron beam which sweeps across the face of the vidicon to detect the light. This electron beam has a roughly Gaussian intensity distribution in the y-direction. Thus, the sampled light intensity is averaged over this Gaussian window. We saw in Section 4.1,B that the effect of this spatial averaging provided by the sample window is to modify the Fourier transform of the unsampled quantity by multiplying it by the Fourier transform of the window function. Thus, the finite size of the scanning electron beam helps to band-limit the measured quantity by rolling off its Fourier transform by a Gaussian function.

Whenever a TV camera system is used to measure the power spectrum, the extent of the aliasing noise must be determined. If it is appreciable, steps must be taken to reduce it. Methods of reducing aliasing noise include (i) increasing the number of raster lines, (ii) decreasing the size of the input object whose power spectrum is being measured, (iii) reducing the resolution of the Fourier transform lens by aperturing or apodizing it, and (iv) defocusing the electron beam of the vidicon tube to increase the spatial averaging before sampling. The first method increases the sample rate and thus the highest allowable spatial frequency, whereas the three other methods are all ways of low-passing the pattern before sampling.

B. Applications

Hybrid systems designed to measure the optical power spectrum have been constructed for a large variety of applications. In this section, we will describe the configuration of representative systems applied to three areas of data processing: (i) pattern recognition, (ii) size measurements, and (iii) time series processing.

1. Pattern Recognition

We saw in Section 4.1,C that the Fourier power spectrum has several characteristics which make it useful for pattern recognition applications. Specifically, the data compression property may allow us to work with fewer power spectrum samples than original image samples. The shift-invariance property makes it possible to detect an object independent of the location of the object in the input scene. It is important to realize, however, that not all pattern recognition tasks are suited to power spectrum sampling. For example, if the location of an input object is important for recognition, power spectrum sampling is inappropriate because the location information is lost. In fact, since the power spectrum is the squared modulus of the Fourier transform, it is the autocorrelation of the input scene that is used for recognition. If position information is important, the statistics which govern the image are generally not stationary, since the mean and variance of pixels in one region of the image are likely to differ from those in another region. Thus, one of the assumptions on which the data compression property was based is invalid in this case (see Section 4.1,C), and a coarse sampling of the Fourier power spectrum may prove to be insufficient. With this precaution, we now turn to some applications which have made effective use of power spectrum sampling for pattern recognition.

The first application of ring and wedge sampling was performed by Lendaris and Stanley [18]. Using aerial photographs as inputs, they noticed that the *signature* provided by the set of intensities from a wedge and ring detector could be used as a description of the image, and that certain regions of this signature provided information about the presence or absence of a certain pattern. Figure 4.2-4 illustrates this effect very dramatically. Parts (a) and (b) are two different aerial pictures of agricultural scenes, with (c) and (d) their two-dimensional Fourier transforms. With ring sampling, the signatures of pictures (a) and (b) are shown in (e) and (f), where the rings increase in radius from left to right in the graph. The signatures provided by the wedge detectors are shown in (g) and (h). In this example, there is a dramatic difference in both the ring and wedge signatures of the two images. It is easy to monitor the power in a few specific regions of these signatures to discriminate between the two scenes.

A more complicated pattern recognition experiment performed by Lendaris and Stanley is shown in Fig. 4.2-5. Here the two scenes are of a railroad, with and without cars. The signatures given by ring sampling show the effect of railroad cars in the image. The presence of the cars adds some spatial frequencies in the middle of the spectrum. Notice that the track-to-track separation is present in both pictures, and that the effect of the cars is to add frequencies lower than this spacing, indicating that the cars were larger than the tracks. Also, since ring sampling was used, the signatures are independent of the orientation of the original images.

Once the signatures have been obtained corresponding to wedge or ring sampling, the hybrid computer may then use an algorithm to extract specific information from the signature. George and Kasdan [19] have suggested two algorithms which work particularly well for sorting handwriting. Their experiment consisted of using the wedge–ring detector to measure the power spectrum from handwriting. Two pages of handwritten text were obtained from each of 21 individuals. The first page was used to establish the characteristic features of each writer. The second page was used as the data base to be sorted.

Wedge signatures were processed using the wedge rank difference sum method, which is best described by the illustration in Fig. 4.2-6. The wedges

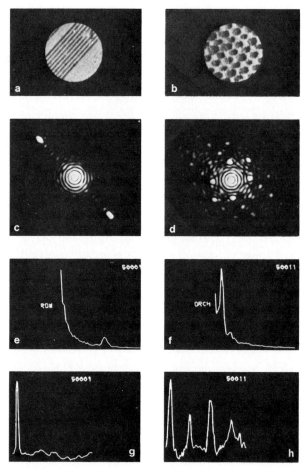

FIG. 4.2-4. Wedge and ring sampling of the optical Fourier power spectrum. (a) Agricultural scene with one-dimensional periodicity. (b) Agricultural scene with two-dimensional periodicity. (c) Optical power spectrum of (a). (d) Optical power spectrum of (b). (e) Ring sampling of (c). (f) Ring sampling of (d). (g) Wedge sampling of (c). (h) Wedge sampling of (d). (From Lendaris and Stanley [18].)

are grouped into six sectors covering one half of the diffraction pattern. This reduces the original 32 data points from the 32 wedges to only 6 data points. Next, the 6 sectors are ranked according to intensity and assigned rank numbers 1–6. This reduces the number of bits needed to express one data point to 3, because $6 < 2^3$. Thus, the features of the 21 handwriting samples can be stored in only (21) (6 × 3) = 378 bits. The features from 2 samples can be compared by summing the absolute values of the differences in rank

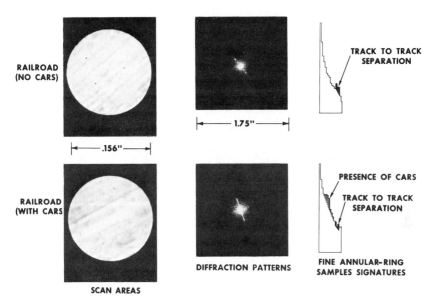

RAILROAD
(NO CARS)

|← .156" →|

|← 1.75" →|

TRACK TO TRACK
SEPARATION

RAILROAD
(WITH CARS)

PRESENCE OF CARS

TRACK TO TRACK
SEPARATION

DIFFRACTION PATTERNS

FINE ANNULAR-RING
SAMPLES SIGNATURES

SCAN AREAS

FIG. 4.2-5. Ring sampling of aerial scenes for detecting the presence of railroad cars. The presence of the cars adds some frequencies which are lower than the track-to-track separation frequency. Since ring sampling is angularly insensitive, detection is independent of track location and orientation. (From Lendaris and Stanley [18].)

number for all 6 sectors. The example in Fig. 4.2-6 shows this sum $\sum$ to be 14 when sample I is compared to sample II. Since wedge sampling is *not* independent of orientation, it may be necessary to compute the wedge rank difference sum for rotated versions of the wedges, which can be performed with the digital computer quite easily by simply shifting the rank numbers of one of the samples and computing a difference sum again. A two-wedge shift is shown in Fig. 4.2-6, resulting in a sum $\sum$ of 0. The lowest value of $\sum$ for the 6 shifts is used as a similarity measure. $\sum$ is computed for all 21 stored samples, with the lowest value corresponding to the highest similarity. This algorithm is a good example of how the digital computer can be used to provide rotational invariance in a hybrid system.

One algorithm used for extracting a feature from the ring signatures was to measure the amount of power in the high spatial frequencies (rings 24–32). A normalized high frequency parameter R^x was defined as

$$R^x = \frac{1}{A_1^x} \sum_{k=24}^{32} A_k^x, \qquad (4.2\text{-}1)$$

where A_k^x is the amplitude of the photodetector current of ring k from hand-

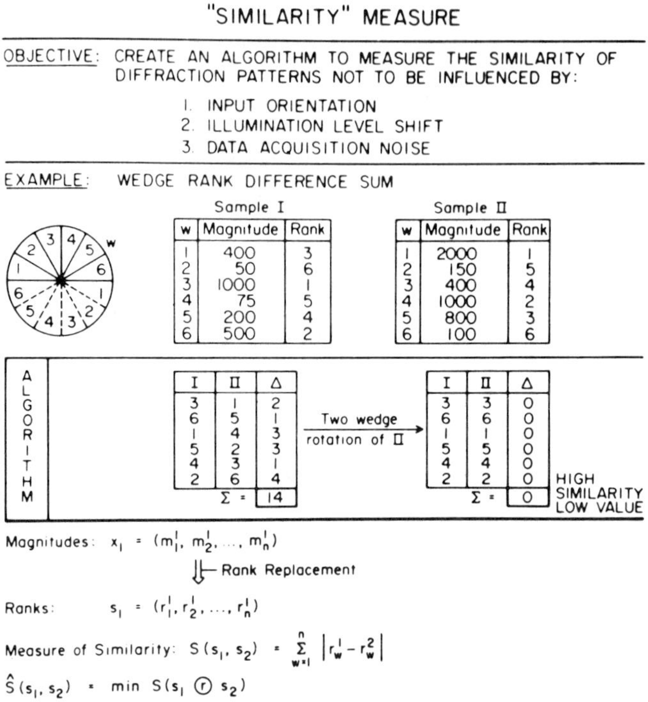

"SIMILARITY" MEASURE

OBJECTIVE: CREATE AN ALGORITHM TO MEASURE THE SIMILARITY OF
DIFFRACTION PATTERNS NOT TO BE INFLUENCED BY:
  1. INPUT ORIENTATION
  2. ILLUMINATION LEVEL SHIFT
  3. DATA ACQUISITION NOISE

EXAMPLE: WEDGE RANK DIFFERENCE SUM

FIG. 4.2-6. Wedge rank difference sum method for analyzing wedge sampling of the optical Fourier power spectrum. Since wedge sampling is affected by orientation, a search must be performed in the computer to calculate the sum for all possible rotations. (From George and Kasdan [19].)

Magnitudes: $x_i = (m_1^i, m_2^i, \ldots, m_n^i)$

Ranks: $s_i = (r_1^i, r_2^i, \ldots, r_n^i)$

Measure of Similarity: $S(s_1, s_2) = \sum_{w=1}^{n} |r_w^1 - r_w^2|$

$\hat{S}(s_1, s_2) = \min S(s_1 \circlearrowright s_2)$

written page x. The similarity between two different pages could then be determined by comparing their respective high frequency parameters.

Using the power spectrum data from the wedge–ring detector and appropriate algorithms, it was found that a sorting accuracy of greater than 95% could be achieved. Thus, it appears that handwriting analysis is well suited to power spectrum analysis.† This is not surprising in view of the analysis of the data compression property discussed in Section 4.1,C. In that section, it was determined that data compression was good in the minimum mean square error sense *if* the statistics were wide-sense stationary. For this condition to hold, the power spectrum of one region of the image must be the same as in another region. (This results from the fact that a wide-sense

---

† We point out, in passing, that another (nonoptical) technique that yields excellent results in handwriting analysis is based on the curvature of symbols. This technique is invariant to scale change.

stationary process has an autocorrelation which is independent of absolute position.) In the case of handwriting, intuition tells us that the statistics are stationary, since a given letter is equally likely to occur anywhere on the page, and the slant and spacing do not vary in different regions of the page.

Another hybrid optical–electronic technique for handwriting classification has been investigated by Duvernoy [20]. In this experiment, the problem consisted of separating pages of handwritten text into two categories: (a) original manuscripts written by the German author Heinrich Heine, and (b) copies written by his secretaries. As in the previous example, the Fourier power spectrum of the handwritten page was performed optically to provide data compression. In this scheme, however, the spectrum was sampled along a single line. The direction of the line was chosen to be perpendicular to the lines of the text [21], and the transform was sampled in ten places along this line. Each page of text (i) then gave rise to a ten-component signature vector with components of the ith vector given by C_{ki}, $k = 1, 10$. A principle component analysis was then performed on the signature vectors resulting from sampling the Fourier power spectrum. This technique, often referred to as a Karhunen–Loève (K-L) analysis, consists of transforming the original signature vectors into a new space whose basis vectors are the eigenvectors of the convariance matrix

$$\mathbf{\Phi}_{kl} = \overline{(C_{ki} - \bar{C}_k)(C_{li} - \bar{C}_l)^*} \qquad (4.2\text{-}2)$$

composed of signatures from both classes (the upper bar denotes an ensemble average). The K-L transformation is identical to the technique described by Eq. (4.1-41) Section 4.1,C for finding the optimal basis set for data compression. Since the K-L transformation maps the maximum amount of information into the fewest dimensions, data which are statistically insignificant and therefore useless for classification purposes are ignored by the transformation. Duvernoy employed a digital computer to solve the eigenvalue equation

$$\mathbf{\Phi}\psi_i = \lambda_i\psi_i \qquad (4.2\text{-}3)$$

for the eigenvectors ψ_i and eigenvalues λ_i, where $\mathbf{\Phi}$ is given by Eq. (4.2-2). The utility of the Fourier transform in the first step in reducing the dimension of the data is fully appreciated from Eq. (4.2-3), as matrix diagonalization is extremely time-consuming for large matrices, the computational time increasing according to the cube of the matrix dimension.

Since the first few eigenvectors of the K-L transform extract the dominant parameters controlling the most important part of the variation among the signatures in the two classes, it was found that a decomposition into the first two eigenvectors was sufficient for class separation. Figure 4.2-7 shows the transformation onto these first two eigenvectors of 14 pages of text written by Heine and 5 pages written by his secretaries. The dotted lines separate this two-dimensional space into two regions containing the two separate classes.

It should be noted that an ideal classification transformation should maximize the interclass differences while minimizing the intraclass ones. A strict application of the K-L transform simply extracts dominant features from both classes without regard to class distinction. If the selected features differ sufficiently between the two classes, clustering occurs as in Fig. 4.2-7. Otherwise, one must resort to more exotic transformations such as those suggested by Fukunaga and Koontz [22].

Briefly summarizing, the previous three applications have all used the Fourier power spectrum to characterize an image for pattern recognition purposes. We saw in Section 4.1,C that these characteristics are independent of the position of the input image. Since the power spectrum is sensitive to rotation and scale change, however, these changes have had to be compensated for when rotation and scale-invariant characteristics were desired. We have thus far used specially designed detectors (wedge–ring detectors) and search algorithms in the digital computer to accomplish this compensation. Another method which is especially applicable with hybrid systems is to perform a coordinate distortion before applying the Fourier transform.

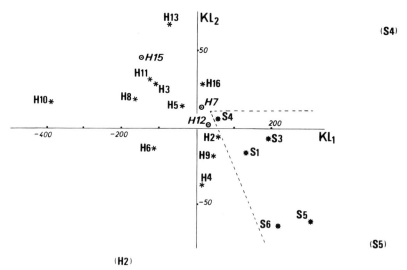

FIG. 4.2-7. Classification of handwriting using optical Fourier and digital Karhunen–Loève transforms. The coefficients of the first two K-L basis functions are plotted for handwriting from the German author Heine and his secretaries. The dotted lines separate these two classes. (From Duvernoy [20].)

We saw in Eq. (4.1-33) of Section 4.1,C that the Fourier transform of an image with a coordinate distortion of $x = e^{\xi}$, $y = e^{\eta}$ gives rise to the two-dimensional Mellin transform. Since the Mellin power spectrum was shown to be invariant to a scale change, it can be used to provide characteristics of an image which are independent of the size of the image.

Casasent and Psaltis [23] demonstrated a hybrid system which performed a two-dimensional Mellin transform. A TITUS tube was used as a spatial light modulator (see Section 4.1,A). The electrical input to the TITUS tube was furnished by a TV camera. A modification of the sweep circuitry of the TITUS tube was made to include logarithmic amplifiers in both the x- and y-sweeps. Thus, the original image detected by the TV camera $f(x, y)$ was recorded on the TITUS tube as $f(e^{\xi}, e^{\eta})$. The intensity in the back focal plane of the Fourier transform lens then became the Mellin power spectrum.

An experiment was performed to test the scale invariance of the hybrid system. Apertures of two different sizes were used as input functions. An input function of the form

$$f(x, y) = \text{rect}\left[\frac{x - (x_1 + x_2)/2}{x_2 - x_1}\right] \text{rect}\left[\frac{y - (y_1 + y_2)/2}{y_2 - y_1}\right] \quad (4.2\text{-}4)$$

corresponds to a rectangular aperture extending from x_1 to x_2 and y_1 to y_2. The Mellin power spectrum associated with this function is

$$|M(f_x, f_y)|^2 = \left|\frac{4}{f_x f_y} \sin\left[f_x \ln\left(\frac{x_2}{x_1}\right)\right] \sin\left[f_y \ln\left(\frac{y_2}{y_1}\right)\right]\right|^2. \quad (4.2\text{-}5)$$

It is apparent that a scale change of $x' = ax$ and $y' = by$ has no effect on the Mellin power spectrum, since the spectrum depends only on the ratios x_2/x_1 and y_2/y_1. Figure 4.2-8a shows the output of the hybrid system corresponding to a square aperture input of width W, and Fig. 4.2-8b that corresponding to a square aperture of width $2W$. The two patterns are identical, as predicted by Eq. (4.2-5).

Notice that the Mellin spectrum is not invariant to shift. Equation (4.2-5) indicates that, as the aperture is shifted in the x-direction, the spacing changes between the nulls in the f_x-direction of the Mellin power spectrum (given by the zeros of the sine function). The shift invariance of the Fourier transform has been replaced by the scale invariance of the Mellin transform. Also notice that the two-dimensional Mellin power spectrum is invariant to changes in scale independently in the x- and y-coordinates. This is evident from Eq. (4.2-5) which indicates that a rectangular aperture extending over $2 < x < 4$ and $5 < y < 10$ will have the same Mellin power spectrum as a square aperture, where $2 < x < 4$, $2 < y < 4$. Many pattern recognition applications do not require this independence of the two coordinates,

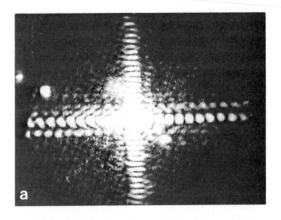

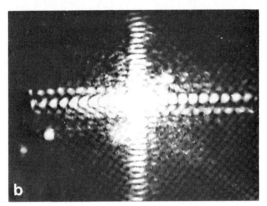

FIG. 4.2-8. Optically produced Mellin power spectrum of a square input. (a) Width of square is W. (b) Width of square is $2W$. (From Casasent and Psaltis [23].)

however. Specifically, if the distance from the object is unknown, the x- and y-coordinates will be scaled by the same unknown amount.

Casasent and Psaltis [24] have suggested a hybrid system which replaces the two-dimensional scale invariance of the two-dimensional Mellin transform with a one-dimensional size invariance and a one-dimensional rotational invariance. This results from applying a Cartesian (x, y) to polar (r, θ) coordinate transformation to the input image $f(x, y)$ and distorting the resulting radial coordinate of $f(r, \theta)$ by a log function to form $f(e^r, \theta)$. A two-dimensional optical Fourier transform of this distorted image produces a Mellin transform in the radial direction and a Fourier transform in the angular direction. A change in size corresponds to a scaling of the radial coordinate. A rotation corresponds to a shift in the angular coordinate. Thus, the resultant intensity distribution is invariant to changes in both size and orientation. A possible hybrid implementation of the system is shown in Fig. 4.2-9.

2. *Size Measurement*

The Fourier power spectrum offers two potential advantages over direct methods of size measurement. First, the smaller the object, the larger its diffraction pattern. This facilitates measurements of small objects. Second, the power spectrum and size measurements derived from it are independent of the spatial position of the object. This property is extremely useful when the location of the object is unknown.

A hybrid system has been developed by Stark *et al.* [25, 26] for particle size determination. Their method is based on the inverse scattering approach discussed by Anderson in Chapter 3. The basic system shown in Fig. 4.2-1 was used, where the spatial light modulator consisted of photographic film and a TV camera was used as a detector. The input consisted of a collection of circularly symmetric particles of three different sizes, randomly arranged in the input. The hybrid computer's task was to estimate the number of particles of each size.

To understand how a collection of randomly positioned particles influences the power spectrum, consider for the moment the power spectrum produced by two circular particles of the same size (radius α) which are

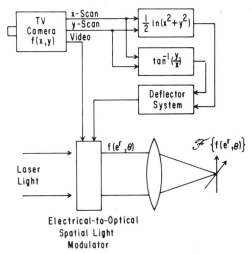

FIG. 4.2-9. Block diagram of scale- and rotation-insensitive hybrid system. The scanning system of an electrical-to-optical spatial light modulator is changed to provide a coordinate distortion. The light intensity in the back focal plane of the lens produces an angular Fourier transform in the y-direction and a radial Mellin transform in the x-direction. (From Casasent and Psaltis [24].)

located at points (a, b) and (c, d), respectively. The Fourier power spectrum is then given by

$$\left| \frac{\alpha J_1(2\pi\alpha\sqrt{f_x^2 + f_y^2})}{\sqrt{f_x^2 + f_y^2}} e^{-i2\pi(af_x + bf_y)} + \frac{\alpha J_1(2\pi\alpha\sqrt{f_x^2 + f_y^2})}{\sqrt{f_x^2 + f_y^2}} e^{-i2\pi(cf_x + df_y)} \right|^2$$

$$= 2\left[\frac{\alpha J_1(2\pi\alpha\sqrt{f_x^2 + f_y^2})}{\sqrt{f_x^2 + f_y^2}} \right]^2 + 2\left[\frac{\alpha J_1(2\pi\alpha\sqrt{f_x^2 + f_y^2})}{\sqrt{f_x^2 + f_y^2}} \right]^2$$

$$\times \cos\{2\pi[(a - c)f_x + (b - d)f_y]\}. \tag{4.2-6}$$

The first term in Eq. (4.2-6) is independent of the particles' positions (a, b) and (c, d). It can thus be used to determine the number of particles with radius α. The second term of Eq. (4.2-6) depends on position, however, and tends to degrade the measurement. When many randomly positioned particles are present, this cross-term takes on a noiselike appearance as a result of the effect of summing many cosines of random frequency. Thus, the power spectrum of a scene with many particles consists of a signal term composed of a simple superposition of single-particle power spectra, and a noise term from the cross-term in Eq. (4.2-6). If the scene consists of N_1 particles of size d_1, N_2 of size d_2, etc., up to N_L particles of size d_L, we can express this distribution by the vector $\mathbf{N} = (N_1, \ldots, N_L)^+$. If $G_i(f_x)$, $i = 1, \ldots, L$ is the diffraction pattern produced by a single particle of size d_i, then we can express the power spectrum of a collection of many particles as

$$\mathscr{H}(f_x) = \sum_{i=1}^{L} N_i G_i(f_x) + n(f_x), \tag{4.2-7}$$

where $n(f_x)$ is the zero mean cross-term noise. If we sample f_x at discrete spatial frequencies $f_{x_1}, f_{x_2}, \ldots, f_{x_N}$, Eq. (4.2-7) can be written in matrix form as

$$\mathscr{H} = \mathbf{GN} + \mathbf{n}. \tag{4.2-8}$$

Consider for the moment the noiseless case, given by

$$\mathbf{H} = \mathbf{GN}. \tag{4.2-9}$$

If $\mathbf{G}$ were a square matrix of full rank, we could solve for $\mathbf{N}$ exactly by multiplying both sides of Eq. (4.2-9) from the left by $\mathbf{G}^{-1}$. In general, however, $\mathbf{G}$ is not square, so it has no inverse and $\mathbf{N}$ has no exact solution. We can obtain a least squares solution by the method of pseudoinversion (see, for example, Ref. [27, p. 130] by multiplying each side of Eq. (4.2-9) from the left by $\mathbf{G}^+$. Since $\mathbf{G}^+\mathbf{G}$ is necessarily square, and often invertible, we can now multiply both sides by its inverse $(\mathbf{G}^+\mathbf{G})^{-1}$. Thus

$$\mathbf{N}_{LS} = (\mathbf{G}^+\mathbf{G})^{-1}\mathbf{G}^+\mathbf{H}. \tag{4.2-10}$$

Returning to the case where cross-term noise is present, the variance of $\mathscr{H}$ in Eq. (4.2-8) caused by the noise term $\mathbf{n}$ cannot be neglected. A plot of $\mathscr{H}(f_x)$ obtained by scanning the Fourier power spectrum of the hybrid processor is shown in Fig. 4.2-10 where the input consisted of a large number of particles of uniform radii.

We recall from Section 4.1,C that the stability of the power spectrum estimate could be increased by smoothing the sample spectrum $\mathscr{H}(f_x)$ with a spectral window. This operation can be expressed in matrix form as

$$\mathscr{H}_W = \mathbf{W}_0 \mathscr{H}, \qquad (4.2\text{-}11)$$

where $\mathbf{W}_0$ is a toeplitz matrix which convolves $\mathscr{H}$ with a specific window function. Applying this spectral window to Eq. (4.2-8) we have

$$\mathscr{H}_W = \mathbf{W}_0\mathbf{GN} + \mathbf{W}_0\mathbf{n}. \qquad (4.2\text{-}12)$$

A plot of $\mathscr{H}_W(f_x)$ is shown in Fig. (4.2-11). Much of the noise has been removed, so that a better estimate of the true power spectrum can be made. If we assume that the noise has been smoothed out in Eq. (4.2-12), then we can perform a pseudoinversion as in Eq. (4.2-10) to solve for the number distribution vector $\mathbf{N}$:

$$\hat{\mathbf{N}}_{LS} = (\mathbf{G}^+\mathbf{G})^{-1}\mathbf{G}^+\mathbf{W}_0\mathscr{H}. \qquad (4.2\text{-}13)$$

The results of using this hybrid processor with and without spectral smoothing are shown in Table 4.2.1. Particles of three different sizes were randomly arranged in the input scene. For the unsmoothed case the results

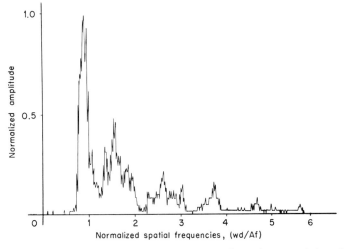

FIG. 4.2-10. Radial scan of the optical power spectrum for an input consisting of particles of uniform size. No smoothing has been applied. (From Stark *et al.* [25].)

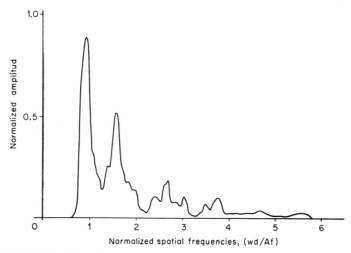

FIG. 4.2-11. Scan from Fig. 4.2-10, where smoothing has been applied to reduce the effects of the cross-term noise. (From Stark *et al.* [25].)

exhibit both a high variation between runs and an overestimation of the number of small particles. Smoothing the power spectrum is seen to reduce both these effects.

3. *Time Series Processing*

The previous applications have all dealt with some form of image processing. The prospect of high speed Fourier transforms has made the hybrid system an attractive candidate for time series processing as well. Indeed, one

TABLE 4.2-1

Summary of Results[a]

Run no.	Actual number of particles			Estimated no. of particles from unsmoothed data			Estimated no. of particles from smoothed data		
	S	M	L	S	M	L	S	M	L
1	50	25	15	76	19	11	52	27	13
2	50	25	15	64	17	13	51	27	14
3	50	25	15	63	17	13	51	27	14
4	50	25	15	67	16	13	49	25	15
5	50	25	15	70	14	12	52	26	13

[a] S, small; M, Medium; L, large.

of the first hybrid systems (although non-real-time) to make extensive use of the Fourier properties of coherent light was built to process the temporal signals from synthetic aperture radar [28]. More recently, hybrid processors have been applied to a large variety of signal processing tasks in radar, acoustics, and communications [29]. Many of these applications are discussed in Chapters 5, and 11 of this book. In this discussion, we shall limit ourselves to a single hybrid system for processing phased array radar.

The hybrid system designed by Casasent and co-workers was previously introduced in connection with the Mellin transform of images. This same system can be used to perform various signal processing operations. Since a TITUS tube is used as a spatial light modulator, an electrical signal can be input to the optical system very easily. For example, phased array radar data can be processed [30, 31] by recording the heterodyned output of each antenna of the array on a separate horizontal line of the TITUS tube, as in Fig. 4.2-12. The x-coordinate of the optical processor becomes proportional to time, and the y-coordinate becomes proportional to distance along the antenna array. The hybrid processor (see Fig. 4.2-1) simply performs a two-dimensional Fourier transform of this input data. Since the antenna array is detecting the far-field diffraction pattern of the targets in one dimension, and we know from Chapter 1 that this diffraction pattern is related to the angular distribution of sources by the Fourier transform, the output of the hybrid processor in the f_y-direction must correspond to the angular dis-

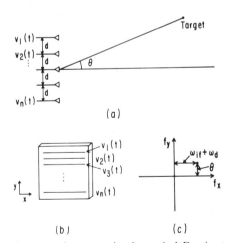

(a)

(b)   (c)

FIG. 4.2-12. Phased array radar processing by optical Fourier transform. (a) Antenna array for phased array radar. (b) Format for recording antenna array data on TITUS tube. (c) Output of optical processor. The f_y-coordinate of an intensity peak is related to the azimuth angle θ of the target, and the f_x-coordinate to its Doppler shift or velocity.

tribution of the targets θ. The output of the processor in the f_x-direction corresponds to the temporal Fourier transform of the antenna outputs.

Consider the effect of a single moving target. If the transmitted radar signal is a sinusoid of frequency ω_t, the return signal is a Doppler-shifted sinusoid $\omega_t + \omega_d$ due to the target's motion. This signal is heterodyned by the receiver and written on the TITUS tube as a sinusoid of frequency $\omega_{if} + \omega_d$, where ω_{if} is the intermediate frequency resulting from the heterodyning in the absence of a Doppler shift. The Fourier transform of this sinusoid produces two delta functions at locations proportional to $\pm(\omega_{if} + \omega_d)$ in the f_x-direction. Since ω_{if} is known, the Doppler frequency, hence the velocity of a given target, can be determined from the f_x-coordinate of the output of the hybrid processor. The position of the point in the f_y-direction is proportional to target angle θ.

The output of the optical system is detected by a television camera. The resultant video signal is processed by some electronics, and input to a digital computer. Electronic processing is very important, since the amount of raw data generated by the TV camera is quite large and the digital computer operates at a much slower data rate. Electronic processing must reduce the amount of data sent to the digital computer by eliminating all unimportant data. In the phased array radar application, the absolute magnitude of all points in the output is of little interest. Instead, the locations of points which exceed a certain threshold are the important data. Also, the resolution of the TV system is frequently higher than necessary. If an array contains 100 antennas, the maximum number of resolvable angles is limited to 100. A typical television system, however, provides almost 500 useable raster lines when sampling the y-coordinate in Fig. 4.2-12c. To reduce the number of data points sent to the digital computer, the electronic processor converts the output video signal into a bit pattern by using the following scheme. A frame of video data is partitioned into a matrix of $M \times N$ cells, where $8 < M < 96$ and $8 < N < 240$. The values of M and N are chosen to conform to the resolution requirements of the particular processing task and are controlled by the digital computer. A single threshold level can also be set to any of 32 levels by the digital computer. The video data are then compared with this threshold level. If the level is exceeded anywhere within a cell, that cell is assigned a value of 1. Otherwise the value of the cell is zero. The resultant $M \times N$ bit pattern is input to the digital computer. It is apparent that this electronic processing has both eliminated redundant data (by reducing unnecessary resolution) and performed the first step of a target identification algorithm (by performing a threshold). Subsequent processing can be conveniently performed by the digital computer.

4.3 HYBRID SYSTEMS BASED ON SPATIAL FILTERING

Hybrid systems based on spatial filtering have been applied to many problems in pattern recognition, image processing, and measurement. Rather than simply detecting the Fourier transform as in the previous section, spatial filtering gives us the opportunity to modify the transform. A second transformation converts the signal back into an image with an altered spatial frequency content. A description of the optical system for performing spatial filtering is given in Chapter 2.

A hybrid implementation of spatial filtering usually involves interfacing to the optical system in three places. The first two, the input and the output, have already been discussed in connection with power spectral measurements (Section 4.2). The third interface connects the electronics to the spatial filter plane of the optical system. By changing the spatial filter, the optical processor can perform different functions. Thus, this interface enables an operator to "program" the optical processor. We will start this section by describing three different spatial filter plane interfaces. Several applications of these hybrid systems will then be reviewed.

A. Hybrid Systems for Generating and Controlling Spatial Filters

In this section, we briefly review some of the materials and devices which have been used as spatial filters. We then discuss the different types of spatial filters and their characteristics. Finally, some specific hybrid spatial filter interfaces are reviewed.

1. *Devices and Materials for Recording Spatial Filters*

Many spatial light modulators are capable of modulating the amplitude or the phase of a coherent light wave but cannot control both amplitude and phase simultaneously. The TITUS tube and PROM devices, for example, modulate the amplitude when the polarization of the coherent beam is at 45 deg with respect to the fast and slow axes of the crystal and an analyzer is used. They modulate the phase when the polarization is parallel to either the fast or slow axis. Photographic films can be used as amplitude or phase modulators also. A conventionally processed film in a liquid gate modulates the light amplitude. If this film is bleached to remove the absorption of the emulsion, the thickness variation between exposed and unexposed parts of the emulsion act to modulate the phase of the light. Although photographic films are not real-time modulators, they have a much higher resolution than the PROM and TITUS tube spatial light modulators and have been found to be useful in parts of a hybrid system where frequent change is unnecessary.

2. Classification of Spatial Filters

Spatial filters which modulate amplitude can be grouped into two classes. The first is the on-axis filter, which is capable of multiplying the Fourier transform by a positive real number. Since this filter cannot alter the phase of the Fourier transform, it can be thought of as modifying only the power spectrum of the input. The second class of amplitude filters is the holographic filter. By amplitude- and frequency-modulating the spatial carrier of the filter, the Fourier transform of the input is effectively multiplied by a complex function, thus modifying both the amplitude and the phase of the input transform. This additional flexibility is generally necessary for performing convolutions and correlations with arbitrary functions, as is required for pattern recognition.

The class of holographic amplitude modulating spatial filters can be further subdivided into Vander Lugt filters and computer-generated filters. Vander Lugt filters are made by recording the interference pattern of the optical Fourier transform of an object and a coherent reference wave. The impulse response of this filter (or its reconstruction), properly apertured to retain only the $+1$ order of the carrier, is the object itself. If the original object were a photographic transparency, it is clear that the impulse response of this type of filter would be restricted to a positive real function. To eliminate this restriction, the computer-generated hologram uses a digital computer to design and construct the filter. The "object" in this case does not physically exist and can be any complex function. Thus, a computer-generated hologram can be designed to have any desired complex impulse response.

3. Examples of Hybrid Spatial Filter Interfaces

A near real-time programmable on-axis spatial filter has been developed by Iwasa [32] and co-workers using a PROM device (see Section 4.1,A). The system uses a scanning blue laser (krypton, 476 nm) to write the filter function on the PROM. The horizontal scanning is provided by an acousto-optic deflector with a resolution of 1000 Rayleigh-limited spots and a dwell time of 10 μsec per spot. The vertical scanning is provided by a galvanometer-driven mirror. The intensity of the laser beam is controlled by an acousto-optic modulator. The deflectors and intensity modulator are driven by a programmable electronic controller, which applies sinusoids of varying amplitudes and phases to the deflectors, causing them to scan the laser beam in a variety of Lissajous patterns. By combining appropriate Lissajous patterns, low-pass, high-pass, and bandpass filters can be constructed, as well as wedges, spirals, and ellipses. These filters are simple enough to be scanned in a fraction of a second, allowing the operator to use a variety of filtering operations in near real time when analyzing an image.

In general, on-axis spatial filters have positive real transmittance functions. The PROM, however, can make use of the baseline subtraction feature described in Section 4.1,A to record both positive and negative transmittance values. Negative transmittance recording is accomplished by adding a bias to the transmittance function to make the entire function positive, scanning this new function onto the PROM and then applying the appropriate amount of baseline subtraction to eliminate the bias. All transmittance levels below this bias will be given a π phase shift. As an example, to perform the x-derivative operation, a real (but not all positive) on-axis spatial filter is needed, according to the Fourier transform relationship [33]

$$\mathscr{F}\left\{\frac{dh(x)}{dx}\right\} = i2\pi f_x H(f_x), \tag{4.3-1}$$

where $H(f_x)$ is the Fourier transform of $h(x)$. Ignoring the complex constant $(i2\pi)$, the spatial filter must multiply the Fourier transform of the input by a linear transmittance function $t(f_x, f_y) = f_x$. If the maximum spatial frequency of the optical system is $(f_x)_{max}$, the transmittance function runs from $-(f_x)_{max}$ to $+(f_x)_{max}$. We scan the biased function $t(f_x, f_y) = f_x + (f_x)_{max}$ onto the PROM and then apply the proper amount of baseline subtraction to subtract $(f_x)_{max}$ from the resulting transmittance.

As a second example, we consider a spatial filter interface designed by Almeida and Eu [34] for controlling a bank of Vander Lugt filters in an optical correlator. In general, optical systems can be designed to correlate high SBP imagery with a single filter function or lower SBP imagery with many multiplexed filter functions simultaneously. The ultimate degree of parallelism for imagery with a certain SBP is determined by the SBP of the optical system. If many filter functions with high SBPs are needed, an alternative is to construct a bank of spatial filters and perform the correlation with each filter sequentially by moving the filters in and out of the optical system. This is equivalent to expanding the system in time rather than space or spatial frequency. For Almeida's interface, a 10×10 matrix of Vander Lugt filters is positioned automatically in the optical system by a pair of stepping motors under digital computer control. The same interface is used to create the filter bank. Since photographic film is used as a recording material, it was necessary to design a special filter holder to ensure accurate repositioning after development. In the following paragraphs, the position accuracy and total travel distance requirements on the filter holder are analyzed.

The position tolerance of the spatial filter is an extremely important design criterion. Since the filter plane is constantly being moved in both the ξ- and η-directions, stepping motors must be used which have sufficient accuracy so that little correlation degradation results. The effect of mispositioning the spatial filter by an amount $\Delta\xi$ is seen by the shift theorem to introduce a phase distortion of $e^{-i2\pi\Delta\xi x/\lambda f}$ in the impulse response of the filter, where f is the focal length of the lenses in the correlator, and λ is the wavelength of the light. The correlation between the input and the impulse response of the filter becomes (in one dimension)

$$c(x') = \int g(x)g^*(x - x')e^{i2\pi\Delta\xi(x-x')/\lambda f} \, dx, \qquad (4.3\text{-}2)$$

where $g(x)$ is the input function and $g^*(-x)$ is the ideal impulse response of the filter. It is evident that the effect of mispositioning the spatial filter is to introduce a phase error in the correlation integral, where the amount of error is related to the displacement $\Delta\xi$. If $g(x)$ is of finite size such that $g(x) = 0$ for $x > |X/2|$, Eq. (4.3-2) evaluated at $x' = 0$ can be rewritten with finite limits as

$$c(0) = \int_{-X/2}^{X/2} g(x)g^*(x)e^{i2\pi\Delta\xi x/\lambda f} \, dx. \qquad (4.3\text{-}3)$$

One common criterion for acceptable phase error is given by the Rayleigh limit, which states that all phase errors should be less than or equal to $\pi/2$. Since the maximum phase error occurs when $x = X/2$, this criterion is ensured if

$$\Delta\xi \le \lambda f/2X. \qquad (4.3\text{-}4)$$

The accuracy requirements for the computer-controlled bank of matched filters in Almeida's hybrid system can now be computed. The first Fourier transform lens has a focal length of 60 cm, and the wavelength of the light is 6.328×10^{-5} cm (helium–neon). The input to the correlator consists of a 35-mm slide, making $X = 3.5$ cm. Thus, the required positioning accuracy in the Fourier plane is $\Delta\xi < 5.4$ μm. The stepping motors used in this system are accurate to 2.5 μm. Thus, the maximum phase distortion caused by mispositioning in one dimension is less the one-half the Rayleigh limit, or $\pi/4$ radians.

It is evident from Eq. (4.3-4) that the positioning accuracy requirement can be relaxed by demagnifying the input image (and thus reducing X) or increasing the focal length of the first transform lens. However, when either of these is done, the size of the filter increases, requiring the positioning mechanism to travel over a longer distance to change filters. The size of a single Vander Lugt filter Ξ is related to the bandwidth of the input function B_x (assuming that the filter is matched to the input as before) by the relation

$$\Xi = 2B_x \lambda f. \tag{4.3-5}$$

Thus, a translation stage must be able to move a spatial filter a distance Ξ to remove it completely from the optical system. An important specification for the translation stage is given by the ratio of the total travel distance to the position accuracy (equivalent to the number of resolvable positions of the stage). We will call this number the space–bandwidth product of the translation stage, in analogy with the SBP of an image (the SBP of an image can be considered to be the number of resolvable points in the image). The required SBP of the translation stage for moving a single filter is given by dividing Eq. (4.3-5) by Eq. (4.3-4), which results in

$$(\text{SBP})_{\text{translation stage}} = \Xi/\Delta\xi \geq 4XB_x. \tag{4.3-6}$$

Note that this specification is independent of the focal length of the Fourier transform lens. We recognize $2XB_x$ as the SBP of the input image. Thus, this parameter is also independent of the scale of the image, since magnifying the image increases the size but reduces the bandwidth. The SBP of the translation stage must simply be greater than or equal to twice the SBP of the input image. The factor of 2 is a result of our use of the Rayleigh limit for defining acceptable phase error and would change for a different error limit.

When a bank of N filters (matched to N different inputs with identical SBPs) must be moved, the total travel distance increases by a factor of N over that stated in Eq. (4.3-5). Equation (4.3-6) then has the natural extension

$$(\text{SBP})_{\text{translation stage}} \geq 2N(\text{SBP})_{\text{input}}. \tag{4.3-7}$$

The required SBP of the translation stage in one dimension of Almeida's spatial filter interface is given by Eq. (4.3-7), where $N = 10$, $X = 35$ mm, $B_x = 6$ mm^{-1}, and $(\text{SBP})_{\text{input}} = 420$. Thus, $(\text{SBP})_{\text{translation stage}}$ must be greater than 8.4×10^3. Since the stage has an accuracy of 2.5 μm and can move a total distance of 5 cm, its SBP is 2×10^4, and the requirement is met.

The third example of a spatial filter interface was designed for generating computer holograms [35, 36, 37, pp. 237–288] as complex spatial filters. A block diagram of the interface is shown in Fig. 4.3-1. A television camera and video digitizer can be used to input the desired impulse response to the digital computer if this information is available in optical form. If the impulse response can be better described mathematically, it can be entered via the teletype.

Once the impulse response is in the computer, it can be modified in

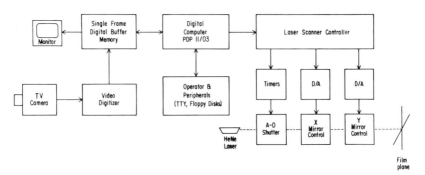

FIG. 4.3-1. Automatic system for generating computer holograms. Image data are input to the digital computer through the TV camera–video digitizer system, Fourier-transformed and coded by the digital computer, and written onto a recording material by the computer-controlled laser scanner.

several ways. For instance, many impulse responses in the form of images can be input through the TV digitizer port and coherently averaged by the digital computer. Nonlinear distortions of the amplitude as well as co-ordinate distortions can be applied. Complex terms can also be incorporated into the impulse response. When the desired response is achieved, the computer performs a two-dimensional discrete Fourier transform on the data and encodes each complex number of the result into three positive real values as required for a computer hologram (for details, see Refs. [38, 39]).

The computer hologram is produced by a laser scanner system under microcomputer control. The laser scanner consists of a pair of servocon-trolled galvanometer scanning mirrors for positioning a helium–neon laser beam in an xy film plane, and an acousto-optic shutter for controlling the film exposure time. A matrix of small rectangular apertures $10 \times 20 \ \mu$m is produced on a piece of photographic film.

The transmittance of each rectangle becomes proportional to a particular positive real number calculated by the computer. Figure 4.3-2 shows a spatial filter at approximately four times its actual size (actual size is $1.7 \times 1.7 \ cm^2$) and a photomicrograph of the center of a filter for displaying the rectangular apertures of varying transmittance.

The computer holograms are usually made with a space–bandwidth product of 128×128 pixels, because of the amount of time needed to perform digital Fourier transforms on larger arrays. The SBP of the scanner system is 10^7 Rayleigh-resolvable points, however. The remaining SBP of the scanner system can be utilized by replicating the hologram up to 16 times, to increase the intensity of the impulse response and reduce noise. Figure 4.3-3 shows the impulse response (or reconstruction) of a computer hologram made with this system. The original data were entered using the

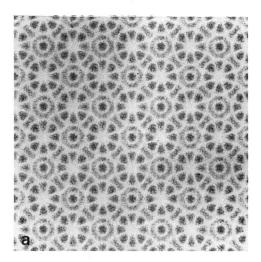

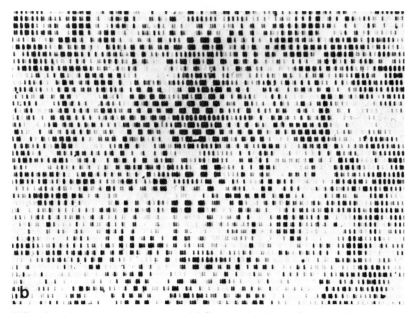

FIG. 4.3-2. A computer-generated spatial filter, made with the laser scanning system. (a) 4× magnification of filter containing 786,432 rectangular apertures of varying transmittance. The actual size is 1.7 × 1.7 cm. (b) 60× magnification of the center of the filter, showing the individual rectangular apertures. The actual size of the apertures is 10 × 20 μm.

FIG. 4.3-3. Reconstruction of a computer hologram created by the laser scanning system. Reconstruction occurs in the -1 diffraction order in the x-direction. Part of the conjugate image in the -2 diffraction order can be seen on the left-hand side of the picture.

TV digitizer port, modified by the computer to reduce noise and enhance certain features, and scanned onto a piece of 649F film using the laser scanner. The resultant developed hologram was placed in front of a Fourier transform lens and illuminated with coherent light. The light distribution shown in Fig. 4.3-3 occurs in the first diffraction order in the back focal plane of the lens.

B. Applications of Hybrid Systems Employing Spatial Filtering

In this section, we will briefly describe five hybrid systems which perform spatial filtering. The first two employ on-axis spatial filters to perform different types of image processing. The second two use Vander Lugt filters and coherent-optical correlators to perform pattern recognition. The final system demonstrates the increased flexibility of the computer-generated hologram as a spatial filter.

Iwasa [32] and co-workers have used the PROM-based programmable on-axis spatial filter described in the previous section in the Fourier plane of a telecentric imaging system (see Fig. 4.3-4). A second PROM was used

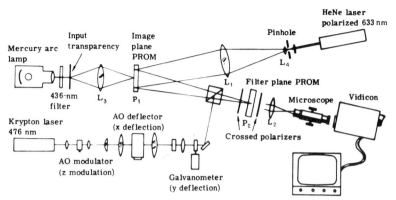

FIG. 4.3-4. Two PROM hybrid systems for image processing. The image plane PROM is used to convert an incoherent input image into a coherent one. The filter plane PROM is used as a programmable spatial filter. (From Iwasa [32].)

as input to the optical system. High- and low-pass filtering, wedge filtering, and differentiation were all demonstrated.

A less conventional hybrid system employing on-axis spatial filtering was constructed by Stark [40] and co-workers for performing optical texture analysis. A texture signature is provided by the variance of an object transparency defined by

$$\sigma^2(x, y) = \overline{[f(x, y) - \overline{f(x, y)}]^2}, \tag{4.3-8}$$

where $f(x, y)$ is the object transmittance function and the over bars indicate ensemble averages. The hybrid system for computing the variance is shown in Fig. 4.3-5. The input transparency is placed directly in front of lens L_1. A spatial filter is placed in the back focal plane of lens L_1 to remove the center of the Fourier transform. A ground glass screen is placed in the back focal plane of L_2, and a TV camera is focused on a plane a small distance behind the screen.

If the input $f(x, y)$ is statistically stationary in the mean, the average transmittance is independent of position, and $\overline{f(x, y)} = \bar{f}_0$. If, in addition, $f(x, y)$ is ergodic in the mean, the ensemble average can be replaced by a spatial average

$$\bar{f}_0 = \frac{1}{A} \int\int_A f(x, y) \, dx \, dy, \tag{4.3-9}$$

where the object is of area A. Comparing Eq. (4.3-9) with the Fourier transform of $f(x, y)$, we see that $\bar{f}_0$ is given by the transform evaluated at its

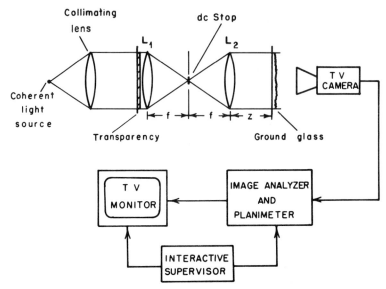

FIG. 4.3-5. Hybrid system for optical texture analysis. The mean of the transparency is removed by a dc stop in the filter plane. The resultant image detected by the TV camera is proportional to the variance of the original transparency. (From Stark [40].)

center. The small dc stop in the Fourier plane of L_1 removes this average from the image, as required in Eq. (4.3-8).

If the dc stop is removed, the image of an object placed in the input plane (directly in front of L_1) will not be formed on the ground glass screen because the image plane is actually one focal length behind the screen. The Fourier power spectrum on the other hand is imaged by lens L_2 to infinity. Thus, the screen must be in a Fresnel transform region of the object. Garcia *et al.* [41] have shown that in this region, when the dc stop is replaced, the expected value of the intensity at the diffusing screen is approximately $\sigma^2(x, y)$, as given by Eq. (4.3-8). The variance of the intensity is quite large, however. By focusing the camera on a plane a small distance behind the diffusing screen, a small amount of blurring is introduced, providing a local spatial average of the estimate of $\sigma^2(x, y)$. As we saw in Section 4.1,C, this averaging tends to reduce the variance and thus increase the stability of the result [42].

The video signal from the TV camera is electronically processed and then is either displayed or sampled, digitized, and fed into a digital computer. The electronic processing consists of a gray scale analyzer for quantizing and false color-encoding the image, and a planimeter for measuring the area of objects in an image. The gray scale analyzer can be adjusted to display on a TV monitor only a specific range of gray levels. When used in conjunction with the above optical system, the display consists of areas in the image of constant variance. Since variance is often a good criterion for classification, this hybrid system can be used as the first part of a pattern recognition routine. This is illustrated in Fig. 4.3-6. The input transparency consisted of an x-ray photograph of the chest cavity. The optical system produced an output whose intensity was proportional to the local variance in the transparency. The gray scale analyzer was then adjusted to display only a small range of gray levels corresponding to a specific variance. It is quite evident from the figure that the lungs have the proper variance, whereas other regions in the chest cavity do not. Thus, the lungs have been identified and separated from the rest of the image. The resultant video signal of this image can be sent to the planimeter to measure the lung area, or it can be sampled, digitized, and stored in the digital computer. The digitized image serves to furnish the computer with the addresses associated with the lung region, effectively rejecting all image points outside the region of interest. Further processing by the digital computer on a reduced number of points from the original transparency is now possible.

A hybrid system for recognition of machine parts on a conveyor belt has been constructed by Gara [43–45] and is shown in Fig. 4.3-7. A liquid crystal light valve (see Section 4.1,A) is used as an image transducer to provide a coherent input to a conventional coherent-optical correlator, and a bleached Vander Lugt filter is used as the matched filter. (Bleaching helps to improve the diffraction efficiency of the filter.) An image rotator presents all possible orientations of the input object to the correlator. The output plane of the correlator consists of the two-dimensional cross-correlation of the input and the impulse response of the spatial filter. If the output plane contains $N \times N$

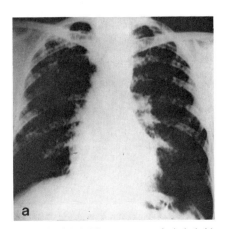
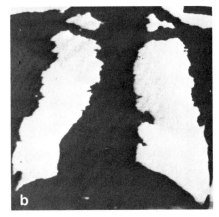

FIG. 4.3-6. The texture analysis hybrid processor is used to locate the lungs in an x-ray photograph of the chest. (a) Original x-ray photograph. (b) Final output from the hybrid processor. Only regions with a specific variance are displayed. (From Stark [40].)

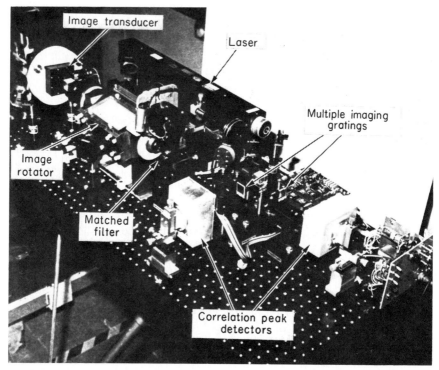

FIG. 4.3-7. Experimental setup of the hybrid correlator for machine part recognition. A liquid crystal light valve is used as an image transducer. The correlation peak detectors consist of two linear arrays of photodiodes oriented at right angles to each other. (From Gara [45].)

resolvable points, a complete description of the cross-correlation would require N^2 measurements. In this system however, the important information consists of the presence or absence of a *single* correlation peak and its x- and y-coordinates. To reduce the number of correlation space measurements to $2N$, two images of the correlation space are formed using a lens and a beam splitter. A multiple-imaging grating is placed in each imaging system to produce a line of correlation peaks, where the peaks are replicated vertically in one image and horizontally in the other. A linear diode array is then oriented at right angles to the line of correlation peaks in each image. Thus, the x- and y-coordinates of the correlation peak can be obtained independently.

The two linear arrays are scanned 360 times, once for every degree of rotation of the input image. An electronic peak detector determines the scan number giving the largest signal, and the photodiode address of the peak signal for each array. By synchronizing the photodiode scan with the image rotator, the peak scan number becomes proportional to the orientation of the object. The peak photodiode addresses of the x- and y-arrays are proportional to the coordinates of the object. A microprocessor calculates these parameters and transfers them to a robot-controlling minicomputer for part manipulation.

The overall accuracy and reproducibility of the system is 1 part in 130 over a 0.6-m object search distance, when the conveyor belt is limited to a speed of less than 25 cm/sec. At speeds greater than this, the strength of the correlation signal would be substantially reduced, resulting in lower accuracy, because the finite response time of the liquid crystal light valve would cause a decrease in edge definition of the coherent object. Thus the filter is no longer accurately matched to the input object.

A hybrid system for water pollution monitoring has been constructed by Almeida and co-workers (see Chapter 2 and Ref. [34]). The pollution level is determined by the number of diatoms of different species found in the water. In this application, a pattern recognition system is required which can differentiate between many classes of objects but be relatively insensitive to variations within a class caused by size difference, misfocus, rotation, etc. The hybrid system consists of a coherent-optical correlator whose output is interfaced to a television camera and digitizer. The input consists of photomicrographs of the diatoms on a computer-controlled film transport. The filter plane interface, consisting of a bank of spatial filters on a computer-controlled translation stage, was described in the previous section.

Intraclass variations are caused by orientation and size differences among a single species of diatom and make conventional matched filtering difficult (see Section 4.1,C). In addition, the depth of focus of the phase-contrast microscope used to create the input transparency can cause two diatoms of the same species to have different appearances. The sensitivity of the correlation to these variations has been reduced by using averaged spatial filters consisting of an incoherent sum of several matched filters, produced by multiple exposure. The resulting impulse response of the filter is the average of several input functions. By using a bank of averaged spatial filters, an input can be examined for the presence of many different species of diatoms simply by moving the appropriate spatial filters into the correlator with the computer-controlled translation stage.

Many other hybrid approaches have been applied to solve the orientation and scale sensitivity problem of the matched spatial filter. One solution to the orientation problem has been to rotate the input via a dove prism and monitor the output of the correlator over an entire 360-deg rotation. Another has been to use a frequency-multiplexed matched filter whose impulse response is a bank of patterns corresponding to all possible orientations of an input. In the first case, the time it takes to complete the pattern recognition operation has been increased. The second case requires an increase in the space–bandwidth product of the optical system. An alternative method for reducing the orientation and scale sensitivity which does not increase the SBP or processing time consists of synthesizing a special computer-generated spatial filter to detect a set of patterns simultaneously [46]. The patterns can be rotated, scaled, or shifted versions of a single image, as well as completely different images. With this flexibility, a spatial filter can be customized for a particular application with the minimum SBP and process time.

The synthesis of a special spatial filter depends on the important facts that pattern recognition can be performed by computing the inner product of a test pattern $f(x, y)$ and the input image $g(x, y)$ and that the correlation function yields this inner product at its center. The rest of the correlation function can be eliminated by using random phase masks. By multiplying the input $g(x, y)$ by a random phase array, and the impulse response of a spatial filter by the complex conjugate of the phase array, the correlation integral becomes

$$c(x', y') = \int \int g(x, y)e^{i\phi_r(x,y)}f^*(x - x', y - y')e^{-i\phi_r(x-x',y-y')}\, dx\, dy$$

$$= \left[\int \int g(x, y)f^*(x, y)\, dx\, dy\right]\delta(x', y'), \qquad (4.3\text{-}10)$$

where $\phi_r(x, y)$ is a random function uniformly distributed from 0 to 2π. The result that only the central value of the correlation remains is due to the fact that an ideal random phase array has an autocorrelation function of a delta function $\delta(x', y')$. By using this technique, we have expressed an inner product as the amplitude of a single point. To perform many different inner products simultaneously (corresponding to different pattern matching operations) we need a filter which has an impulse response containing the sum of many test patterns. To separate the different inner products in the output, the complex conjugate of each test pattern is multiplied by the conjugate random phase mask and a sum is formed from shifted versions of these products. Thus the impulse response of the spatial filter we need is

$$h(-x, -y) = \sum_{p,q} f^*_{pq}(x + p\Delta, y + q\Delta)e^{-i\phi_r(x+p\Delta,y+q\Delta)}, \qquad (4.3\text{-}11)$$

where $f_{pq}(x, y)$ are the test patterns, and Δ is a shift constant. The correlation integral corresponding to Eq. (4.3-10) becomes

$$\int \int g(x. y)e^{i\phi_r(x,y)} \sum_{p,q} f^*_{pq}(x - x' + p\Delta, y - y' + q\Delta)$$

$$\times e^{-i\phi_r(x-x' + p\Delta,y-y + q\Delta)}\, dx\, dy$$

$$= \sum_{p,q} \left[\int \int g(x, y)f^*_{pq}(x, y)\, dx\, dy\right]\delta(x' - p\Delta, y' - q\Delta). \qquad (4.3\text{-}12)$$

The output of the optical system consists of a set of points, where the amplitude of a specific point corresponds to a particular pattern matching operation.

The hybrid system using this optical processor is shown in Fig. 4.3-8. An expanded and filtered beam from an argon laser is passed through a computer-generated hologram (CGH #1). This light is polarized vertically and Fourier-transformed by lens L_3 so that the reconstruction of CGH #1 occurs at the face of the liquid crystal light valve. This reconstruction, consisting of a random phase array, is spatially modulated by the LCLV according to the incoherent object projected onto the left side of the light valve. The coherent light leaving the right side of the light valve thus has a horizontal polarization component which is proportional to the product of an input (from the incoherent object) and a random phase array. This horizontal component is transmitted by the polarizing beam splitter, and the Fourier transform of the light field appears in the back focal plane of L_4. A second computer-generated hologram (CGH #2) with an impulse response given by Eq. (4.3-11) is placed in this plane, and the resultant light field Fourier-transformed by lens L_5. The intensity of the final Fourier transform corresponding to the squared magnitude of Eq. (4.3-12) is detected by a vidicon,

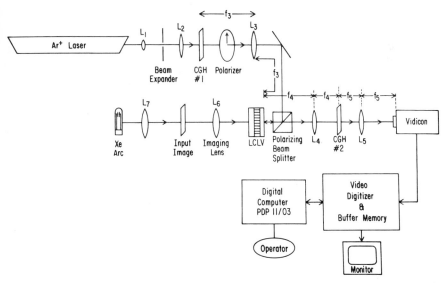

FIG. 4.3-8. A hybrid optical processor for performing a generalized set of pattern matching operations. A liquid crystal light valve (LCLV) performs incoherent-to-coherent image conversion. The output of the optical processor is detected by a TV camera, digitized, analyzed by the digital computer, and displayed on a TV monitor. CGH #1 and CGH #2 are computer-generated holograms.

sampled (256 × 512 pixels), digitized, and stored in a buffer memory. This process takes place at TV frame rates ($\frac{1}{30}$ sec). The buffer memory can then be displayed on a TV monitor or analyzed by a digital computer.

The interface diagrammed in Fig. 4.3-1 for generating spatial filters is also part of this hybrid system. Both computer holograms (CGH #1 and CGH #2) are fabricated automatically using this interface. Although CGH #1

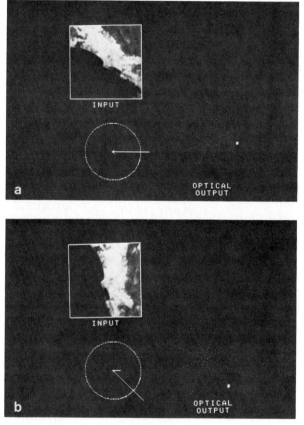

FIG. 4.3-9. TV monitor output of the hybrid processor for detecting several rotated versions of an image. The input image is displayed at the upper left. The output of the optical processor is detected by a TV camera and displayed at the lower right as a series of points in a circular pattern. At the lower left is a graph of the intensities of these points, drawn on the monitor by the digital computer. Two different orientations of the input pattern are shown in (a) and (b).

needs only to be fabricated once, CGH #2 must be changed when a different set of patterns is needed. The advantages of using computer holograms as spatial filters are apparent. First, a complicated impulse response such as that represented by Eq. (4.3-11) is difficult to achieve using Vander Lugt filters. Second, the pattern functions $f_{pq}(x, y)$ need not exist physically, or can be modifications of existing objects. Third, the summation is performed in the digital computer and therefore is coherent. A coherent sum offers higher filter diffraction efficiency, and therefore more pattern matching operations can be performed simultaneously than with an incoherent summation from a multiply exposed Vander Lugt filter.

Examples of the output of the hybrid system, displayed on a TV monitor, are shown in Fig. 4.3-9. At the lower right is a video display of the output of the optical processor. A graphed version of this display (lower left) shows the intensity of an inner product point as the length of a given line. The input object is displayed at the upper left. The filter used in this experiment was designed to detect a LANDSAT photograph of the coast of France in eight different orientations. The inner product points were designed to occur in a circle at the output of the correlator, where the position in the circle corresponded to the orientation. Part (a) shows an input with a rotation of 90 deg, the corresponding optical output, and a graph indicating that the pattern corresponding to a 90 deg rotation matches the input. Part (b) shows the result of rotating the input pattern to 135 deg.

4.4 HYBRID SYSTEMS USING INCOHERENT LIGHT

Although the majority of this chapter has been devoted to coherent hybrid systems, we would like to discuss briefly some hybrid systems which use incoherent-optical processors. Numerous systems have been designed for a wide variety of applications [47, Chapter 3]. However, our attention will be confined to three representative systems. A thorough description of incoherent-optical systems is also contained in Chapter 12 of this book, as well as in Ref. [48].

Incoherent-optical processors are of interest for several reasons. First, the input to the processor is very easy to generate, since it no longer must be coherent. As we saw in Section 4.1,A, a television monitor serves as a convenient electrical-to-optical interface. An LED array can also serve this purpose. Second, coherent noise and artifacts resulting from dust and blemishes on optical components are suppressed because of the multiplicity of paths that light takes in traveling from input to output in an incoherent system. Third, the output information of an incoherent system is conveyed by the real-valued intensity distribution rather than a complex amplitude distribution, making detection easier since the phase does not have to be recovered. The price paid for these advantages is generally one of flexibility. Since incoherent processors manipulate light intensities rather than amplitudes, negative and complex numbers are not as easy to represent. Also, the relationship between the input transparency and the light field at the back focal plane of a lens is no longer described by a simple Fourier transform. The hybrid approach often can relieve many of these constraints and is therefore very important. The next few examples illustrate how television technology and incoherent light sources can be used to extend the flexibility of incoherent systems.

An incoherent-optical processor has been developed by Bromley *et al.*, which is capable of performing a broad variety of general linear filtering operations on a one-dimensional signal [49]. The processor, diagrammed in Fig. 4.4-1, consists of a light emitting diode (LED). a two-dimensional mask, and a charge-coupled device (CCD). A sampled electrical signal f_n is applied to the LED to generate an illumination which is temporally proportional to the signal voltage and spatially uniform across the mask. The mask consists of an $M \times K$ array of apertures of varying size which are aligned with the photosite array of the CCD. The light received by the m,kth element of the CCD at time n is thus proportional to the product of the signal f_n and the aperture size $h_{m,k}$.

By using the "shift-and-add" mode of a CCD, the photogenerated charge of one cell can be shifted to an adjacent cell. Suppose that the appropriate clocking waveforms are applied to the CCD in Fig. 4.4-1 so that rows of charge are shifted up on the structure, and that this shifting is synchronized

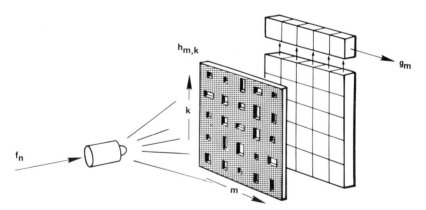

FIG. 4.4-1. Diagram of an optical processor using incoherent light. A light emitting diode is driven by a sampled electrical signal f_n. By applying various masks to the CCD, a multi-channel cross-correlation is performed on the one-dimensional time signal. (From Bromley [49].)

with the input sampling. As these charge packets travel upward, more and more charge is added to them as a result of the time-varying pattern illuminating the CCD. The accumulated charge in the top row of the CCD after K shifts is then given by

$$g_m = \sum_{n=1}^{K} f_n h_{m,n}. \tag{4.4-1}$$

A parallel-to-serial converter on the CCD is used to read out the M values of g_m between row shifts. We recognize Eq. (4.4-1) as a general vector–matrix product between a vector $\mathbf{f} = (f_1, f_2, \ldots, f_n)$ and a matrix

$$\mathbf{H} = \begin{bmatrix} h_{11} & h_{12} & h_{13} & \cdots & h_{1n} \\ h_{21} & h_{22} & h_{23} & & \\ h_{31} & h_{32} & h_{33} & & \\ \vdots & & & & \\ h_{m1} & & & & h_{m,n} \end{bmatrix}, \tag{4.4-2}$$

where all the elements of the vector and the matrix are nonnegative real numbers. Bipolar quantities can be handled either by adding a bias to make the resulting numbers nonnegative or by dividing the quantities into positive and negative parts and handling each part separately.

The ability to perform a high speed vector–matrix multiply has many applications in signal processing. A few of these applications are multichannel cross-correlation, multichannel FIR filtering, and general linear transformations. Using a CCD with 512 rows ($K = 512$) and 320 columns ($M = 320$), and using an input sample rate of 32,000 samples/sec, this processor performs the equivalent of 5×10^9 multiplications/sec.

A fully parallel hybrid incoherent-optical system has been developed by Goodman et al. [50] for performing a discrete Fourier transform (DFT) on one-dimensional data. The input in this case consists of a linear array of LEDs and thus provides a means of entering the data into the optical processor in parallel. A CCD is used as an integrating detector. The hybrid aspect in this case makes it possible to handle complex input data by using electronic circuits to convert a single complex signal into three positive real ones. Similarly, electronic circuits can be employed to convert the output of three positive real channels into the real and imaginary parts of the complex result. If high speed LEDs were used in this system along with avalanche photodiodes, a 32-point DFT could be performed every 10 nsec [50, p. 3]. This corresponds to a total throughput rate of about 3×10^9 complex samples per second.

A hybrid system which makes use of television technology and feedback has been investigated by Häusler and Lohmann [51]. Electronic processing in this hybrid system makes nonlinear and space-variant operations possible. Gain can also be introduced by electronic amplification of the video signal. The system has been applied to solving the diffusion equation in two spatial coordinates and time, including a time-dependent diffusion rate [52].

An extension of this work to a television optical operational amplifier has been performed by Götz et al. [53]. A block diagram of the operational amplifier and its incoherent-optical implementation are shown in Fig. 4.4-2. The feedback loop consists of a TV monitor for converting the output video signal into incoherent light and a TV camera for detecting this light and converting it into a video signal. The output of the camera is electronically subtracted from the input video signal to provide negative feedback. Optical processing can be performed in the feedback loop by placing optical elements between the monitor and the camera. The hybrid nature of the system makes gain and signal subtraction easy to perform.

The response of the system can be analyzed with the help of Fig. 4.4-2a. The output $\mathcal{O}'(x, y)$ can be written as

$$\mathcal{O}'(x, y) = \mathcal{G}[\mathcal{O}(x, y) - \mathcal{O}_F(x, y)]\beta, \qquad (4.4\text{-}3)$$

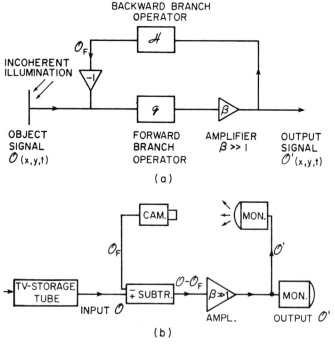

FIG. 4.4-2. The incoherent-optical operational amplifier. (a) Block diagram. (b) Experimental setup. (From Götz et al. [53].)

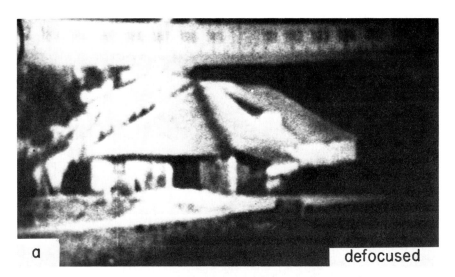

a · defocused

b · restored

FIG. 4.4-3. Image restoration using the optical operational amplifier. (a) Input image produced by a defocused camera. (b) Restored image using the same defocused camera in the feedback branch. (From Götz *et al.* [53].)

where $\mathcal{O}(x, y)$ is the two-dimensional input function, $\mathcal{O}_F(x, y)$ is the feedback function, $\mathcal{G}$ is the forward branch operator, and β is an amplification provided by the electronics with the property $\beta \gg 1$. The feedback term $\mathcal{O}_F(x, y)$ is given by

$$\mathcal{O}_F(x, y) = \mathcal{H}[\mathcal{O}'(x, y)], \tag{4.4-4}$$

where $\mathcal{H}$ is the feedback branch operator. In part (b) of the figure, $\mathcal{G}$ corresponds to any electronic processing in the forward branch, and $\mathcal{H}$ corresponds to the combined effects of the TV monitor, camera, and optical elements placed in between the two, in the feedback branch. Substituting Eq. (4.4-4) into Eq. (4.4-3) we have

$$\mathcal{O}'(x, y) = \beta\mathcal{G}\{\mathcal{O}(x, y) - \mathcal{H}[\mathcal{O}'(x, y)]\}. \tag{4.4-5}$$

The operators $\mathcal{G}$ and $\mathcal{H}$ can generally be nonlinear or space-variant. They may depend on intensity, spatial coordinates, or temporal coordinates. If we consider for simplicity a space-invariant system, the operators can then be represented as transfer functions multiplying the Fourier transforms of the images. In this case, Eq. (4.4-5) becomes

$$\tilde{\mathcal{O}}'(f_x, f_y) = \beta G(f_x, f_y)[\tilde{\mathcal{O}}(f_x, f_y) - H(f_x, f_y)\tilde{\mathcal{O}}'(f_x, f_y)], \tag{4.4-6}$$

where G and H are the transfer functions of the forward and feedback paths, and the tilde denotes a Fourier transform. Solving for $\tilde{\mathcal{O}}'(f_x, f_y)$ yields

$$\tilde{\mathcal{O}}'(f_x, f_y) = \frac{\beta G(f_x, f_y)\mathcal{O}(f_x, f_y)}{1 + \beta G(f_x, f_y)H(f_x, f_y)}. \tag{4.4-7}$$

By assuming that $\beta \gg 1$, this can be approximated by

$$\tilde{\mathcal{O}}'(f_x, f_y) = [H(f_x, f_y)]^{-1}\tilde{\mathcal{O}}(f_x, f_y). \tag{4.4-8}$$

The action of the operational amplifier is then equivalent to that of an operator $\mathcal{H}^{-1}$ and is independent of the forward operator $\mathcal{G}$. This remains true if the operators are space-variant and nonlinear also.

An example of the space-invariant case is shown in Fig. 4.4-3. Part (a) shows a defocused image, where the aberration is described by a transfer function $\tilde{H}(f_x, f_y)$. If this same aberration is introduced in the feedback path in Fig. 4.4-2, the output of the operational amplifier will consist of a corrected image [part (b)].

4.5 SUMMARY

We have tried to introduce the basic concepts of hybrid system design in this chapter. Various aspects of electrical-to-optical and incoherent-to-coherent interfaces were studied, with an emphasis on design requirements. The capabilities of optical processors were studied by investigating several properties of the Fourier transform. The results of these studies helped to define which tasks the optical processor could best perform and which should be handled by electronics. Finally, these points were exemplified in several representative hybrid systems. Each example was chosen to make a specific point, and no attempt was made to list all the hybrid systems now in existence. It is hoped that this chapter has served as an introduction to hybrid systems, and that the basic concepts may be of use to future hybrid system designers.

APPENDIX

We asserted in Eq. (4.1-41) that the orthonormal transformation which minimizes the mean square error of an n-component vector when only m components are measured was given by the eigenvectors of the covariance matrix $\mathbf{R}_x$. The proof provided here is essentially that derived by Fukunaga (see Ref. [54, pp. 229–231]). We start by considering another orthonormal transformation provided by a matrix $\boldsymbol{\Phi}$ of n column vectors $\boldsymbol{\phi}_i$. Applying this transformation to the covariance matrix $\mathbf{R}_x$ we have

$$\mathbf{H} = \boldsymbol{\Phi}^{+}\mathbf{R}_x\boldsymbol{\Phi}. \tag{4.A-1}$$

The diagonal terms of $\mathbf{H}$ are given by

$$h_{ii} = \boldsymbol{\phi}_i^{+}\mathbf{R}_x\boldsymbol{\phi}_i. \tag{4.A-2}$$

We can partition the matrix $\mathbf{H}$ such that

$$\mathbf{H} = \begin{array}{c} \overbrace{}^{m} \overbrace{}^{n-m} \\ \left[\begin{array}{c|c} \mathbf{H}_{11} & \mathbf{H}_{12} \\ \hline \mathbf{H}_{21} & \mathbf{H}_{22} \end{array}\right] \begin{array}{l} \}m \\ \}n\text{-}m \end{array} \end{array}. \tag{4.A-3}$$

From Eq. (4.1-40) the mean square error which results from measuring only m components of an n-component vector when an orthonormal transformation has been applied is given by

$$E\{\|\boldsymbol{\varepsilon}\|^2\} = \sum_{i=m+1}^{n} \boldsymbol{\phi}_i^{+}\mathbf{R}_x\boldsymbol{\phi}_i = \operatorname{tr}\mathbf{H}_{22}, \tag{4.A-4}$$

where tr is the trace of the matrix. Let $\boldsymbol{\Psi}$ and $\boldsymbol{\Lambda}$ be the eigenvector and eigenvalue matrices of $\mathbf{R}_x$, where the eigenvalues in the diagonal matrix $\boldsymbol{\Lambda}$ are ordered from largest to smallest. Then $\mathbf{H}$ can be expressed as

$$\mathbf{H} = \boldsymbol{\Phi}^{+}\mathbf{R}_x\boldsymbol{\Phi} = \boldsymbol{\Phi}^{+}\boldsymbol{\Psi}\boldsymbol{\Lambda}\boldsymbol{\Psi}^{+}\boldsymbol{\Phi}. \tag{4.A-5}$$

If we let $\mathbf{K} = \boldsymbol{\Phi}^+\boldsymbol{\Psi}$, then we can partition Eq. (4.A-5) as

$$\mathbf{H} = \begin{matrix} m\{ \\ n-m\{ \end{matrix} \overbrace{\left[\begin{array}{c|c} \mathbf{K}_1 & \mathbf{K}_2 \\ \hline \mathbf{K}_3 & \mathbf{K}_4 \end{array}\right]}^{m \quad n-m} \left[\begin{array}{c|c} \boldsymbol{\Lambda}_1 & 0 \\ \hline 0 & \boldsymbol{\Lambda}_2 \end{array}\right] \overbrace{\left[\begin{array}{c|c} \mathbf{K}_1^+ & \mathbf{K}_3^+ \\ \hline \mathbf{K}_2^+ & \mathbf{K}_4^+ \end{array}\right]}^{m \quad n-m}. \tag{4.A-6}$$

It is apparent from Eq. (4.A-6) that the mean square error is given by

$$E\{\|\boldsymbol{\varepsilon}\|^2\} = \text{tr } \mathbf{H}_{22} = \text{tr } (\mathbf{K}_3\boldsymbol{\Lambda}_1\mathbf{K}_3^+ + \mathbf{K}_4\boldsymbol{\Lambda}_2\mathbf{K}_4^+). \tag{4.A-7}$$

If in Eq. (4.A-1) we had used the eigenvectors of $\mathbf{R}_x$ to perform the transformation, we would have arrived at the expression

$$\mathbf{H}' = \boldsymbol{\Psi}^+\mathbf{R}_x\boldsymbol{\Psi} = \boldsymbol{\Lambda}, \tag{4.A-8}$$

where $\boldsymbol{\Psi}$ and $\boldsymbol{\Lambda}$ are the eigenvector and eigenvalue matrices as defined in Eq. (4.A-5). In this case, the mean square error given by Eq. (4.A-4) simply reduces to

$$E\{\|\boldsymbol{\varepsilon}\|^2\} = \text{tr } \boldsymbol{\Lambda}_2, \tag{4.A-9}$$

where $\boldsymbol{\Lambda}$ is partitioned as in Eq. (4.A-6). Therefore, our assertion will be proven if we can show that

$$\text{tr } \boldsymbol{\Lambda}_2 \leq \text{tr}(\mathbf{K}_3\boldsymbol{\Lambda}_1\mathbf{K}_3^+ + \mathbf{K}_4\boldsymbol{\Lambda}_2\mathbf{K}_4^+). \tag{4.A-10}$$

To show this, we recall that $\mathbf{K}$ is the product of two orthonormal matrices $\mathbf{K} = \boldsymbol{\Phi}^+\boldsymbol{\Psi}$. This implies that $\mathbf{K}$ is orthonormal, since

$$\mathbf{K}^+\mathbf{K} = (\boldsymbol{\Phi}^+\boldsymbol{\Psi})^+(\boldsymbol{\Phi}^+\boldsymbol{\Psi}) = \boldsymbol{\Psi}^+(\boldsymbol{\Phi}\boldsymbol{\Phi}^+)\boldsymbol{\Psi} = \mathbf{I}.$$

Thus,

$$\mathbf{K}^+\mathbf{K} = \mathbf{K}\mathbf{K}^+ = \mathbf{I}. \tag{4.A-11}$$

Expressing $\mathbf{K}$ in the partitioned form of Eq. (4.A-6), Eq. (4.A-11) implies

$$\left[\begin{array}{c|c} \mathbf{K}_1 & \mathbf{K}_2 \\ \hline \mathbf{K}_3 & \mathbf{K}_4 \end{array}\right]\left[\begin{array}{c|c} \mathbf{K}_1^+ & \mathbf{K}_3^+ \\ \hline \mathbf{K}_2^+ & \mathbf{K}_4^+ \end{array}\right] = \left[\begin{array}{c|c} \mathbf{I} & 0 \\ \hline 0 & \mathbf{I} \end{array}\right] \tag{4.A-12}$$

and

$$\left[\begin{array}{c|c} \mathbf{K}_1^+ & \mathbf{K}_3^+ \\ \hline \mathbf{K}_2^+ & \mathbf{K}_4^+ \end{array}\right]\left[\begin{array}{c|c} \mathbf{K}_1 & \mathbf{K}_2 \\ \hline \mathbf{K}_3 & \mathbf{K}_4 \end{array}\right] = \left[\begin{array}{c|c} \mathbf{I} & 0 \\ \hline 0 & \mathbf{I} \end{array}\right]. \tag{4.A-13}$$

From Eq. (4.A-12) we have the result

$$\mathbf{K}_1\mathbf{K}_1^+ + \mathbf{K}_2\mathbf{K}_2^+ = \mathbf{I}, \tag{4.A-14}$$

and from Eq. (4.A-13) we have the results

$$\mathbf{K}_1^+\mathbf{K}_1 + \mathbf{K}_3^+\mathbf{K}_3 = \mathbf{I} \tag{4.A-15}$$

and

$$\mathbf{K}_2^+ \mathbf{K}_2 + \mathbf{K}_4^+ \mathbf{K}_4 = \mathbf{I}. \tag{4.A-16}$$

Using these three results along with the matrix identities

$$\mathrm{tr}(\mathbf{AB}) = \mathrm{tr}(\mathbf{BA}) \tag{4.A-17}$$

and

$$\mathrm{tr}(\mathbf{A} + \mathbf{B}) = \mathrm{tr}(\mathbf{A}) + \mathrm{tr}(\mathbf{B}), \tag{4.A-18}$$

we can simplify the expression for the mean square error given by Eq. (4.A-7):

$$\begin{aligned}
\mathrm{tr}&(\mathbf{K}_3 \boldsymbol{\Lambda}_1 \mathbf{K}_3^+ + \mathbf{K}_4 \boldsymbol{\Lambda}_2 \mathbf{K}_4^+) \\
&= \mathrm{tr}(\boldsymbol{\Lambda}_1 \mathbf{K}_3^+ \mathbf{K}_3 + \boldsymbol{\Lambda}_2 \mathbf{K}_4^+ \mathbf{K}_4) = \mathrm{tr}\left[\boldsymbol{\Lambda}_1 \mathbf{K}_3^+ \mathbf{K}_3 + \boldsymbol{\Lambda}_2 (\mathbf{I} - \mathbf{K}_2^+ \mathbf{K}_2)\right] \\
&= \mathrm{tr}(\boldsymbol{\Lambda}_1 \mathbf{K}_3^+ \mathbf{K}_3 + \boldsymbol{\Lambda}_2 - \boldsymbol{\Lambda}_2 \mathbf{K}_2^+ \mathbf{K}_2) = \mathrm{tr}(\boldsymbol{\Lambda}_2) + \mathrm{tr}(\boldsymbol{\Lambda}_1 \mathbf{K}_3^+ \mathbf{K}_3 - \boldsymbol{\Lambda}_2 \mathbf{K}_2^+ \mathbf{K}_2) \\
&\geq \mathrm{tr}(\boldsymbol{\Lambda}_2) + \left[\lambda_m \mathrm{tr}(\mathbf{K}_3^+ \mathbf{K}_3) - \lambda_{m+1} \mathrm{tr}(\mathbf{K}_2^+ \mathbf{K}_2)\right],
\end{aligned} \tag{4.A-19}$$

since λ_m is the smallest eigenvalue in $\boldsymbol{\Lambda}_1$ and λ_{m+1} is the largest eigenvalue in $\boldsymbol{\Lambda}_2$. The right side of the inequality in Eq. (4.A-19) can be further simplified as

$$\begin{aligned}
\mathrm{tr}(\boldsymbol{\Lambda}_2) &+ \left[\lambda_m \mathrm{tr}(\mathbf{K}_3^+ \mathbf{K}_3) - \lambda_{m+1} \mathrm{tr}(\mathbf{K}_2^+ \mathbf{K}_2)\right] \\
&= \mathrm{tr}(\boldsymbol{\Lambda}_2) + \left[\lambda_m \mathrm{tr}(\mathbf{I} - \mathbf{K}_1^+ \mathbf{K}_1) - \lambda_{m+1} \mathrm{tr}(\mathbf{I} - \mathbf{K}_1 \mathbf{K}_1^+)\right] \\
&= \mathrm{tr}(\boldsymbol{\Lambda}_2) + (\lambda_m - \lambda_{m+1}) \mathrm{tr}(\mathbf{I} - \mathbf{K}_1^+ \mathbf{K}_1) \\
&= \mathrm{tr}(\boldsymbol{\Lambda}_2) + (\lambda_m - \lambda_{m+1}) \mathrm{tr}(\mathbf{K}_3^+ \mathbf{K}_3) \geq \mathrm{tr}(\boldsymbol{\Lambda}_2),
\end{aligned} \tag{4.A-20}$$

since $\lambda_m \geq \lambda_{m+1}$ and $\mathrm{tr}(\mathbf{K}_3^+ \mathbf{K}_3) \geq 0$. Combining Eq. (4.A-19) and (4.A-20), we obtain

$$\mathrm{tr}(\mathbf{K}_3 \boldsymbol{\Lambda}_1 \mathbf{K}_3^+ + \mathbf{K}_4 \boldsymbol{\Lambda}_2 \mathbf{K}_4^+) \geq \mathrm{tr}(\boldsymbol{\Lambda}_2), \tag{4.A-21}$$

which proves the relationship stated in Eq. (4.A-10) and thus the assertion that the minimum mean square error results when a vector is expanded in terms of the eigenvectors of its covariance matrix.

REFERENCES

[1] H. Kasden and D. Mead (1975). *Proc. Electron. Opt. Syst. Des. Conf.*, Anaheim, Calif. pp. 248–258.
[2] G. Knight (1981). *In* "Optical Information Processing Fundamentals" (S. H. Lee, ed.), Vol. 48, Chap. 4. Springer-Verlag, Berlin and New York.
[3] B. A. Horwitz and F. J. Corbett (1978). *Opt. Eng.* **17**, 353–364.
[4] W. P. Bleha *et al.* (1978). *Opt. Eng.* **17**, 371–384.
[5] D. Casasent (1978). *Opt. Eng.* **17**, 344–352.
[6] J. Goodman (1968). "Introduction to Fourier Optics." McGraw-Hill, New York.

[7] W. K. Pratt (1978). "Digital Image Processing." Wiley, New York.

[8] R. Lesault (1973). *In* "Perception of Displayed Information" (L. Biberman, ed.), Chap. 7. Plenum, New York.

[9] J. L. Harris (1966). *J. Opt. Soc. Am.* **56,** 569–574.

[10] C. W. Helstrom (1967). *J. Opt. Soc. Am.* **57,** 297–303.

[11] B. R. Frieden (1972). *J. Opt. Soc. Am.* **62,** 511–518.

[12] A. V. Oppenheim and R. W. Schafer (1975). "Digital Signal Processing." Prentice-Hall, Englewood Cliffs, New Jersey.

[13] G. M. Jenkins and D. G. Watts (1968). "Spectral Analysis and Its Applications." Holden-Day, San Francisco, California.

[14] N. Ahmed and K. R. Rao (1975). "Orthogonal Transforms for Digital Signal Processing." Springer-Verlag, Berlin and New York.

[15] H. C. Andrews and B. R. Hunt (1977). "Digital Image Restoration." Prentice-Hall, Englewood Cliffs, New Jersey.

[16] A. Vander Lugt, F. B. Rotz, and A. Klooster (1965). *In* "Optical and Electro-Optical Information Processing" (J. T. Tippett, D. A. Berkowitz, L. C. Clapp, C. J. Koester, and R. Vanderburgh, Jr., eds.), Chap. 7. MIT Press, Cambridge, Massachusetts.

[17] D. Casasent and A. Furman (1977). *Appl. Opt.* **16,** 1652–1661.

[18] G. Lendaris and G. Stanley (1970). *Proc. IEEE* **58,** 198–216.

[19] N. George and A. L. Kasdan (1975). *Proc. Electron. Opt. Syst. Des. Conf., Anaheim, Calif.* pp. 494–503.

[20] J. Duvernoy (1976). *Appl. Opt.* **15,** 1584–1590.

[21] J. Duvernoy (1973). *Opt. Commun.* **7,** 14.

[22] K. Fukunaga and W. L. G. Koontz (1970). *IEEE Trans. Comput.* **C-19,** 311–318.

[23] D. Casasent and D. Psaltis (1977). *Proc. IEEE* **65,** 77–84.

[24] D. Casasent and D. Psaltis (1976). *Appl. Opt.* **15,** 1795–1799.

[25] H. Stark, D. Lee, and B. Dimitriadis (1975). *J. Opt. Soc. Am.* **65,** 1436–1442.

[26] H. Stark and B. Dimitriadis (1975). *J. Opt. Soc. Am.* **65,** 425–431.

[27] G. Strang (1976). "Linear Algebra and its Applications." Academic Press, New York.

[28] L. J. Cutrona, E. N. Leith, L. J. Porcello, and W. E. Vivian (1966). *Proc. IEEE* **54,** 1026–1032.

[29] D. Casasent (1977). *In* "Optical Signal Processing" (D. Casasent ed.), Topics in Applied Physics, Vol. 23. Chap. 8. Springer-Verlag, Berlin and New York.

[30] L. Lambert, M. Arm, and A. Aimette (1965). *In* "Optical and Electro-Optical Information Processing" (J. T. Tippett, D. A. Berkowitz, L. C. Clapp, C. J. Koester, A. Vanderburgh, Jr., eds.), Chap. 38. MIT Press, Cambridge, Massachusetts.

[31] D. Casasent and F. Casasayas (1975). *IEEE Trans. Aerosp. Electron. Syst.* **AES-11,** 65–75.

[32] S. Iwasa (1976). *Appl. Opt.* **15,** 1418–1424.

[33] R. Bracewell (1965). "The Fourier Transform and Its Applications." McGraw-Hill, New York.

[34] S. P. Almeida and J. K. T. Eu (1976). *Appl. Opt.* **15,** 510–515.

[35] J. R. Leger and S. H. Lee (1979). *Opt. Eng.* **18,** 518–523.

[36] J. R. Leger, J. Cederquist, and S. H. Lee (1978). *J. Opt. Soc. Am.* **68,** 1414.

[37] J. R. Leger (1980). The coded-phase optical processor and its hybrid implementation, Ph.D. Thesis, Univ. of California, San Diego.

[38] W. H. Lee (1970). *Appl. Opt.* **9,** 639–643.

[39] C. B. Burckhardt (1970). *Appl. Opt.* **9,** 1949.

[40] H. Stark (1974). *Opt. Eng.* **13,** 243–249.

[41] E. Garcia, H. Stark, and R. C. Barker (1972). *Appl. Opt.* **11,** 1480–1490.

[42] H. Stark and E. Garcia (1974). *Appl. Opt.* **13,** 648–658.

[43] A. D. Gara (1979). *Appl. Opt.* **18,** 172–174.
[44] A. D. Gara (1977). *Appl. Opt.* **16,** 149–153.
[45] A. D. Gara (1979). *Res. Publ.—Gen. Mot. Corp., Res. Lab.* **GMR-2841,** 1–46.
[46] J. R. Leger and S. H. Lee (1982). *Appl. Opt.* **21.**
[47] W. T. Rhodes and A. A. Sawchuk (1981). *In* "Optical Information Processing Fundamentals" (S. H. Lee, ed.), Vol. 48, Chap. 3. Springer-Verlag, Berlin and New York.
[48] M. A. Monahan, K. Bromley, and R. P. Bocker (1977). *Proc. IEEE* **65,** 121–128.
[49] K. Bromley, M. A. Monahan, R. P. Bocker, A. C. H. Louie, and R. D. Martin (1977). *Proc. SPIE* **118.**
[50] J. W. Goodman, A. R. Dias, and L. M. Woody (1978). *Opt. Lett.* **2,** 1–3.
[51] G. Häusler and A. Lohmann (1977). *Opt. Commun.* **21,** 365–368.
[52] J. Götz, G. Häusler, A. W. Lohmann, and M. Simon (1979). *Angew. Opt. Annu. Rep. 1977/78* p. 7.
[53] J. Götz, G. Häusler, and R. Sesselmann (1979). *Appl. Opt.* **18,** 2754–2759.
[54] K. Fukunaga (1972). "Introduction to Statistical Pattern Recognition." Academic Press, New York.

Chapter 5

Fourier Optics and
Radar Signal Processing

MARVIN KING

RIVERSIDE RESEARCH INSTITUTE
NEW YORK, NEW YORK

5.1 INTRODUCTION

In the introduction to Chapter 4, J. Leger and S. Lee mentioned the application of Fourier optics to radar signal processing. This is the subject of this chapter.

Fourier-optical configurations have been applied to a diverse family of radar signal processing functions, including the detection of radar returns, the estimation of radar target range and range rate, the estimation of target angular position for phased array radars, and the mapping of distributed-target scattering distributions for synthetic aperture radar. Fourier-optical processors are successful as radar signal processors because they supply a flexible linear filter implementation which can be adapted to the particular processing needs of radar.

Section 5.2 discusses the linear filtering operations required to process signals for radar range and range rate estimation, estimation of target angular position in phased array radars, and mapping the radar cross-section distribution for a synthetic aperture radar.

In addition to providing a discussion of radar signal processors for these operations, Section 5.3 presents a brief discussion of real-time spatial light modulators which convert received radar signals into spatially modulated optical wavefronts instantaneously for these processors. The discussion of processors for radar range and range rate is somewhat lengthy, reflecting the rich literature which began with conventional Fourier-optical configurations

209

and has evolved to the current interest in thin-film optics processors. The discussion of Fourier-optical processors for phased array antennas is particularly interesting, because it describes configurations which may be regarded as optical analog simulations of the phased array antennas whose signals they process. Perhaps the most striking and well-known application of Fourier optics to radar signal processing is their use for synthetic aperture radar image formation; although the principal years of discovery in this area of optical processing are apparently well in the past, the elegance of these processors and the striking nature of their outputs never fail to impress one with the potentiality of Fourier optics in general.

The discussions of specific optical processors presented in this chapter are designed to familiarize the reader with the various approaches taken to the particular signal processing problems of interest; not all published and proposed Fourier-optical approaches are reported here. Similarly, the references on Fourier-optical processors support the particular processors described and are not intended to be exhaustive. The aim of the discussions of the various processors is to communicate an understanding of the principles underlying them; in some cases, factors which limit the performance or successful application of the processors are discussed. However, a complete critical assessment of the processors and their principal design tradeoffs requires a much more lengthy discussion than is possible in this space.

5.2 RADAR SIGNAL PROCESSING

A. Processing for Range and Range Rate

Suppose that a radar transmits a pulsed signal to the target, which scatters a portion of the signal toward the receiver diagrammed in Fig. 5.2-1. The radar antenna senses the return, and the receiver amplifies this signal and

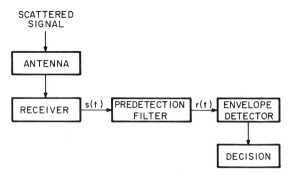

FIG. 5.2-1. Radar reception system.

translates its energy band to the intermediate frequency (IF) of the radar. The IF signal, which is a time function $s(t)$, is operated on linearly by the predetection filter, which has an impulse response $h(t)$. The predetection filter output signal $r(t)$ is simply given by the convolution relation

$$r(t) = \int_{-\infty}^{\infty} s(\tau)h(t - \tau) \, d\tau, \tag{5.2-1}$$

where the integration is actually carried out over the duration of $s(t)$. The output signal $r(t)$ then passes to an envelope detector, which contains a square law device and a low-pass filter. The output of the envelope detector is then used to decide something about the radar target, e.g., whether a (point) target is present at a particular range, what the (extended) target radar cross section is at a particular position, or what the most probable range and range rate are for a point target. The problem treated by radar signal processing lies in specification of the IF filter impulse response $h(t)$ that leads to making these determinations effectively, unambiguously, or in some sense optimally. The subject of Fourier-optical processing is concerned with realizing the predetection IF filtering operation in a coherent-optical configuration. In our discussions, it will be assumed that the IF signal $s(t)$ and the filter output $r(t)$ are in electrical form; the Fourier-optical system converts $r(t)$ to optical form, performs the required operation, and converts the output back to electrical form.

The present discussions assume that the radar transmitter uses a pulsed carrier waveform given by

$$g(t) = a(t) \cos[2\pi f_0 t + \phi(t)], \tag{5.2-2}$$

where t is time, $a(t)$ is the amplitude waveform, f_0 is the transmitted carrier frequency, and $\phi(t)$ is the phase modulation of the waveform. If the point target is situated at range R from the radar and moving at a range rate of v, then the received radar waveform resembles $g(t)$ except for a time delay given by

$$t_R = 2R/c, \tag{5.2-3}$$

where c is the speed of light, and a Doppler frequency offset given by

$$f_D = 2f_0 v/c. \tag{5.2-4}$$

When the time-delayed and possibly Doppler-shifted signal is sensed at the radar receiver, it is desired to perform the optimum predetection processing. It has been shown that the optimum predetection processing for a linear system receiving a signal of known form in the presence of additive white Gaussian noise is a matched filter [1, 2]. A matched filter forms the correlation between the expected (noise-free) waveform and the actual noise-corrupted received signal. Thus, if the received signal at the IF is of the form

$$s(t) = g_r(t) + n(t), \tag{5.2-5}$$

where $g_r(t)$ is due to the transmitted radar waveform and $n(t)$ is due to white Gaussian noise, then the predetection filter must be matched to $g_r(t)$. This is accomplished by setting

$$h(t) = g_r(-t). \tag{5.2-6}$$

Substituting Eqs. (5.2-5) and (5.2-6) into Eq. (5.2-1), it is found that the matched filter output is given by

$$r_m(t) = \int_{-\infty}^{\infty} [g_r(\tau) + n(\tau)] g_r(\tau - t) \, d\tau. \tag{5.2-7}$$

Equation (5.2-7) shows that the matched filter output is the sum of two terms: the first term is the autocorrelation of the noise-free version of the received signal waveform, and the second term is the cross-correlation of the expected signal waveform and the additive white Gaussian noise.

The matched filter is optimum in at least three senses [1, 2]. First, the matched filter optimizes the signal-to-noise ratio achieved at one instant of time. Second, when statistical decision theory is applied in choosing between the hypotheses that a radar target is present at a particular position and that only noise exists at the location, it is found that the matched filter forms the desired statistic, which is called the likelihood ratio. Third, from the viewpoint of parameter estimation theory, which regards the radar reception problem as estimating the parameters (such as time delay and Doppler shift) of a signal of known form in the presence of additive white Gaussian noise, the matched filter output supplies the optimum maximum likelihood estimates.

It is instructive to apply the matched filter notion to the problems of reception of signals from targets which cause a simple time delay and a combination of time delay and frequency shift. For a simple radar point target at a range R the transmitted signal waveform given by Eq. (5.2.-2) yields the radar receiver IF signal given by

$$s(t) = g(t - t_R) + n(t)$$
$$= a(t - t_R) \cos[2\pi f_0(t - t_R) + \phi(t - t_R)]. \tag{5.2-8}$$

The output of the matched filter whose impulse response is given by Eq. (5.2.1-6) is of the form

$$r_m(t) = \int_{-\infty}^{\infty} \{a(\tau - t_R) \cos[2\pi f_0(\tau - t_R) + \phi(\tau - t_R)] + n(\tau)\}$$
$$\times a(\tau - t) \cos[2\pi f_0(\tau - t) + \phi(\tau - t)] \, d\tau. \tag{5.2-9}$$

Notice that, when $t = t_R$, the filter output is given by

$$r_m(t_R) = \int_{-\infty}^{\infty} a^2(t) \cos^2[2\pi f_0 t + \phi(t)] \, dt$$

$$+ \int_{-\infty}^{\infty} n(\tau) a(\tau - t_R) \cos[2\pi f_0(\tau - t_R) + \phi(\tau - t_R)] \, d\tau. \qquad (5.2\text{-}10)$$

The time t_R is the instant at which the output signal-to-noise ratio is maximum. The signal component of $r_m(t_R)$ is of the form given in the first term of Eq. (5.2-10), which is recognized as a measure of the energy in the transmitted radar pulse.

The envelope of $r_m(t)$ has significance for the parameter estimation and decision theory approaches to radar reception as well. With regard to parameter estimation theory, it can be shown that the envelope of $r_m(t)$ peaks at the time which equals the maximum likelihood estimate of range delay. With regard to decision theory, the envelope of $r_m(t)$ is the desired likelihood ratio which, when subjected to threshold detection, provides the maximum probability of detecting a target when it is present for a fixed probability of false detections (false alarms) when a target is absent.

Figure 5.2-2 shows a typical matched filter output, as a function of time, after envelope detection. This figure shows that the matched filter output is generally small and noiselike except for a brief interval of time near the time of the true range delay t_R. The position in time of the peak of this function is the optimum estimate, in the senses mentioned above, of the actual range delay. The width of the peaked function is indicative of the resolution with which range delay may be measured; under rather broad conditions, the full width of this function at half-maximum is roughly equal to the inverse of the radar-transmitted signal bandwidth B. The exact location of the peak of the envelope function can be perturbed slightly by the noise term which is included in Eq. (5.2-10). As the relative strength of the noise component of

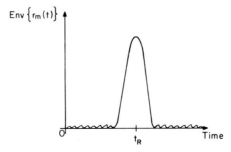

FIG. 5.2-2. Matched filter output.

the output of the predetection filter increases, it becomes more likely that the peak of the envelope function will be displaced from t_R. Consequently, the ultimate accuracy of the estimate of range delay depends upon the signal-to-noise ratio of the radar receiver.

A similar discussion applies to the case when both the target range and Doppler shift are uncertain. In this case, the Doppler-shifted received signal at IF is expressed by the form

$$s(t) = a(t - t_R) \cos[2\pi(f_0 - f_D)(t - t_R) + \phi(t - t_R)] + n(t), \quad (5.2\text{-}11)$$

where this received signal waveform is corrupted by additive white Gaussian noise. In this situation, the receiver must supply a matched filter for each possible target Doppler shift. The IF signal must be fed to all such filters, and each filter must be followed by an envelope detector, as shown in Fig. 5.2-3. Comparison of the outputs of all the envelope dectectors will show that only the channel which is matched to the actual target Doppler shift will produce a characteristic response like the one shown in Fig. 5.2-2. The other outputs will have noiselike signal outputs at the envelope detectors.

It is more difficult to implement a matched filter receiver for an unknown target range and Doppler than for the simple case of an unknown range, because the receiver must have a filter matched to each and every possible Doppler shift which the target may impose on the radar signal. Therefore, in the interest of economy, a practical question arises as to exactly how few Doppler filters can cover a given Doppler uncertainty bandwidth adequately. This question is treated from the viewpoint of Doppler resolution, which considers the allowable frequency mismatch between the actual target Doppler shift and the center frequency of the matched filter. A small frequency mismatch results in a modest decrease in the strength of the peak of the output of the envelope detector, much in the same way that small differences from t_R in Fig. 5.2-2 cause a modest decrease in the strength of the peak of the

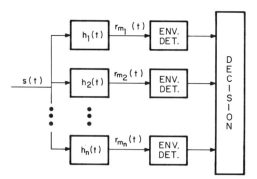

FIG. 5.2-3. Matched filter bank for unknown range delay and Doppler shift.

output of the envelope detector. It can be shown under rather broad conditions that the spacing in center frequency of the Doppler filters may be roughly equal to the inverse of the duration T of the pulsed radar signal. For example, radar signal transmission with duration T equal to 1 msec may be tested for target Doppler by matched filters which have center frequencies spaced by 1 kHz; if the target-induced Doppler shift is possibly located anywhere over a 100-kHz frequency band, then 1000 Doppler-matched filters must be realized in the optimum receiver. The outputs of all envelope detectors in Fig. 5.2-3 are connected to the decision apparatus which determines when a radar return occurs and at what Doppler shift it is detected.

The output of each predetection filter in Fig. 5.2-3 consists of a signal term and a noise term. If we neglect the noise components of the predetection filter outputs, it is important to recognize that the family of outputs consists of the cross-correlation between the IF version of the transmitted waveform and Doppler-shifted versions of this signal. The family of outputs, considered as a function of time shift and frequency shift, is known as the ambiguity function of the transmitted waveform, as defined by Woodward [3]. Formally, the ambiguity function for the complex waveform $g(t)$ is defined by

$$X(t, f) = \int_{-\infty}^{\infty} g(\tau)g^*(t + \tau)e^{j2\pi f\tau} \, d\tau. \qquad (5.2\text{-}12)$$

This function peaks at $t = f = 0$; the output of a real matched filter for $g(t)$ is given by $\text{Re}[X(t, f)]$. In a very real sense, realization of a matched filter receiver which seeks a Doppler- and time-shifted version of $g(t)$ is equivalent to realizing an ambiguity function operator; if the received signal is $f(t)$, then the output of the ambiguity function processor is given by $\text{Re}[X(t, f)]$, where

$$X(t, f) = \int_{-\infty}^{\infty} f(\tau)g^*(t + \tau)e^{j2\pi f\tau} \, d\tau. \qquad (5.2\text{-}13)$$

If $f(t)$ is a delayed (by t_0) and Doppler-shifted (by f_0) version of $g(t)$, then the envelope of the matched filter output of Eq. (5.2-13), $|X(t, f)|$, will reach a maximum at $t = t_0, f = f_0$. The ambiguity function description of matched filtering will be used in Section 5.3.

The ambiguity function is a powerful tool for examining the similarity of the transmitted waveform to time-shifted and frequency-shifted versions of itself. A transmitted waveform which is highly similar to time- and frequency-shifted versions of it will not be particularly valuable for resolving or distinguishing multiple targets spaced closely in range and Doppler, respectively. On the other hand, if a transmitted waveform is selected which, in the cross-correlation sense, does not resemble itself for modest shifts of time and carrier frequency, then it is expected that the output of the matched filter will peak up at a very brief instant of time and only for a particular Doppler frequency, thereby providing excellent resolution capability for these parameters. Consequently, it is recognized that the problem of estimating both range and Doppler delay for radar target returns reduces to that of forming the ambiguity function on the received signals.

In the remaining portion of this section, we consider two widely used radar waveforms that will be discussed in the section on optical processing for range and range rate; these waveforms are known as the linear FM or chirp signal, and the coherent pulse train or pulse-Doppler signal. The linear FM or chirp signal was devised in order to permit radar transmitters that are peak power-limited to transmit high energy, wide bandwidth pulses. The form of the chirp signal [1] is given by

$$g(t) = p_T(t) \cos 2\pi(f_0 t + Bt^2/2T), \tag{5.2-14}$$

where $p_T(t)$ represents a simple amplitude envelope of effective duration T. The argument of the cos function contains terms that are linear and quadratic in time; the quadratic term adds a time-variable frequency to the nominal carrier frequency of the waveform. This is evident in Fig. 5.2-4, which shows that the instantaneous frequency of the chirp waveform varies linearly with time, spanning a total bandwidth B during pulse duration T. The bandwidth of a chirp waveform is roughly equal to B, with minor variations depending upon the precise shape of the amplitude envelope. The autocorrelation of the chirp waveform in time depends on the product BT (see Fig. 5.2-5 for $BT = 10$ and 50) but generally contains a narrow main lobe of half-amplitude width equal to $1/B$. The ratio of the initial chirp waveform duration to the output of the envelope detector is thus equal to BT, and the shortening of the duration of the transmitted pulse by the matched filter process is usually referred to as pulse compression; BT is known as the pulse compression ratio. It will be seen in Section 5.3 that a particularly simple optical configuration can be used for pulse compression for linear FM signals.

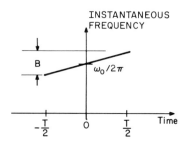

FIG. 5.2-4. Chirp frequency versus time.

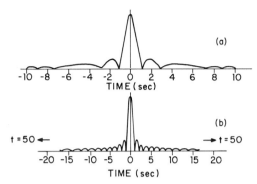

FIG. 5.2-5. Linear FM autocorrelation waveforms. (a)$BT = 10$. (b) $BT = 50$.

The basic pulse-Doppler radar waveform [4] consists of a sequence of equally spaced, pulsed samples of a coherent, continuously running carrier frequency, which may be expressed as

$$g(t) = \sum_{n=0}^{M} p_T(t - n\tau) \cos 2\pi f_0 t, \qquad (5.2\text{-}15)$$

where T is the pulse duration, τ is the period of the equally spaced pulses, and $N\tau$ is the duration of the burst of $(N + 1)$ pulses. The pulse-Doppler waveform provides range delay measurement resolution of T and Doppler shift resolution of $1/N\tau$. Although such resolutions are available for single-point targets, the periodicity of the amplitude envelope leads to ambiguous radar returns for multiple-point or extended targets with a range delay extent comparable to or larger than τ, or Doppler shift spreads comparable to or larger than $1/\tau$. This ambiguity problem may be appreciated by inspecting Fig. 5.2-6, which illustrates the key features of the waveform. The temporal autocorrelation, shown in Fig. 5.2-6a, is seen to be contain nearly equal-amplitude resolution peaks of width T spaced by time τ, which make it difficult to distinguish between range delays of t_R and $t_R + \tau$. Ambiguity in Doppler shift may be anticipated by inspecting the power spectrum in Fig. 5.2-6b, where it is noted that the structure of nearly equal-amplitude spectral lines of width $1/N\tau$ spaced by $1/\tau$ makes it difficult to distinguish between a Doppler shift of f_D and $f_D + 1/\tau$. Since the unambiguous range delay extent of the pulse-Doppler waveform is τ, and the range delay resolution is T, there are only τ/T distinct range delays or range resolution cells associated with this waveform. Similarly, the unambiguous Doppler extent of $1/\tau$, coupled with the Doppler resolution $1/N\tau$, leads to N Doppler resolution cells for the waveform. This situation is sketched in Fig. 5.2-7, which easily evokes the interpretation that the pulse-Doppler waveform seeks to deter-

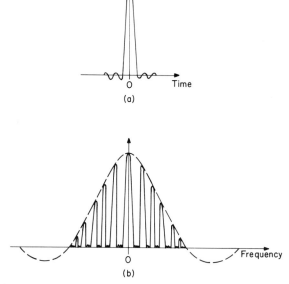

FIG. 5.2-6. Pulse-Doppler autocorrelation (a) and power spectrum (b).

mine the target position unambiguously within $N\tau/T$ range-Doppler resolution cells.

The ideal receiver for the pulse-Doppler waveform [4] may be best understood as a combination of matched filtering [the filter is matched to $p_T(t)$] and Fourier analysis of each range cell of interest. Figure 5.2-8 is a diagram of such a receiver, which is seen to gate the sequence of $(N + 1)$ returns from each of τ/T range cells into a separate N-channel Doppler

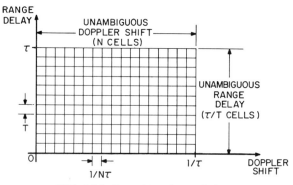

FIG. 5.2-7. Range-Doppler resolution.

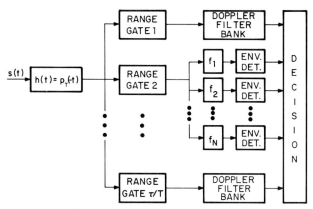

FIG. 5.2-8. Pulse-Doppler receiver.

filter bank; this design assumes that the actual change in target range delay during the burst duration $N\tau$ is small compared with the range delay resolution cell size T. These notions of resolution and ambiguity apply as well to a coherent, regularly spaced sequence of modulated pulses, such as chirps; in this case, the matched filter would be tailored to the single-pulse waveform, and the range delay resolution T would correspond to the compressed pulse width. It will be the goal of the optical processor to provide the equivalent of an N-channel Doppler frequency analysis for each range cell of interest for pulse-Doppler waveforms.

B. Processing for Phased Array Antennas

Phased array antennas are employed to transmit and receive radar signals from an extended angular segment of space without mechanical motion of the aperture. This section is concerned only with phased arrays composed of equally spaced antenna elements lying in a plane (planar array) or along a line (linear array) and the problem of determining the angular direction from which the radar return is being received. Consider the linear phased array and receiver in Fig. 5.2-9, where it is seen that the $(N + 1)$ element linear array consists of elements separated by distance d. The diagram and calculations will assume that the target range is large enough so that only the plane wave component of the radar return needs to be calculated. If the radar transmission to the target is of the form

$$g(t) = p_T(t) \cos 2\pi f_0 t, \tag{5.2-16}$$

then the received signal in the nth element of the linear array is given by

$$s_n(t) = p_T(t - t_R) \cos 2\pi f_0 [t - t_R - (nd \sin \theta)/c]. \tag{5.2-17}$$

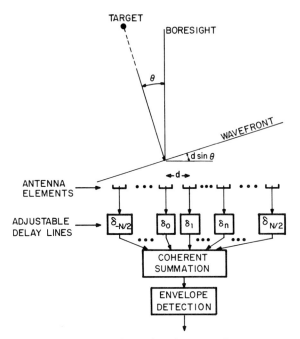

FIG. 5.2-9. Linear phased array receiver.

This formula assumes that the differential signal delay between adjacent elements of the linear array is small compared with the radar pulse duration T. Consequently, the differential delay time shows only in the argument of the cos function; the phase difference between adjacent elements in the linear array is proportional to $d \sin \theta$. The receiver processing in Fig. 5.2-9 consists of an adjustable delay line in each channel followed by a simple coherent summation of the signals in each channel. It is quite apparent that, if the delay is adjusted in each channel precisely to cancel out the differential phase shift caused by the wave front tilt, then the coherent summation output will be at a maximum. This is the principle behind receiver beam forming in a linear phased array; the receiver controller determines what angle in the antenna field of view is desired to be addressed by the receiver, adjusts the delay lines in each channel to cancel precisely the differential phase shift in a wave front arriving from that angle, and then forms a coherent summation of the resulting returns. In a pulsed radar, the delay lines must be adjusted before the radar return begins to appear at the antenna elements. When the linear phased array receiver addresses more than one angular position simultaneously, the signal generated in each antenna element is divided among a number of delay lines equal to the number of beams to be formed, the delay lines in each element and throughout the array are adjusted to compensate for the phase differences for the various beam angles, and the resulting signals are coherently summed separately for each beam direction.

Quite analogously with the ambiguity problem discussed above for pulse-Doppler waveforms, there is a finite number of distinctive angular resolution elements available from a linear array with a finite number of antenna elements. This can be appreciated in Fig. 5.2-10 which shows the response of a linear array forming a single beam for a distant point target. The comblike structure under the single-element response envelope can be shown to be expressed by the form

$$v(\theta) = \sin(N\gamma) \cdot e(\theta)/(N \sin \gamma), \qquad (5.2\text{-}18)$$

where $\gamma = N\pi(f_0 d/c) \sin \theta$, and $e(\theta)$ is the single-element beam pattern. The key features of the angular response of a linear phased array are the periodically spaced grating lobes (separated by λ/d) and the lobe width of λ/Nd; this leads to an unambigious angular position coverage of N beam positions. Consequently, it is necessary to have N possible delays implementable in each channel of the linear array processor.

The reciprocity of propagation of electromagnetic waves leads to a rather interesting way to regard the problem of processing phased array data. Reciprocity suggests that it should be possible to determine the nature of the source which creates the received wave by simply retransmitting from the same antenna elements the received waves and measuring how the transmitted wave forms beams at long distances. It will be shown in Section 5.4 that optical processors configured for phased array antenna beam forming can be regarded as devices which create an optical wave front which is the analog of the wave front received by the antenna elements and then propagates that wave front back to the source.

In the case of planar array processing, it is only necessary to extend the notions developed above. Imagine that a planar array consists of a rectangular array of elements separated by d_x and d_y, and that the array size is $(M + 1)$ elements in the x-direction and $(N + 1)$ elements in the y-direction.

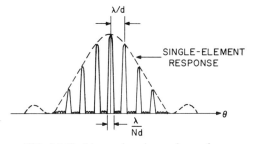

FIG. 5.2-10. Linear phased array beam shape.

Once again, the array receiver consists of antenna elements followed by variable delay lines and coherent summation operations prior to signal power detection. A direct extension of the notions of resolution discussed above shows that the planar phased array can, at most, distinguish unambiguously among M angular positions in the x-direction and N angular beam positions in the y-direction; this implies that in two-dimensional angular space there are MN distinguishable target positions. As suggested in Fig. 5.2-10, the focused lobe width and the grating lobe spacings depend on the interelement separation and the wavelength of operation of the radar. The receiver signal processing must now supply delays in each antenna channel in order to provide the signals required for coherent summation all of the possible beam directions. Once again, this processing is equivalent to reconstructing the target by transmitting the detected wave front back toward the target space. It will be seen later that optical processors for beam forming in two-dimensional planar phased arrays form an optical analog of the received radiofrequency wave front and propagate it to the far field, where the target distribution is formed.

C. Synthetic Aperture Radar Processing

Synthetic aperture radar obtains fine-angular resolution images by monitoring a sequence of pulsed radar returns obtained while there is relative motion between the target and the radar antenna. Typically, the relative motion between target and radar is linear or simply rotational, and the synthetic aperture receiver processing forms a linear superposition of the sequence of pulsed returns in order to provide a target image well resolved in angle. The radar pulse duration or bandwidth determines resolution in the second direction, that of range. The synthetic aperture radar is intended to provide a radar reflectivity map of an extended scatterer distribution which is resolved in the range and line-of-flight directions.

The angular resolution realized by the synthetic aperture in the linear translational motion case is equivalent to that obtained by a linear antenna array whose length equals twice the line-of-flight distance traveled during the integration period of the receiver signal processing.

Consider the synthetic aperture configuration shown in Fig. 5.2-11. The aircraft carries an antenna which illuminates a wide swath of the nominally flat ground in a band parallel to the flight of the aircraft; it is sufficient to discuss the radar signal processing by studying the response of the radar to a single scattering point on the ground, and by superposition extending the results to extended target situations. In particular, assume that the point target is located at position x_1 along the ground path, at a range r_1 from the aircraft flight path, which is parallel to the x-direction. the synthetic aperture

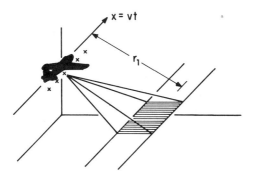

FIG. 5.2-11. Synthetic aperture radar.

radar transmitter sends out coherent pulses at regular intervals of time τ. The nth pulse is given by

$$g_n(t) = p_T(t - n\tau) \cos 2\pi f_0 t. \tag{5.2-19}$$

If the aircraft travels at constant speed v along the x-direction, then at time t the radar range to the scattering point is given, within second order, by the form

$$r_1 + \frac{(x - x_1)^2}{2r_1} = r_1 + \frac{x_1^2}{2r_1} - \frac{2x_1 vt - v^2 t^2}{2r_1}. \tag{5.2-20}$$

Consequently, the sequence of return pulses from N pulse transmissions can be expressed by

$$s(t) \cong \cos\left\{2\pi\left[\left(f_0 + \frac{x_1 v}{cr_1}\right)t - \frac{v^2 t^2}{2cr_1}\right] + \theta\right\} \sum_{n=-(N-1)/2}^{(N-1)/2} p_T\left(t - n\tau - \frac{r_1}{c}\right),$$

where

$$\theta = -2\pi f_0(r_1/c + x_1^2/2r_1 c). \tag{5.2-21}$$

This formulation assumes that the quadratic term in the expression for range delay is small compared with the width of an individual pulse T. Inspection of the argument of the cos function shows that the phase modulation resulting from the point target leads to a function very similar to the chirp waveform of Eq. (5.2-14). In the present case, however, the radar return is Doppler-shifted by an amount which depends on the target range and the along-track position of the point. In addition, the quadratic phase (chirp) modulation depends only upon the known aircraft speed and the measured range at the point of closest approach between the aircraft track and the target point.

The azimuth (x-direction) position of the target point is reclaimed from the phase information carried on the cos function. The required processing in the azimuth direction can be regarded as a matched filter operation or a cross-correlation operation. In this situation, the filter must match both the Doppler shifts and the quadratic phase components. It will be seen in Section 5.3 that a coherent-optical system is strikingly well suited to realizing the required filtering operation in a simple and elegant fashion.

In general, the scene illuminated by the synthetic aperture radar transmitter contains a large number of scatterers distributed over the "plane" of the scene. The result of matched filter processing of the return obtained from this extended target distribution is a map of radar scattering cross-section distribution in the target plane. The resolution of the map in the ground range direction is controlled simply by the pulse width T, and the resolution in the azimuth direction is controlled by the synthetic aperture length $nv\tau$.

5.3 OPTICAL PROCESSORS FOR RADAR SIGNALS

A. Real-Time Light Modulators

Radar signal processing is perhaps distinguished from other applications of Fourier optics by the need, under many circumstances, to process the radar signal in real time. Consequently, real-time devices have been applied to the problem of converting an IF electrical signal instantaneously into a spatial modulation of an optical wave front for processing in a Fourier-optical configuration. Optical processors that do not require strictly real-time operation often employ photographic film as a medium for storing the signal in a format useful for modulating an optical wave front. This section is devoted to a discussion of two particularly useful classes of real-time devices for converting temporally modulated carrier waveforms into spatially modulated optical wave fronts.

Perhaps the simplest device, conceptually, for converting time functions into space functions is the acousto-optic (AO) light modulator [5, 6], which is shown schematically in Fig. 5.3-1. The AO light modulator uses a piezo-electric transducer to couple the electrical signal into a traveling acoustic wave in a transparent delay medium. The dimensions of the transducer surface are selected so that the acoustic disturbance propagates essentially in a geometric column; an absorber is placed at the remote end of the AO light modulator to prevent reflection of the incident acoustic wave. The sound wave travels through the light modulator in the negative x-direction at speed s, and the aperture of the modulator is limited for practical reasons to length P; note that the AO light modulator can only store a signal in its aperture of duration P/s. Practical reasons for the limits on the modulator length are acoustic wave absorption and dispersion. The bandwidth B of the electrical signal coupled into the medium is limited essentially by the piezoelectric transducer and matching network. Suppose that a simple single-frequency

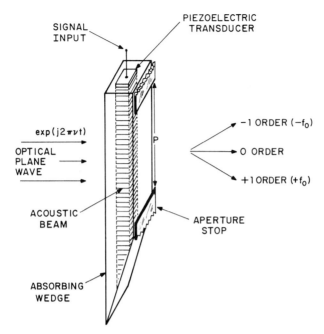

FIG. 5.3-1. Acousto-optic light modulator.

sine wave is coupled into the AO light modulator by the transducer. The traveling acoustic beam would perturb the local optical index of refraction. Therefore, an optical plane wave passing through the delay medium would be phase-modulated differently at different points across the wave front when it emerges from the modulator. If the AO light modulator is driven by the electric signal expressed in Eq. (5.2-2), and if a unit amplitude optical plane wave is incident on the modulator, then the light amplitude emerging from the modulator is of the form

$$\exp(j2\pi vt)\exp[-j\psi(t, x)]\,\mathrm{rect}(x/P)$$

$$= \exp(j2\pi vt)[1 - j\psi(t, x)]\,\mathrm{rect}(x/P), \qquad \text{if} \quad \psi_0 \ll 1, \quad (5.3\text{-}1)$$

where v is the optical carrier frequency, and

$$\psi(t, x) = \psi_0 a(t + x/s)\cos[2\pi f_0(t + x/s) + \phi(t + x/s)].$$

Note that functional dependences on the (transverse) y-direction in the modulator have been suppressed for simplicity in this formula. This expression shows that for small peak phase modulation ψ_0, the AO light modulator imposes a traveling replica of a P/s-second segment of the input electric signal on the transmitted optical wave front. The temporal carrier frequency of the input signal f_0 is associated with a spatial carrier frequency f_0/s in the modulated wave front. Essentially, when ψ_0 is small, the AO light modulator imposes a spatial modulation term that is in optical quadrature with the incident plane wave.

This, and other interesting properties of the AO light modulator can be seen by expanding Eq. (5.3-1) to the form given by

$$\exp(j2\pi vt)[1 - j\psi(t, x)]\,\text{rect}(x/P)$$

$$= \text{rect}(x/P)\{\exp(j2\pi vt)$$

$$- \tfrac{1}{2}[j\psi_0 a(t + x/s)]\exp j[2\pi(v + f_0)t + 2\pi f_0 x/s + \phi(t + x/s)]$$

$$- \tfrac{1}{2}[j\psi_0 a(t + x/s)]\exp j[2\pi(v - f_0)t - 2\pi f_0 x/s - \phi(t + x/s)]\}.$$

$$(5.3\text{-}2)$$

When the AO modulator is illuminated with wavelength λ, the second and third terms of Eq. (5.3-2) propagate at angles $\pm\lambda f_0/s$ with respect to the direction of the first term; the second and third terms are, respectively, the $+1$ and -1 diffraction orders associated with the phase grating in the AO light modulator and will be focused, respectively, below and above the optical axis in the back focal plane of a Fourier-transform lens placed after the light modulator. Inspection of the arguments of the two diffraction terms also shows that the optical carrier frequency of the ±1 order is shifted by $\pm f_0$. This implies that, if a portion of the incident light wave is heterodyned with either diffracted first-order wave, then a beat frequency f_0 will be observed in the photodetection process. Heterodyne detection can make available carrier phase information in the detection of the outputs of optical radar signal processors.

The light valve family of coherent optical wave front modulators employs electrostatically induced thickness variations in a transparent fluid film or the electro-optic effect in a transparent crystal (such as deuterated potassium dihydrogen phosphate, DKDP, or bismuth silicon oxide). Figure 5.3-2 shows an electron beam-addressed light valve [7]. The electro-optic plate has a transparent electrode on one face, may be addressed by a write gun and an erase gun on the other face, and is typically maintained in a vacuum enclosure near the phase transition temperature ($-52°C$ for DKDP) in order to maximize the electro-optic interaction at high resolution. The write gun current is modulated with the time-varying signal as the electron beam sweeps along a line on the crystal. This deposits a static charge distribution which perturbs the crystal index ellipsoid locally. Either phase- or amplitude-modulated wave fronts can be obtained by passing linearly polarized mono-chromatic waves through the crystal by way of the transparent windows. If the output light is passed through a crossed polarizer, amplitude modulation

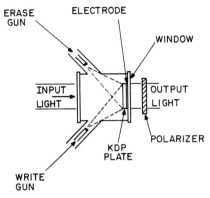

FIG. 5.3-2. Electro-optic light valve.

may be obtained; when the output polarizer is parallel to the input polarization, phase modulation is obtained. Although transition-temperature
operation retains the deposited charge distribution for long periods of time,
use of the erase gun to discharge the pattern by secondary emission permits
real-time operation. Noble [8] has used an electron beam-addressed fluid
film in a vacuum tube to achieve real-time phase modulation of optical wave
fronts at standard video resolutions and frame rates; phase modulation of
transmitted light results when the charge deposited by the electron beam on
the fluid film causes localized relief distortion of the fluid surface.

Suppose that a light valve is driven with the real-time IF signal of Eq.
(5.2-2). If the electron beam writes along the negative x-direction, traveling at
speed s, and the aperture size of the light valve is P, then a P/s-second segment
of the signal will be borne in the charge pattern. Succeeding time segments of
the signal may be written adjacent to the first one, along a parallel path in
the x-direction; although a single electron beam trace stores only P/s seconds
of signal, it is evident that N successive traces can store NP/s seconds of
signal. The maximum bandwidth of the signal that can be stored on the face
of the light valve is limited by the spatial frequency response of the electro-
optic crystal. Electro-optic light valves have been operated with 1000
pixels/line and 1000 lines, at a rate of 30 frames/sec.

When driven by the signal in Eq. (5.3-2) and used in the phase modulation
mode, the form of the wave front transmitted by the light valve is quite
similar to the form obtained for the acousto-optic light modulator, except
that the wave front modulation does not travel in time. Consequently, the
$+1$ grating order may be expressed as

$$j(\psi_0/2) \, \mathrm{rect}(x/P)a(x/s) \exp j[2\pi f_0 x/s + \phi(x/s)] \qquad (5.3\text{-}3)$$

when illuminated by a unit amplitude monochromatic plane wave. The expression for the -1 order has $-x$ wherever x occurs in this equation. The strength of the modulation depends on the charge density deposited on the crystal, the electric field induced in the crystal by this charge, and the electro-optic coefficient of the crystal. When the light valve is operated in the amplitude mode, the forms of the diffracted orders are the same as in the phase modulation case, except that the orders are in phase (not phase quadrature) with the zero order.

Closely related to electron beam-addressed light valves are optically addressed light valves, which modulate a coherent-optical wave front via the static charge distribution generated photoelectrically on one face of an electro-optic crystal when it is exposed to an optical pattern representing the radar signal of interest. When the charge distribution has been formed, amplitude- or phase-modulated waves couple the radar signal into the Fourier-optical system. In the PROM device [9], the crystal plate (bismuth silicon oxide) is operated at room temperature and exhibits the photoelectric effect in the blue region of the spectrum; the optical pattern representing the input radar signal may be created by modulating the intensity of a scanned spot of light with the radar IF signal. The stored signal is read with coherent light in the red region of the spectrum, where bismuth silicon oxide exhibits the electro-optic effect. The photo-DKDP light valve [7], which houses a transparent electro-optic crystal in a vacuum enclosure near the transition temperature, uses a photoconductive layer sandwiched between two transparent electrodes on one face to generate the static charge pattern in response to incident light; these electrodes control when the modulator is exposed, read out, and erased. The photo-DKDP is read out in reflection and is erased by flooding the photoconductor with a uniform light pattern. Radar signals may be written into the photo-DKDP light valve in the same way they are written into the PROM. Schneeberger et al. [10] have recently fabricated a photoaddressed light valve based on room temperature oil film deformation by variation of surface tension; although it is sensitive only to infrared light, it appears to promise resolution and speed comparable to other light valves in a compact, low cost package.

B. Optical Processors for Radar Range and Range Rate

Optimum radar reception requires the Fourier-optical processor to correlate the received radar IF signal with the waveform expected to be received, thereby producing the cross-correlation between the (noisy) received and the sought-after signal, as discussed in Section 5.2. The real-time optical processor of Fig. 5.3-3 performs this matched filtering operation and detects the envelope of the matched- filter output. The AO light modulator is driven with the noisy IF signal, and the amplitude transmittance of the signal

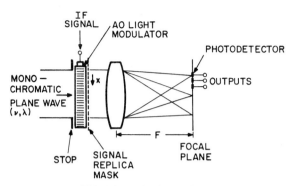

FIG. 5.3-3. Simple correlator.

replica mask contains a spatial replica of the sought-after signal. The length of the AO light modulator P equals the length of the signal on the replica mask sT, so that at any time a T-second sample of the IF signal modulates the incident plane wave. If the signal replica mask is close enough to the plane of the AO light modulator, then diffraction effects are insignificant and the mask is illuminated by the zero-order and first-order components discussed above. If the mask transmittance is given by

$$\text{rect}(x/P)\{b_0 + ba(x/s)\cos[2\pi(f_0 + f_B)x/s + \phi(x/s)]\}, \qquad (5.3\text{-}4)$$

then the light amplitude in the focal plane of the lens is given by

$$
\begin{aligned}
\int_{-P/2}^{P/2} & \{b_0 + ba(x/s)\cos[2\pi(f_0 + f_B)x/s + \phi(x/s)] \\
& - j\psi_0 b_0 a(t - t_0 + x/s)\cos[2\pi(f_0 + f_D)(t - t_0 + x/s) \\
& + \phi(t - t_0 + x/s)] + n(t + x/s)\}e^{-j2\pi xx'/\lambda F}\,dx \\
& - j\psi_0 b \int_{-P/2}^{P/2} a(x/s)\cos[2\pi(f_0 + f_B)x/s + \phi(x/s)] \\
& \times [a(t - t_0 + x/s)\cos[2\pi(f_0 + f_D)(t - t_0 + x/s) \\
& + \phi(t - t_0 + x/s)] + n(t + x/s)]e^{-j2\pi xx'/\lambda F}\,dx. \qquad (5.3\text{-}5)
\end{aligned}
$$

The first integral contains the sum of the spatial Fourier transforms of four terms which tend to be strong at different points in the Fourier plane. The first term creates a sinc function centered at $x' = 0$. The second and third terms form the spatial Fourier transforms of the signal replica on the mask and the time segment of the signal in the AO light modulator; the carrier frequencies of these components cause these transforms to concentrate around the positions $x = \pm(f_0 + f_B)\lambda F/s$ and $x' = \pm f_0\lambda F/s$, respectively. The transform of the last component of the first integral, which is a white noise component, is spread over the same frequency band and same region of x' as the received signal.

The second integral of Eq. (5.3-5) contains the matched filter response to signal plus noise; the form of this integral is the same as the ambiguity functions of Eqs. (5.2-12) and (5.2-13). Consequently, this processor realizes the optimum range-Doppler predetection filter. For the radar parameters of the present example, the modulus of the matched filter term (in the absence of noise) reaches a peak at time $t = t_0$ and position $x' = \pm(f_B + f_D)\lambda F/s$; the light amplitude arriving adjacent to these points provides the frequency-mismatched filter response. The frequency offset f_B on the mask signal may be selected to ensure that, even for zero Doppler shift, the matched filter response falls at a point in the Fourier plane which is separate from all the other terms that may cause unwanted interference; other ways to separate the desired term include polarization discrimination for shear mode AO light modulators and heterodyne detection near $x' = 0$, between the matched filter term and the zero-order light.

The output current of the photodetector at the point where the matched filter response forms is proportional to the squared modulus of the light amplitude; this quantity, which is the squared output of the envelope detector in Fig. 5.2-1, may be used for target detection and parameter estimation. The other detector element outputs are the filter outputs for frequency-mismatched Doppler channels. Neighboring Doppler channels are separated in temporal frequency by $1/T$, and in the x-plane by distance $\lambda F/sT$.

The matched filter shown in Fig. 5.3-3 was implemented for chirp and pseudorandom phase-modulated signals without Doppler shifts. The spatial light modulator was an AO device using distilled water between optical flats as the transparent delay medium and a quartz piezoelectric transducer to couple the IF signal into the optical system. The matched filter signals were stored on specially prepared optical transmission gratings; the received IF signal was generated by detecting with a photomultiplier tube the light intensity transmitted by the grating when it was scanned by a flying-spot scanner. Figure 5.3-4 shows the received waveforms and the optical matched filter outputs near the time of signal correlation obtained by Arm *et al.*, [11], who worked with 10-MHz-bandwidth, 100-μsec-duration pulses; Slobodin [12] obtained comparable results with 60-μsec-duration, 2-MHz-bandwidth pulses.

Several researchers have considered the effects of Fresnel diffraction, which tends to blur the AO modulator output arriving on the reference mask of Fig. 5.3-3. Meltz and Maloney [13] analyzed the sensitivity of the strength of the correlation peak to separation of the AO light modulator and replica and to signal parameters, especially the fractional modulation bandwidth.

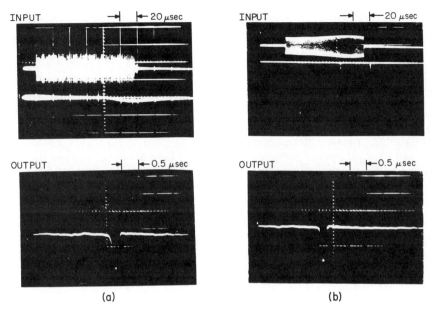

FIG. 5.3-4. Matched filter results: (a) pseudorandom signal and (b) chirp signal.

They considered the heterodyne-detected output available at the origin of the frequency plane when $f_B = 0$ and there are no Doppler shifts. Generally, they found that this output consisted of a pair of symmetrically time-shifted matched filter responses possessing negligible quadratic phase distortion. They confirmed their analysis, which included specific calculations of response for zero-Doppler shift, single-frequency rectangular pulses and for chirped pulses, with laboratory experiments which measured the heterodyne-detected outputs at the origin of the frequency plane. Atzeni [14] analyzed the chirp signal matched filter response with direct detection for various AO modulator-to-replica spacings and noted that in the near diffraction field the correlation of Fresnel images formed in a nearly periodic succession of planes parallel to the AO light modulator was very close to the ideal signal correlation function; the number of these planes and the quality of the correlator response were found to depend essentially on the signal fractional bandwidth. Stark et al. [15] proposed to accommodate Fresnel diffraction effects by modifying the signal stored on the mask to match the expected Fresnel diffraction pattern. Although target range and Doppler were to be determined from the time and position of the correlation output, the authors found these parameters to be coupled for a coherent-pulse burst waveform in a fashion similar to the range-Doppler estimates of a chirp signal; they proposed a modified optical configuration to exploit this peculiarity but envisioned potentially serious challenges to implementing the configuration.

The three-lens correlator configuration of Fig. 5.3-5 permits simple spatial filtering in the x'-plane to eliminate all potentially degrading signals and does away with Fresnel diffraction problems as well [16]; the penalties associated with this configuration are the need for two additional high quality lenses and the burden of keeping the system aligned. The three lenses are separated by their common focal length F from all planes of interest. Lens 1 separates the three diffraction orders in the x'_1-plane, and the bandpass filter allows the $+1$ diffraction order to be inversely transformed by lens 2 and to fall on the replica mask in the x-plane. Lens 3 functions similarly to the single lens of Fig. 5.3-3, and the optimum receiver outputs are available for the photodetector array in the x'-plane. The replica mask used in this configuration differs from the previous one because an offset frequency f_B is not needed to separate the various terms in the spatial frequency plane. The light amplitude falling on the x'-plane in Fig. 5.3-5 can be shown from Eqs. (5.3-2) and (5.3-1) to be of the same form as the optimum predetection filter; in fact, if the light falling on the detector plane is coherently detected by means of optical heterodyning with a portion of the unmodulated light which illuminates this configuration, then the outputs of the detectors will correspond totally to the outputs of the bank of predetection filters shown in Fig. 5.2-3 [17].

Another configuration for matched filtering, which is useful for zero-Doppler returns and a limited range interval of interest, is the AO time integrating correlator of Fig. 5.3-6 [18]. In this system, which uses no replica mask, the radar IF signal is fed into the AO modulator, and the temporal reference signal being sought is fed into the laser intensity modulator; the intensity modulator output waveform in time consists of a constant optical bias upon which is superposed the temporal reference signal. The light transmitted by the AO modulator at every point in the aperture includes the product of the temporally varying illumination beam and the local component of the radar return signal. The two lenses and spatial filter reimage the

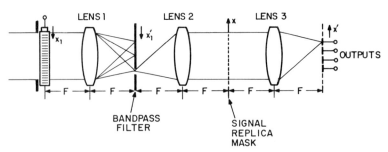

FIG. 5.3-5. Three-lens correlator.

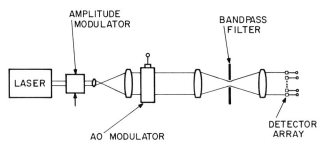

FIG. 5.3-6. AO time-integrating correlator.

first diffraction order of this output distribution onto the array of photo-detector cells; photodetector element sizes are chosen to match the linear extent of a single-range cell in the image of the AO modulator aperture. It can be shown that, if the output of each individual cell is integrated over the time for which the reference waveform exists, then the result is equivalent to the matched filter output for that particular range cell. This configuration offers the advantage of handling very long duration reference signals without the need for a very long storage time AO modulator. On the other hand, the length of the range interval which can be examined is limited by the length of time that can be stored in the AO modulator. Apparently, however, this configuration has been found to be attractive for implementation with integrated optics. In a recent experiment [19], a temporally modulated laser illuminated a wideband surface acoustic wave transducer mounted on a lithium niobate optical waveguide structure which included lens and spatial filters. This experiment demonstrated matched filtering for signals with 20-MHz-bandwidth, 7-msec-duration radar signals. Workers in integrated optics argue that they are preferable to bulk Fourier optics because of reduced size, weight, power consumption, cost, and susceptibility to environmental effects.

The Fourier-optical matched filtering of a chirp radar waveform with an AO light modulator may do away with the signal replica mask by taking advantage of the quadratic phase modulation imposed on the transmitted optical wave front. Gerig and Montague [20] note that one of the first-order diffraction wave fronts emerging from the AO light modulator (when a chirp signal lies within) is inherently convergent in a way that is analogous to the action of a one-dimensional Fresnel zone plate. In particular, if the chirp signal bandwidth and duration are B and T, respectively, and if the speed of sound in the light modulator is s, then the focal length of the convergent wave is $s^2 T/\lambda B$, where λ is the wavelength of the monochromatic illuminating light. The experimental configuration proposed by Gerig and Montague was adapted to the layout shown in Fig. 5.3-7 by Collins et al., [21]; they

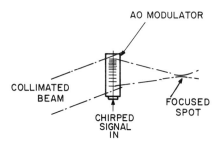

FIG. 5.3-7. Chirp filter with collimated illumination.

compressed a 240-nsec-duration, 75-MHz-bandwidth chirp pulse down to 15 nsec with a sapphire AO light modulator operated at 1.4-GHz carrier frequency. (The off-normal incidence angle of the collimated illumination beam in the figure allowed more efficient AO coupling over the full signal bandwidth.) The measured and predicted focal lengths of the chirp pulse were 17.2 cm, and the envelope of the compressed pulse was detected by a wideband photomultiplier placed behind a 65-μm slit situated at the expected focal point. These writers also noted that this configuration could be adjusted simply to accommodate various chirp rates (B/T) and center frequencies; they also suggested that the amplitude envelope of the chirp pulse could be matched by controlling the distribution of intensity across the illuminating laser beam. Nearly simultaneously, Schulz et al. [22] exploited the Fresnel zone plate property for chirp signals in a somewhat different configuration; Fig. 5.3-8 shows that they not only inclined the illuminating wave but also used divergent illumination to increase the AO efficiency further. Using an yttrium gallium garnet AO modulator operated at a center frequency of 1.16 GHz, and a wideband photomultiplier behind a 25-μm slit, they demonstrated the compression of a 60-MHz-bandwidth, 2-μsec pulse down to 18 nsec. More recently, Tsai et al. [23] adapted the arrangement of

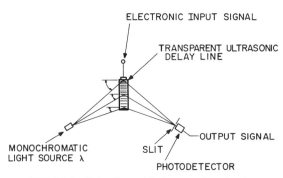

FIG. 5.3-8. Chirp filter with divergent illumination.

Fig. 5.3-7 to an integrated optics configuration, where they demonstrated the compression of a 2.5-μsec-duration, 130-MHz-bandwidth chirp pulse down to 83 msec.

Processing of the returns from a pulse-Doppler waveform requires the Fourier analysis of the sequence of returns obtained from each range resolution cell within the range coverage of interest. Typically, this processing enables the receiver to distinguish among targets which may occupy the same range cell or Doppler cell simultaneously. Figure 5.3-9 shows two wave front modulation formats which may be used for the required optical processing. The space domain format displays the return from the first pulse in the sequence in the x-direction at the left edge of the format; the second pulse return is placed immediately to the right, and so on, with equal pulse-to-pulse spacing in the y-direction. One-dimensional Fourier transformation of the space domain format in the y-direction provides Fourier analysis of each range cell independently. The alternative format, employed by King and Arm [24], imposes the frequency spectrum of the range extended return from each pulse in the burst on the optical wave front in the processor. In this format, a simple two-dimensional Fourier transformation provides the required range-Doppler analysis. Figure 5.3-10 shows the reconstruction arrangement for the frequency domain format. The detailed sketch of the

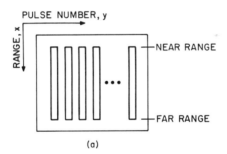

(a)

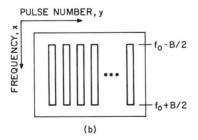

(b)

FIG. 5.3-9. Pulse-Doppler formats: (a) space–domain format and (b) frequency–domain format.

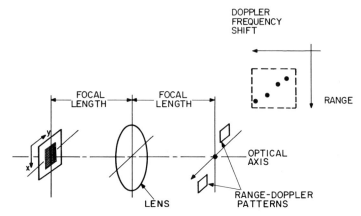

FIG. 5.3-10. Processing configuration for frequency domain pulse-Doppler format.

output plane indicates four point targets which are observed to occupy distinct range cells and Doppler frequency cells.

This processor was demonstrated [24] for simulated pulse-Doppler radar returns. The simulated transmitted radar waveform was a train of fifty 1-μsec pulses with a 1-sec interpulse period. Five point-target returns were simulated, three in adjacent range cells and the remaining two spaced by 1 μsec as shown in Fig. 5.3-11. Note that each of the five simulated target returns included a distinctive simulated Doppler shift. The sequence of 50 returns was recorded in a special holographic recorder [25] as a photographic transparency in the frequency domain format of Fig. 5.3-9b. The exposed and developed photographic film record was then reconstructed in the configuration of Fig. 5.3-10 to provide the Fourier plane patterns shown in Fig. 5.3-12. The upper photograph in this figure shows the zero-order light

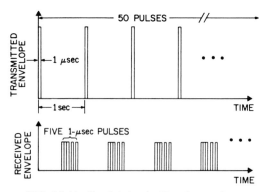

FIG. 5.3-11. Simulated pulse-Doppler envelopes.

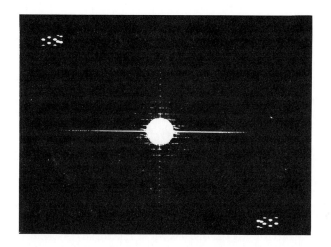

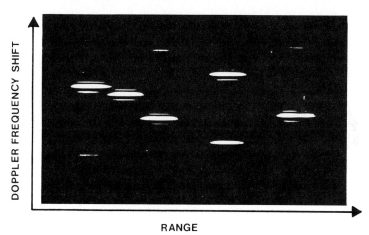

RANGE

FIG. 5.3-12. Optically processed pulse-Doppler outputs.

and the two first-order diffraction patterns; the lower figure shows the detailed structure of one of the diffraction orders. Note that in the range direction, starting from the left, the three targets are lying in adjacent range cells; the next two targets are each spaced by 1 msec. In the Doppler frequency shift direction, note that the targets are separated and that the Doppler frequency ambiguity at 1 Hz is also reconstructed so that frequency ambiguities are visible in this pattern. In practice, it would be required to mask out the ambiguous Doppler frequency region for processing. Had the pulse-Doppler returns been recorded on a light valve rather than photographic film, these results could have been obtained in real time; however, real-time processing of pulse-Doppler returns for an extended range gate has not been performed to our knowledge. Nevertheless a promising approach to real-time processing of pulse-Doppler and other radar waveforms is discussed in Chapter 7. There the interaction between coherent light and surface acoustic waves is used to realize matched filter operations and the generation of ambiguity functions.

C. Optical Processors for Phased Array Antennas

The Fourier-optical processor for phased array antennas determines target angular position by reconstructing the received radar wave front at an optical wavelength and transmitting it to the far field. Perhaps the conceptually simplest real-time processing configuration couples the $(N + 1)$ signals from the typical linear array in Fig. 5.2-9 to an $(N + 1)$-channel AO light modulator resembling the one shown in Fig. 5.3-13. As a consequence of the AO modulation process, the microwave antenna wavelength λ_m is simulated with optical illumination wavelength λ, the linear array interelement spacing d is simulated by acoustic channel-to-channel spacing l, and microwave aperture size Nd is simulated by optical aperture size Nl: therefore, the diffraction-limited microwave lobe width λ_m/Nd is simulated by optical processor lobe width λ/Nl, and the N angular beam positions of the microwave array are contained within the angle λ/l in the optical simulation.

The length of time stored by the AO modulator in the x-direction should

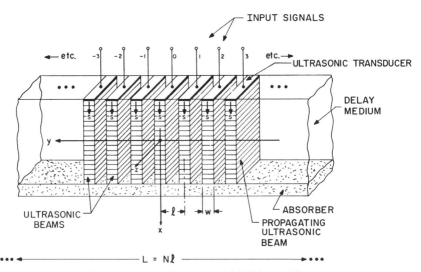

FIG. 5.3-13. Multiple-channel AO light modulator.

be at least as great as the received radar pulse duration; longer signal storage time causes the far-field diffraction to persist for a correspondingly longer period of time. If the received IF signal from the nth channel of the linear array is given by Eq. (5.2-17), then the $+1$ diffraction order component of the light transmitted by the nth channel of the AO modulator at the received signal time t_R is given by the form

$$(j\psi_0/2)\,\text{rect}[(y - nl)/W)]\,p_T(x/s)\exp\{j2\pi f_0[x/s - (nd\sin\theta)/c]\}. \quad (5.3\text{-}6)$$

This form shows that the $+1$ order from each AO channel contains the desired element-to-element phase shift which characterizes the target angular position off boresight.

The far field of the simulated linear array is obtained in real time at the back focal plane of a Fourier transform lens operating on the AO modulator output as sketched in Fig. 5.3-14a. Lambert *et al.* [26] showed that in the region of the $+1$ diffraction order, at $t = t_R$, the form of the optical irradiance distribution was essentially the same as the linear array pattern given by Eq. 5.2-18. (The exception is that $e(\theta)$ of the microwave antenna element is replaced by the element radiation pattern of the rectangularly shaped acoustic beam.) Figure 5.3-14b shows how details of the light distribution in the back

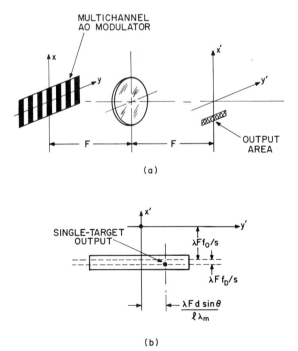

(a)

(b)

FIG. 5.3-14. Spatially multiplexed processor: (a) processor and (b) output area.

focal plane of this spatially multiplexed processor are related to the significant parameters of the optical processor and the received radar signal. Lambert *et al.* [26] noted the following properties of the spatially multiplexed processor for linear arrays: Target angular position may be determined unambiguously and with the full angular resolution of the microwave array; all antenna beam positions are formed simultaneously over the continuous angle space observed by the array; the location of the beam formed in the optical processor also indicates target Doppler shift with a resolution limited by the transmitted radar pulse duration; the time at which the beam is formed corresponds to the range delay of the target.

Lambert *et al.* [26] also showed that the linear array signals may be processed in a single-channel AO modulator by temporally multiplexing the element signals so they appear end to end in the processor aperture. Temporal multiplexing is accomplished with fixed delay lines which provide a constant incremental delay T_D between adjacent elements; typically, T_D is chosen to be slightly larger than the inverse of the radar pulse bandwidth. The far field of the wavefront transmitted by the AO modulator is obtained in real time with a Fourier transform lens. Figure 5.3-15 shows the relationship between the location of the beam formed by the temporally multiplexed processor and the target parameters. Notice that all information is obtained in the x'-direction with $y' = 0$ in the back focal plane of the lens. The x'-position corresponds to the target angular position 0, according to the factors shown in Fig. 5.3-15; target Doppler is manifested in the location of the sinc function

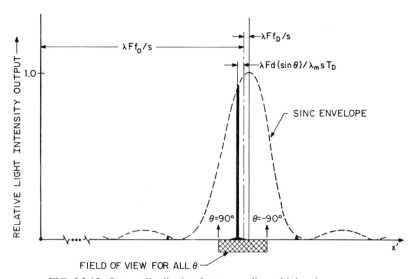

FIG. 5.3-15. Output distribution for temporally multiplexed processor.

which multiplies the (sin Na)/sin a array factor. It has been concluded that this temporally multiplexed processor provides the same information as the spatially multiplexed approach, except that target Doppler does not influence the location of the beam that is formed but appears as a weighting function on the output.

Lambert *et al.* [26] showed that the design concept of a Fourier-optical processor for plane array antenna beam forming in real time was a straightforward combination of the above linear array processing concepts. If the plane array antenna consists of a rectangular array of $(N + 1)$ rows by $(M + 1)$ columns of elements, then the received wave front may be reconstituted in real time with an $(M + 1)$-channel AO modulator; the mth channel of the modulator is driven by the time-multiplexed outputs of the $(N + 1)$ antenna elements in the mth column of the antenna array. The optical output of the AO modulator is brought to the far field by a Fourier transform lens, and all targets are consequently reconstructed optically. Figure 5.3-16 shows the nature of the information available in the output region of this planar array processor. Among the interesting properties of this processor are: All possible two-dimensional beam angles are obtained simultaneously, at the full angular resolution of the phased array antenna; multiple targets and continuous extended targets will produce multiple and continuous outputs with all angles properly associated; target Doppler shift may affect the apparent amplitude of the radar return; side lobe reduction by aperture weighting may be accomplished by weighting the illuminating beam or by adjusting amplifier gains appropriately; only one fixed delay line (for time multiplexing) is needed for each antenna element with this real-time processor.

A 24-channel, spatially multiplexed, real-time, linear array processor was tested in the laboratory with signals that simulated radar for on- and off-boresight point target returns [26, 27]. The signals which simulated the

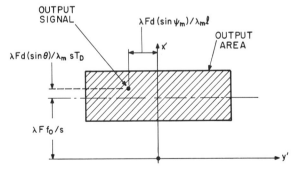

FIG. 5.3-16. Output scheme for planar array processor.

returns in a uniformly weighted 24-element linear array were run continuously, rather than pulsed, so that the beam formed in real time could be
scanned by a small slit and photomultiplier tube. Figure 5.3-17a is a logarithmic plot of measured intensity as a function of slit position; in accord
with the phased array literature, the slit position axis (y' in Fig. 5.3-14) is
calibrated in sin θ, on the assumption that the antenna elements of the array
being simulated were spaced by $\lambda_m/2$. Comparison of the main lobe width and
side lobe levels with the ideal distribution showed that the simulated antenna
angle resolution was preserved in the optical processor; the departures in the
side lobe levels from the ideal were found, by supplementary measurements,
to be caused by residual errors in the setting of the phases of the 24 separate
simulated antenna element signals. Figure 5.3-17b shows the logarithmic
intensity distribution in the beam when an off-boresight signal (19.5 deg) was
simulated; significantly, the beam forming performance did not deteriorate

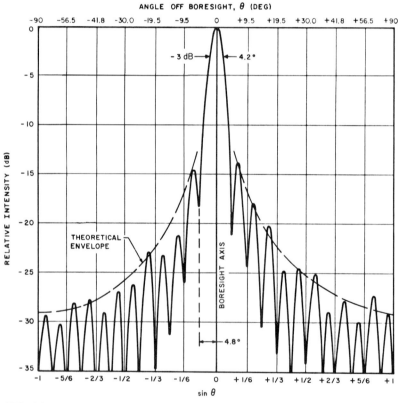

FIG. 5.3-17a. Spatially multiplexed array processor outputs with boresight target. Input
signals simulate a uniform linear array antenna with 24 elements spaced at one-half wavelength.

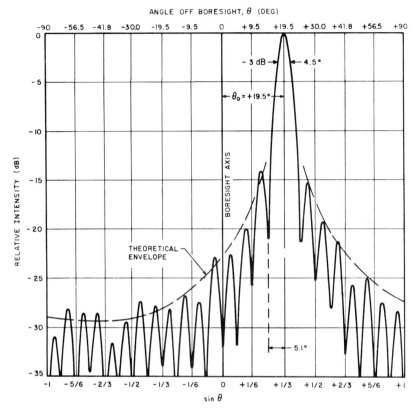

FIG. 5.3-17b. Spatially multiplexed array processor output with off-boresight target. Input signals are the same as in Fig. 5.3-17a.

for the off-boresight signal. Excellent results were also obtained for a simulated 48-element linear array with a time-multiplexed processor configuration. These experiments also processed simulated returns for a Hamming-weighted linear array in real time yielding the logarithmic intensity distribution shown in Fig. 5.3-18; it was determined that the above-expectation near-lying side lobes were caused by residual figure errors in the optical components of about $\lambda/20$. Finally, Fig. 5.3-19 shows the results of real-time processing for simulated off-boresight returns in a 576-element (24 × 24) planar array radar; these excellent results demonstrate vividly the potential value of optical processing for phased array radar processing.

Casasent and Casasayas [28] used an electro-optic light valve to reconstruct optically and to process simulated linear and planar array returns. With this device, they were able to store the IF signals from adjacent elements of a linear array (or from one column of a planar array) along a

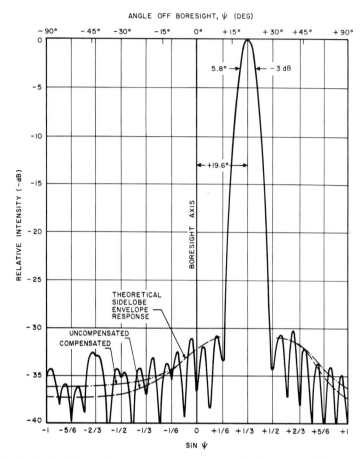

FIG. 5.3-18. Spatially multiplexed processor output with Hamming weighting, pedestal amplitude h = 0.2, and optical aperture 28.8 × 8 mm. The input signal simulates a weighted linear array of 24 elements spaced at $\lambda/2$.

single trace by modulating the write gun electron beam sequentially with the IF signals from the antenna; fixed delay lines would be needed to implement this approach, which is equivalent to time-multiplexed processing. Planar array signals were stored by writing adjacent array columns on adjacent traces with the light valve. Their results demonstrated the full resolution of the antennas that were simulated (100-element linear array and a 70 × 70 element planar array) and showed that the measured beam positions corresponded to simulated target positions within experimental error. It appears that the light valve-based array processor may be at a disadvantage with respect to the AO approach, because the former does not display the con-

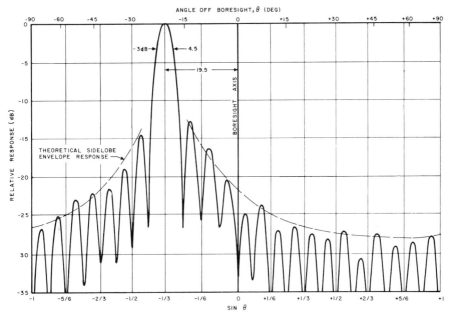

FIG. 5.3-19a. Azimuth cut through planar array processor outputs. Input signals simulate a uniform planar array with 576 elements (24 × 24) spaced at $\lambda/2$. Optical aperture is 28.8 × 72 mm.

tinuously changing pattern that results as returns are received from larger target ranges.

Williams and von Bieren [29] proposed an optical configuration that would be useful for phased array radars which employ coded pulses that require matched filtering at the receiver to realize full radar performance. This configuration would perform, in successive planes, matched filtering of the returns in each antenna element and beam forming from the ensemble of predetection filter outputs. Casasent and Klimas [30] demonstrated a non-real-time system which simultaneously performed predetection filtering and beam forming for simulated returns from a linear array that were recorded holographically, side by side, on a photographic transparency. Their results showed that the AO modulator-based, spatially multiplexed processor of Lambert and co-workers could be adapted to perform simultaneous predetection filtering and beam forming in real time.

In radar applications, the phased array beam former must process a completely new set of wave fronts for each range resolution cell of interest. However, it is not at all clear that light valves have the capacity to write in completely new data frames in time scales of 1 μsec or less. Of all the approaches mentioned for phased array processors, only the spatially

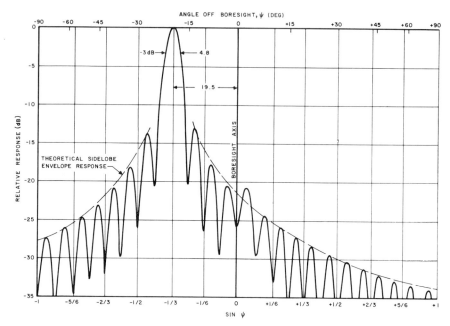

FIG. 5.3-19b. Elevation cut through planar array processor outputs. Input signals and optical aperture are the same as in Fig. 5.3-19a.

multiplexed AO approach appears to operate strictly in real time without qualification. One- and two-dimensional AO modulator-based systems using time multiplexing keep the signals received from one range cell within the optical processor aperture for NT_D/s seconds, where N is the number of antenna elements multiplexed in time; as a consequence, beams formed from a single range cell will be seen to narrow and build up in amplitude for the NT_D/s seconds it takes for the signals to fill the AO modulator and to fade away in the same fashion over the next NT_D/s seconds. However, M. Arm and L. Lambert have conjectured that the persistence phenomenon in the time-multiplexed approach may be eradicated by imposing a fixed, pseudo-random phase mask on the modulator and adding complementary phases to the signals from the individual elements so that efficient beam formation is only possible when the temporally multiplexed signal is centered in the modulator aperture; this idea appears sound but has not been analyzed or reduced to practice.

D. Optical Processors for Synthetic Aperture Radar

The development of practical coherent-optical processors for synthetic aperture radar signals is beautifully summarized in two technical papers

[31, 32], and the present discussion is based largely on them. The synthetic aperture radar processor records the return from each transmitted pulse as a strip hologram on photographic film according to the format shown in Fig. 5.3-20. The hologram spatial carrier bears the amplitude and phase of the radar return as a function of range. Successive pulses are recorded side by side, as indicated in the figure, on the photographic film. The hologram exposures are made with a cathode ray tube used as a flying-spot scanner; the bias current of the cathode ray tube beam is modulated with the radar IF signal. Further, the photographic film being exposed is moved at a constant speed such that adjacent strip holograms are closely packed. As a result, radar range is recorded along the x-direction and radar azimuth (or increasing pulse number) is recorded along the y-direction of the photographic film format. Consequently, a point target contained at a particular range will have the phase history of its returns contained along a single horizontal (y-direction) strip of the film record. The optical processing essentially sorts out and compresses signal histories from all point scatterers in the radar field of view simultaneously, thus yielding an image with an azimuth resolution controlled by the distance the aircraft has traveled during the effective integration time.

To a rather significant extent, the processing of this film format resembles the linear phased array processing discussed in the preceding section of this chapter. However, separate point targets in the radar field of view in the present case do not produce tilted plane waves in the aperture but rather produce tilted spherical waves that have a distinct quadratic phase history across the aperture. From another viewpoint, the signal history recorded from each point scatterer in the y-direction on the film is a cylindrical Fresnel zone plate. It was shown in Section 5.2 that the curvature of the optical wave front emerging from a Fresnel zone plate or chirplike signal depends on the target range only; consequently, all scattering points lying along a line parallel to the aircraft track produce a Fresnel zone plate with the same focal length but with a different angle of arrival. In addition, targets at different ranges produce zone plates of different focal lengths. When the

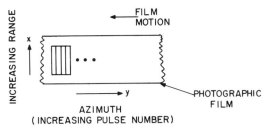

FIG. 5.3-20. Signal storage format for synthetic aperture radar.

film record is developed and illuminated with a plane wave of coherent light, the transmitted wave front contains three components; these are the zero-order and the virtual- and real-image terms of holography. However, the situation is somewhat complicated, because the azimuth positions of radar scatterers lie in a different plane than the range positions. Figure 5.3-21 illustrates this situation for a simple radar pulse; a virtual azimuth image plane lies to the rear (left side) of the film record, and a real azimuth image plane lies to the front of the film record. The azimuth planes are tilted because the zone plates at different ranges have different focal lengths. It is the purpose of optical processing to cause the range and azimuth image planes to coincide. Figure 5.3-21b shows the situation when a chirp-coded radar pulse is transmitted; in this case, the range focal planes lie off the film plane and are parallel to the film plane. In this latter case as well, it is the purpose of the coherent-optical signal processor to operate on a pair of range and azimuth planes so that they coincide in the output image plane. Figure 5.3-22 shows schematically the operation performed by the optical system in order to bring a tilted azimuth plane, A, and a range plane, R, into near coincidence. Kozma *et al.* [32] have noticed that anamorphic telescopic

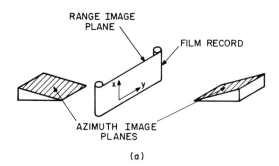

(a)

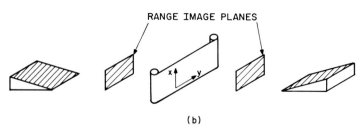

(b)

FIG. 5.3-21. Image planes for signal formats: (a) simple radar pulse and (b) chirped radar pulse.

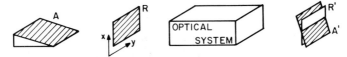

FIG. 5.3-22. Anamorphic optical system for bringing azimuth and range planes into coincidence.

optical systems can perform this function; further, they pointed out that the final coincidence of the R' and A' planes in Fig. 5.3-22 could be achieved by tilting the photographic film record. For the case illustrated in this figure, the photographic film should be tilted such that the upper edge of the R plane leans forward; the upper edge of the R' plane consequently leans backward, falling into coincidence with the A' plane.

Figure 5.3-23 shows top and side views of a typical anamorphic telescopic system used by Kozma and his colleagues to achieve the required transformations. Lenses L_1 and L_2 are circular, and lenses L_3 and L_4 are cylindrical. Because L_1 and L_2 are confocal, the tilted range plane is imaged into the tilted image plane as indicated in the side view. The top view shows how a typical azimuth line lying somewhere to the left of the film plane is imaged jointly by all four lenses (the cylindrical lenses are confocal, as indicated) onto the corresponding line on the tilted image plane. Space does not permit a complete discussion of the optical relationships which must be satisfied in order to achieve the coincidence of the range and azimuth image planes and the compensation of the severely distorted metric scale.

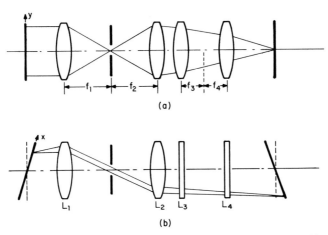

FIG. 5.3-23. Anamorphic telescopic system for synthetic aperture images: (a) top view and (b) side view.

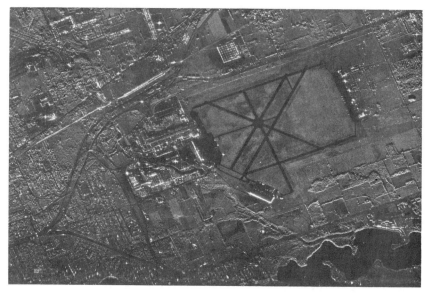

FIG. 5.3-24. Synthetic aperture radar processor output.

The reader is referred to the paper of Kozma *et al.* [32] for complete details of the discussion.

This optical system replaces an earlier one [31] which required a conical lens to be placed near the photographic film record so that the various focal length cylindrical zone plates would fall into focus in the same plane as the image of the range information. The present anamorphic system uses simpler optical components and also corrects the image scale so that the optical processing of lengthy synthetic aperture radar film records can be performed continuously; the radar signal film record and the output image film move at the same speed. Another advantage of this processor over the earlier one is that a larger area of photographic film record may be reconstructed simultaneously, and this tends to suppress speckle noise. An example of a radar image formed with this coherent-optical processor is shown in Fig. 5.3-24. This striking image and the many others that have been produced by the optical processing of synthetic aperture radar signals have for years been among the most highly appreciated outputs of coherent-optical processing of radar signals.

REFERENCES

[1] C. E. Cook and M. Bernfeld (1967). "Radar Signals: An Introduction to Theory and Application." Academic Press, New York.
[2] C. W. Helstrom (1960). "Statistical Theory of Signal Detection." Pergamon, New York.

[3] P. M. Woodward (1953). "Probability and Information Theory, with Applications to Radar." Pergamon, New York.

[4] D. K. Barton (1965). "Radar System Analysis." Prentice-Hall, Englewood Cliffs, New Jersey.

[5] E. I. Gordon (1966). *Proc. IEEE* **54**, 1391–1401.

[6] G. A. Alphonse (1974). *Appl. Opt.* **14**, 201–207.

[7] G. Marie and J. Donjon (1973). *Proc. IEEE* **61**, 942–958.

[8] M. Noble (1979). *Proc. Int. Conf. Lasers, 1978* pp. 574–578.

[9] R. A. Sprague (1978). *Appl. Opt.* **17**, 2762–2767.

[10] B. Schneeberger, F. Laeri, and T. Tschudi (1979). *Opt. Commun.* **31**, 13–15.

[11] M. Arm, L. Lambert, and I. Weissman (1964). *Proc. IEEE* **52**, 842–843.

[12] L. Slobodin (1963). *Proc. IEEE* **51**, 1782.

[13] G. Meltz and A. T. Maloney (1968). *Appl. Opt.* **7**, 2091–2099.

[14] C. Atzeni (1971). *Appl. Opt.* **4**, 863–872.

[15] H. Stark, F. B. Tuteur, and M. Sayar (1971). *Appl. Opt.* **10**, 2728–2733.

[16] M. King (1966). Heterodyne-type electro-optical signal processors, Ph.D.Thesis, Columbia Univ., New York.

[17] M. King, W. R. Bennett, L. B. Lambert, and M. Arm (1967). *Appl. Opt.* **6**, 1367–1375.

[18] R. A. Sprague and K. L. Koliopoulos (1976). *Appl. Opt.* **15**, 89–92.

[19] C. S. Tsai (1980). *SPIE Semin. Proc.* **209.**

[20] J. S. Gerig and H. Montague (1964). *Proc. IEEE* **52**, 1753.

[21] J. H. Collins, E. G. H. Lean, and H. J. Shaw (1967). *Appl. Phys. Lett.* **11**, 240–242.

[22] M. B. Schulz, M. G. Holland, and L. Davis (1967). *Appl. Phys. Lett.* **11**, 237–240.

[23] C. S. Tsai, L. T. Nguyen, B. Kim, and I. W. Yao (1977). *SPIE Semin. Proc.* **128**, 68–74.

[24] M. King and M. Arm (1969). *IEEE Trans. Quantum Electron.* **QE-5**, 332–333.

[25] M. Arm and M. King (1969). *Appl. Opt.* **8**, 1413–1419.

[26] L. B. Lambert, M. Arm, and A. Aimette (1965). *In* "Optical and Electro-Optical Information Processing" (J. T. Tippett *et al.*, eds.), pp. 715–748. MIT Press, Cambridge, Massachusetts.

[27] L. B. Lambert (1965). Electro-optical signal processors for array antennas, Ph.D. Thesis, Columbia Univ., New York.

[28] D. Casasent and F. Casasayas (1975). *Appl. Opt.* **14**, 1364–1372.

[29] R. E. Williams and K. von Bieren (1971). *Appl. Opt.* **10**, 1386–1392.

[30] D. Casasent and E. Klimas (1978). *Appl. Opt.* **17**, 2058–2063.

[31] L. J. Cutrona, E. N. Leith, L. J. Porcello, and W. E. Vivian (1966). *Proc. IEEE* **54**, 1026–1032.

[32] A. Kozma, E. N. Leith, and N. G. Massey (1972). *Appl. Opt.* **11**, 1766–1777.

Chapter 6

Application of Optical Power Spectra to Photographic Image Measurement

R. R. SHANNON

OPTICAL SCIENCES CENTER
UNIVERSITY OF ARIZONA
TUCSON, ARIZONA

and

P. S. CHEATHAM

PACIFIC-SIERRA RESEARCH CORPORATION
SANTA MONICA, CALIFORNIA

6.1 INTRODUCTION

Many of the techniques discussed by Almeida and Indebetouw in Chapter 2 and by Leger and Lee in Chapter 4 involve the use of photographic film either as input mask, Fourier plane mask, or output storage. In King's discussion (Chapter 5) of synthetic aperture radar (SAR) he noted that photographic film is the primary medium of storage for the strip hologram that contains the amplitude and phase of the radar echo. Given then the importance of photographic film in Fourier optics, what can we say about how it affects and/or limits the performance of the system? What are the parameters of importance to the designer of signal processing systems and how are they measured? This is the subject matter of this chapter. The work described herein includes the investigations of the coauthors and Stephen Sagan and David Honey. All of the work reported here was completed at the Optical Science Center, University of Arizona.

A photographic image is a two-dimensional recording of the light distribution in an aerial image formed by a lens. The image is recorded by

253

random samples of developed silver grains in the processed emulsion. Both the image information content, or signal, and the description of the noise content, or granularity, may be obtained through application of optical power spectrum (OPS) analysis [1]. Use of an appropriate detector in the Fourier transform plane permits the simultaneous collection of power spectral data in several spatial frequency bands. This type of spectrum analysis is useful for quantifying photographic image and noise characteristics on a statistical basis but requires care in its application [2, 3].

Our work has been concerned with measurement of the image content that can eventually be used for image quality judgments. Toward this end, we have defined several quantities which sometimes lead to minor conflicts with the standard nomenclature. We have used the symbol PS to represent the power spectrum, S to represent the power spectrum of the signal, N to represent the power spectrum of the noise, and $SNPS$ to represent the signal-to-noise power spectrum.

These measurements of image content are based on samples in the frequency domain of an input image. The standard coherent-optical processor is probably the best device for this purpose. The processor (Fig. 6.1-1) consists of a laser, spatial filter, collimator, object, and transform lens. The function of the processor can easily be shown by simple optical principles. The laser gives us a temporally and spatially coherent beam which is expanded and collimated by the spatial filter. The spatial filter acts to broaden the spatial coherence of the laser. Leaving the collimator we have a plane wave of a single wavelength λ which is incident upon an object, traditionally a photographic transparency. The transparency, of course, diffracts the light according to the distribution of spatial frequencies in the object; i.e., each frequency acts as a sinusoidal grating producing plane waves diffracted at angles proportional to that frequency. This diffraction arises from both amplitude modulation and phase effects of the relief image on the film. We have made no attempt to isolate the separate contributions but rather have chosen to deal with the total image content regardless of the source. Note that the relief image can be exploited to enhance an image during printing

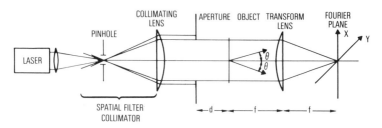

FIG. 6.1-1. Coherent-optical processor schematic.

and therefore represents information content. Our work has been directed toward a system which can be easily used. An index-matching oil (a so-called liquid gate) could be used to minimize the phase effects but would not be feasible in routine application.

The angle of diffraction θ from the initial direction of propagation can be determined for any spatial frequency u by the grating equation

$$\sin \theta = \lambda u, \tag{6.1-1}$$

and the resulting diffracted beam can be expressed as

$$a(x) = e^{-j2\pi ux}. \tag{6.1-2}$$

The amplitude distribution leaving the transparency will be a sum or, in the limit, a continuum, of plane waves,

$$f(x) = \int_{-\infty}^{\infty} F(u)e^{-j2\pi ux}\, du, \tag{6.1-3}$$

where $F(u)$ is a weighting function of the relative strength of each spatial frequency in the object. The effect of the Fourier transform lens is obvious. Each of the plane waves will be brought to a focus in the focal plane of the lens. The focal point is, of course, determined by the angle of the entering beam, so that for any particular spatial frequency the distance from the optical axis d is given by the equation

$$d = f \tan \theta, \tag{6.1-4}$$

where f is the focal length of the lens (Fig. 6.1-1). If we assume small angles so that

$$\tan \theta = \sin \theta = \lambda u, \tag{6.1-5}$$

then the distance d is related to the spatial frequency u by

$$d = \lambda f u. \tag{6.1-6}$$

The net effect is a distribution of light in the focal plane where the linear distance is directly proportional to the spatial frequency in the original object, and the amplitude at any point is proportional to the value of $F(u)$, the relative strength of the component at the desired frequency. In other words, the optical processor linearly transforms a spatial frequency into a physical distance in the focal plane of the lens. A detector placed in the Fourier plane will be sensitive not to the amplitude but to the intensity, or power, of the light distribution, which is proportional to $|F(u)|^2$. Therefore measurements taken in this plane will yield the values of the relative power in the diffracted beam from each spectral component, or a power spectrum. Equation (6.1-3) is, of course, a Fourier transform relationship, and the intensity distribution in the detector plane is the squared modulus of the Fourier transform, where the relationship between spatial frequency and physical distance in the detector plane is given by Eq. (6.1-6).

This discussion of the power spectrum has been of a general nature with the intention of showing the basic principles. Several rigorous derivations of the Fourier relationships involved in the coherent processor can be found in the literature [3–6], as well as in this book. These derivations are based on analysis of the amplitude and phase of the incoming beam and a free-space propagation operator. When the same approximations are used that we assume here, i.e., those of small angles and perfect lenses, these derivations yield the same result with one addition: In our discussion we ignored the effects of a limiting aperture in the system, whereas a rigorous derivation includes this effect. If this aperture were placed in contact with the object, then the intensity in the Fourier image plane $I(x)$ would be

$$I(x_f) = a|\mathcal{F}\{t(x) \cdot p(x)\}|^2_{u = x_f/\lambda f}, \qquad (6.1\text{-}7)$$

where a is a constant, $t(x)$ is the transmittance of the object, $p(x)$ is the pupil function describing the limiting aperture, and $\mathcal{F}$ is the Fourier transform operator. We can express $I(x)$ in terms of Fourier transforms as

$$I(x_f) = a|T(u) * P(u)|^2_{u = x_f/\lambda f}, \qquad (6.1\text{-}8)$$

where $T(u)$ and $P(u)$ are the Fourier transforms of $t(x)$ and $p(x)$, respectively. This convolution between $T(u)$ and $P(u)$ can normally be ignored if the pupil function is sufficiently large. Its transform will then approximate a delta function, and the transmission function will be largely unaltered. The expression for the intensity then becomes [4–6]

$$I(x_f) = a|\mathcal{F}\{t(x)\}|^2_{u = x_f/\lambda f}. \qquad (6.1\text{-}9)$$

Since the system is composed of optical elements, there will naturally be contributions to the intensity distribution from the aberrations in the system. The importance of the system aberrations depends, of course, on the intended use of the processor. For our purposes we require only coarse samples of the power spectrum and can therefore tolerate a moderate amount of aberration. We estimated the aberration blurs which would be expected from simple singlets bent for minimum spherical aberration. The detector used in our apparatus (Section 6.3) was an array of elements oriented for power spectrum measurements. The radius of the expected blurs was well within the size of these elements both on-axis and at the field edge. We chose, after this analysis, to use commercially available doublets corrected for spherical aberration. The aberration effects from these lenses will be less than those of the singlet used in our analysis. This allows us to neglect the effect of the optical aberrations of the processor as long as reasonable angular fields are used.

Several properties of the power spectrum $PS(u)$ become readily apparent with the realization of the Fourier relationship between the object and the focal plane of the lens. For a sum of signals

$$f(x) = g(x) + h(x), \tag{6.1-10}$$

the power spectrum becomes

$$PS(u) = |G(u)|^2 + |H(u)|^2 + G^*(u)H(u) + G(u)H^*(u), \tag{6.1-11}$$

where $G(u)$ and $H(u)$ are the Fourier transforms of $g(x)$ and $h(x)$, respectively. The cross-terms arise from interactions between the two signals.

Random or noiselike functions are usually described in terms of an ensemble average of many samples of the function. However, one sample can be used to characterize the function when the sample is large enough to provide satisfactory statistics. Often this can be obtained from the first joint moment of the probability function, which is given by the autocorrelation function [7]

$$R_f(\alpha) = \lim_{x \to \infty} \int_{-x}^{x} f^*(x) f(x + \alpha) \, dx. \tag{6.1-12}$$

For the sum of two functions, the autocorrelation function is

$$R_f(\alpha) = R_g(\alpha) + R_h(\alpha) + R_{gh}(\alpha) + R_{hg}(\alpha). \tag{6.1-13}$$

Again we have cross-terms from the cross-correlation of the two functions. The Fourier transform of the autocorrelation is normally defined as the Wiener spectrum, after N. Wiener who pioneered the methods of analyzing random processes. The Wiener spectrum is given by

$$W(u) = \lim_{X \to \infty} \left\langle \frac{1}{2X} \left| \int_{-X}^{X} f(x) e^{-j2\pi u x} \, dx \right|^2 \right\rangle, \tag{6.1-14}$$

where $\langle \; \rangle$ denotes an ensemble average. The power spectrum and the Wiener spectrum are really equivalent terms. However, the Wiener spectrum has traditionally been computed from digital data, for instance, micro-densitometer measurements, rather than direct measurement in a coherent-optical processor. For this reason we maintain a distinction between the terms *power spectrum* and *Wiener spectrum*, the former being obtained from a coherent processor and the latter from digitized data. The relationship between the autocorrelation function and the power spectrum can be confirmed, by starting with Parseval's theorem,

$$\int_{-\infty}^{\infty} |f(x)|^2 \, dx = \int_{-\infty}^{\infty} |F(u)|^2 \, du \tag{6.1-15}$$

and using the shift property of the Fourier transform to obtain

$$\int_{-\infty}^{\infty} f^*(x)f(x + x')\,dx = \int_{-\infty}^{\infty} F^*(u)F(u)e^{j2\pi ux'}\,du. \qquad (6.1\text{-}16)$$

From Eq. (6.1-12) we see that the left-hand side is the autocorrelation function. Taking its transform we find

$$\mathscr{F}\{R_f(x')\} = \int_{-\infty}^{\infty} \int_{-\infty}^{\infty} F^*(u)F(u)e^{j2\pi ux'}\,du\, e^{-j2\pi u'x'}\,dx'$$

$$= \int_{-\infty}^{\infty} \int_{-\infty}^{\infty} |F(u)|^2 e^{j2\pi(u-u')x'}\,dx'\,du$$

$$= \int |F(u)|^2 \delta(u - u')\,du = |F(u')|^2. \qquad (6.1\text{-}17)$$

The power spectrum of a noiselike function will be taken as

$$PS(u) = \mathscr{F}\{R_f(x)\}$$

$$= \mathscr{F}\{R_g(x)\} + \mathscr{F}\{R_h(x)\} + \mathscr{F}\{R_{gh}(x)\} + \mathscr{F}\{R_{hg}(x)\}. \qquad (6.1\text{-}18)$$

6.2 POWER SPECTRAL MEASUREMENTS

Our concern is the measurement of image quality based on power spectral samples. Consider an image, on film, whose transmittance function is $f(x)$. Its power spectrum will contain effects from the film and measuring apparatus, which must be separated to allow evaluation of the image. For a constant image signal we can write the amplitude transmittance function as

$$f(x) = a + b(x), \qquad (6.2\text{-}1)$$

where $b(x)$ describes the random fluctuations about the constant transmittance a.

For a more complex image signal, we would expect these fluctuations to depend not only on the noise but on the signal as well. Thus, in this case, we obtain

$$f(x) = a(x) + b(x). \qquad (6.2\text{-}2)$$

Now the function $a(x)$ is merely the contribution of the recorded image, and $b(x)$ is a description of the variations from this input signal. The Fourier transform for this function is

$$F(u) = A(u) + B(u), \qquad (6.2\text{-}3)$$

and the intensity in the Fourier plane will be

$$I(x_f) = |F(u)|^2 = \langle|A(u)|^2\rangle + \langle|B(u)|^2\rangle$$
$$+ \langle A^*(u)B(u)\rangle + \langle A(u)B^*(u)\rangle, \qquad (6.2\text{-}4)$$

where $\langle \ \rangle$ again refers to ensemble averages. If the signal and noise are uncorrelated or independent, then the cross-terms become negligible and the equation reduces to

$$PS(u) = |A(u)|^2 + |B(u)|^2. \qquad (6.2\text{-}5)$$

The terms $A(u)$ and $B(u)$ are the signal and noise power spectra and can be replaced by $S(u)$ and $N(u)$, giving

$$PS(u) = S(u) + N(u). \qquad (6.2\text{-}6)$$

The independence of noise and signal can be examined by measuring the power spectra of uniformly flashed density film samples. Here the signal is constant, and its power spectrum becomes a delta function. Equation (6.2-6) reduces to

$$PS(u) = S_1 \, \delta(u) + N(u), \qquad (6.2\text{-}7)$$

where S_1 is a constant indicating the bias level. Now the noise curve for different density levels can be compared by ignoring the zero-frequency component. If the noise curve is independent of signal level, it can then be subtracted, leaving just the signal power spectrum.

The signal power spectrum, although useful, is not the best form for our purposes. Power spectral measurements at different frequencies differ by several orders of magnitude and can lead to erroneous conclusions about the relative values of image content. In communications theory, channel capacity of information content is expressed, for a Gaussian channel, as

$$C = \Delta u \log_2[(P + N)/N], \qquad (6.2\text{-}8)$$

where u is the frequency bandwidth and P and N are the power in the signal and noise. In terms of the optical power spectrum, this can be written

$$IC = \int_0^\infty \log_2[1 + IPS(u)/NPS(u)] \, du \qquad (6.2\text{-}9)$$

or

$$IC = \int_0^\infty \log_2[1 + SNPS(u)] \, du, \qquad (6.2\text{-}10)$$

where $SNPS(u)$, the signal-to-noise power spectrum, is the ratio of the signal power spectrum to the noise power spectrum $S(u)/N(u)$. Our work has shown that the $SNPS(u)$ is indeed a useful and reliable measure of information content and as such should prove useful in image quality determinations.

6.3   EXPERIMENTAL CONFIGURATION

The power spectrum is normally measured by one of two methods. Either the image is digitized and Fourier-transformed by computer, or the power spectrum can be measured directly with a coherent-optical processor [8]. The use of the $SNPS$ does not favor the use of either method, but the first method defeats the main advantage of optical power spectrum measurements, namely, the ability to process numerous parallel data samples over a large region of the image. Since the direct measurement of the OPS is an intensity reading, the spatial phase information is lost and reconstruction of the image is impossible, but our purposes only require judgments based on samples of the image power spectrum. Thus, the second method is preferable and was used in our investigations of the $SNPS$. For the measurements of the granularity function, however, the coherent optical processor results were compared with the computed spectrum obtained from a Mann microdensitometer.

We used a sensor (designed for OPS measurement by Recognition Systems, Inc. [9]) at the focal plane of our coherent-optical processor. The sensor is a silicon wafer consisting of 64 photoactive elements (p-n junction type) is a semiannular ring-and-wedge pattern. This geometry is an approximately constant frequency bandpass, thus permitting image spectra errors, if any, to be larger at the edge of the field. For this reason, we can use simple doublets, corrected only for spherical aberration, instead of expensive Fourier transform lenses.

The sensor is provided with controlling electronics and gain amplifiers to give a dynamic range of 10^4. These electronics have been supplemented by an interface designed and built at the University of Arizona. It allows synchronization of detector operation with a Texas Instruments remote computer terminal for data recording.

Software has been written for the manipulation of the raw power spectral readings. Included are programs that average two readings of the same image, compact the data for easy storage, and normalize the readings by both the area of each detector element and the zero-frequency reading. The signal-to-noise power spectrum is computed from the image power spectrum and a stored noise spectrum derived from earlier measurements. The noise curve is scaled so that its values are identical to the image data in the frequency band where it is known that the image contains no information. Display programs have also been written to provide plots of the power spectral data as well as the signal-to-noise power spectrum.

As part of a general investigation into the optical processor operation, we made measurements of the effects of the various parameters in the system (Fig. 6.1-1). We found the system to be linear throughout its dynamic range. Film placement has no effect on the power spectrum as long as vignetting is not introduced by a large distance between the image and the transform lens. Power spectral readings show some sensitivity to aperture. This is probably a result of the convolution between the aperture and the image transmittance function [Eq. (6.1-8)]. Since the dynamic range of the power spectrum is so large, the side lobes of the Airy disk seem to have an effect even when the Airy disk size is smaller than the individual detector element areas. This effect does not seem to influence the *SNPS*, however, probably because the noise and image curves are measured with the same aperture, and the aperture effects are removed with the rest of the system noise. We also found

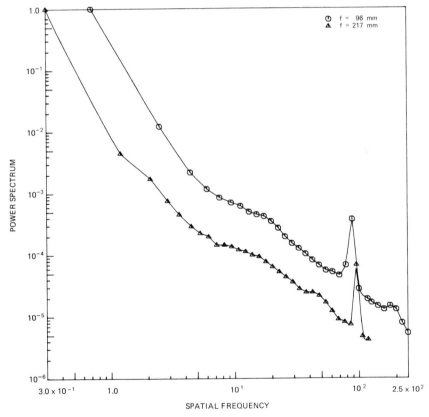

FIG. 6.3-1. Optical power spectrum of sample grating recorded with two different Fourier-transform lens focal lengths.

that power spectral readings of the same image taken with different transform lenses could vary greatly because of differences in processor noise. However, the *SNPS* is identical for the images (Figs. 6.3-1 and 6.3-2).

We experienced some difficulty with reflections in the system from the detector. Although the detector is anti-reflection-coated, there is still a reflection of about 4%. We were able to eliminate most of this effect by tilting the detector slightly so that the reflection was off-axis. This tilt introduced some additional blur in the outer rings, but it was well within the detector element size.

Although the *SNPS* is independent of system perturbations, we used the same aperture size, film placement, and transform lens for our experiments, to maintain continuity throughout the data.

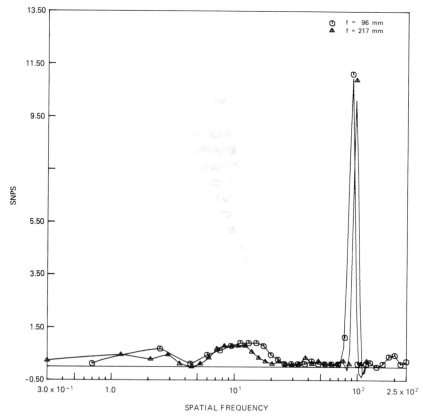

FIG. 6.3-2. Signal-to-noise power spectrum of sample grating recorded with two different Fourier-transform lens focal lengths.

6.4 GENERATION OF THE *SNPS*

Although the signal-to-noise concept offers, in theory, an excellent measure of image content, its practicality must be investigated. Several experiments have been carried out that demonstrate the validity and application of the concept. In most cases, it has been shown to function as expected with no significant surprises. In a few cases, as seen in the experiments with fine-grained material in Section 6.5, there is some ambiguity in the definition of noise spectra with use of the concept. However, in the case of imagery, it appears to be a useful measure of image content.

All the measurements to be described here were carried out on a coherent-optical processor using film images obtained from various sources. In no case was the film immersed in an **index-matching** fluid for the measurement.

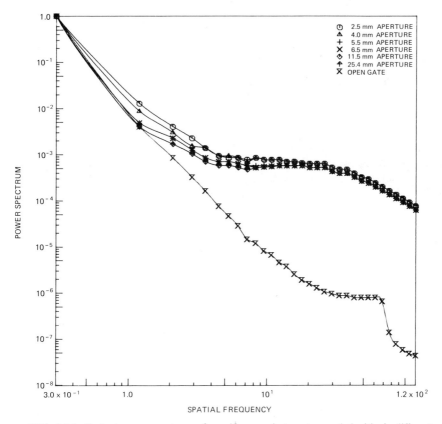

FIG. 6.4-1. Optical power spectrum of a uniform grain target recorded with six different apertures.

We know that most emulsions exhibit a relief image which will have an effect on the coherent power spectrum. This relief image can be suppressed or eliminated by immersion of the material in an appropriate index-matching oil. However, the practical application of coherent measurement methods to the measurement of imagery does not easily permit the use of an immersion gate. The reason is that the benefits of coherent-optical measurements of images lie in the high speed processing capability of the spectrum analyzer. A necessity for immersing each sample would greatly increase the expense and problems involved in evaluating the film. It is for this reason that we avoided suppressing the relief images and chose to examine the effects relative to image and noise structures in the absence of oil immersion techniques.

The effects that can be considered noise in the signal-to-noise power spectrum arise not only from the film but from the optical processor noise.

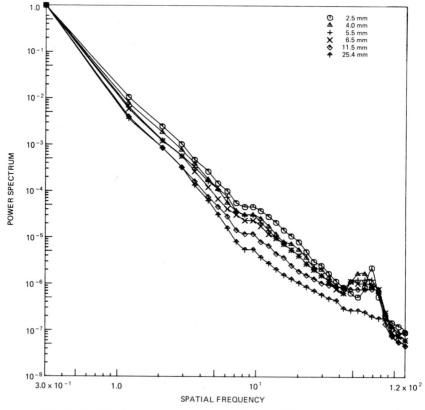

FIG. 6.4-2. Optical power spectrum recorded for six different empty apertures.

This noise is due to scattered light, diffraction, and other problems inside
the optical system. Figures 6.4-1 and 6.4-2 demonstrate some of the characteristics of this noise. These figures show the logarithm of the power spectrum
as obtained from a coherent sensor versus the logarithm of spatial frequency
over the range of approximately 0.5–100 cycles/mm. The power spectrum
values occupy almost seven decades in magnitude. The lower curve shows
the effect of a 25.4-mm-diameter empty aperture. The upper curves show the
effect of loading the aperture with a uniform target for six different-size
apertures. Note that, in the high frequencies, there are about four orders of
magnitude between the power spectrum level of the grain noise and the
values from the empty aperture. In the low frequency region, however, values
from the empty aperture and uniform targets are approximately the same,
indicating that processor noise is the dominant component in the total

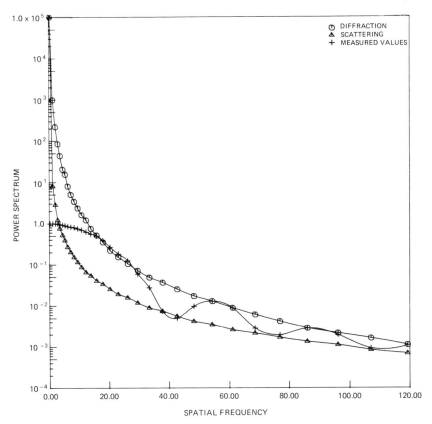

FIG. 6.4-3. Optical power spectrum of a 25-μm pinhole plotted with theoretical diffraction
and scattering curves.

power spectrum. Figure 6.4-2 shows the effect of size variation in the aperture-diffracted light. The 25.4-mm open gate is similar to that in Fig. 6.4-1. The variations in the curves indicate differing amounts of effective power spectra all normalized at the lower spatial frequency for the various apertures. The results indicate that there is a difference in the scattered light content of the processor, as for different aperture sizes. As for evaluations of the image structure or the grain noise structure, these variations constitute a background "artifact" noise that must be removed from the measurement. The lumps appearing in these curves were found to be due to reflections in the processor. The detector was tilted to minimize these reflections for all future measurements. In the work discussed here, usually no differentiation was made between the processor noise and the noise introduced by granularity or

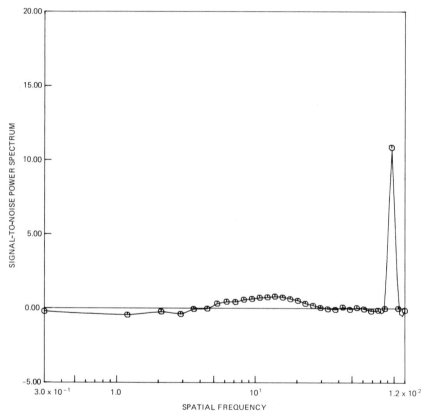

FIG. 6.4-4. Signal-to-noise power spectrum for a grating of density 0.88 and input modulation of 1:1.

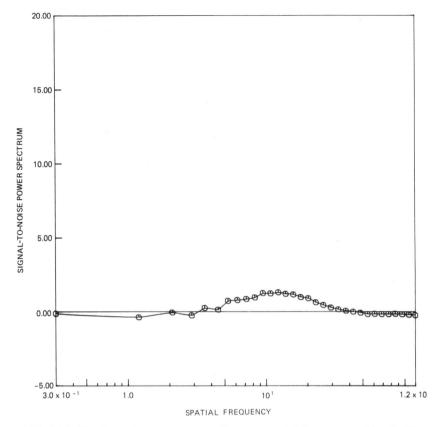

FIG. 6.4-5. Signal-to-noise power spectrum for a zero-modulation grating of density 0.88.

film base errors. Figure 6.4-3 shows an experimental attempt to localize the source of processor noise. In this case, measured values were compared to the theoretical diffraction falloff and the hypothesized scattering characteristics of the optical surface. The diffraction effects should fall off as f^3, where f is the frequency, and the scattering should fall of as f^2. A 100-μm aperture was used for this measurement. The pattern detected is obviously an Airy disk pattern, but the graph shows that the scattering effects become comparable to the contributions from the image power spectrum as frequency increases, and areas of the pattern which contain zero information from the original object (the diffraction pattern minima) have contributions due only to processor noise.

There are some additional sources of instrument noise in coherent measurement that must be considered. Many high resolution films have a pelloid backing applied to the base, which consists of a rippled surface generally made up of a suspension of small plastic beads in a matrix. These beads will cause coherent diffraction and scattering, thereby increasing system noise. Experimental demonstrations showed that the effect of the pelloids could be removed by dissolving them from the back of the film base. However, we can take the added noise of the pelloids into account in calculating the signal-to-noise power spectrum.

To obtain the required background noise data versus spatial frequency, we take a uniformly dense sample approximating that of the sample images to be measured and use several average measurements of the sample as the noise curve for construction of the signal-to-noise power spectrum. Measurement of the film granular structure, of course, requires noise measure-

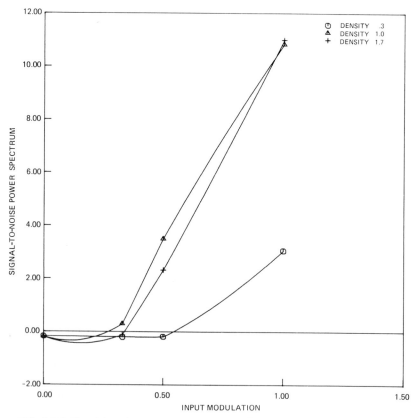

FIG. 6.4-6. Signal-to-noise power spectrum at grating peak versus input modulation for constant density gratings.

ments of the film base alone with the granular structure removed, but in the case of an image, the power spectrum of a uniform density with the same average transmittance as the image serves for the noise background measurements. Structure at an average density serves as a sample in obtaining the noise background measurements [10].

A series of sinusoidal gratings were produced to investigate the consistency of the *SNPS* measurements. The gratings were fabricated using interference patterns to give a sinusoidal grating of a single known frequency with variations in intensity and modulation. The power spectra for these gratings were matched to the noise curve at frequencies near the peak. The *SNPS* for two of the gratings appear in Figs. 6.4-4 and 6.4-5. These curves show some low frequency information between 5 and 30 cycles/mm that we

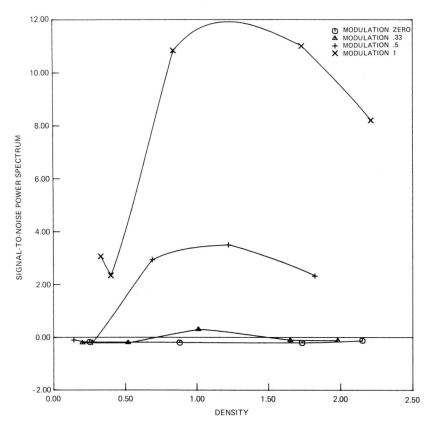

FIG. 6.4-7. Signal-to-noise power spectrum at grating peak versus density for constant input modulation gratings.

had not expected. However, after the *SNPS* computation, we went back and visually confirmed the presence at these frequencies of small diffraction peaks, probably caused by low frequency noise produced in the grating manufacture.

The gratings were manufactured by recording the interference pattern of two laser beams allowing control of the grating density and input modulation. Input modulation is the ratio of the intensities of the two interfering laser beams. We also attempted to measure the recorded signal on the film by microdensitometer trace. The quantity of interest in the optical processor is the variation in amplitude transmittance across the image, but unfortunately we can only measure variations in the transmitted intensity. The measurements can be used to estimate the variation in amplitude transmittance b by the formula

$$b = (f_{max} - f_{min})/2 = (\sqrt{I_{max}} - \sqrt{I_{min}})/2. \tag{6.4-1}$$

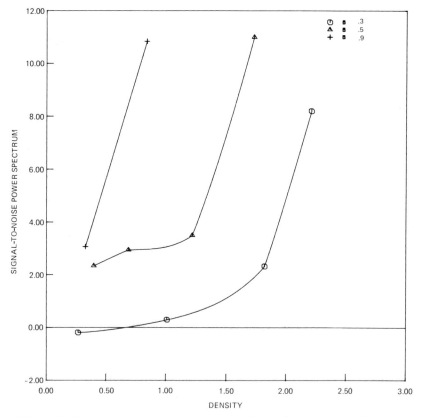

FIG. 6.4-8. Signal-to-noise power spectrum at grating peak versus density for constant amplitude transmission gratings.

On some of the lower density gratings the low frequency noise was much larger than the grating modulation, preventing calculation of the quantity *b*.

The results for the grating experiments appear in Figs. 6.4-6, 6.4-7, and 6.4-8. The curves show the variations of the *SNPS* at the grating frequency, 119 cycles/mm, versus input modulation, density, and variation in amplitude transmittance. The constant modulation results confirm our expectations. The same input modulation produces lower *SNPS* values at high and low densities where we are operating at the edges of the log exposure-versus-density (*H-D*) curve. In the case of constant amplitude transmittance and constant density, however, some discrepancies occur. The *SNPS* is higher at the higher densities than the input signal indicates. The results are probably caused by phase effects of the film. In order to produce the same intensity variations at higher density levels, the density difference must increase. This increase can result in a larger relief image and, therefore, in increased corre- lated phase effects, even though the amplitude transmittance remains con-

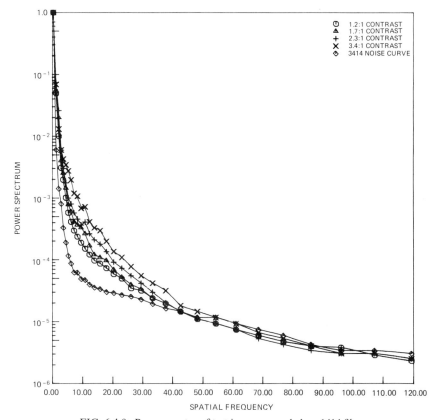

FIG. 6.4-9. Power spectra of test images recorded on 1414 film.

stant. This phase effect is actually consistent with our ideas about the *SNPS*. Phase effects can be used to enhance an image depending on the printing method, and therefore increased phase effects indicate increased information content on the film, provided that the phase effects are induced by the image, i.e., strongly correlated with it. The *SNPS* should be higher for images with larger emulsion relief, even though the transmission is the same.

Now, the major purpose in defining this power spectrum is to work with the images of realistic scenes, that is, objects whose spatial frequency distribution extends over a broad spatial frequency range. Figure 6.4-9 shows the power spectral data obtained from a series of simulated images. These images are all simulated photography of the same object but were made at several different contrast levels. As can be seen, there appear to be significant differences among the power spectra curves.

The images, obtained from an outside source, were available only on 1414 film, and no large area flashed grain sample of this material was avail-

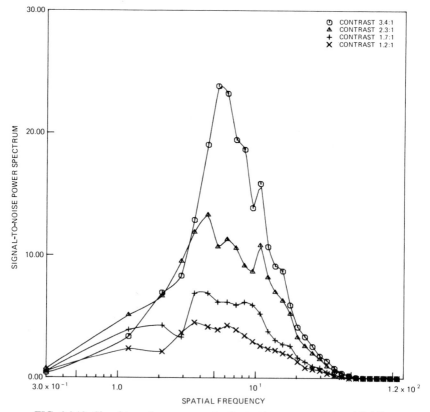

FIG. 6.4-10. Signal-to-noise power spectra for test images recorded on 1414 film.

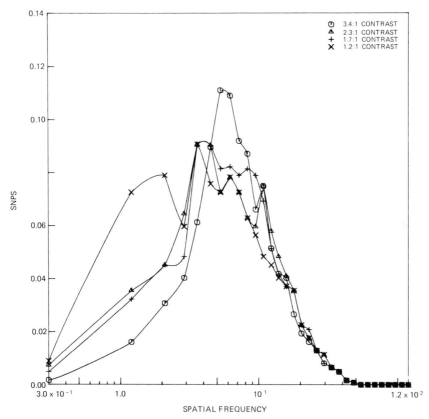

FIG. 6.4-11. Signal-to-noise power spectrum for test images. *SNPS* has been normalized so that enclosed area equals unity.

able for establishing the background noise curve. Since 3414 film is similar but with a different thickness of base, it was decided to use the granular characteristics of the flashed sample of 3414 film to establish the background noise curve. This required removal of the 3414 pelloid backing, since 1414-type film does not contain such a backing material.

As shown in Fig. 6.4-9, the similar emulsion was matched at the high spatial frequency region to obtain a noise calibration curve. The signal-to-noise spectrum curves obtained from using this noise background curve, which includes the processor plus film grain noise, are shown in Fig. 6.4-10. In general, the curves have the same shape. Their level varies monotonically with the contrast of the original image. Since the ratios seem reasonable and plausible, the signal-to-noise power spectrum concept does indeed allow separation of the image spectral content and contrast from the film, processor, and base noise even when a nonimmersed processor is used.

The curves shown in Fig. 6.4-10 are presented again in Fig. 6.4-11 when the area under each is normalized to unity. The shapes of each of the curves match those of the other curves extremely well. The spatial frequency content of this simulated scene is consistently obtained, and the use of the signal-to-noise power spectrum concept is appropriate for quantifying image content.

6.5 EFFECT OF FILM TYPE AND SAMPLING AREA

The experimental work to be described in this section and the following section is based upon Master's thesis work [11] carried out by Stephen Sagan at the Optical Sciences Center, University of Arizona, in 1978. He produced a series of sample images on two different types of film which varied in several controlled respects. Four architectural scenes were photographed, giving different contrast images of the type found in technical photography. In addition, some samples were made of images of random pebble patterns as would be observed in a driveway or garden plot. The resulting imagery displays a random set of spatial frequencies but no significant straight-line object content. The images were all photographed at relatively small numerical apertures to ensure that the effect of lens aberrations in the taking lens were held at a minimum. A number of focal positions were chosen in order to vary the spatial frequency content on the image. Two different types of film, Kodak Plus X and Panatomic X, were used in collecting the imagery. This provided large quantities of different types of information which permitted cross-checking for consistency of the signal-to-noise power spectrum results under a number of different conditions. Figure 6.5-1 is a composite of the six scenes chosen for the study, represented here merely to give some idea of the difference in the types of scenes that will be discussed. The conclusions are, in fact, found to be generally independent of the type of scene observed. This result is expected, since the coherent power spectrum measurement does in fact give an average result over the area of measurement.

Figure 6.5-2 shows a measurement of the signal-to-noise power spectrum of the random distribution of pebbles. In general, the shape is approximately the same and indicates the use of this type of object as a possible calibrating device for an optical system. When the first four scenes were examined using a sampling aperture with a 20-mm diameter, the indications in Fig. 6.5-3 were observed. They showed a general similarity between the scenes but a much broader and lower spatial frequency content than that illustrated by the random pebble samples shown in Fig. 6.5-2. This is to be expected, because the scenes are similar in nature and were produced by the same optical system. The broader spatial frequency content implies that the random distribution of pebbles probably contained a high distribution of small pebbles rather than very large pebbles.

FIG. 6.5-1. Random test scenes recorded on Pan-X Film: (a) Arizona State Museum, (b) Old Main, (c) Mission Church (view one), and (d) Mission Church (view two). Test scenes containing random pebble distributions: (e) medium-sized pebbles and (f) small-sized pebbles.

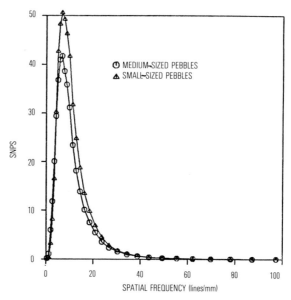

FIG. 6.5-2. Signal-to-noise power spectra of random pebble distributions.

It was noted that the signal-to-noise power spectrum characteristics of the random scene were not significantly dependent on the size of the sampling area or the placement of this area on the photographic frame. This would be expected, because of the isotropic nature of the image. However, the other scenes were seen to be dependent on sampling size and placement. One extreme example of this is shown in Fig. 6.5-4. In this case a 6-mm aperture was placed at seven different locations on the frame of "Old Main." The signal-to-noise spectrum obtained at these seven different locations is shown. Examination of the original "Old Main" scene shows that there is obviously a significant variation in object content with location in the frame. The other scenes showed a somewhat smaller variation within the frame. Figure 6.5-5 shows in another way the magnitude of this varying distribution. The means of the signal-to-noise power spectrum values obtained in Fig. 6.5-4 are plotted along with the standard deviations, indicated by the vertical bars. Finally, for this same scene, Fig. 6.5-6 shows a comparison of the seven measurements which were made independently and covered approximately the same area as a 20-mm-diameter aperture measurement. Quite reasonable agreement is observed between the averaged power spectrum from a single large aperture measurement and the averaging of the power spectral distribution in the small areas.

Similar results were obtained for all of the other scenes. We conclude from this observation that—at least for photography of the type being

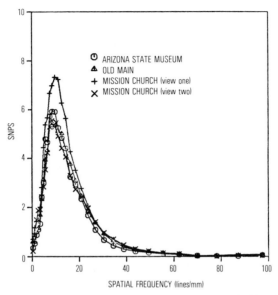

FIG. 6.5-3. Signal-to-noise power spectra of random test scenes in Figs. 6.5-1a–6.5-1d.

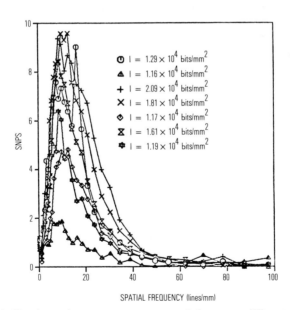

FIG. 6.5-4. Signal-to-noise power spectrum recorded at seven different aperture placements on the "Old Main" test scene. An aperture size of 6 mm was used, and the information content computed for each curve.

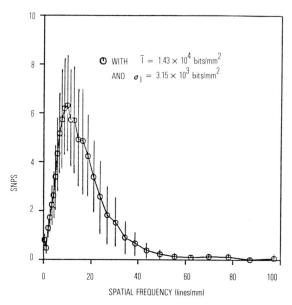

FIG. 6.5-5. Average signal-to-noise power spectrum curve computed from curves of Fig. 6.5-4 with mean and standard deviation.

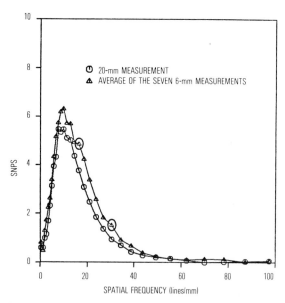

FIG. 6.5-6. Average signal-to-noise power spectrum computed from seven measurements with a 20-mm aperture of the "Old Main" test scene.

discussed—the averaging effects of optically processing a large area sample of an imagery to obtain a statistical evaluation of the scene content is indeed a feasible and sensible way of describing the image content. When the calculation of information content as obtained from the signal-to-noise power spectrum definition is used, this curve indicates an information content of about 14,000 bits/mm^2, with the averaging of the six individual measurements indicating a standard deviation of about 3000 bits/mm^2 across the image. A factor of about 2 in fact existed between the very lowest measurement of information content and the highest measurement of information content noted with the six apertures.

Other measurements were made with different aperture sizes on the images, and in general the conclusion was the same. The averaging of information content over small apertures or the ranking of the information content measure over a single large aperture gave the same average results.

Another area of interest was the effect of different materials on the signal-to-noise power spectrum. The two materials chosen were Plus X and Panatomic X, and both were processed to different gammas. The large amount of data expected meant that not every scene was evaluated for every contrast on each film. In order to determine whether the theory in fact worked, the images obtained at medium contrast were used to predict what the change in the level of the signal-to-noise power spectrum would be for the high and low contrast images. A comparison also was made of high and low contrast development.

Figure 6.5-7 shows a result of this experiment on Plus X film. A single medium contrast image was evaluated, as were those from low and high contrast development. As can be seen, the prediction of the contrast change for the low and high contrast images is quite good, indicating again that the concept of signal-to-noise power spectrum seems to be workable and matches experimental characteristics. Figure 6.5-8 shows a similar experiment carried out with Panatomic X film. In this case the results, although still impressive, show larger differences between the predicted and measured values, especially in the 20 cycles/mm range. It is to be noted that relatively higher spatial frequencies were represented in the Panatomic X material, as would be expected. However, the prediction obtained by the simple process of scaling by gamma did not appear to work very well, probably because of the different shapes of the characteristic curves obtained from the different development times. No attempt was made to make a detailed correction for the change in shape of this curve. This correction could probably be obtained by evaluating a histogram of density differences. However, even without this correction, the agreement is fairly good and indicates that, as long as a reasonable choice of development characteristics is used, the signal-to-noise power spectrum concept should be applicable for comparing photographic imaging systems.

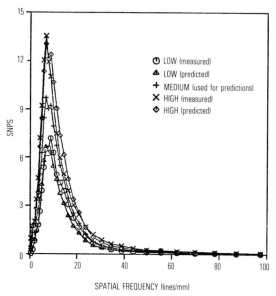

FIG. 6.5-7. Comparison of measured and predicted signal-to-noise power spectrum curves for the contrast samples on Plus-X film.

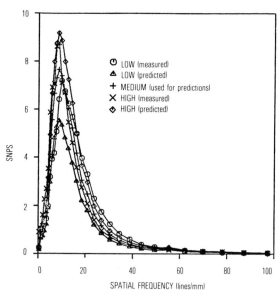

FIG. 6.5-8. Comparison of measured and predicted signal-to-noise power spectrum curves for contrast samples on Pan-X film.

As a final important example in this regard, Sagan also investigated the effect of defocus. A large number of sample images were obtained and measured. A theoretical defocus modulation transfer function for the lens was then used to correct the signal-to-noise power spectrum at each spatial frequency. The in-focus signal-to-noise power spectrum curve was used as a standard, and then the theoretical adjustments made throughout the focus range.

Many examples were presented in the thesis [11]. One example is given as Fig. 6.5-9. The agreement between the measured and predicted values is seen to be excellent. This is typical of the results obtained over all the sample sets. One can then conclude that, in general, the technique of signal-to-noise power spectrum analysis of different scene contents provides useful average information which can be corrected for changes in the optical system. Although changes in the *H-D*, or characteristic, curve caused by different processing presented some difficulty, a reasonable level of agreement was demonstrated. In all the examples referred to here the significant spatial frequency components of the imagery were below 100 cycles/mm. Such imagery was chosen in order that the optics used in providing the sample imagery could image at close to the diffraction limit. This made development of the theoretical predictions of image formation somewhat easier.

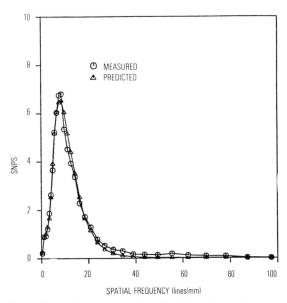

FIG. 6.5-9. Comparison of measured and predicted signal-to-noise power spectrum curves for the medium contrast $f/16$ Pan-X sample with defocus = 1.79 waves.

6.6 DESCRIPTION OF GRANULARITY

The discussions so far have dealt with a description of the information portion, or the image portion, by optical power spectrum techniques. In the preceding work it was necessary to remove the effect of granularity noise and processor noise. This section deals with the measurement and description of the granularity of photographic materials by optical power spectrum techniques. It is entirely possible that the granularity of other processes that have been photographically recorded, such as the output of electro-optical systems, may be described by these techniques as well.

A difficulty arises in that it is desirable to obtain the granularity in terms of an absolute quantity. Historically, the silver image on a film has been described in terms of density, as it was believed that the latter was proportional to the amount of developed silver that was retained in the emulsion. Optical measures of the opacity of the emulsion may be related through a proportional coefficient to the amount of deposited silver per unit area [12]. Granularity and the perceived quantity graininess are a consequence of the statistical variation in the size and distribution of the individual silver grains that make up the photographic image. Since silver is a carrier for the image, it is desirable to be able to quantify the magnitude of this fluctuation in absolute units rather than relative units. In this case the direct use of signal-to-noise power spectrum techniques must be approached with a bit of caution.

The unit to be used is, of course, that of optical transmission. In fact, it is not absolute but generally relative to the transmission aperture. The power spectrum description of the granularity is then obtained by comparing the granularity spectrum to the spectrum of the light diffracted by the aperture plus base (but no emulsion), as was previously done in defining the signal-to-noise power spectrum curve. The difference is that it is necessary to be able to state an absolute level for the granularity in order to permit comparison with theoretical models of the granularity.

David Honey, for his Master's degree thesis at the Optical Sciences Center, University of Arizona, studied the question of predicting the granularity spectrum from parameters associated with the emulsion [13]. In part of this work he measured the granularity spectrum associated with the emulsion. Honey measured the granularity of Tri-X and 3414 emulsions both with an optical power spectrum analyzer and with a microdensitometer. Computation of the Wiener spectrum from the measurements on the microdensitomer were then compared with the optical power spectrum measurement. In addition, he compared liquid-gated and non-liquid-gated samples. Some of the data obtained will now be presented and their relationship to optical power spectrum measurements discussed. Figures 6.6-1 to 6.6-3 show the Wiener spectrum developed from microdensitometer data compared

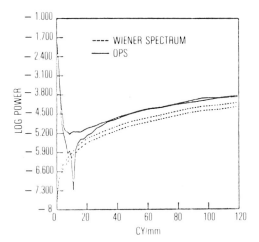

FIG. 6.6-1. Optical power spectrum and phase-compensated Wiener spectrum for shoulder or high exposure area of the Tri-X characteristic curve.

with the optical power spectrum measurement of the granularity for different sections of the *H-D*, or characteristic, curve of Tri-X film. The curve is analyzed for three sections, the toe or low exposure area, the midrange area, where the film is essentially linear, and the shoulder or high exposure area, where film saturation occurs. A reasonable level of agreement can be noted here except in the case of the toe region. The optical power spectrum data show a significantly lower granularity spectrum level for this case. The general shapes of the curves are noted to be consistent.

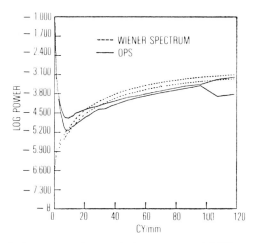

FIG. 6.6-2. Optical power spectrum and phase-compensated Wiener spectrum for mid-range area of the Tri-X characteristic curve.

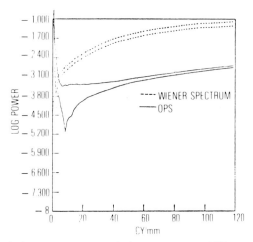

FIG. 6.6-3. Optical power spectrum and phase-compensated Wiener spectrum for toe or low exposure area of the Tri-X characteristic curve.

Figures 6.6-4 to 6.6-6 show a similar behavior for the non-phase-compensated comparisons. Again, reasonable agreement can be noted, but we note that the levels of the spectra are higher in both cases for the non-liquid-gated than for the liquid-gated values. These data show reasonable agreement between the two methods of measurement. A reasonable level of agreement

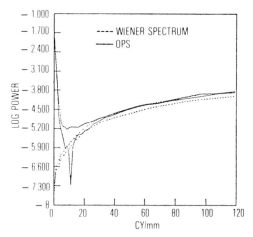

FIG. 6.6-4. Optical power spectrum and non-phase-compensated Wiener spectrum for shoulder or high exposure area of the Tri-X characteristic curve.

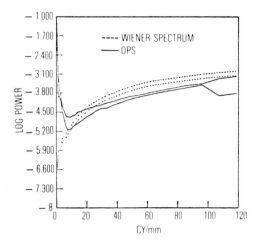

FIG. 6.6-5. Optical power spectrum and non-phase-compensated Wiener spectrum for mid-range area of the Tri-X characteristic curve.

was also noted between these spectra and predictions made on the basis of Monte Carlo simulations of the grain distributions in the emulsion.

Figures 6.6-7 to 6.6-9 show an attempt to carry through the same measurements with the 3414 film. The data were not in as good agreement, perhaps because of the pelloid backing on the 3414 film, which was not

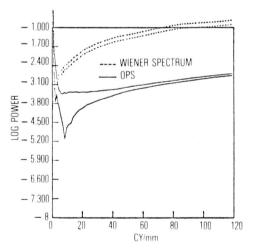

FIG. 6.6-6. Optical power spectrum and non-phase-compensated Wiener spectrum for toe or low exposure area of the Tri-X characteristic curve.

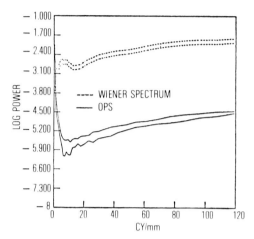

FIG. 6.6-7. Optical power spectrum and phase-compensated Wiener spectrum for shoulder or high exposure area of the 3414 characteristic curve.

removed during the measurement. Another source of error could have been the small size of the grains in the 3414 material, which may in fact exhibit considerably different scattering properties when observed through the large angle aperture of the microdensitometer rather than the small numerical aperture optics of the optical power spectrum device. At present, the detailed cause of this is left as a matter of some conjecture.

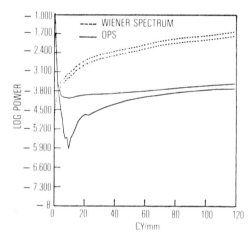

FIG. 6.6-8. Optical power spectrum and phase-compensated Wiener spectrum for midrange area of the 3414 characteristic curve.

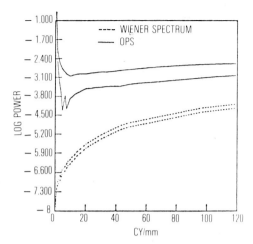

FIG. 6.6-9. Optical power spectrum and phase-compensated Wiener spectrum for toe or low exposure area of the 3414 characteristic curve.

6.7   CONCLUSION

The theory as developed in this chapter proposes a method for calculating the signal-to-noise power spectrum which is related to the information content of the image on film. Several examples with different objects and different imaging systems illustrate that indeed this concept serves as a valid measure of the image content of what is on the film.

Attempts to quantify the granularity characteristics of the film were quite successful with the large coarse-grained emulsion Tri-X. Results were not as successful with the very fine-grained emulsion. The conclusion, based on this work, is that the optical power spectrum technique is not satisfactory for absolute quantification of granularity levels on very fine-grained materials.

Nevertheless, we have demonstrated that, on both fine- and coarse-grained materials, the power spectrum characteristics of the recorded image are reasonably accurately portrayed, even in the absence of any elaborate phase compensation of the film to eliminate surface relief effects. This demonstration indicates that this concept may be used for the screening and evaluation of technical photography.

Future work should examine the application of this concept to different types of imagery to determine whether the power spectrum is related to the perceived image quality. Since it appears at least to date that no good correlation exists between subjective image quality and information content, it is speculative to suppose that an evaluation of the pictorial characteristics of photography other than the technical or information content is possible through this route. Use of the power spectral data for entry into various image quality criteria is of course possible but is beyond the scope of the work presented here.

REFERENCES

[1] K. R. Hessel (1974). *Appl. Opt.* **13,** 1023.
[2] H. Stark, W. R. Bennet, and M. Arm (1969). *Appl. Opt.* **8,** 2165.
[3] H. Stark (1971). *Appl. Opt.* **10,** 333.
[4] J. W. Goodman (1968). "Introduction to Fourier Optics." McGraw-Hill, New York.
[5] J. D. Gaskill (1978). "Linear Systems, Fourier Transforms, and Optics." Wiley, New York.
[6] P. S. Cheatham (1976). Calibration of an optical power spectra measurement device, Ph.D. Thesis, Opt. Sci. Cent., Univ. of Arizona, Tucson.
[7] J. C. Dainty and R. Shaw (1974). "Image Science." Academic Press, New York.
[8] G. G. Lendaris and G. L. Stanley (1970). *Proc. IEEE* **58,** 198.
[9] WRD 6400 Data Sheet, Recognition Systems, Inc., Van Nuys, California.
[10] J. W. Goodman (1967). *J. Opt. Soc. Am.* **57,** 493.
[11] S. Sagan (1980). Power spectrum analysis of photographic images, M.S. Thesis, Opt. Sci. Cent., Univ. of Arizona, Tucson.
[12] H. Thiry (1963). *J. Photogr. Sci.* **11,** 69.
[13] D. A. Honey (1978). A comparison of measured and predicted photographic noise power spectrum, M.S. Thesis, Opt. Sci. Cent., Univ. of Arizona, Tucson.

Chapter 7

Fourier Optics and SAW Devices

*P. DAS AND F. M. M. AYUB**

DEPARTMENT OF ELECTRICAL, COMPUTER,
  AND SYSTEMS ENGINEERING
RENSSELAER POLYTECHNIC INSTITUTE
TROY, NEW YORK

7.1 INTRODUCTION

An elastic wave propagating in a solid or liquid medium induces a periodic perturbation in the medium's refractive index via the photoelastic effect. As a result, light propagating through such a medium can be diffracted by the wave. This diffraction phenomenon has been studied extensively since the first observations of Brillouin [1] and the pioneering studies of Raman and Nath [2] and Debye and Sears [3]. More recently the interaction has been utilized in many applications, including several of those mentioned in the earlier chapters of this book.

The acousto-optic interaction is shown schematically in Fig. 7.1-1, in which the diffracted light is modulated by the signal used to generate the ultrasound. (Ultrasound is an elastic wave of frequency typically above ~ 50 KHz.) An electrical signal is fed to the ultrasound transducer, usually a piece of piezoelectric material, which generates ultrasound in the medium to which the transducer is bonded. The modulated light emerges at an angle from the incident beam. Although very wideband bulk transducers can be easily fabricated, the bonding of the transducer to the modulating medium is rather inconvenient from the practical point of view. This can be avoided if the same material is used for the transducer and the modulating medium.

* Present address: The Bendix Corporation, Advance Technology Center, Columbia, Maryland.

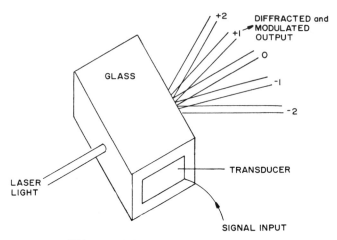

FIG. 7.1-1. Ultrasound laser light modulator.

This integration is easily accomplished using surface acoustic waves (SAWs) instead of bulk ultrasound. An added advantage of these surface acousto-optic devices is their compatability with guided optical waves in integrated optical configurations [4, 5].

The operations that can be performed with coherent-optical processors [6, 7] are Fourier, Fresnel, and Mellins transforms, convolution, correlation, and linear space-invariant filtering. These operations rely mostly on the Fourier transform property of the lens and can be performed in the frequency domain. There is another class of optical processors [8] which use an incoherent source. In this class, no Fourier transform relation exists between the front and rear focal planes of a lens, and therefore all the operations are done in the spatial domain. The advantages of the optical processor over a digital processor are (i) the ability of the optical system to operate in parallel on all inputs in a two-dimensional input space and (ii) the speed with which these operations are performed. To realize these advantages of optical processing in real time, efficient transducers are required to introduce the input data in real time. Acousto-optic cells are widely used as transducers in coverting real-time electrical signals to optical signals for optical signal processing. The first application of these devices for television projection system was developed by Okolicsanyi [9] in 1937. With the discovery of the laser and the development of SAW devices, acousto-optic signal processing has become relatively simpler and cheaper. The acousto-optic interaction between SAW and optically guided waves is a rapidly developing area of integrated optics. It should be mentioned that other SAW devices [10, 11] have been developed which can perform useful signal processing functions by themselves without the acousto-optic interaction, and in a sense these devices and SAW optic devices compete with each other.

In Section 2 the properties of SAW and the interaction of SAW with light are reviewed. Recently, the application of surface acousto-optic devices in real-time signal processing and image scanning has become increasingly popular because of their wide bandwidth and high dynamic range capabilities. In Section 3 some of these devices, such as the spectrum analyzer, real-time correlator or convolver, and electronically focusable lens are discussed. The final section suggests future developments and summarizes the main achievements of these devices to date.

7.2 SURFACE ACOUSTO-OPTIC INTERACTION

Surface acoustic waves have been well known and well studied by seismologists since Lord Rayleigh's discovery of this mode of wave propagation in 1895 [12]. Only in the last two decades [10], however, has the potential of SAWs in the electronics industry been fully realized. This has been due to two main factors, namely, the availability of piezoelectric substrates such as lithium niobate ($LiNbO_3$), and the ease with which SAW signals can be generated and detected on these materials through the use of interdigital transducers (which require microcircuit techniques for their fabrication, as explained later in this section). Thus, one can easily fabricate a delay line with an insertion loss of only a few decibels, tens of microseconds of delay, and a center frequency anywhere from 10 MHz to a few gigahertz. It is important to mention that a delay line using bulk ultrasound can also be made, but there are major differences between the two types of delay lines which make the SAW delay line more attractive. For example, the SAW signal can be easily tapped from the surface of a piezoelectric substrate by sets of interdigital transducers placed between the main input and output transducers, resulting in a tapped delay line structure. Additional flexibility is easily achieved by assigning independent tapping weights at each tap through apodization (variable finger overlap) or by other techniques making it possible to realize transversal filters with any desired characteristics within certain limits.

A SAW is ultrasound with surface confinement. Compared to bulk wave propagations in solids a SAW is rather complex, as shown in Fig. 7.2-1. Figure 7.2-1a shows the propagation of a bulk sound wave in a solid represented by grid points. Each of the grid points in the figure is uniformly spaced before the wave propagates. While the wave is propagating (in the z-direction), one observes compression and elongation along the z-axis and uniformity along the y-axis. Figure 7.2-1b shows the SAW propagation. Observe that, for large values of y, the wave barely exists. In an isotropic solid, the

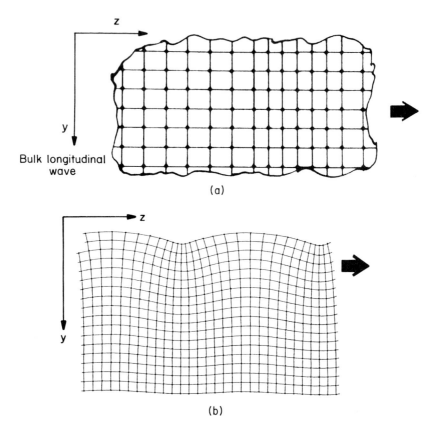

FIG. 7.2-1. Cross section of the displacements of a rectangular grid of material points for a longitudinal bulk wave in (a) and in (b) a surface acoustic wave on an isotropic material.

SAW displacements are given by

$$u_y = \bar{u}_y(y) \cos k(vt - z), \qquad (7.2\text{-}1)$$

$$u_z = \bar{u}_z(y) \sin k(vt - z), \qquad (7.2\text{-}2)$$

where $k = 2\pi/\lambda$, λ is the SAW wavelength, v is the SAW velocity, and $\bar{u}_y$ and $\bar{u}_z$ are the amplitudes of u_y and u_z, respectively. The amplitudes, $\bar{u}_y$ and $\bar{u}_z$, are sketched in Fig. 7.2-2 for a specific material. The mechanical displacements are mostly confined to within a wavelength from the top of the surface and have both shear and longitudinal components propagating with a velocity slightly less than the bulk shear wave velocity.

  In an anisotropic solid, the complexity is further increased. Unless the surface wave is propagating along some particular orientation of the crystal, the propagation vector and the direction of energy flow may not be coinci-

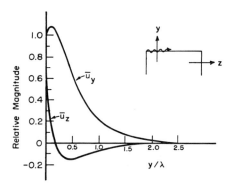

FIG. 7.2-2. Particle displacements for surface acoustic waves in an isotropic medium.

dent. For a piezoelectric crystal, one must also include the propagating electric field associated with the mechanical displacements. Although the mechanical deformation is confined to the solid itself, the electric field exists both inside and outside the crystal. The expression for the resultant inner potential is rather complex, however, the outer potential can be written as

$$\Phi(t, y, z) = \Phi_0 e^{-y/\lambda} \cos k(vt - z), \tag{7.2-3}$$

where Φ_0 is related to the input power. Thus, a SAW carries with it an electric field that exists in air, decays within a wavelength λ from the free surface, and has both a y- and a z-component. These electric fields can interact with electrons in metals or semiconductors brought into close proximity of the surface. This interaction is basic to the understanding of how interdigital transducers (IDTs) [13] can be used for the easy and efficient generation of SAW on piezoelectric substrates such as $LiNbO_3$ or ST-cut quartz. By using these IDTs as input/output ports in a SAW device, energy can be transformed from an electrical signal to ultrasound or from ultrasound to an electrical signal in a controlled manner.

The heart of the SAW optic device is the SAW delay line. A typical SAW delay line, shown in Fig. 7.2-3a, uses two interdigital transducers where one is the input and the other is the output. By applying a voltage to the input at frequency $f = 1/T$, where T is the time required for the surface to travel one wavelength λ with velocity v, one generates an electric field between the fingers. A typical interdigital transducer is shown in Fig. 7.2-3b. The width of each metal finger is $\lambda/4$. The alternate fingers are connected to one another by a summing buss. When an rf voltage is applied across these two summing busses, an electric field is set up between adjacent fingers. Recall that this electric field alternates with a time period $\frac{1}{2}T = 1/2f = \pi/\omega$. Thus the piezoelectric material is alternately compressed and elongated between the fingers, and these elastic deformations propagate with a velocity v, the surface

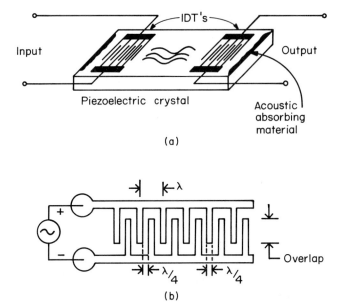

FIG. 7.2-3. (a) SAW delay line. (b) Interdigital transducer.

wave velocity. Thus for the SAW to propagate a distance of $\lambda/2$, it takes a time $T/2$, so that all the excitations reinforce and efficient surface waves are generated. For detection, the reverse is true; i.e., as the elastic deformation passes beneath the fingers, a potential is induced between them.

A basic advantage of the SAW devices discussed above is that they use planar technology and thus can be cheaply mass-produced. For the tapped delay line applications, one major handicap is the rather large temperature coefficient of $LiNbO_3$ or other high coupling material which has a low insertion loss. If one is willing to sacrifice efficiency, a substance known as ST-cut quartz is available, which has a zero first-order temperature coefficient. However, for SAW optic devices $LiNbO_3$ is most often used, since the variation of SAW velocity with temperature plays an insignificant role in this case.

SAW technology includes not only Rayleigh waves, but all the waves which can propagate in a solid and be somewhat confined near the surface. For Rayleigh waves the confinement is on the order of one wavelength (for $LiNbO_3$ at 100 MHz, $\lambda \simeq 35~\mu m$), but for other waves such as Blustein–Gulayev waves, the confinement width may be much larger, say $\epsilon_r \lambda$, where ϵ_r is the effective relative dielectric constant which, for example, is approximately 30 for $LiNbO_3$. Although some acousto-optic interaction experiments have been performed for other SAW waves [14], most of the devices use Rayleigh wave propagation.

A. Photoelastic Effect

The change in the refractive index of the medium due to externally applied stress is known as the photoelastic effect or sometimes as the elasto-optic or piezo-optic effect. The photoelastic effect provides the interaction between the elastic waves (bulk or surface) and the light, usually known as acousto-optic interaction. The photoelastic effect was first discovered by Sir David Brewster [15] in 1815. The first phenomenological formulation of the photoelastic effect was put forward by Pockels [16] for isotropic materials.

The optical property of a crystal is generally described by the indicatrix [17]. The indicatrix is an ellipsoid which describes the refractive index of a crystal as a function of propagation direction and polarization of the light. In the absence of stress, the general equation of the indicatrix in terms of the impermeability tensor B_{ij} is given by

$$B_{ij}x_ix_j = 1 \qquad i, j = 1, 2, 3, \tag{7.2-4}$$

where by definition

$$B_{ij} = (1/n^2)_{ij} \tag{7.2-5}$$

(summation convention for repeated indices is used). Spatial coordinates are represented by x_i, and n is the refractive index. This is a sphere for isotropic materials and an ellipsoid for anisotropic materials.

The photoelastic effect then results in a change in the size, shape, and orientation of the indicatrix, and Eq. (7.2-4) is modified as

$$(B_{ij} + \Delta B_{ij})x_ix_j = 1. \tag{7.2-6}$$

Pockels formulated that this change in impermeability ΔB_{ij} is linearly related to strain S_{kl} by the relation

$$\Delta B_{ij} = P_{ijkl}S_{kl}, \tag{7.2-7}$$

where P_{ijkl} is the fourth-rank photoelastic tensor and

$$S_{kl} = \frac{1}{2}\left(\frac{\partial u_k}{\partial x_l} + \frac{\partial u_l}{\partial x_k}\right). \tag{7.2-8}$$

Since the strain and indicatrix tensors are unaffected by changing the order of the ij or kl subscripts, Eq. (7.2-7) can be conveniently written in reduced notation as

$$\Delta B_m = P_{mn}S_n, \tag{7.2-9}$$

where m, n varies from 1 to 6. When ultrasound propagates in a piezoelectric medium, it generates an electric field propagating with it, which in turn can cause a change in the refractive index through the electro-optic effect. This indirect contribution is given by

$$\Delta B_{ij} = r_{ijr}E_r, \qquad (7.2\text{-}10)$$

where r_{ijr} are Pockels's electro-optic tensors and E_r are the electric fields generated by ultrasound. This indirect effect was ignored until its importance was demonstrated in the $LiNbO_3$ crystal which displays very large piezo-electric coupling. Recently it has been found that the contribution of the indirect effect dominates the guided acousto-optic interaction in $LiNbO_3$ [18]. The change in the dielectric constant $\Delta\epsilon_{ij}$ follows from Eq. (7.2-10):

$$\Delta\epsilon_{ij} = -\tfrac{1}{2}\epsilon_{ii}^2 P_{ijkl}S_{kl} + r_{ijr}E_r. \qquad (7.2\text{-}11)$$

It has been recently shown that the antisymmetric part of the strain tensor contributes to the photoelastic tensor. This contribution of the so-called roto-optic effect [19] is not negligible in anisotropic material. Thus Eq. (7.2-11) should be modified to include these second-order effects. The most important material for SAW optic devices is $LiNbO_3$, and its important properties are listed in Table 7.2-1.

TABLE 7.2-1

Optical Properties of $LiNbO_3$[a]

Symbol	Value	Description
$\epsilon_{11}^0 = \epsilon_{22}^0$	5.24	Unperturbed relative permittivity
ϵ_{33}^0	4.84	
$P_{11} = P_{22}$	-0.026	Elasto-optic tensor elements
$P_{12} = P_{21}$	0.091	
$P_{13} = P_{23}$	0.133	
$P_{14} = -P_{24} = P_{65}$	-0.075	
$P_{31} = P_{32}$	0.179	
P_{33}	0.071	
$P_{41} = -P_{42} = P_{56}$	-0.151	
$P_{44} = P_{55}$	0.146	
r_{33}	$30.8 \times 10^{-2}e^{j-0.125\pi}$ m/V	Electro-optic tensor elements
$r_{22} = -r_{12} = -r_{61}$	3.4×10^{-12} m/V	
$r_{13} = r_{23}$	8.6×10^{-12} m/V	
$r_{51} = r_{42}$	28×10^{-12} m/V	

[a] From [20].

B. Bulk Acousto-Optic Interactions: Raman–Nath and Bragg Scattering

For a monochromatic ultrasonic wave traveling in the z-direction in an isotropic, homogeneous material, we can represent the refractive index in the region of the wave as

$$n(z, t) = n_0 + n_1 \sin(\omega t - kz), \tag{7.2-12}$$

where n_0 is the unperturbed refractive index and n_1 is the maximum change in refractive index arising from the presence of the sound wave. (Note that, in SAW, n_1 depends on y.) In terms of the elastic strain S, the elasto-optic constant p, and the unperturbed refractive index, it follows from Eqs. (7.2-5) and (7.2-7) that n_1 is given by

$$n_1 = \tfrac{1}{2}n_0^3 pS. \tag{7.2-13}$$

The velocity of the acoustic wave is normally about five orders of magnitude less than the velocity of light in the material. Consequently an optical beam incident on the sound wave effectively encounters a static phase grating of periodicity λ, the acoustic wavelength. It is well known that a spatially and temporally coherent, monochromatic, collimated light beam normally incident on such a phase grating is diffracted in the far field into several orders, as shown schematically in Fig. 7.1-1. The angle of the nth order from the optic axis θ_n is inferred by

$$\sin \theta_n = n\lambda_{0f}/\lambda, \tag{7.2-14}$$

in which λ_{0f} is the free-space wavelength of the incident light. The intensity of the nth-order diffracted light, normalized to that of the incident light, is given by

$$I_n/I_0 = J_n^2(\phi). \tag{7.2-15}$$

In Eq. (7.2-15) J_n is the nth-order Bessel function of the first kind, and ϕ is the phase retardation introduced in the light as a result of the presence of the sound wave and is given by

$$\phi = (2\pi/\lambda_0)n_1 L, \tag{7.2-16}$$

in which λ_0 is the wavelength in the medium and L is the interaction length. The interaction length is the distance over which the light and sound overlap, measured along the direction of optical propagation.

This description of the acousto-optic interaction in which the acoustic wave is considered a phase grating was first given by Raman and Nath [2]. Under certain circumstances to be discussed, the effect of the acoustic wave is better described by considering the wave a series of partially reflecting mirrors corresponding to the maxima of the refractive index wave, Eq. (7.2-12). By analogy to x-ray diffraction, when the light is incident at an angle θ_i inferred by

$$\sin \theta_i = \tfrac{1}{2}\lambda_{0f}/\lambda, \tag{7.2-17}$$

the reflections interfere constructively in one direction, giving rise to a single diffraction order of intensity

$$I_1/I_0 = \sin^2 (\tfrac{1}{2}\phi). \tag{7.2-18}$$

This type of diffraction is called Bragg diffraction and is shown schematically in Fig. 7.2-4a, while Fig. 7.2-4b shows the momentum diagram for Bragg diffraction. The momentum diagram corresponds to a particle model of the interaction in which a photon and phonon are annihilated with the simultaneous creation of a new photon [17]. Conservation of momentum provides the result that scattered light exists at an angle

$$\theta_d = \sin^{-1} \tfrac{1}{2}(\lambda_{0f}/\lambda), \tag{7.2-19}$$

and conservation of energy requires that the frequency of the scattered light be upshifted with respect to the incident light by the acoustic frequency. The shift in frequency can also been arrived at by considering the Doppler frequency shift of the light wave reflected from moving mirrors.

A unified theory treating both Raman–Nath and Bragg scattering has

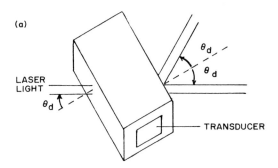

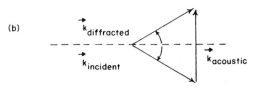

FIG. 7.2-4. Bragg diffraction: (a) acousto-optic interaction with bulk wave, (b) momentum conservation for Bragg interaction.

been given by Klein and Cook [21], in which it was shown that the quantity Q, defined by

$$Q = 2\pi\lambda_{0f}L/n_0\lambda^2, \qquad (7.2\text{-}20)$$

describes the nature of the acousto-optic interaction. For $Q \ll 1$, Raman–Nath diffraction occurs; for $Q \gg 1$, Bragg diffraction occurs. In the region $Q \sim 1$ an intermediate type of diffraction occurs.

For anisotropic crystals several complications arise. For example, the diffracted light might have a polarization perpendicular to that of the incident light. This happens when there are nonzero off-diagonal elements in the impermeability tensor of Eq. (7.2-6). Equations (7.2-17) and (7.2-19) describing the Bragg conditions are then modified as follows [22]:

$$\theta_i = \sin^{-1}(\lambda_{0f}/2n_d\lambda)[1 + (\lambda/\lambda_{0f})^2(n_i^2 - n_d^2)], \qquad (7.2\text{-}21a)$$

$$\theta_d = \sin^{-1}(\lambda_{0f}/2n_d\lambda)[1 - (\lambda/\lambda_{0f})^2(n_i^2 - n_d^2)], \qquad (7.2\text{-}21b)$$

where n_i and n_d are the indices for the incident and diffracted wave, respectively.

C. Surface Acousto-Optic Interaction

Four interaction configurations [23] have been used to achieve an acousto-optic interaction between a surface acoustic wave and an unguided optical wave. These are shown in Fig. 7.2-5. In the top reflection configuration, light is reflected from the surface on which the acoustic wave propagates. Diffraction occurs because of the slight surface corrugation associated with

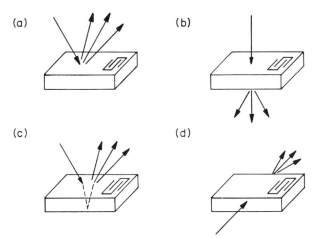

FIG. 7.2-5. Interaction configurations: (a) top reflection, (b) top transmission, (c) back reflection, and (d) side entry.

the propagation of the acoustic wave. Although the interaction is weakest in this configuration, it is a useful arrangement for noncontact probing of surface waves.

A second interaction geometry is shown in Fig. 7.2-5b, the top transmission configuration. The incident laser light is directed nearly normal to the acoustic propagation surface, and the diffracted light exits through the bottom of the crystal. Because the strain components of the surface acoustic wave decay with distance from the crystal surface, this configuration has a very short interaction length, on the order of an acoustic wavelength. A twofold improvement in the interaction length is achieved through the scheme shown in Fig. 7.2-5c, the back-reflection configuration. In this case the light is incident, as in the top transmission case, but after exiting from the backside of the crystal, it is reflected back through the crystal by a mirror for a second pass through the acoustic wave. The periodic corrugation of the crystal surface due to the acoustic wave provides a slight phase contribution in addition to that due to the internal effect. The top transmission and back-reflection configurations have received considerable attention [24, 25].

The side-entry technique, Fig. 7.2-5d, provides a vast improvement in interaction length over the other configurations. The incident light in this case enters from the side of the crystal, near the crystal top surface, and propagates parallel to the surface on which the acoustic wave is propagating. The interaction length for the side-entry mode of operation is given, therefore, not by the depth of penetration of the acoustic wave below the crystal surface but rather by the acoustic beam width which is determined by the aperture of the interdigital transducer. For a transducer on y-cut z-propagating LiNbO$_3$ designed to match to 50 Ω at the center frequency, this is an improvement of two orders of magnitude over the top transmission case. Recent work suggests that a total internal reflection condition is achievable with the side-entry technique which provides additional enhancement of the interaction in this device configuration [26]. Although optimum optical alignment is more difficult to achieve with the side-entry technique than with the other configurations, it is generally quite easy to exceed the interaction strength possible with the other interactions even with a nonoptimal alignment.

One feature common to all acousto-optic devices employing surface acoustic waves is that the figures of merit are larger than those for bulk devices employing the same material. This is due to the fact that the surface acoustic wave velocity is invariably smaller than the corresponding bulk wave velocity, on the order of a factor of 1.5. Another advantage of surface acoustic waves over bulk waves is the very small acoustic beam waist aspect ratio, allowing very efficient diffraction. However, of the four interaction configurations shown in Fig. 7.2-5, only the side-entry technique takes advantage of this fact. It is also easy to achieve the Bragg condition $Q \gg 1$ in this configuration.

The diffraction of light by SAWs is more complex than that by bulk ultrasound because of the presence of both the longitudinal and shear strain components. In general, the SAW particle displacement can be written as

$$U_i(t, y, z) = A_i \sum_{n=1}^{3} a_{in} e^{\alpha_n k y + j(\omega t - kz)}, \qquad i, n \rightarrow x, y, z, \qquad (7.2\text{-}22)$$

where A_i and a_{in} are amplitude constants and α_n are decay constants. The numerical values of the constants are given in reference [27]. These displacements cause a change in refractive index of the medium as well as a ripple on the surface. This surface ripple, a unique property of SAW is given by

$$\delta(t, z) = \delta_0 \cos(\omega t - kz), \qquad (7.2\text{-}23)$$

where $\delta_0 = |U_y|_{y=0}$ for a y-cut z-propagating SAW. This ripple introduces additional phase modulation of the reflected light. Thus, for a complete theory of surface acousto-optic interaction, one should consider both the changes in refractive index and the surface ripple. However, only the surface corrugation need be considered for the top reflection configuration, and this is discussed below.

The top reflection interaction configuration has been extensively used as a diagnostic tool to determine the SAW parameters, i.e., attenuation, beam steering, diffraction profile, velocity, and generator efficiency of SAW and harmonic generation due to elastic nonlinearities. The different approaches in solving the SAW acousto-optic interaction for the top reflection configuration and the different optical probing techniques have been dealt with in detail by Stegeman [28]. For the configuration shown in Fig. 7.2-6a, the diffracted light intensity is given by [29]

$$I(\theta, t) = A(\theta, t)A^*(\theta, t), \qquad (7.2\text{-}24)$$

where $A(\theta, t)$ is the amplitude of the light scattered into angle θ and is given by

$$A(\theta, t) = (1/L)e^{j\omega t} \int_0^L \exp[j(kz\theta \cos \theta_i + \phi_1 \sin kz)] \, dz$$

$$= \sum_{n=0}^{\infty} J_n(\phi_1) \left[\sin \frac{L}{2}(nk - k_1 \cos \theta_i) \right] \bigg/ \frac{L}{2}(nk - k_1 \cos \theta_i), \qquad (7.2\text{-}25)$$

where $\phi_1 = (4\pi/\lambda_0)\delta_0 \cos \theta_i$, L is the width of the optical beam incident at the the surface, and $k_1 = 2\pi/\lambda_0$.

For the other three configurations there are two different approaches in calculating the light intensity diffracted by SAW. In the first case, to obtain

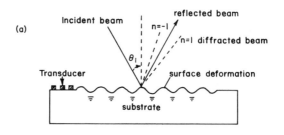

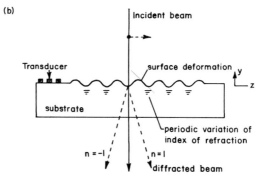

FIG. 7.2-6. Schematic of light diffraction by a SAW: (a) reflection case, $\phi_1 = (2\pi/\lambda_0)\delta_0 \cos\theta_i$, (b) transmission case. $\phi_2 = (2\pi/\lambda_0)[(n-1)\delta + \int_0^\infty \Delta n \, dy]$.

the diffracted electric field, the following wave equations in different regions have to be solved satisfying proper boundary conditions:

$$-\nabla^2 E_i + [\nabla(\nabla E)]_i = \mu_0 \frac{\partial^2}{\partial t^2} \epsilon_{ij} E_j, \qquad i = 1, 2, 3, \quad j = 1, 2, 3. \quad (7.2\text{-}26)$$

In the above equation E_i is the component of the electric field inside the diffracting media and ϵ_{ij} is obtained from Eq. (7.2-6). The coupling of the light and the acoustic field, due to a periodical change in the electrical susceptibility ϵ_{ij} of the material caused by the acoustic wave, gives rise to diffracted light of different orders. The amplitudes of this diffracted light can be obtained by matching the SAW modified boundary conditions. This approach has been used for front reflection transmission and side entry with total interval reflection [30]. Alipi et al. [24] used continuity of the tangential electric field at the front and back reflecting and transmitting surfaces, and Schumer and Das [31] used coupled-mode equations for the side-entry configuration. In the latter approach, the diffracted electric field is obtained for both the reflection and transmission case by integrating the Helmholtz integral with proper Green's functions [25]. The results of two particular analyses [25, 31] are detailed below. They have been selected because of their simplicity and validity for different angles of incidence and polarizations.

As pointed out earlier, the interaction in the transmission configuration can be considered a two-step process: first, the interaction brought about by the surface ripple and, second, the interaction brought about by the refractive index variation. The phase modulation of the light transmitted in the y-direction for the y–z LiNbO$_3$ is given by

$$\phi_j = \frac{2\pi}{\lambda_{0f}}\left[(n-1)\delta_0\cos(\omega t - kz) + \int_0^\infty \Delta n_{jj}\, dy\right], \qquad (7.2\text{-}27)$$

in which the first term inside the brackets is due to the surface ripple, the second term is due to the refractive index variation and Δn_{jj} is obtained from Eq. (7.2-11). The contribution due to the periodic variation in the refractive index of the medium may add or subtract from that of the surface ripple, depending on the incident light polarization. For example, when they add, Eq. (7.2-27) reduces to

$$\phi_j = \phi_2\cos(\omega t - kz), \qquad (7.2\text{-}28)$$

where

$$\phi_2 = \frac{2\pi}{\lambda_0}\left[(n-1) + \left|\int_0^\infty \Delta n_{ij}\, dy\right|\right]$$

and is shown in Fig. 7.2-6b. The intensity of the nth-order diffracted light and the angular relation are the same as given in Eq. (7.2-25), with ϕ_1 replaced by ϕ_2. The experimental results obtained by Alippi *et al.* [24] agree quite well with the theoretical expression. They have studied the first-order diffracted light for different polarizations as a function of the incidence angle for LiNbO$_3$ and quartz substrates. The normalized diffraction efficiency as a function of the incidence angle is shown in Fig. 7.2-7 for LiNbO$_3$. They have shown that the contributions due to surface corrugation and refractive index variation add constructively for S-polarized incident light (electric field vector parallel to z) and add destructively for O-polarized light (electric field vector parallel to y).

The theory of the interaction between SAWs and light for the side-entry configuration adopted here is similar to that of the phase lattice theory developed by Raman–Nath. Figure 7.2-8 defines the interaction geometry in which the light propagation direction is along the x-axis of the y-cut z-propagating LiNbO$_3$. The relative permittivity can be written as

$$\Delta\epsilon_{ij} = R_{ij}(y)e^{jh_{ij}(y)}e^{j(\omega t - kz)} + \text{c.c.}, \qquad (7.2\text{-}29)$$

where R_{ij} and h_{ij} are real-valued functions of y which can be obtained through Eqs. (7.2-6) and (7.2-22) and c.c. is complex conjugate. In order to

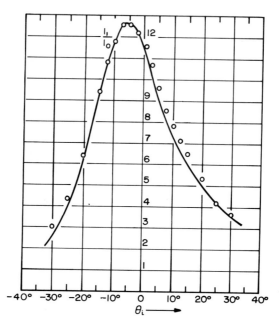

FIG. 7.2-7. Comparison between theory and experiment for z-polarized light diffracted on transmission by surface waves propagating on y–z lithium niobate. The angle of optical incidence is θ_i, and the theory is denoted by the solid line. (After Alippi et al. [24].)

solve the equation, the electric field $E(\cdot)$ can be expanded in a Fourier series

$$E_{(i,y,z,t)} = \sum_{n=-\infty}^{\infty} \phi_n(n, y) \exp[j(\omega + n\omega)t - \mathbf{k}_n^{(i)} \cdot \mathbf{r}], \qquad (7.2\text{-}30)$$

where

$$\mathbf{k}_n^{(i)} \cdot \mathbf{r} = k_n^{(i)}(\cos \theta_n^{(i)} y + \sin \theta_n^{(i)} z). \qquad (7.2\text{-}31)$$

The superscript i on the mode amplitude and wave vector indicates that these depend on the electric field polarization. By substituting these equations in the wave equation [Eq. (7.2-26)] one obtains a set of coupled-mode equations which can be solved for Raman–Nath and Bragg diffraction under proper boundary conditions. Thus the nth-order diffracted intensity for normal incidence is given by

$$I_n/D^2 = \delta_{ip}J_n^2(\tfrac{1}{2}R_{ii}(y)(k_0/n^{(i)})L), \qquad (7.2\text{-}32)$$

where θ_n satisfies Eq. (7.2-14), $n^{(i)}$ is the refractive index for light polarized along the principal axis x_i, and D is a constant related to the incident light intensity.

For the incident light polarized in the x_p-direction ($p = 2$ or 3) the

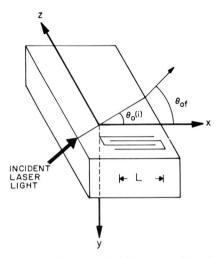

FIG. 7.2-8. Geometry for side-entry configuration.

diffracted light intensity at the Bragg angle is given by

$$I_0^{(p)} = D^2 \cos^2\left(\tfrac{1}{4} R_{pp}(y) \frac{k_0}{n^{(p)}} L\right) \tag{7.2-33a}$$

and

$$I_1^{(p)} = D^2 \sin^2\left(\tfrac{1}{4} R_{pp}(y) \frac{k_0}{n^{(p)}} L\right), \tag{7.2-33b}$$

where $n^{(p)}$ is the refractive index for light polarized along axis x_p and D is the amplitude for the p-polarized light. The total diffracted intensity is obtained by integrating over the y-direction.

The theoretical results obtained by Schumer and Das [31] for the intensity of the first-order diffracted light as a function of depth y, taking into account both elasto-optic and electro-optic effects, are shown in Fig. 7.2-9 for z-polarization. It is observed that the theoretical results agree well with the experimental results obtained by Kramer and Das [32].

D. SAW-Guided Wave Interaction

Thus far we have discussed the situation where light propagates inside the solid as a plane wave with a finite aperture. However, light can be guided at the surface of the LiNbO$_3$ or quartz crystal by making a thin surface layer having a refractive index different from that of the substrate. This can be done by out-diffusing lithium ions or in-diffusing other ions such as titanium [33–34] in a heated furnace. Another way to make an optical waveguide is

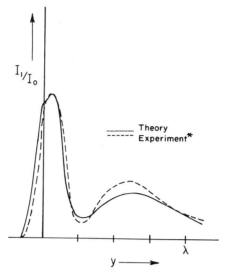

FIG. 7.2-9. Comparison of experimental and theoretical results for first-order diffracted light beam intensity as a function of depth from the surface; z-polarized light is used.

to sputter a glass film on the piezoelectric substrate [35]. A somewhat different approach is becoming of great interest as the technique of depositing reproducible piezoelectric ZnO film becomes better understood. This technique allows the optical waveguide to be placed on a regular glass plate, permitting surface acoustic waves to be generated because of the piezoelectric property of ZnO film [36].

In general the guided wave is coupled through a prism or a grating coupler. This guided wave is rather complex and can have different mode structures such as TE and TM. Thus the complexity of SAW interaction increases by an order of magnitude, although the basic concepts as embodied in the Bragg diffraction angle results given by Eqs. (7.2-21) remain the same. Consider Fig. 7.2-10 where the optical waveguide is of thickness $2d$ and the incident and diffracted lights are both guided TE modes. The intensity I_d of the diffracted light, having a TE mode structure for an interaction length L, has been shown by Lean [25] to be given by

$$I_d = I_i |\Gamma_{di} \cos \theta_d|^2 L^2 (\sin^2 TL)/(TL)^2, \qquad (7.2\text{-}34)$$

where I_i is the incident intensity of the input TE_m mode,

$$T^2 = |\Gamma_{di}\Gamma_{id}(\cos \theta_i \cos \theta_d)| + \tfrac{1}{4}\Delta K_2^2, \qquad (7.2\text{-}35)$$

θ_d and θ_i are the incident and diffracted angles, respectively, $\Delta K_2 = K_d \cos \theta_d$

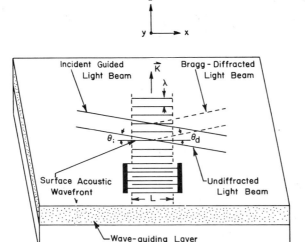

FIG. 7.2-10. Schematic layout for guided wave Bragg diffraction by a SAW.

$- K_i \cos \theta_i$,

$$\Gamma_{di} = \frac{\pi u_n}{2\lambda} \frac{(\sin \theta_i \sin \theta_d \epsilon_{yy} I_{yy} + \cos \theta_i \cos \theta_d \epsilon_{zz} I_{zz})}{\displaystyle\int_0^\infty |u_d|^2 \, dy} \qquad (7.2\text{-}36)$$

$$\Gamma_{id} = \Gamma_{di}(u_m/u_n) \left(\int_0^\infty |u_d|^2 \, dy \Big/ \int_0^\infty |u_i|^2 \, dy \right) \qquad (7.2\text{-}37)$$

$$I_{22} = \int_0^\infty p_{22q3} B_{q3} V_{q3} u_i u_i^* \, dy \qquad (7.2\text{-}38)$$

$$I_{33} = \int_0^\infty p_{33q3} B_{q3} V_{q3} u_i u_d^* \, dy, \qquad (7.2\text{-}39)$$

and u_n corresponds to the propagation velocity of the nth mode. Here u_y is the transverse dependence of the E-field for the TE_n mode and V_{q3} is the y-dependence of the SAW displacement vector given by Eq. (7.2-22). For an isotropic medium, Eq. (7.2-34) simplifies to

$$I_d = I_i \sin^2 \{(\pi \cos \theta_i)/\lambda_{0f}] L\sqrt{(M/2)P_{ac}F^2}\}, \qquad (7.2\text{-}40)$$

where $M = n^6 p^2/\rho v^3$, P_{ac} is the acoustic power density, and F is given by

$$F = \int_0^\infty u^2 V \, dy \Big/ \int_0^\infty |u|^2 \, dy. \qquad (7.2\text{-}41)$$

SAW can diffract a TE mode into a TM mode, and vice versa, if the crystal orientation is chosen such that proper photoelastic constants, which rotate the polarization of the incident mode to that of the diffracted mode, come into play.

7.3 APPLICATIONS

Some of the potential applications of the surface acousto-optic interaction are in real-time signal processing and image scanning. Specifically, in signal processing one can use this interaction to make light modulators, spectrum analyzers, and correlator–convolvers. Using the convolver as a basic building block, one can also perform different transformations such as real-time Fourier transformations. To understand the basic principles involved in all the above-mentioned devices, consider the schematic shown in Fig. 7.3-1. The signals $f(t)$ and $g(t)$ are electrical input signals to the SAW delay line either as it is or after electronic processing. The second input $g(t)$ is zero except for the case of correlation–convolution application. An incident laser beam can be uniform or modulated by a transparency or lenses. The output light—after interacting in the SAW acousto-optic cell—is incident on a photodetector. This photodetector can be a point, an array, or a charge-coupled photodetector device (CCPD) [37]. The electrical output of the photodetector, either as is or after electronic postdetection processing, is the desired output.

The acousto-optic interaction can be either in the Raman–Nath or in the Bragg region. In the Raman–Nath region the diffraction efficiency is in general lower than that in the Bragg region.

A. Raman–Nath Region

For a typical device in the Raman–Nath region, a detailed version of Fig. 7.3-1 is shown in Fig. 7.3.2. With reference now to Fig. 7.3-2, a uniform plane wave is amplitude-modulated by the transmittance function $T(\zeta)$ which is imaged via lens L_1 onto the z-plane, where it appears as $T(z)$. For simplicity we assume unity magnification but, depending on the placement of L_1, the imaging process could involve magnification. Since $T(z)$ is spatially limited, we can consider an aperture of width l, where $l \leq d$, positioned in the z-plane of the diffraction-limited system. Voltages $f_1(t)$ and $f_2(t)$ amplitude-modulate rf carriers at the acoustic frequency ω, which excite transducers at opposite ends of the ultrasonic delay line. It is assumed throughout that the time rates of change in $f_1(t)$ and $f_2(t)$ are small compared to that in the rf carrier. These signals produce traveling optical phase gratings which diffract the incident light. The Fraunhofer diffraction pattern is formed in the z'-plane

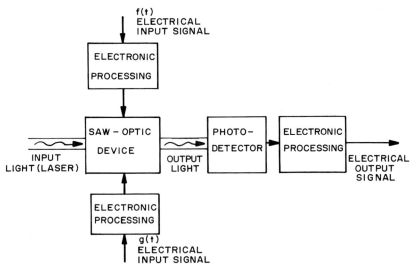

FIG. 7.3-1. Block diagram of a SAW optic signal processor.

which is the focal plane of lens L_2. Based on our discussion in Section 7.2 it can be shown that the light intensity in the z'-plane can be approximately given by

$$I = \sum_{n=0}^{\infty} I_n \delta(z' - z_n'),$$

where

$$I_n = C^2 \sum_r \sum_p$$

$$\times e^{j2(r-p)\omega t} \left| \int_{-l/2}^{l/2} T(z) J_r[V_1(vt - z - d/2)] J_{r-n}[V_2(vt + z + d/2)] \, dz \right|^2$$

(7.3-1)

is the intensity of the nth-order diffracted light, $z_n' = nF\lambda_{0f}/\lambda$, C is related to the incident light intensity, and F is the focal length of L_2. It has been assumed that $l \to \infty$. In the physical case in which the aperture l is finite, diffraction spreading of these spots occurs, the width of each spot being given approximately by $F\lambda_{0f}/l$. The functions V_1 and V_2 are defined as

$$V_i(vt) = (2\pi Lb/\lambda_{0f})f_i(t) = Kf_i(t),$$

(7.3-2)

where b is the constant which takes into account material photoelastic constants and transducer electromechanical coupling.

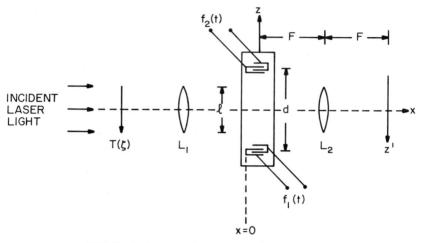

FIG. 7.3-2. Geometry for acousto-optic device model.

B. Modulator

The modulator is the simplest acousto-optic device. By varying the input voltage $f_1(t)$, the time-averaged light intensity of the various orders is controlled. With $f_2(t) = 0$ and $T(z) = 1$ in the interval $l/2 \le z \le l/2$, from Eq. (7.3-1) the intensity in the nth-order light is given by

$$I_n = I_0 J_n^2 [K f_1(t)]. \tag{7.3-3}$$

It should be recalled that Eq. (7.3-1) was derived under the assumption that V_1 was constant in the optical aperture region. Thus we require that the aperture l be small compared to the spatial rate of change in V_1. In terms of the angular frequency ω_{max} of the highest frequency component of $f_1(t)$, we can express this requirement as

$$l < \pi v / \omega_{max}, \tag{7.3-4}$$

where v is the acoustic velocity in the SAW device.

C. Spectrum Analyzer

For a particular order and $f_2(t) = 0$, the diffracted light at z' will be proportional to the input frequency of the signal $f_1(t)$. However, if $f_1(t)$ contains many frequencies, each diffracted order will have many spots at the positions given by

$$z'(f) = (nF\lambda/v)f. \tag{7.3-5}$$

The intensity of the light at $z'(f)$ will be proportional to the power spectra of the input signal. Thus replacing a single photodiode by a photodiode array or a charge-coupled photodetector device one can easily build a spectrum analyzer. The spacing S between the detector elements, the focal length F of the transform lens, and the optical beam diameter D must satisfy

$$S \leq (F\lambda_0/n_{\text{eff}}v) \, \Delta f \qquad \text{(rf resolution)},$$

where Δf is resolution in $f_1(t)$ and

$$S \leq aF/n_{\text{eff}} D \qquad \text{(optical resolution)},$$

where a is the aperture shape factor.

D. Image Scanner

With $f_2(t) = 0$, the configuration shown in Fig. 7.3-1 can be used to transform an intensity transmission function $\tau(\zeta) \equiv |T(\zeta)|^2$ into a voltage for the purpose of facsimile transmission. From Eq. (7.3-1) we obtain

$$I_n = C^2 \left| \int_{-l/2}^{l/2} T(z)J_n[V_1(vt - z - d/2)] \, dz \right|^2. \tag{7.3-6}$$

For small arguments of the Bessel function, the first-order light is given by

$$I_1 = C^2 \left| \int_{-l/2}^{l/2} T(z)V_1(vt - z - d/2) \, dz \right|^2. \tag{7.3-7}$$

Equation (7.3-7) can be interpreted as the square of the convolution of the spatial mask $T(z)$ with the acoustic signal V_1. If $f_1(t)$ is a very short rectangular pulse which amplitude-modulates the rf carrier, then V_1 approximates the delta function $\delta(v_a t - z - d/2)$. Thus Eq. (7.3-7) becomes

$$I_1 = C^2|T(vt - d/2)|^2 = C^2\tau(vt - d/2). \tag{7.3-8}$$

As a function of time, the intensity replicates the spatial information $\tau(\zeta)$. In practice V_1 need not be made small in amplitude. The pulse height is adjusted to maximize the first-order Bessel function.

E. Convolver

It is apparent from Eq. (7.3-1) that the time-dependent intensity in each diffraction order is made up of a sum of temporal frequency components occurring at even integral multiples of the acoustic rf frequency. Any given frequency component can be electronically isolated by proper filtering of the output signal of a photodetector placed to collect the light of a single diffraction order. The resultant signal, after filtering, provides the real-time convolution of $f_1(t)$ and $f_2(t)$. We illustrate this for the case of light of the zero order at twice the acoustic frequency, but a similar analysis can be applied to obtain real-time convolution in any order at any even multiple of the acoustic frequency.

In the special case where $f_1(t)$ and $f_2(t)$ are time-limited signals,

$$f_1(t), f_2(t) = 0 \qquad \text{for} \quad t < 0, t > d/v, \qquad (7.3\text{-}9)$$

and $T(z)$ is a simple window function,

$$T(z) = \begin{cases} 1, & |z| \le d/2, \\ 0, & |z| > d/2, \end{cases} \qquad (7.3\text{-}10)$$

the limits of integration for all integrals appearing in Eq. (7.3-1) can be taken from $-\infty$ to ∞ without affecting the value of the integrals. When we consider the zero-order diffracted light (i.e., $n = 0$), and $r - p = 1$ which corresponds to the component at twice the acoustic frequency, Eq. (7.3-1) provides, after some rearrangement of terms,

$$I_0(2\omega) \simeq \tfrac{1}{2} C^2 d \int_{-\infty}^{\infty} V_1(vt - z - d/2) V_2(vt + z + d/2)\, dz. \qquad (7.3\text{-}11)$$

With the substitution $v\tau = vt - z - d/2$, Eq. (7.3-11) reduces to the familiar convolution integral. In terms of the voltage signals $f_1(t)$ and $f_2(t)$,

$$I_0(2\omega)_a \simeq \frac{2\pi^2 L^2 b^2}{\lambda_{0f}^2} C^2 d \int_{-\infty}^{\infty} f_1(\tau) f_2(2t - \tau)\, d\tau. \qquad (7.3\text{-}12)$$

An interesting feature of Eq. (7.3-12) is the time compression factor of 2 (i.e., the argument $2t - \tau$ instead of the usual argument $t - \tau$), which arises from the fact that both functions $V_1(\cdot)$ and $V_2(\cdot)$ propagate in opposite directions along the acoustic track. In the usual convolution integral, one function is considered fixed with respect to the coordinate system, while the other is shifted with respect to it.

The analog convolver is a powerful signal processor because of its great flexibility. If the two input carrier frequencies are not identical, and if $f_2(t)$ is an even function of time, then the modulating factor of the output frequency $2(\omega_a - \Delta\omega_a)$ is proportional to the cross-ambiguity function of f_1 and f_2:

$$I_0[2(\omega_a - \Delta\omega_a)] \propto \int_{-\infty}^{\infty} f_1(\tau) f_2(\tau - 2t) e^{j\Delta\omega_a \tau}\, d\tau. \qquad (7.3\text{-}13)$$

In the case where $\Delta\omega_a = 0$, Eq. (7.3-13) reduces to the correlation integral. Equation (7.3-13) also provides a means of producing the Fourier transform of the function $f_1(t)$ by making $f_2(t)$ equal to a constant and varying $\Delta\omega_a$. Table 7.3.1 indicates the application capabilities of the optical system of Fig. 7.3-2. In this table, a δ-function optical input $T(z)$ refers to a narrow

TABLE 7.3-1

Optical Processor Functions[a]

Function	Time-limited voltage inputs $f_1(t)$	$f_2(t)$	Space-limited optical input, $T(z)$	Output voltage signal	Comment
Modulation (power)	$f_1(t)$	0	$\delta(z_0)$	$\lvert f_1(t - z_0/v)\rvert^2$	Detect time-averaged intensity. Time delay controlled by position of optical window
Image scanniag (intensity)	$\delta(t)$	0	$T(z)$	$\lvert T(vt)\rvert^2$	Detect time-averaged intensity
Convolution	$f_1(t)$	$f_2(t)$	1	$\int f_1(\tau)f_2(2t - \tau)\,d\tau$	Detect component at $2\omega_a$
Modulation (voltage)	$f_1(t)$	$\delta(t)$	1	$f_1(2t - d/2v)$	—
Image scanning (amplitude)	$\delta(t)$	1	$T(z)$	$T(v_a t)$	—
Ambiguity function generation[b]	$f_1(t)$	$f_2(t)$	1	$\int f_1(\tau)f_2(\tau - 2t)e^{j\Delta\omega_a\tau}\,dt$	Input signals shifted in frequency by ω_a. Detect component at $2(\omega_a - \Delta\omega_a)$. Input signals must exhibit time symmetry
Fourier transformation[b]	$f_1(t)$	1	1	$\int f_1(\tau)e^{j\Delta\omega_a\tau}\,d\tau$	Input signals shifted in frequency by ω_a. Detect component at $2(\omega_a - \Delta\omega_a)$.

[a]From [31].

[b] Sweeping of the frequency $\Delta\omega_a$ is required to implement function.

window function; a δ-function voltage input refers to a short rectangular pulse.

F. Experimental Results: Raman–Nath Diffraction

The basic optical processor [38, 39] is shown schematically in Fig. 7.3-3. Light from a laser is brought to a line focus by lenses L_1 L_4, producing a ~ 15-μm-wide line in the surface of the LiNbO$_3$ delay line. Lens L_5 recollimates the light, and L_6 produces the Fraunhofer diffraction pattern of the light amplitude distribution in its focal plane. Amplitude-modulated signals are applied to inputs 1 and 2, and the final optical signal is detected by a photodiode positioned in the focal plane of L_6 where it collects the light of a single diffracted order.

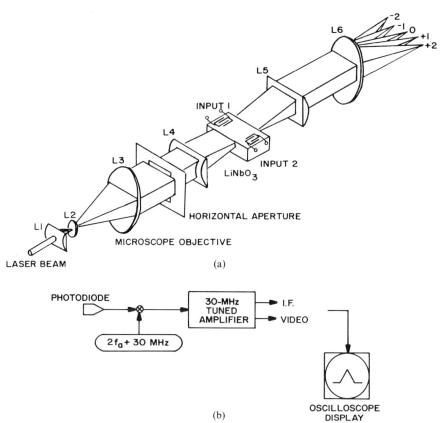

FIG. 7.3-3. Optical processor for real-time convolution: (a) optical arrangement. (b) electronic processing for convolution. Electronic heterodyning isolates the component of the photodiode signal at twice the acoustic frequency. (From Das and Schumer [39].)

The output of the photodiode, after electronic processing, is displayed on an oscilloscope. A typical processing scheme for convolution is shown in Fig. 7.3-3b. Heterodyning of the photodiode signal and subsequent filtering with a tuned amplifier isolate the voltage component at the desired frequency.

An acoustical delay line with the interdigital transducers employs y-cut z-propagating $LiNbO_3$. The use of the surface acousto-optic interaction as a modulator is illustrated by Fig. 7.3-4. This figure was obtained using 45 MHz as the center frequency, and the output of a commercial closed-circuit television camera as the input to the modulator. The output from the modulator was detected by a p-i-n diode and displayed on a regular television screen.

Figure 7.3-5 indicates the results of the different methods of obtaining the output of the optical convolver using a helium–neon laser as the light source. There is a choice as to the diffracted order in which the photodiode is positioned, as well as a choice of how the signal from the photodiode is processed. The input square pulses of 500-mW power and 2-μsec duration applied to both transducers are shown in Fig. 7.3-5a. In Fig. 7.3-5b, on the same time scale, is the total intensity detected in the zero order, displayed on a high frequency oscilloscope after amplification. The signal is composed of a dc component and high-frequency components containing the desired

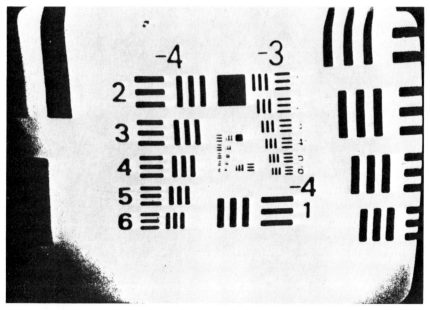

FIG. 7.3-4. Television image recovered from an intensity-modulated first-order diffracted beam using a $LiNbO_3$ SAW modulator. (From Kramer and Das [32].)

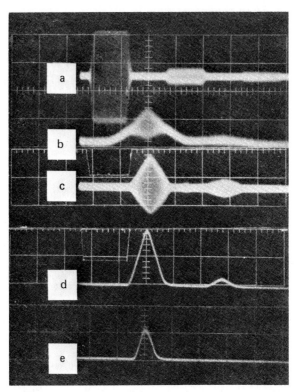

FIG. 7.3-5. Experimental results of real-time convolution: (a) rectangular pulse which is applied to both inputs (15 V p-p), (b) total intensity detected in the zero order, (c) autoconvolution signal at twice the acoustic frequency detected in the zero order, (d) envelope of autoconvolution at twice the acoustic frequency in the first order, and (e) envelope of autoconvolution at four times the acoustic frequency in the first order. Horizontal scale = 2 μsec/division. (From Schumer and Das [31].)

autoconvolution. The component as $2\omega_a$ is obtained by heterodyning and filtering and is displayed in Fig. 7.3-5c, which is the if signal of the tuned amplifier. Figure 7.3-5d shows the tuned amplifier video output, which is the desired envelope of the autoconvolution signal at 2ω in the first order.

Figure 7.3-6 shows results for the autoconvolution of a signal composed of two rectangular pulses. Shown in the figure are (a) the input signal, (b) the autoconvolution detected at 2ω in the zero order, and (c) the autoconvolution envelope.

For a rectangular acoustic pulse of fixed length, it is expected that the peak height of the triangular autoconvolution depends on the magnitude of the acoustic signal. For an ideal convolver, the peak height would be proportional to the input signal voltage. However, as discussed earlier, this situation

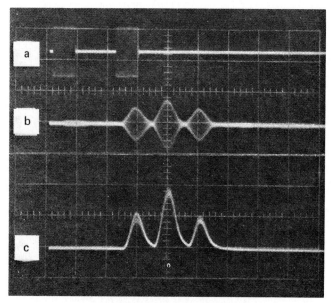

FIG. 7.3-6. Autoconvolution of signal composed of two rectangular pulses.

only pertains in the small signal limit for the acousto-optic convolver. For larger input signals the convolution height follows a Bessel function dependence. It is also sensitive to polarization and is shown in Fig. 7.3-7. The y and z in the figure refer to cases where the incident light has an E-field vector perpendicular and parallel to the SAW propagation vector, respectively.

Figure 7.3-8 shows the input signal and detected autocorrelation for a single rf pulse, a 3-bit Barker code, a 7-bit Barker code, and a 13-bit Barker code. In obtaining these results a 100-MHz acousto-optic convolver and an argon laser were employed, using the side-entry configuration. The argon laser light source operated at 5145 Å. The Barker code was impressed upon

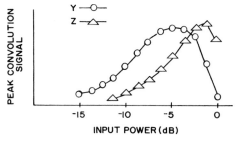

FIG. 7.3-7. Peak convolution signal versus input power of both the signals. Results are for y- and z-polarized light.

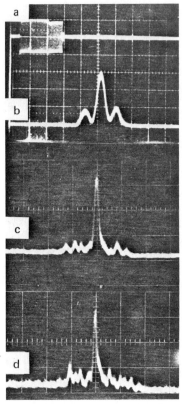

FIG. 7.3-8. Autocorrelation of 100 MHz signals. Results for Barker encoded data: (a) input rf signal and (b) autocorrelation of 3-bit Barker code (1 μsec/division), (c) autocorrelation of 7-bit Barker code (2 μsec/division), (d) autocorrelation of 13-bit Barker code (2 μsec/division). (From Schumer and Das [31].)

the rf carrier by introducing a phase shift of π in the electrical cw signal used to launch the acoustic wave. In this way $+1$ was distinguished from -1.

As was pointed out, under certain conditions the analog convolver can be used to generate the ambiguity function of time-limited signals. Figure 7.3-9 shows the electronic arrangement for ambiguity function generation. Figure 7.3-10 shows results for (a) a rectangular pulse and (b) a signal consisting of two rectangular pulses. The multiexposure photographs provide a three-dimensional plot of the ambiguity function as a function of t and $\Delta\omega$.

The convolver structure discussed above is called space integrating because the integration is performed over the length of the delay line by the lens collecting the diffracted light and bringing it to a focus. Thus the time–bandwidth (TB) product of a space-integrating correlator is limited by the

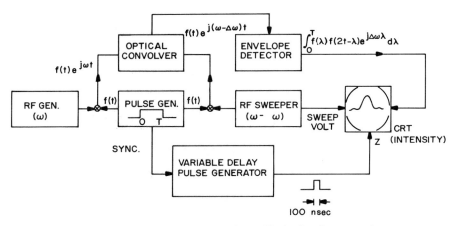

FIG. 7.3-9. Electronic arrangement for ambiguity function generation.

bandwidth of the SAW transducers and the interaction time of the SAW device. Also, a reference signal is required to perform a correlation which results in pulse compression. However, it is possible to obtain pulse compression by using the self-focusing property of the linear FM signal. The optical processor which uses this property is shown in Fig. 7.3-11. Consider the grating set up by the propagating SAW excited by rf chirp signal. Ray 1 is diffracted by an angle θ_1, and ray 2 is diffracted by a larger angle θ_2. These rays will converge at a distance F, known as the focal length of the acousto-optic lens. The focal length is given by

$$F = K_1 v^2/\beta, \tag{7.3-14}$$

where v is the SAW velocity, K_1 is the propagation constant of light, and β is the chirp rate or rate of change in input frequency. As the SAW propagates along the delay line, the focused spot travels in the same direction with the same speed. The amplitude of the first-order diffracted light at the focal plane and $z_0 = \omega v/B$, ω being the acoustic frequency, can be shown to be given by

$$A(F, z_0) \propto \int_0^L J_1(\phi)e^{j(\beta t/v)z}\,dz, \tag{7.3-15}$$

where ϕ is given by Eq. (7.2-16) and L is the interaction length of the SAW delay line.

A photodiode placed behind a stationary narrow slit gives an output signal

$$I = |A(x_0, z_0)|^2 \propto J_1^2(\phi)\left[\frac{\sin(\beta\Delta Tt)}{\beta\Delta Tt}\right]^2, \tag{7.3-16}$$

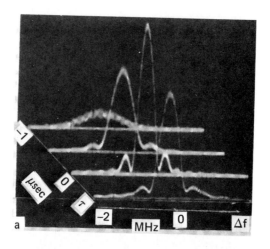

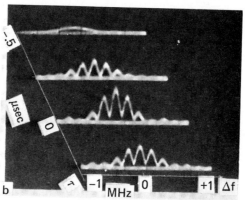

FIG. 7.3-10. Experimental results for ambiguity function: (a) multiexposure photograph showing slices of the ambiguity function surface of a 3-μsec pulse, (b) ambiguity function for a signal comprised of two rectangular pulses each 0.8 μsec and separated by approximately 1.5 μsec. (From Schumer and Das [31].)

which shows pulse compression in accordance with well-known chirp radar theory [40]. Here $\Delta T = L/v$ and ΔT is the time duration of the applied chirp pulse. The pulse compression ratio achieved using SAW device is limited by the bandwidth of the transducer. A typical example [41] is shown in Fig. 7.3-12 where a compression ratio of 60 is achieved using a 4-μsec-wide pulse with a 15-MHz bandwidth at 100 MHz compressed to 70 nsec. The focal distance F can be varied by varying the chirp rate, but this requires slits or a number of slits. The convergent property is analogous to that of a Fresnel lens. The advantage of an acousto-optic lens achieved by a chirp signal is that it is free of aberrations and the focal length can be adjusted

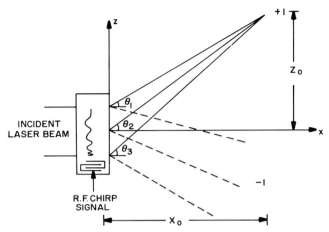

FIG. 7.3-11. Schematic of the acousto-optic lens.

electronically. It should be mentioned that this electronic lens is not centro-symmetric and that most of the light is not diffracted by the lens.

If the incident light is modulated with intensity transmission function $\tau(Z)$, which is placed near the SAW delay line, then Eq. (7.3-15) gives the Fourier transform of the transparency. Thus, one can achieve a one-dimensional Fourier transform of the given transparency without the use of a lens in an optical processor.

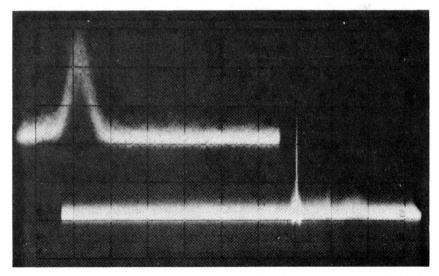

FIG. 7.3-12. Pulse compression using acousto-optic lens. Upper: compressed optical pulse (1 μsec/division); lower: expanded (10 nsec/division). (From Ayub and Das [41].)

Other optical transformations, such as the Mellin transform, can also be performed using an acousto-optic lens. The one-dimensional Mellin transform is given by

$$M(j\omega) = \int_0^\infty f(x)x^{-j\omega-1}\, dx, \qquad (7.3\text{-}17)$$

which, through use of the transformation $x = e^\xi$, can be written as

$$M(j\omega) = \int_{-\infty}^{+\infty} f(\exp \xi)e^{-j\omega\xi}\, d\xi. \qquad (7.3\text{-}18)$$

We see from Eq. (7.3-18) that the Mellin transform of $f(x)$ is a Fourier transform of $f(e^\xi)$. To realize this, one has to generate transparencies of $\xi = \ln x$. It is of interest to note that the acousto-optic lens can be used to demultiplex light of many wavelengths, as the light intensity associated with different wavelengths will be focused at different points.

Another implementation of a real-time correlator, called a time integrating correlator [42], is shown in Fig. 7.3-13. Although several different configurations are possible, this one is suitable for SAW delay lines. The light source is modulated by the signal $s_1(t)$ using a SAW delay line modulator. The modulated light is expanded and collimated to form a thin sheet of beam, which illuminates the SAW device. The acoustic signal $s_2(t - z/v)$ propagating on the surface of the SAW delay line diffracts the light into many orders. The image of the doubly diffracted light is integrated in time by the charge-coupled photodiodes. The output of the CCPD is proportional to the required correlating function

$$\int_0^T s_1(t)s_2(t - z/v)\, dt, \qquad (7.3\text{-}19)$$

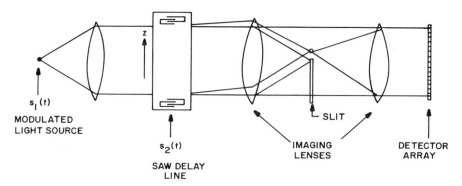

FIG. 7.3-13. Time-integrating correlator.

where T is the detector integration time. There is distinct difference between this correlation signal and the one discussed in connection with the spatially integrating correlator. The complete correlation signal is obtained if one uses a spatially integrating correlator, whereas only a portion of the correlating signal is obtained if one uses a time-integrating correlator. In the time-integrating correlator the time–bandwidth product is limited by the integration time and dynamic range of the CCPD array.

To implement the real-time Fourier transform [11] the chirp transform algorithm is used. The Fourier transform $S(f)$ of a signal $s(t)$ is given by

$$S(f) = S(\beta t) = \int_{-\infty}^{\infty} s(\tau)e^{-j2\pi(\beta t)\tau}\, d\tau$$

$$= (e^{-j2\pi\beta t^2}) \int_{-\infty}^{\infty} (s(\tau)e^{-j\pi\beta\tau^2})(e^{j\pi\beta(t-\tau)^2})\, d\tau. \tag{7.3-20}$$

The basic operation consists of multiplying the input signal $s(\tau)$ by a chirp $e^{-j\pi\beta\tau^2}$, then convolution with a chirp $e^{-j\pi\beta\tau^2}$, and then multiplying the convolution product by a chirp $e^{-j\beta\pi t^2}$. The complete complex Fourier transform $S(f)$ is obtained at the output of these operations. The frequency variable f will be a linear function of time; i.e., $f = \beta t$, so that $S(f)$ will be generated at the device output in real time.

Thus far it has been assumed that the light source is coherent. However, incoherent sources have been used successfully [43], although they furnish poorer signal-to-noise ratios. The coherency requirements are similar to that of the bulk wave case discussed in Ref. [44].

The lenses used in the processors discussed so far make it rather difficult to make the device compact. However, a lensless version is possible [45] and is shown in Fig. 7.3-14. The basic principal of this device is conversion of the

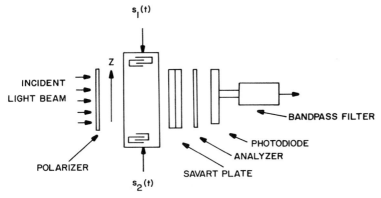

FIG. 7.3-14. Schematic arrangement of a lensless convolver.

phase modulation into polarization modulation by applying polarization interferometry techniques. The use of a Savart plate or a Wollaston prism for this purpose is very well known. In both cases the incident light ray is split into an ordinary (O) and an extraordinary ray (E). For the Savart plate these rays emerge from the plate parallel to each other with a lateral displacement, and for the Wollaston prism the two rays travel in different directions with variable lateral displacement.

It is well known that the light emerging from an acousto-optic device is phase-modulated. If this phase-modulated light is passed through the Savart plate, the two emerging E and O wavefronts interfere to give rise to polarization modulation. To convert the polarization modulation into intensity modulation, a polarizer and an analyzer are used. A lensless surface acousto-optic processor using the side-entry technique is shown in Fig. 7.3-14. This device offers several advantages: (a) large diffraction efficiency is obtained because of the side-entry technique; (b) an incoherent source such as a high power light emitting diode (LED) can be used; (c) the background radiation can be discriminated easily by the polarizer; (d) since no lens is required following the delay line, the device becomes very compact. The feasibility of using an LED source to make this processor a packaged device is a very attractive one.

G. Image Scanning

A black-and-white image scanner is shown in Fig. 7.3-15a. Incident laser light is spatially modulated by a one-dimensional transparency or a movie imaged onto the $LiNbO_3$ crystal. A short acoustic pulse scans this image and provides a light intensity in the first-order diffracted light, which replicates the intensity distribution as a function of time. Scanning of a two-dimensional transparency is achieved by mechanical translation of the transparency in the vertical direction after each horizontal scan of the acoustic pulse [46, 47].

If we recall that in Raman–Nath diffraction the diffraction angle of the first order is proportional to the light wavelength λ_{0f}, we see that it is a natural extension of this system to provide scanning of multicolor transparencies. The incident light is now composed of several colors, such as red, green, and blue laser light, and the intensity transmission function is a color transparency. The first-order diffracted light exits at a different angle for each color, as shown in Fig. 7.3-15b. Three photodiodes, one for each color, are used to collect the light. The three photodiode signals, after amplification, can then be multiplexed for transmission. Because diffraction rather than color filters is used to separate color components, the system is capable of color separation with high spectral purity.

An important factor which characterizes any image scanner is resolution.

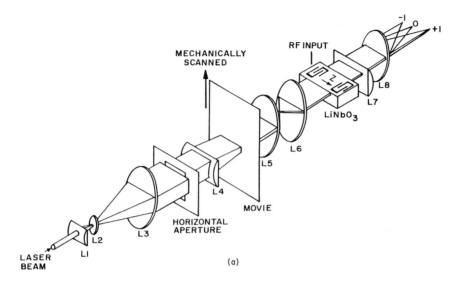

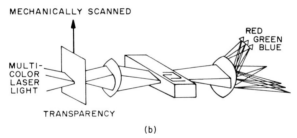

FIG. 7.3-15. Acousto-optic image scanner: (a) monochromatic operation, (b) color image scanner.

For an acoustic pulse of time duration T, the spatial resolution obtained is given by

$$R = Tv, \qquad (7.3-21)$$

where v is the SAW velocity.

The amount of light diffracted is proportional to the pulse length T. Therefore, as resolution is improved, the signal is decreased, other factors being equal. For this reason, in a practical scheme, one cannot simply improve resolution without bound by going to higher acoustic frequencies resulting in a smaller T. Ultimately, signal-to-noise considerations will determine the optimal pulse width, and therefore the resolution. The signal-to-noise ratio at the photodetector output is improved by increasing the incident light intensity, but optical damage to the $LiNbO_3$ places a limit on useful intensity.

In addition to determining resolution, the distance Tv also determines the spot size of the diffracted orders, since this is the diffraction-limiting aperture. If a lens of focal length F is used to form the Fraunhofer diffraction pattern of the diffracted light, the spot size s (i.e., the width of the main lobe) in each order will be on the order of

$$s = F\lambda_{0f}/Tv. \qquad (7.3\text{-}22)$$

It is therefore quite possible that, if T is small enough, the spots corresponding to the first-order diffraction for each color may overlap. In this case, a dispersive element such as a prism can be used to enhance the spatial separation of the different color components. A typical result is shown in Fig. 7.3-16 where a black-and-white 8-mm movie was used. The output signal

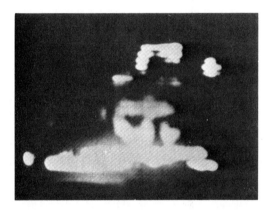

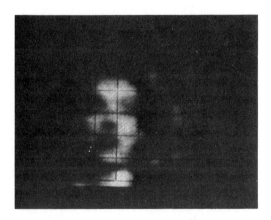

FIG. 7.3-16. Demonstration of two-dimensional image scanning. (From Ayub and Das [47].)

from the photodiode modulated the z-axis of the oscilloscope from which this picture was obtained.

H. Bragg Region

All the applications discussed in connection with the Raman–Nath region can be implemented using Bragg diffraction. In Bragg diffraction the angle of the incident light must change for peak efficiency as the SAW wavelength is varied. For this reason Bragg processors are somewhat bandwidth-limited. However, this restriction can be reduced significantly by using multiple tilted interdigital transducers [48], as shown in Fig. 7.3-17. The transducers are staggered in their center frequency and tilted in their propagation axis. It is clear that the multiple tilted SAWs generated by such a transducer array can individually satisfy the Bragg condition in each frequency band and thus permit a wide bandwidth to be realized. With the use of tilted transducer structure, spectrum analysis, time integrating and space correlating correlation has been performed with a bandwidth as high as 900 MHz and time–bandwidth product on the order of 10^6.

An integrated optic spectrum analyzer [49] is shown in Fig. 7.3-18. It consists essentially of a solid state laser, waveguide lenses for beam focusing, an optical surface waveguide with a SAW transducer, and a detector array coupled to the waveguide. In comparison with the bulk device, the integrated optic spectrum analyzer offers low fabrication cost, small size, and greater reliability. The laser beam is coupled to the waveguide on the surface of the substrate and is collimated by a waveguide lens L_1. The first-order diffracted light is passed through the second waveguide lens L_2, which focuses it onto the linear photodiode array. The intensity at the back focal plane of

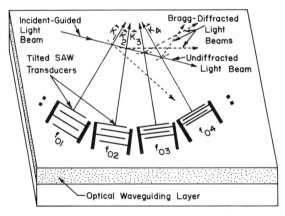

FIG. 7.3-17. Guided wave Bragg diffraction from multiple tilted surface acoustic waves. (From Tsai [48]. © 1975 IEEE.)

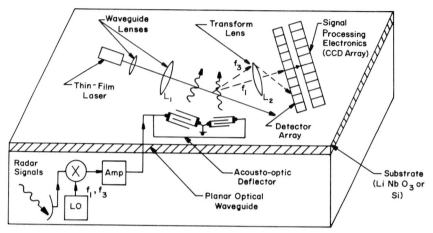

FIG. 7.3-18. Schematic of an integrated acousto-optic rf spectrum analyzer. (After Tsai *et al.* [52].)

the transform lens is proportional to the power spectrum of the input signal.

The side-entry technique discussed in connection with the Raman–Nath region can be easily extended to the Bragg region. When side entry is used in the Bragg region, a signal processing capability with a time–bandwidth product of up to 10,000 has been reported [50]. A typical result, where a real-time Fourier transformation of a rectangular pulse is performed, is shown in Fig. 7.3-19.

Another interesting device, a so-called memory correlator, has also been implemented using the acoustophoto refractive effect to store an image of a SAW in a $LiNbO_3$ delay line. This image is stored in the form of semi-permanent changes in the index of refraction of the acousto-optic medium

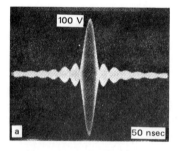
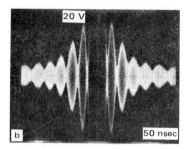

FIG. 7.3-19. (a) Normal (left) and (b) vertically expanded (right) side-lobe pattern $[\sin(x)/x]$ of the acousto-optic Fourier transform using a uniform intensity laser beam profile. (After Berg *et al.* [50].)

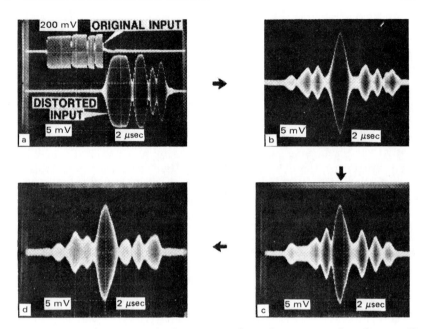

FIG. 7.3-20. Changes with time of output wave form of memory correlator for seven-bit Barker code. (After Berg *et al.* [50].)

and is caused by the interaction of a high intensity, short duration laser pulse with a propagating SAW. The storage time, strongly dependent on the energy density and wavelength of incident light, can be as long as a few months. The signal to be stored, say $f_1(t)$, is applied at the transducer, while the properly synchronized writing pulse of light is passed through the delay line. To achieve correlation with a signal $f_2(t)$ at a later time, it is applied to the same transducer. However, if convolution is desired, the $f_2(t)$ signal is applied to transducer on the other side of the delay line. One particular advantage of the memory correlator is that no synchronization of the signals is needed to perform a correlation. The performance of a memory correlator is shown in Fig. 7.3-20 where the writing light was obtained by the second harmonic conversion of a neodymium–YAG laser and the reading light was a helium–neon laser. The stored signal was a seven-bit Barker code, and the correlation signal was obtained using the same coded signal at a later time.

7.4 CONCLUSION

In this chapter the basic properties of the SAW acousto-optic interaction and many of its applications have been discussed. Although there are many practical applications, especially in the field of real-time signal processing, most are highly specialized. For this reason, no successful commercial product has been developed as yet. However, as the technological problems of manufacturing integrated optics products are solved, the wideband real-time spectrum analyzer will probably be among the first of the marketable items. In the meantime the SAW interaction will find many specialized applications, such as in Doppler radar requiring a very high bandwidth and time–bandwidth product coupled with a large dynamic range. Also, with the maturing of the optical communication field, SAW optic devices will play an increasingly useful role, since the signal can be processed with the optical carrier frequency, thereby avoiding its repeated detection and modulation [51]. In this respect SAW optic signal processing devices will have a significant advantage over other signal processing devices such as SAW electronic devices [10] and charge-coupled devices [37].

REFERENCES

[1] L. Brillouin (1921). *Ann. Phys. (Paris)* **17**, 103.
[2] C. W. Raman and N. S. N. Nath (1935). *Proc. Ind. Acad. Sci.* **2A**, 406–413.
[3] P. Debye and F. W. Sears (1932). *Proc. Natl. Acad. Sci. U.S.A.* **18**, 409
[4] M. K. Barnoski, ed. (1974). "Introduction to Integrated Optics." Plenum, New York.
[5] P. K. Tien (1971). *Appl. Opt.* **10**, 2395–2413.
[6] J. W. Goodman (1977). *Proc. IEEE* **65**, 29–38.
[7] D. Casasent and D. Psaltis (1977). *Proc. IEEE* **65**, 77–83.
[8] G. L. Rogers (1977). "Noncoherent Optical Processing." Wiley, New York.
[9] F. Okolicsanyi (1937). *Wireless Eng.* **14**, 527–536.
[10] H. Matthews, ed. (1977). "Surface Wave Filters." Wiley, New York.
[11] L. B. Milstein and P. Das (1979). *IEEE Commun. Mag.* **17**, 25–33.
[12] Lord Rayleigh (1885). *Proc. London Math. Soc.* **17**, 4–11.
[13] W. R. Smith, H. M. Gerard, J. H. Collins, T. M. Reeder, and J. H. Shaw (1969). *IEEE Trans. Microwave Theory Tech.* **MTT-17**, 856–864.
[14] F. Freyre and W. C. Wang (1978). *J. Appl. Phys.* **49**, 3629–3631.
[15] D. Brewster (1815). *Philos. Trans. R. Soc. London* p. 60.
[16] F. Pockels (1903). *Ann. Phys. (Leipzig)* **11**, 726.
[17] A. Yariv (1971). "Introduction to Optical Electronics," Chap. 9. Holt, New York.
[18] J. M. White, P. F. Heidrich, and E. G. Lean (1975). *Electron. Lett.* **10**, 510–511.
[19] D. F. Nelson (1979). "Electric, Optic and Acoustic Interactions in Dielectrics," Chap. 13. Wiley, New York.
[20] W. R. Cook, R. F. S. Hearman, H. Jaffe, and D. F. Nelson (1979). *In* "Landolt–Bornstein" (K. H. Hellwege, ed.), Vol. 11, Chap. 5. Springer-Verlag, Berlin and New York.
[21] W. R. Klein and B. D. Cook (1967). *IEEE Trans. Sonics Ultrason.* **SU-14**, 123–134.
[22] R. W. Dixon (1967). *IEEE J. Quantum Electron* **QE-3**, 85–93.
[23] P. Das, D. Schumer, and F. M. M. Ayub (1976). *Proc. Soc. Photo-Opt. Instrum. Eng.* **84**, 91–96.
[24] A. Alippi, A. Palma, L. Palmieri, and G. Socino (1974). *J. Appl. Phys.* **45**, 1492–1497.
[25] E. G. Lean (1973). *Prog. Opt.* **11**, 123–166.
[26] C. J. Kramer, M. N. Araghi, and P. Das (1976). *Dig. Tech. Pap. Conf. Laser Electro-Opt. Syst.* Pap. WD5.

[27] R. N. Spaight and G. G. Koerber (1971). *IEEE Trans. Sonics Ultrason.* **SU-23**, 33–63.

[28] G. I. Stegeman (1976). *IEEE Trans. Sonics Ultrason.* **SU-23**, 33–63.

[29] W. G. Mayer, G. B. Lamers, and D. C. Auth (1967). *J. Acoust. Soc. Am.* **42**, 1255–1257.

[30] D. Sarid and G. I. Stegeman (1978). *J. Appl. Phys.* **49**, 2301–2305.

[31] D. Schumer and P. Das (1979). Tech. Rept. MA-ONR-30. Rensselaer Polytechnic Inst., Troy, New York.

[32] C. J. Kramer and P. Das (1976). Tech. Rep. RADC-TR-76-104. Griffiss Air Force Base, New York.

[33] R. V. Schmidt, I. P. Kaminow, and J. R. Carruthers (1973). *Appl. Phys. Lett.* **23**, 417–419.

[34] R. V. Schmidt and I. P. Kaminow (1975). *IEEE J. Quantum Electron.* **QE-11**, 57–59.

[35] L. Kuhn, M. L. Dakss, P. F. Heidrich, and B. A. Scott (1970). *Appl. Phys. Lett.* **17**, 265–267.

[36] T. Shiosaki (1978). *Ultrason. Symp. Proc.* pp. 100–110.

[37] C. H. Sequin and M. F. Tompsett (1975). "Charge Transfer Devices," Chap. 4. Academic Press, New York.

[38] C. J. Kramer, M. N. Araghi, and P. Das (1974). *Appl. Phys. Lett.* **25**, 180–183.

[39] P. Das and D. Schumer (1976). *Ferroelectric* **10**, 77–80.

[40] C. E. Cook and M. Bernfield (1967). "Radar Signals." Academic Press, New York.

[41] F. M. M. Ayub and P. Das (1978). *Opt. Commun.* **26**, 161–164.

[42] R. A. Sprague (1977). *Opt. Eng.* **16**, 467–474.

[43] P. Das and F. M. M. Ayub (1979). *Proc. Soc. Photo-Opt. Instrum. Eng.* **185**, 110–117.

[44] E. B. Felstead (1967). *IEEE Trans. Aerosp. Electron. Syst.* **AES-3**, 907–914.

[45] A. Alippi, A. Palma, L. Palmieri, G. Socino, and E. Verono (1978). *Electron. Lett.* **14**, 525–526.

[46] A. Alippi, A. Palmo, L. Palmieri, and G. Socino (1975). *Appl. Phys. Lett.* **26**, 357–360.

[47] F. M. M. Ayub and P. Das (1979). *Opt. Laser Technol.* **11**, 87–90.

[48] C. S. Tsai (1975). *Ultrason. Symp. Proc.* pp. 130–125.

[49] M. K. Barnoski, B. Chen, H. M. Gerad, E. Marom, O. G. Ramer, W. R. Smith, Jr., G. L. Tangonan, and R. D. Weglien (1978). *Proc. IEEE Ultrason. Symp.* Cat. No. 78 CH 1344-1 SU, pp. 75–79.

[50] N. J. Berg, J. N. Lee, M. W. Casseday, and B. J. Udelson (1979). *Appl. Opt.* **18**, 2767–2774.

[51] P. Das, D. Schumer, and H. Estrada-Vazquez (1977). *In* "Optical Communications using Surface Acoustic Waves" (E. Marom *et al.*, eds.), pp. 447–455. Pergamon, Oxford.

[52] C. S. Tsai, C. C. Lee, and I. W. Yao (1978). *Proc. Soc. Photo-Opt. Instrum. Eng.* **154**, 60–63.

Chapter 8

Space-Variant Optical Systems and Processing

WILLIAM T. RHODES

SCHOOL OF ELECTRICAL ENGINEERING
GEORGIA INSTITUTE OF TECHNOLOGY
ATLANTA, GEORGIA

8.1 INTRODUCTION

The study of optical systems, especially those designed for imaging and signal processing, has been aided greatly during the past several decades by the analytical methods of linear systems theory. The application of Fourier transform techniques in particular had led to a special emphasis on space-invariant characteristics in optical systems analysis. The input–output relationship for an imaging system, for example, is often modeled by a convolution integral, and transfer function concepts are applied to its analysis. Implicit in these models is the assumption that the system is space-invariant. In fact almost all the image and signal processing systems discussed in earlier chapters were assumed to be space-invariant.

From a physical standpoint, space invariance in optical systems means that a lateral shift in a given input distribution results in a proportional lateral shift in the output—but nothing more. To the extent possible, most imaging systems are by design space-invariant: There is negligible geometrical distortion of the image; the image of a point object does not change shape as the object is moved about in the input plane. Examples of imaging systems that are *not* space-invariant include ultrawide angle camera lenses, for which a shift in the location of the object is not matched by a proportional shift in the location of its image, and lenses exhibiting coma. In the latter case, an object point on the optic axis produces an image distribution, also on-axis, that is compact and symmetric. An off-axis object point, on the other hand, produces the well-known comalike image distribution. Systems of this type are called *space-variant*.

333

Space variance in optics is important to us in several ways. First, as already noted, some imaging operations cannot be modeled as space-invariant. And even if an imaging system is itself space-invariant, it may be used in a space-variant manner. As a simple example, consider the blurred image that results from making a photograph with a camera that rotates about the lens axis during exposure. The image point on-axis is unblurred; all other image points are blurred, by an amount proportional to their distance off-axis. If such image blurring effects are to be modeled analytically, space-variant representations must be used. In addition, there are optical information processing operations of interest to us that are space-variant in nature. These include, naturally enough, deblurring of imagery obtained with space-variant imaging systems, but extend to such signal processing operations as the optical implementation of Walsh–Hadamard transforms. (The Fourier transform operation performed so naturally by a spherical lens is, in fact, itself a space-variant operation.) In what follows, we describe a variety of space-variant systems and operations, both one- and two-dimensional. We begin by considering the basic analytical tools used for describing such systems and look at some analytical examples.

8.2 REPRESENTATION AND ANALYSIS
OF SPACE-VARIANT LINEAR SYSTEMS

A. Basic Mathematical Description

Consider a system that operates on a two-dimensional (2-D) input distribution $f(x, y)$ to produce a 2-D output distribution $g(x, y)$. If the system is linear, the input–output operation can be characterized analytically by a superposition integral:

$$g(x, y) = \int_{-\infty}^{\infty} \int_{-\infty}^{\infty} f(\xi, \eta) h_1(x, y; \xi, \eta) \, d\xi \, d\eta. \tag{8.2-1}$$

The meaning of this integral and the function $h_1(x, y; \xi, \eta)$ is clarified if we assume the input to be a spatial impulse (a 2-D Dirac delta function) located at coordinates $x = x_0, y = y_0$:

$$f(x, y) = \delta(x - x_0, y - y_0). \tag{8.2-2}$$

Substituting in Eq. (8.2-1) and exploiting the sifting property of the delta function [1], we obtain

$$g(x, y) = \int_{-\infty}^{\infty} \int_{-\infty}^{\infty} \delta(\xi - x_0, \eta - y_0) h_1(x, y; \xi, \eta) \, d\xi \, d\eta$$

$$= h_1(x, y; x_0, y_0). \tag{8.2-3}$$

The function $h_1(x, y; x_0, y_0)$ is thus seen to be the output response as a function of spatial coordinates x and y to a spatial input consisting of an impulse located at coordinates x_0, y_0. We refer to $h_1(x, y; \xi, \eta)$ as the impulse response of the system. An arbitrary input function $f(x, y)$ can be thought of as the superposition of delta function components in accord with the identity

$$f(x, y) = \int_{-\infty}^{\infty} \int_{-\infty}^{\infty} f(\xi, \eta)\, \delta(x - \xi, y - \eta)\, d\xi\, d\eta. \qquad (8.2\text{-}4)$$

Each such component is operated on by the system to produce, on superposition of all responses, the distribution $g(x, y)$.

If the system is space-invariant, a shift in the input results in a corresponding shift—but no other change—in the output distribution. Stated analytically, this means that, if input $\delta(x, y)$ results in output $h_1(x, y; 0, 0)$, then the shifted input $\delta(x - \xi, y - \eta)$ results in the shifted output $h_1(x - \xi, y - \eta; 0, 0)$. For a space-invariant system, we can thus make the identification $h_1(x, y; \xi, \eta) = h_1(x - \xi, y - \eta; 0, 0)$. Since h_1 depends in this case only upon two rather than four arguments, we define a new function $h(x, y)$ given by

$$h(x, y) = h_1(x, y; 0, 0). \qquad (8.2\text{-}5)$$

The superposition integral of Eq. (8.2-1) then becomes the convolution integral

$$g(x, y) = \int_{-\infty}^{\infty} \int_{-\infty}^{\infty} f(\xi, \eta)h(x - \xi, y - \eta)\, d\xi\, d\eta, \qquad (8.2\text{-}6)$$

which we denote by

$$g(x, y) = f(x, y) * h(x, y). \qquad (8.2\text{-}7)$$

If we take the 2-D Fourier transform of both sides of Eq. (8.2-7), we obtain

$$G(u, v) = F(u, v)H(u, v), \qquad (8.2\text{-}8)$$

where $G(u, v)$ and $F(u, v)$ are the 2-D Fourier transforms of $g(x, y)$ and $f(x, y)$, respectively; i.e.,

$$G(u, v) = \mathscr{F}_2\{g(x, y)\} = \int_{-\infty}^{\infty} \int_{-\infty}^{\infty} g(x, y)e^{-j2\pi(ux + vy)}\, dx\, dy, \qquad (8.2\text{-}9)$$

$$F(u, v) = \mathscr{F}_2\{f(x, y)\} = \int_{-\infty}^{\infty} \int_{-\infty}^{\infty} f(x, y)e^{-j2\pi(ux + vy)}\, dx\, dy, \qquad (8.2\text{-}10)$$

and where $H(u, v)$ is the system transfer function, given by the 2-D Fourier transform of $h(x, y)$:

$$H(u, v) = \mathscr{F}_2\{h(x, y)\} = \int_{-\infty}^{\infty} \int_{-\infty}^{\infty} h(x, y)e^{-j2\pi(ux + vy)} \, dx \, dy. \quad (8.2\text{-}11)$$

The distinction between a space-invariant system and its space-variant counterpart are illustrated in Fig. 8.2-1. Figure 8.2-1a shows two spatial impulses assumed as input, and Fig. 8.2-1b the corresponding responses of a space-invariant system. Figure 8.2-1c illustrates a space-variant response to these impulsive inputs. Note that $h_1(x, y; x_0, y_0)$ not only has a shape different from $h_1(x, y; 0, 0)$ but its center of mass is displaced from (x_0, y_0).

In parts (d) and (e) of Fig. 8.2-1 we illustrate an alternative analytical description of the system response $h_2(x, y; \xi, \eta)$ that is often more convenient to use than h_1. Like $h_1(x, y; \xi, \eta)$, the function $h_2(x, y; \xi, \eta)$ describes the response of the system to an impulse located at (ξ, η). However, h_2 describes the response as a function of the distance from (ξ, η) rather than the distance from the origin. The two functions describe the same distribution, being

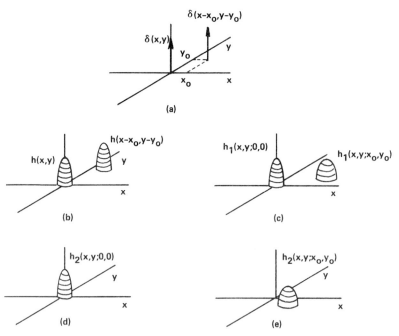

FIG. 8.2-1. Linear system responses: (a) input impulses, (b) response of space-invariant system, (c) response h_1 of space-variant system, (d) response h_2 to impulse at (0, 0), (e) response h_2 to impulse at x_0, y_0.

related by a simple shift:

$$h_2(x - \xi, y - \eta; \xi, \eta) \equiv h_1(x, y; \xi, \eta). \tag{8.2-12}$$

B. Frequency-Variant Systems

The above provides an adequate analytical basis for the study of space-variant systems. Before moving on to specific examples of such systems, however, we introduce one additional concept, that of a *frequency-variant system*. Instead of decomposing $f(x, y)$ into a superposition of weighted impulses, as in Eq. (8.2-4), we represent it, through an inverse Fourier transform integral, by a superposition of complex exponentials or Fourier components:

$$f(x, y) = \mathscr{F}_2^{-1}\{F(u, v)\} = \int_{-\infty}^{\infty} \int_{-\infty}^{\infty} F(u, v)e^{j2\pi(ux + vy)} \, du \, dv. \tag{8.2-13}$$

Substituting for $f(\xi, \eta)$ in Eq. (8.2-1) and interchanging the order of integration, we obtain

$$g(x, y) = \int_{-\infty}^{\infty} \int_{-\infty}^{\infty} F(u, v)\hat{h}_1(x, y; u, v) \, du \, dv, \tag{8.2-14}$$

where $\hat{h}_1(x, y; u, v)$ is given by

$$\hat{h}_1(x, y; u, v) = \int_{-\infty}^{\infty} \int_{-\infty}^{\infty} h_1(x, y; \xi, \eta)e^{j2\pi(u\xi + v\eta)} \, d\xi \, d\eta \tag{8.2-15}$$

and is the *system response to a complex exponential (Fourier component) of spatial frequency u, v*. This spatial frequency domain description of a general 2-D linear system is useful to us because spatial frequency components of an input can be isolated and modified in certain planes within a coherent-optical system just as easily as impulsive spatial components [the delta functions of Eq. (8.2-4)] can be isolated and modified in other planes.

The distinction of a system being frequency-invariant follows naturally from our definition of $\hat{h}_1(x, y; u, v)$: If the system's response to a Fourier component shifts in direct proportion to a shift in frequency of the input component but remains otherwise unchanged, this system is frequency-invariant, and

$$\hat{h}_1(x, y; u, v) = \hat{h}_1(x - u, y - v; 0, 0). \tag{8.2-16}$$

The overall system response to input functions $f(x, y)$ then becomes

$$g(x, y) = \int_{-\infty}^{\infty} \int_{-\infty}^{\infty} F(u, v)\hat{h}(x - u, y - v) \, du \, dv = F(x, y) * \hat{h}(x, y), \tag{8.2-17}$$

where we define $\hat{h}(x, y)$ by

$$\hat{h}(x, y) = \hat{h}_1(x, y; 0, 0). \tag{8.2-18}$$

The notion of frequency-invariant and frequency-variant optical systems becomes particularly important to us in the context of coherent-optical spectrum analysis. The optical spectrum analyzer described in Chapter 1 is, by assumption, frequency-invariant: The shape of the response of the system to a particular Fourier component of the input wave amplitude distribution is independent of the frequency (u, v) of this component; only the position of the response changes, and in direct proportion to u, v. In practice, coherent-optical spectrum analyzers are frequency-invariant only up to a certain point; off-axis aberrations ultimately change the shape of the response $\hat{h}_1(x, y; u, v)$ for sufficiently high spatial frequencies. As before, we can emphasize the change in shape of the impulse response *about* the location (u, v) if we define a related function $\hat{h}_2(x, y; u, v)$ according to

$$h_2(x - u, y - v; u, v) \equiv \hat{h}_1(x, y; u, v). \tag{8.2-19}$$

As we discuss later, spectrum analyzers that are intentionally frequency-variant have been built for special applications.

We have not provided a complete review of space-variant linear systems theory here but have considered the subject in sufficient depth that we can analyze a variety of examples and applications of space-variant systems. Additional analytical details will be provided as we go along. The interested reader is referred to Refs. [2, 3] for descriptions of additional methods for modeling space-variant systems.

8.3 EXAMPLES OF SPACE VARIANCE

Before considering the design and application of space-variant or frequency-variant systems for signal analysis and processing, we analyze three examples of space variance in imaging systems. These examples are associated, respectively, with atmospheric turbulence, rotation blur, and diffusion in scanned (TV-like) imaging systems. In all three cases, our objective is to model the space variance by obtaining an expression for the space-variant impulse response, either $h_1(x, y; \xi, \eta)$ or $h_2(x, y; \xi, \eta)$, whichever is more convenient.

A. Atmospheric Turbulence

If a star is observed with a telescope having an aperture of 10 cm diameter or less, two effects of atmospheric turbulence predominate: The star image moves about in the image plane in response to low level turbulence (which effectively changes the direction of arrival of the light waves at the telescope), and the brightness of the image fluctuates in response to high level turbulence (which introduces a slight focusing or defocusing of the light traveling toward the telescope). Since the light path through the atmosphere to the telescope depends on the location of the star in the sky, the star image—which is essentially the impulse response of the imaging system for short duration exposures—is space-variant. If we denote by $s(x, y)$ the point spread function (PSF) of the imaging system in the absence of turbulence, then $h_2(x, y; \xi, \eta)$ with turbulence present is given, for any instant in time, by

$$h_2(x, y; \xi, \eta) = A(\xi, \eta)s[x - \Delta_x(\xi, \eta), y - \Delta_y(\xi, \eta)]. \qquad (8.3\text{-}1)$$

In this expression, $A(\xi, \eta)$ represents space-variant scintillation effects and $\Delta_x(\xi, \eta)$ and $\Delta_y(\xi, \eta)$ the shift of the impulse response from its turbulence-free location. If the aperture of the telescope is significantly larger than 10 cm, then wavefront curvature as well as wavefront tilt must be taken into account, and $h_2(x, y; \xi, \eta)$ will have a more complicated form. It might be noted that, if the instantaneous space-variant point spread function is time-averaged over several seconds, the result is a PSF that is essentially space-invariant. This time-averaged PSF is somewhat broader than the instantaneous impulse response, corresponding to images of lower spatial resolution.

B. Rotation Blur

Figure 8.3-1 illustrates the particular kind of space-variant blur that occurs when a camera is rotated through an angle $\Delta\theta$ about its lens axis during an exposure. The scene being photographed is assumed to consist of two point sources, one located at coordinates (ξ_1, η_1) at the beginning of the exposure, the other at (ξ_2, η_2). The coordinate system is assumed to be centered on the axis of rotation. The response of the system to a point object is an arc of a circle, the radius and length of which depend on the radial coordinate of the object. Since radially more distant image points move faster during the exposure, the amplitude of the impulse response must fall

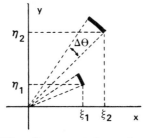

FIG. 8.3-1. Effect of camera rotation on two point objects.

off with distance from the origin. Using polar coordinate notation [1, Chapter 3], we can write

$$h_1(x, y; \xi, \eta) = (1/r_0)\, \delta(r - r_0)\, \text{rect}[(\theta - \theta_0 - \Delta\theta/2)/\Delta\theta], \quad (8.3\text{-}2)$$

where rect ( ) denotes the unit rectangle function and

$$r = \sqrt{x^2 + y^2}, \qquad r_0 = \sqrt{\xi^2 + \eta^2}, \qquad (8.3\text{-}3)$$

$$\theta = \tan^{-1}(y/x), \qquad \theta_0 = \tan^{-1}(\eta/\xi). \qquad (8.3\text{-}4)$$

This expression does not take into account the resolution limitations of the imaging system that exist even in the absence of rotation blur. If we assume the zero-rotation PSF to be radially symmetric and denote it by $s(x, y)$, then the actual rotation blurred PSF assumes the form

$$h_1(x, y; \xi, \eta) = \left[\left(\frac{1}{r_0}\right)\delta(r - r_0)\,\text{rect}\!\left(\frac{\theta - \theta_0 - \Delta\theta/2}{\Delta\theta}\right)\right] * s(x, y). \quad (8.3\text{-}5)$$

In order to evaluate this expression analytically, we would write r and θ as explicit functions of x and y. However, working out the convolution integral would provide us with no additional insight into the form of $h_1(x, y; \xi, \eta)$. Better is simply to interpret Eq. (8.3-5) as representing a 2-D smoothing of the impulsive function of Eq. (8.3-2).

C. Blur by Diffusion

Scanning-type thermal imaging systems generally exhibit some degree of space invariance when used to record events whose duration is short compared to the image scan time. The reason is a spreading or diffusion with time of the image pattern as it is scanned following exposure. This spread is a natural consequence of basic thermal imaging principles. First, an infrared scene is imaged onto a thin layer of a material whose electrical conductivity is varied on a point-by-point basis by local heating effects. Then, the resultant conductivity pattern is measured by scanning the layer from the back side with an electron beam. Immediately after exposure to the infrared pattern, the associated temperature variation pattern is well defined. With time, however, the thermal image diffuses. With isotropic diffusion [4], a pointlike detail in the thermal image blurs into a Gaussian distribution whose width is proportional to the square root of the diffusion time and whose spatial integral remains constant. If significant blurring occurs during the scan time, the resultant displayed image exhibits space-variant resolution.

In representing the impulse response of such a system, we assume the simple noninterlaced scanning pattern of Fig. 8.3-2. The scan starts at the origin at the moment of exposure and progresses left to right, bottom to top.

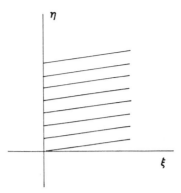

FIG. 8.3-2. Raster scan pattern. Scan moves from left to right, bottom to top.

Each line is scanned with constant velocity; retrace between lines is assumed instantaneous. Under these circumstances, the diffusion time is proportional to η, and $h_2(x, y; \xi, \eta)$ can be written in the form

$$h_2(x; y; \xi, \eta) = (\alpha^2 c/\eta) \exp[-\pi(\alpha^2 c/\eta)(x^2 + y^2)], \qquad (8.3\text{-}6)$$

where α is a constant representing the thermal capacity and thermal resistivity of the scanned material and c is the scan velocity in the η-direction. Note that $h_2(x, y; \xi, \eta)$ does not represent the impulse response of the complete imaging system; it describes only the relationship between the thermal image originally incident on the thermal conducting layer and the resultant conductivity pattern as it is scanned in time.

8.4 SYSTEMS FOR SPACE-VARIANT PROCESSING OF 1-D SIGNALS

We now turn our attention to optical systems designed for space-variant *processing* operations, concentrating in this section on systems for processing one-dimensional (1-D) signal information. Such systems, first discussed in the early sixties by researchers at the University of Michigan [5, 6], can be used in the analysis of radar and sonar signals or in the processing of speech waveforms, to name two areas of application.

The basic 1-D space-variant processing operation is characterized by an input–output relationship of the form

$$g(x) = \int_{-\infty}^{\infty} f(\xi) h_1(x; \xi)\, d\xi, \qquad (8.4\text{-}1)$$

where $g(x)$ is the output, $f(x)$ is the input, and $h_1(x; \xi)$ is the response at coordinate x to a unit impulse applied at ξ. An alternate form of the impulse response given by $h_2(x - \xi; \xi) \equiv h_1(x; \xi)$ can be employed.

There are a number of ways of implementing the operation of Eq. (8.4-1) with a coherent-optical processing system. We look first at four basic methods and then consider possible extensions.

A. Basic Systems and Examples

Fundamental to many systems for space-variant processing of 1-D signals is a cylindrical or spherical–cylindrical lens system that images in one transverse direction while Fourier-transforming in the other. The operation of such a system on an input $f(x, y)$ is characterized by the integral expression

$$g(u, y) = \int_{-\infty}^{\infty} f(x, y)e^{-j2\pi ux}\, dx \equiv \mathscr{F}_x\{f(x, y)\}. \qquad (8.4\text{-}2)$$

The system of cylindrical lenses of Fig. 8.4-1a performs such an operation. We can show this without resorting to a diffraction theory analysis if we

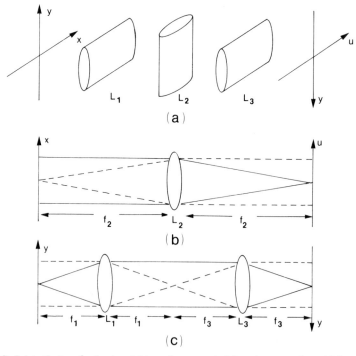

FIG. 8.4-1. System for horizontal transform–vertical imaging operation: (a) lens system, (b) top view, (c) side view.

note that (1) the basic operation of an ideal 2-D Fourier transform system is to map plane waves into points and points into plane waves, and (2) an ideal coherent imaging system maps points into points and plane waves into plane waves. As shown by simple ray drawings, we see from Fig. 8.4-1b that in the x-direction the system performs a Fourier transform operation; in the y-direction, as shown in Fig. 8.4-1c, the system of lenses images. Note that in labeling the horizontal axis in the output plane we have used the rectangular spatial frequency u, which is given by the actual horizontal coordinate normalized by λ times the focal length of the central lens element. Alternate systems can be used, particularly if only the output plane intensity is of interest. The system of Fig. 8.4-2, for example, which effects an imaging operation in the y-direction that includes a multiplicative 1-D quadratic phase factor (parallel rays in the input are mapped into diverging rays in the output) is acceptable if the output plane phase is not of concern. Additional systems along these lines are described in Refs. [7, 8].

The first system we consider for implementing Eq. (8.4-1) is the most direct [9, 10]. Assume that the function $f(x)$ is impressed across the input plane of the system of Fig. 8.4-1 as a complex wave amplitude distribution of the form $f(x)1(y)$. In this expression the function $1(y)$, which equals unity for all y, is included to emphasize that there is no variation in the y-direction (ignoring the effect of a finite input plane aperture). This distribution is incident on a mask in the input plane with complex amplitude transmittance $t(x, y)$ given by

$$t(x, y) = h_1(y; x). \qquad (8.4\text{-}3)$$

It is the product distribution $f(x)1(y)h_1(y; x) = f(x)h_1(y; x)$ that is Fourier-transformed in x and imaged in y, with the resultant output wave amplitude having the form

$$g(u, y) = \int_{-\infty}^{\infty} f(x)h_1(y; x)e^{-j2\pi ux}\, dx. \qquad (8.4\text{-}4)$$

If u is set equal to zero in this expression, the integral of Eq. (8.4-1) results, with the output given as a function of y. The basic system is thus completed by placing a narrow vertical slit in the output plane that restricts attention

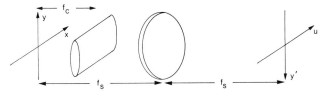

FIG. 8.4-2. Alternative system for horizontal transform–vertical imaging.

to the y-axis. It must be kept in mind, of course, that the output distribution actually observed is proportional to the intensity $|g(0, y)|^2$.

The operation of this system is illustrated in Fig. 8.4-3 with $f(x) = \delta(x - x_1) + \delta(x - x_2)$ and $h_1(y; x) = \text{rect}[(y - x)/w(x)]$, where $\text{rect}(\cdot)$ denotes the 1-D unit rectangle function. We can think of $w(x)$ as the width, in the y-direction, of the rectangle as a function of x. Figure 8.4-3a shows the input $f(x)1(y)$ with dark lines representing the delta functions. Figure 8.4-3b shows this input overlaid by the mask function $h_1(y; x)$. The result of a horizontal transform–vertical imaging operation is illustrated in Fig. 8.3-3c: Short delta function line segments exposed by the mask in Fig. 8.4-3b are smeared out uniformly in the horizontal direction by the Fourier transform. The result of masking off all but the vertical axis is shown in Fig. 8.4-3d. The output is seen to have the expected form:

$$g(y) = \text{rect}[(y - x_1)/w(x_1)] + \text{rect}[(y - x_2)/w(x_2)].$$

An alternate system for implementing Eq. (8.4-1) is shown in Fig. 8.4-4. This system is particularly attractive because it uses only a single spherical lens. (It is easier to obtain high quality spherical lenses at a reasonable price than high quality cylindrical lenses.) As before, the input distribution $f(x)1(y)$ is incident on a mask characterized by complex amplitude transmittance $t(x, y)$. The product distribution $f(x)t(x, y)$ is Fourier-transformed by the lens in *both* directions, and the resultant distribution observed along the vertical or v-axis. (Again, we label output axes in terms of spatial frequency variables; u and v are given by the actual focal plane coordinates x_f and y_f divided by λf.) Along this axis, the output is given by

$$g(v) = \int_{-\infty}^{\infty} \int_{-\infty}^{\infty} f(x)t(x, y)e^{-j2\pi vy}\, dx\, dy = \int_{-\infty}^{\infty} f(x)h_1(v; x)\, dx, \quad (8.4\text{-}5)$$

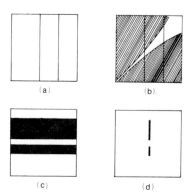

FIG. 8.4-3. Operation of 1-D space-variant processor.

FIG. 8.4-4. Alternate single-lens 1-D space-variant processor.

where

$$h_1(v; x) = \int_{-\infty}^{\infty} t(x, y)e^{-j2\pi vy}\, dy. \qquad (8.4\text{-}6)$$

To obtain the desired impulse response $h_1(v; x)$, the mask must be made proportional to

$$t(x, y) = \int_{-\infty}^{\infty} h_1(v; x)e^{j2\pi vy}\, dv, \qquad (8.4\text{-}7)$$

i.e., a 1-D inverse transform of $h_1(v; x)$. We note that in general $t(x, y)$ will be complex-valued.

For illustration, we again assume that $f(x) = \delta(x - x_1) + \delta(x - x_2)$ and that $h_1(v; x) = \text{rect}[(v - x)/w(x)]$. From Eq. (8.4-7) we obtain

$$t(x, y) = w(x)\,\text{sinc}[w(x)y]e^{j2\pi xy}. \qquad (8.4\text{-}8)$$

Such a mask can be made by Fourier-transforming the mask of Fig. 8.4-3b vertically while imaging it horizontally and recording the resultant distribution holographically. Figure 8.4-5 illustrates the operation of the processing system by decomposing the 2-D Fourier transform operation into a horizontal imaging–vertical transforming operation followed by a horizontal transform–vertical imaging operation. Figure 8.4-5a shows the pair of line delta functions, which are overlaid by mask $t(x, y)$. The line at x_1 is multiplied by $w(x_1)\,\text{sinc}[w(x_1)y]\exp(j2\pi x_2 y)$; similarily for the line at x_2. A horizontal imaging–vertical transform operation results in the distribution of Fig. 8.4-5b: Each mask-modified line is transformed vertically into a short line segment whose position and extent along the vertical axis depend upon x. The attendant horizontal transform–vertical imaging operation

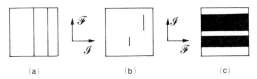

(a)   (b)   (c)

FIG. 8.4-5. Operation of system of Fig. 8.4-4.

yields the horizontal smears of Fig. 8.4-5c, which are viewed through the slit along the vertical (v) axis.

Although the system of Fig. 8.4-4 requires a holographic mask for the particular operation illustrated, this is not always the case. In general the system used to implement Eq. (8.4-1)—whether one of the above or one of the following methods—should be chosen with the particular application in mind.

The two methods described thus far involve illumination of a mask transparency by the input functions $f(x)$. There is a second class of systems for which the mask transparency is illuminated not by $f(x)$ but by its Fourier transform $F(u)$. These latter systems are slightly more complicated to set up but may have advantages under a given set of circumstances. In essence, they evaluate an integral of the form $\int_{-\infty}^{\infty} F(u)H(u, v)\, du$; however, this integral can be related to the desired superposition integral by the power theorem (generalized form of Parseval's theorem) which states that

$$\int_{-\infty}^{\infty} f(x)s^*(x; v)\, dx = \int_{-\infty}^{\infty} F(u)S_x^*(u; v)\, du, \qquad (8.4-9)$$

where $S_x(u; v) = \mathscr{F}_x\{s(x; v)\}$.

The two systems of Fig. 8.4-6 illustrate this second class of processors. In

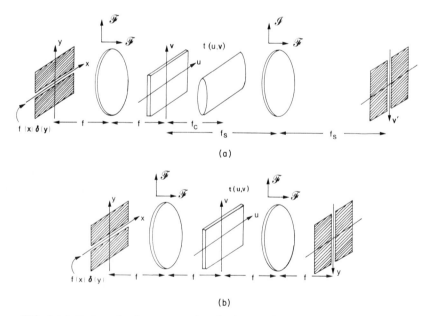

(a)

(b)

FIG. 8.4-6. Systems for frequency-variant implementation of general 1-D superposition integral.

both cases, the distribution $F(u)1(v)$ is projected onto a mask denoted by $t(u, v)$. In the first system, the product distribution $F(u)t(u, v)$ is Fourier-transformed horizontally and imaged vertically. The resultant output, as viewed along the vertical (v) axis, is given by

$$g(v) = \int_{-\infty}^{\infty} F(u)t(u, v) \, du. \tag{8.4-10}$$

If this output is to correspond to Eq. (8.4-1), one can easily show by using Eq. (8.4-9) that $t(u, v)$ must be given by

$$t(u, v) = \int_{-\infty}^{\infty} h_1(v; \xi)e^{j2\pi u\xi} \, d\xi. \tag{8.4-11}$$

In the second system, shown in Fig. 8.4-6b the distribution $F(u)t(u, v)$ is Fourier-transformed both horizontally and vertically. The output along the vertical (y) axis is given by

$$g(y) = \int_{-\infty}^{\infty} F(u)T_y(u, y) \, du, \tag{8.4-12}$$

where

$$T_y(u, y) \equiv \int_{-\infty}^{\infty} t(u, v)e^{-j2\pi vy} \, dv. \tag{8.4-13}$$

If this distribution is to correspond to Eq. (8.4-1), we require that

$$t(u, v) = \int_{-\infty}^{\infty} \int_{-\infty}^{\infty} h_1(\xi; \eta)e^{j2\pi(u\xi + v\eta)} \, d\xi \, d\eta. \tag{8.4-14}$$

In distinguishing between the two classes of processors just discussed, the following terminology is sometimes convenient: If the input function $f(x)$ itself illuminates the 2-D mask, as in the systems of Figs. 8.4-1 and 8.4-4, we speak of a *space-variant implementation*. If, on the other hand, it is the frequency spectrum $F(u)$ that is multiplied by the mask function, as for the systems of Fig. 8.4-6, we speak of a *frequency-variant implementation*. The key word here is implementation. As functions of v, the left and right sides of Eq. (8.4-9) are equal. However, the left side represents a space-variant implementation, whereas the right side represents a frequency-variant implementation. As noted earlier, which implementation is chosen in a particular case depends on the nature of the impulse response h_1. The two classes of processors are summarized in Table 8.4.1. Applications of such processors to various space-variant operations are discussed in Refs. [9–11].

There are various ways in which the basic methods discussed above can be generalized for increased flexibility. One approach is to go from a system

TABLE 8.4-1

Basic Single-Mask Systems for 1-D Space-Variant Processing

Implementation class	Distribution incident on mask	Mask function	Operation performed by lens system following mask	Refer to figure
Space-variant	$f(x)$	$t(x, y) = h_1(y; x)$	$\mathscr{J} \downarrow \mathscr{F}$	8.4-1
Space-variant	$f(x)$	$t(x, y) = \int_{-\infty}^{\infty} h_1(\xi; x)e^{j2\pi y\xi}\, d\xi$	$\mathscr{F} \downarrow \mathscr{F}$	8.4-4
Frequency-variant	$F(u)$	$t(u, y) = \int_{-\infty}^{\infty} h_1(y; \xi)e^{j2\pi u\xi}\, d\xi$	$\mathscr{J} \downarrow \mathscr{F}$	8.4-6
Frequency-variant	$F(u)$	$t(u, y) = \int_{-\infty}^{\infty}\int_{-\infty}^{\infty} h_1(\eta; \xi)e^{j2\pi(u\xi + y\eta)}\, d\xi\, d\eta$	$\mathscr{F} \downarrow \mathscr{F}$	8.4-6

employing a single mask transparency to one that uses two. An example of such a system is shown in Fig. 8.4-7. The input to this system is the wave amplitude $f(x)1(y)$. Perhaps the easiest way to analyze the operation of this system is to note that it essentially *images* the distribution $f(x)t_1(x, y)$ onto the output plane with a coherent 2-D impulse response that is determined by the spatial filter plane transparency $t_2(u, v)$. Denoting this coherent impulse response by $s(x, y)$, where

$$s(x, y) \equiv \mathscr{F}_2^{-1}\{t_2(u, v)\}, \qquad (8.4\text{-}15)$$

the wave amplitude along the y-axis in the output plane is given by

$$g(y) = \left[f(x)t_1(x, y) \right] * s(x, y)\big|_{x=0}$$
$$= \int_{-\infty}^{\infty}\int_{-\infty}^{\infty} f(\xi)t_1(\xi; \eta)s(-\xi; y - \eta)\, d\xi\, d\eta = \int_{-\infty}^{\infty} f(\xi)h(y; \xi)\, d\xi,$$

$$(8.4\text{-}16)$$

where

$$h(y; \xi) = \int_{-\infty}^{\infty} t_1(\xi; \eta)s(-\xi, y - \eta)\, d\eta. \qquad (8.4\text{-}17)$$

Additional systems that operate along these lines can be realized by replacing one or both of the spherical lenses with cylindrical–spherical lens combinations for 1-D transform–1-D imaging operations [11].

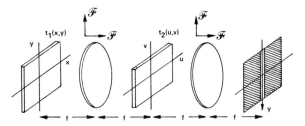

FIG. 8.4-7. Two-mask system for 1-D space-variant processing.

It is not generally clear how best to exploit the characteristics of two-mask systems or how they compare, in terms of light efficiency, ease of mask manufacture, etc., with the single-mask systems described earlier. Again, the best rule appears to be to consider the various possible implementations with a specific processing operation in mind. Choice of system will often be determined by relative simplicity, mask dynamic range requirements, light efficiency, and similar considerations.

B. Frequency-Variant Spectral Analysis

In Section 8.2 we introduced the concept of frequency variance. In this section we consider this concept again in the context of optical systems designed for spectral analysis of 1-D signals.

A conventional 1-D spectrum analysis has the form

$$F(u) = \int_{-\infty}^{\infty} f(x)e^{-j2\pi ux}\, dx. \tag{8.4-18}$$

Such an operation can be performed optically on an input signal record using either a spherical or a spherical–cylindrical lens system. Occasionally something other than a simple display of spectral content versus frequency is desired. In the analysis of musical sounds, for example, harmonic structures are emphasized if spectral content is displayed versus the logarithm of the frequency variable, for then octaves are spaced by equal distances, as on a piano keyboard. A log frequency display is also important in connection with the spectral analysis of radar and sonar signals that have undergone substantial compression or expansion in time scale upon reflection from rapidly moving objects. As an example, compare the conventional linear frequency spectra of a bandpass signal $s(t)$ and its threefold compressed version $s(3t)$, shown in part (a) of Fig. 8.4-8, with the log frequency displays shown in part (b). In the latter case, the effect of the time compression is simply a *shift* in the basic log frequency structure along the display axis rather than a stretching of the spectrum.

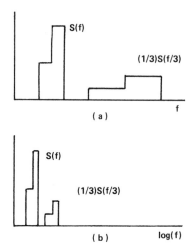

FIG. 8.4-8. Spectra of $s(t)$ and $s(3t)$: (a) linear frequency scale, (b) log frequency scale.

Several systems for implementing a log frequency spectrum analysis optically have been recently described [12, 13]. One example, appropriately described as a "slit-mapping" system, is shown in Fig. 8.4-9. Both lens components in this system are designed to image in the vertical direction and Fourier-transform in the horizontal direction. The mask in plane P_2 consists of a logarithmically shaped slit in an opaque background. The log frequency spectrum is viewed along the vertical axis in the output plane. A component $\exp(j2\pi u_0 x)$ of the input signal is transformed in the x-direction to produce a vertical line of light a distance u_0 along the u-axis in plane P_2. This line of light is passed only at one point by the logarithmic slit mask. The resultant compact spot of light is smeared horizontally by spherical–cylindrical lens combination L_2. Because of the shape of the mapping slit, the vertical distance of this smear from the origin in the output plane is proportional to the logarithm of u_0. Figure 8.4-10a shows the output of such a spectrum analyzer with a Ronchi ruling (square wave) for input. By way of comparison

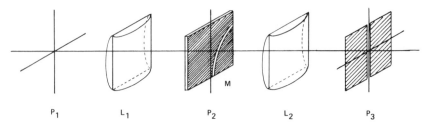

FIG. 8.4-9. Slit mapping system for log frequency spectrum analysis.

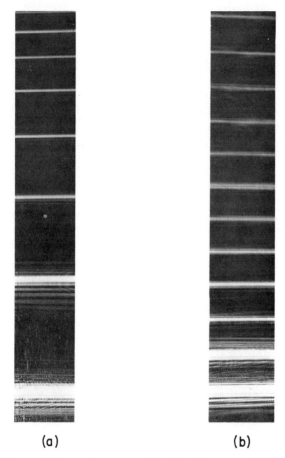

(a) (b)

FIG. 8.4-10. Output of slit mapping system: (a) with log frequency slit, (b) with linear slit.

a linear frequency display of the same spectrum is shown in Fig. 8.4-10b, obtained by replacing the log frequency mapping slit with a straight line slit.

Operation of the system is easily analyzed if we model the slit transmittance function by $t(u, y) = \delta(u - e^y)$, appropriate for a slit following the curve $y = \log(u)$. The output along the vertical axis in plane P_3 is then given by

$$g(y) = \int_{-\infty}^{\infty} F(u)\, \delta(u - e^y)\, du = F(e^y), \qquad (8.4\text{-}19)$$

which represents the desired mapping (increases in y correspond to progressively larger increases in the argument of $F(\cdot)$. It is possible to display signal spectral content versus many different functions of frequency simply by changing the functional form of the mapping slit.

The slit mapping technique, although relatively simple to implement, suffers severe light efficiency limitations since only a small fraction of the incident light finds it way to the output display slit. More efficient is an alternate method where the log frequency mapping is effected by a refractive or diffractive mapping element. In this approach, an input distribution $f(x)1(y)$ is Fourier-transformed in x and imaged in y; the resultant distribution $F(u)1(y)$ is incident on a mask with amplitude transmittance

$$t(u, y) = \begin{cases} \exp\{j2\pi[\log(u)]y\}, & u > 0, \\ 0, & \text{otherwise.} \end{cases} \tag{8.4-20}$$

The product distribution $F(u)t(u, y)$ is Fourier-transformed in both the horizontal and vertical directions, with the resultant output distribution viewed along the vertical output axis. This distribution is given by

$$g(0, v) = \int_{-\infty}^{\infty} \int_{-\infty}^{\infty} F(u)e^{j2\pi[\log(u)]y} e^{-j2\pi yv} \, du \, dy$$

$$= \int_{-\infty}^{\infty} F(u) \, \delta[v - \log(u)] \, du. \tag{8.4-21}$$

Using the relationship [14],

$$\delta[f(x)] = \sum_n \delta(x - x_n)/|f'(x_n)|, \tag{8.4-22}$$

where x_n are the roots of $f(x) = 0$ and $f'(x) = df/dx$ [$f'(x_n)$ is assumed to exist and is nonzero], we can evaluate Eq. (8.4-21) with the result

$$g(0, v) = \int_{-\infty}^{\infty} F(u)e^{-v} \, \delta(u - e^v) \, du = e^{-v}F(e^v). \tag{8.4-23}$$

This output has the desired form. The factor $\exp(-v)$ compensates for the tighter packing of higher frequency components along the display axis.

A holographic mask element can be recorded with the desired mapping characteristic relatively easily [12]. As illustrated in Fig. 8.4-11, the desired logarithmic mapping curve is plotted on white paper above a horizontal line which serves as a line impulse reference. A high contrast negative photograph is made of the pattern and placed in the input plane of a system that images in the horizontal direction and Fourier-transforms in the vertical direction. The resultant distribution in the output plane consists of a space-varying sinusoidal fringe pattern like that of Fig. 8.4-12, which is recorded on high resolution film. The recording should be of high contrast for improved diffraction efficiency when used in the spectrum analyzer. To avoid confusion

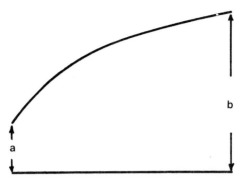

FIG. 8.4-11. Mask for recording holographic mapping element for log frequency analysis.

resulting from multiple diffraction orders, the distance b in Fig. 8.4-11 should be no larger than twice the offset distance a.

In a variation of the single-mask systems discussed above, the input plane distribution $f(x)1(y)$ is modified by mask function $w(x, y)$. The effect of this mask is to "window" the input signal differently for different values of y and, therefore, for different frequencies as viewed in the output plane [13]. A particularly important example is in a constant Q or constant proportional bandwidth analysis in connection with a log frequency display.

A general analysis of this system with both window mask and mapping transparency in place is outlined below. For convenience we assume the simpler slit mapping system. With the input modified by $w(x, y)$, the wave amplitude incident on the mapping slit is given by

$$u_2(u, y) = F(u) * W_x(u, y), \qquad (8.4\text{-}24)$$

where $*$ denotes a 1-D convolution with respect to the variable u and where

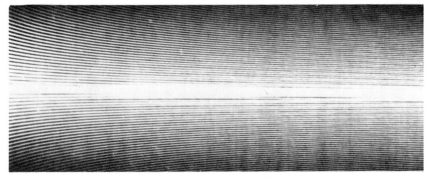

FIG. 8.4-12. Appearance of hologram for log frequency mapping.

$W_x(u, y) = \mathcal{F}_x\{w(x, y)\}$. We assume the mapping slit to follow the curve $y = \beta(u)$ and to have a constant (and satisfactorily small) width in the horizontal, i.e., u-direction. The amplitude transmittance of the mapping slit is then represented by

$$t(u, y) = \delta[u - \gamma(y)], \tag{8.4-25}$$

where γ is the inverse of β:

$$\gamma(y) = \beta^{-1}(y). \tag{8.4-26}$$

Multiplying $u_2(u, y)$ by $t(u, y)$ and integrating in the u-direction, we obtain as the final output distribution

$$u_3(0, y) = \int_{-\infty}^{\infty} f(\xi)w(\xi, y)e^{-j2\pi\gamma(y)\xi} \, d\xi. \tag{8.4-27}$$

Interpretation of this integral is straightforward: $u_3(0, y)$ represents a spectral analysis of a windowed version of $f(x)$, where both the window function $w(\cdot, \cdot)$ and the Fourier frequency variable γ can depend in a more-or-less general way on the display coordinate y. For the specific case of a constant Q, log frequency analysis, the mapping slit follows the curve $y = \log(u)$, as before, and the input window function has the form $w(x, y) = \tilde{w}(x/e^{-y})$, where $\tilde{w}(x)$ is the basic 1-D window function (e.g., rectangular or Gaussian).

It should be emphasized that there are no *fundamental* differences between space-variant signal processing and frequency-variant spectral analysis for 1-D signals. In view of the relationship of Eq. (8.4-9), operations of the form $\int_{-\infty}^{\infty} f(\xi)h_1(y; \xi) \, d\xi$ and $\int_{-\infty}^{\infty} F(u)H_1(y; u) \, du$ can be identical in result. For any one of the systems described in this section, either a single-mask or a two-mask configuration can be used to implement either integral operation. It is worthwhile to recognize, however, that the two expressions lend themselves to different interpretations: In one case the superposition integral represents a general mapping—a distortion or blurring—of time or space domain components of a 1-D signal, whereas in the other case the integral expression represents a general mapping of spectral components. This difference in point of view can be important in the application of these systems to specific signal processing or signal analysis problems.

C. Systems with 1-D Inputs and 2-D Outputs

Thus far in this section we have considered systems for which both the input and output distributions are one-dimensional. We now discuss two systems, shown in Fig. 8.4-13, for which the input is a 1-D signal but the output is two-dimensional. In the figure we have omitted lens components, simply indicating their function in terms of Fourier transform $\mathcal{F}$ and imaging

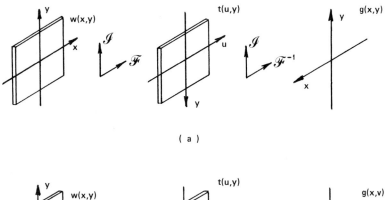

(a)

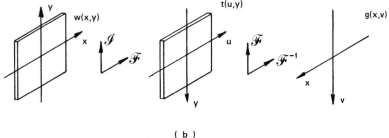

(b)

FIG. 8.4-13. Processors for 1-D inputs with 2-D outputs.

$\mathscr{I}$ operations. The input to each system is the wave amplitude distribution $f(x)1(y)$ [where again $1(y) = 1$ for all y].

The operation of the system of Fig. 8.4-13a is easily analyzed if we note that it is basically a channelized (in y) spatial filtering system operating on the distribution $f(x)w(x, y)$, where $w(x, y)$ is a 2-D window function corresponding to the input plane mask. The output wave amplitude $g(x, y)$ is given by a 1-D convolution integral, with y as a parameter:

$$g(x, y) = \int_{-\infty}^{\infty} f(\xi)w(\xi, y)T_x(x - \xi, y)\, d\xi, \qquad (8.4\text{-}28)$$

where

$$T_x(x, y) = \int_{-\infty}^{\infty} t(u, y)e^{j2\pi ux}\, du, \qquad (8.4\text{-}29)$$

where $t(u, y)$ is the complex amplitude transmittance of the intermediate plane mask. Equation (8.4-28) can be viewed as a special case of the 2-D space-variant superposition integral; i.e.,

$$g(x, y) = \int_{-\infty}^{\infty} \int_{-\infty}^{\infty} [f(\xi)1(\eta)]h_1(x, y; \xi, \eta)\, d\xi\, d\eta, \qquad (8.4\text{-}30)$$

where

$$h_1(x, y; \xi, \eta) = w(\xi, \eta)T_x(x - \xi, \eta)\,\delta(\eta - y). \qquad (8.4\text{-}31)$$

In the most common use of such a system, there is no input plane mask [i.e., $w(x, y) = 1$] and the operation performed is simply a 1-D convolution of $f(x)$ with a y-dependent impulse response:

$$g(x, y) = \int_{-\infty}^{\infty} f(\xi)T_x(x - \xi, y)\,d\xi. \qquad (8.4\text{-}32)$$

Such a system can be used to filter a 1-D input signal with multiple filter functions [5], each one of different scale. Alternatively, it can be used as a multichannel correlator, where the input signal is simultaneously correlated with a large number of 1-D distributions [15].

The system of Fig. 8.4-13b performs a similar operation, but the relationship between the impulse response and the mask transmittance functions is different. The easiest method for finding the system impulse response is to assume $\delta(x - \xi, y - \eta)$ as the input and follow its transformation through the system. The horizontal transform–vertical imaging operation produces the distribution $w(\xi, \eta)\exp(-j2\pi u\xi)\,\delta(y - \eta)$, which is multiplied by $t(u, y)$. The product function is inverse Fourier-transformed in u and forward transformed in y, yielding the distribution

$$h_1(x, v; \xi, \eta) = w(\xi, \eta)T_x(\xi - x, \eta)e^{-j2\pi\eta v}, \qquad (8.4\text{-}33)$$

where $T_x(\cdot, \cdot)$ is defined above. As before, if $w(x, y) = 1$, the superposition integral simplifies to a 1-D convolution in the x-direction with v as a parameter:

$$g(x, v) = \int_{-\infty}^{\infty} f(\xi)\left[\int_{-\infty}^{\infty} T_x(\xi - x, \eta)e^{-j2\pi\eta v}\,d\xi\right]d\eta$$

$$= \int_{-\infty}^{\infty} f(\xi)T(\xi - x, v)\,d\xi, \qquad (8.4\text{-}34)$$

where $T(u, v) = \mathscr{F}_2\{t(x, y)\}$.

D. The Ambiguity Function Processor

An operation of particular interest in radar and sonar signal processing is calculation of the cross-ambiguity function $g(\tau, v)$ for a pair of signals $f(t)$ and $h(t)$, defined by

$$g(\tau, v) = \int_{-\infty}^{\infty} f(t)h^*(t - \tau)e^{-j2\pi vt}\,dt. \qquad (8.4\text{-}35)$$

In this expression, τ represents a delay and v a temporal frequency. The significance of the cross-ambiguity function can be explained in terms of a simple radar example. Consider a radar system that transmits a short electromagnetic pulse denoted by $p(t)$. If the pulse is reflected from a fixed target, the return (echo) signal $r(t)$ has the form $r(t) = kp(t - \Delta t)$, where Δt is the roundtrip delay and k is a constant less than unity. The received pulse, along with accompanying noise, is input to a matched filter receiver which calculates its correlation with the transmitted waveform $p(t)$. The basic pulse shape is chosen to have good autocorrelation characteristics; thus, the receiver output, given by (for $k = 0$)

$$g(\tau) = \int_{-\infty}^{\infty} r(t)p(t - \tau)\,dt = \int_{-\infty}^{\infty} p(t - \Delta t)p(t - \tau)\,dt, \quad (8.4\text{-}36)$$

is strongly peaked at $\tau = \Delta t$.

In practice, the processing operation is complicated in two ways. First, the pulse is conveyed on an rf carrier. Second, the target of interest is generally not stationary but instead moves with some velocity component in the radial (i.e., transceiver target) direction. When the rf carrier is modulated in amplitude only, the transmitted pulse can be represented by

$$s(t) = p(t)e^{j2\pi f_c t}, \quad (8.4\text{-}37)$$

where f_c is the rf carrier frequency. The effect of target motion on the received signal is to introduce a compression or expansion of the time scale of the returned pulse. Thus,

$$r(t) = s[\alpha(t - \Delta t)] = p[\alpha(t - \Delta t)]e^{j2\pi\alpha f_c(t - \Delta t)}, \quad (8.4\text{-}38)$$

where α is the compression–expansion factor. In most cases of interest, $s(t)$ satisfies narrowband conditions: The dominant effect of target motion is to shift the frequency of the carrier, while leaving the envelope essentially the same; i.e.,

$$r(t) \cong p(t - \Delta t)e^{j2\pi\alpha f_c(t - \Delta t)}. \quad (8.4\text{-}39)$$

If, under these circumstances, we replace $f(t)$ and $h(t)$ in Eq. (8.4-35) with our expressions for $r(t)$ and $s(t)$, we obtain

$$g(\tau, v) = \int_{-\infty}^{\infty} p(t)e^{j2\pi f_c t}p(t - \tau - \Delta t)e^{-j2\pi\alpha f_c t}e^{-j2\pi\alpha f_c \Delta t}e^{-j2\pi vt}\,dt$$

$$= e^{-j2\pi\alpha f_c \Delta t}\int_{-\infty}^{\infty} p(t)p(t - \tau - \Delta t)e^{-j2\pi[v - (1 - \alpha)f_c]t}\,dt. \quad (8.4\text{-}40)$$

For $v = (\alpha - 1)f_c$, the integral becomes the simple correlation of Eq. (8.4-36). Thus $|g(\tau, v)|$ has a maximum at $v = (\alpha - 1)f_c$ and $\tau = \Delta t$, and the ambiguity function is seen to give both range and radial velocity (Doppler) information on the target. We can think of the Fourier kernel $\exp(-j2\pi vt)$ as shifting the frequency of the return rf pulse $r(t)$ back to the original carrier frequency f_c, so as to maximize the integral.

An optical system for implementing the cross-ambiguity function employs once again the system of Fig. 8.4-1 (or its equivalent) for Fourier-transforming in x and imaging in y. The signal $f(t)$ and $h(t)$ are recorded on film, scaled so as to produce transparencies governed by amplitude transmittance functions $f(x)1(y)$ and $h^*(\sqrt{2}x)1(y)$† These transparencies are placed in contact in the input plane, the second transparency being rotated clockwise through 45 deg. The result is the input plane transmittance

$$t(x, y) = f(x)h^*(x - y). \tag{8.4-41}$$

If this product distribution is now Fourier-transformed in the x-direction, the result is

$$g(u, y) = \int_{-\infty}^{\infty} f(x)h^*(x - y)e^{-j2\pi ux}\, dx, \tag{8.4-42}$$

which is the desired cross-ambiguity function, with y corresponding to τ and u corresponding to v. Variations on this basic scheme are described in Refs. [16, 17].

8.5 SYSTEMS FOR SPACE-VARIANT PROCESSING OF 2-D SIGNALS

The versatility of optical systems as space-variant processors for 1-D signals can be viewed as a natural consequence of the fundamentally 2-D nature of the optical systems' operation. Since the inputs are only 1-D in nature, there is a dimension or degree of freedom left over that can be exploited to increase the class of operations that can be performed, from space-invariant to space-variant. This is to be contrasted with the case of space-variant optical processing of 2-D signals, where there are no additional spatial degrees of freedom to be exploited. In the 2-D case, there must either be a drastic reduction in processing capability (as measured by the space–bandwidth product), or—as an alternative—some additional parameter such

---

† If $h(t)$ represents a real baseband signal (i.e., no carrier), the conjugation is unnecessary. If, however, $h(t)$ represents a narrowband modulated rf carrier of the form $a(t)\cos[2\pi ft - \theta(t)]$, additional steps must be taken to ensure the correct form of the output. Perhaps the simplest solution is to introduce a spatial filtering operation that removes either all positive or all negative frequencies, as appropriate, of the input signals. Thus, for $h(t)$, the effective input signal would be either $a(t)\exp[j2\pi ft - \theta(t)]$ or its complex conjugate, as desired. The system described in Ref. [16] operates in this manner.

as time, temporal frequency, or wavelength, must be exploited, often at the expense of system simplicity. In this section we describe four fundamentally different kinds of systems for space-variant processing of 2-D signals. Some are more general in their capabilities than others; each suffers from its own characteristic limitations.

A. Implementations Based on 2-D to 1-D Mappings

The success of space-variant processors for 1-D signals suggests an approach to space-variant processing of 2-D signals in which the 2-D distributions to be processed are first converted, via some invertible mapping, into 1-D distributions. The 1-D signal representations can then be processed in space-variant manner by one of the systems described in the preceding section and the resultant 1-D distributions converted to 2-D outputs via a complementary mapping.

The basic idea is illustrated with the following example. Let the desired superposition integral be given, as before, by

$$g(x, y) = \int_{-\infty}^{\infty} \int_{-\infty}^{\infty} f(\xi, \eta) h_1(x, y; \xi, \eta) \, d\xi \, d\eta. \tag{8.5-1}$$

If $f(\xi, \eta)$ and $h_1(x, y; \xi, \eta)$ are both band-limited in ξ and η, Eq. (8.5-1) can be expressed in discrete form as

$$g_{ij} = \sum_{k} \sum_{l} f_{kl} \hat{h}_{ijkl}, \tag{8.5-2}$$

where g_{ij}, f_{kl}, and $\hat{h}_{ijkl}$ are sample values of $g(x, y)$, $f(x, y)$, and $\hat{h}_1(x, y; \xi, \eta)$, respectively, taken on a regular sampling lattice, and where $\hat{h}_1(x, y; \xi, \eta)$ is obtained by low-pass-filtering $h_1(x, y; \xi, \eta)$ with respect to x and y to a bandwidth consistent with the output [18]. So long as the necessary sampling conditions are satisfied, $g(x, y)$ can be recovered without error from samples g_{ij} by an interpolation operation.

To go from a 2-D space-variant operation to a 1-D space-variant operation, we note that in practice the subscripts i, j, k, and l assume only a finite range of values. Thus, we can define an invertible correspondence

$$(i, j) \leftrightarrow n \tag{8.5-3}$$

$$(k, l) \leftrightarrow m \tag{8.5-4}$$

that links the two pairs of integer subscripts (i, j) and (k, l) to a single pair of integer subscripts m and n. With these subscripts, the double summation of Eq. (8.5-2) becomes a single summation over m,

$$g_n = \sum_{m} f_m \hat{h}_{nm}, \tag{8.5-5}$$

which is the discrete version of a space-variant 1-D superposition integral. Since the correspondence between (i, j) and n is invertible, g_{ij} can be determined from g_n and, via interpolation, $g(x, y)$ can therefore be reconstructed.

Figure 8.5-1 shows schematically the operation of an optical system for performing this kind of operation. First, the input distribution $f(x, y)$ (Fig. 8.5-1a) is sampled by a mask containing a regular array of pinholes (Fig. 8.5-1b). The resultant array of light spots is smeared out vertically by a vertical transform–horizontal imaging operation (Fig. 8.5-1c). Because of the staggering of the pinholes in the x-direction, each sample line is separated from its neighbors: The 2-D to 1-D mapping has been effected. This distribution of vertical lines is now multiplied by mask function $t(x, v)$, and a horizontal transform–vertical imaging operation performed. Along the vertical or v-axis, the output is masked by a slit. The wave amplitude along this slit is given by a weighted sum of the sample values of $f(x, y)$, where the weights depend on ordinate v. The final 1-D to 2-D mapping is achieved by smearing this v-axis distribution in the horizontal direction and sampling the result with another skewed sampling array (Fig. 8.5-1f).

This scheme for 2-D space-variant processing has several drawbacks. First, a train of multiple optical subsystems is required for the different transform-imaging operations, each subsystem contributing noise and reducing the signal level at the output. Of a more fundamental nature, both input and output distributions must be of a sufficiently small 2-D space–bandwidth product that they can be represented by 1-D distributions of space–bandwidth products consistent with the optical system's limitations. Taking 3000 as an achievable 1-D space–bandwidth product for a coherent-optical processor, we conclude that it must be possible to represent the input

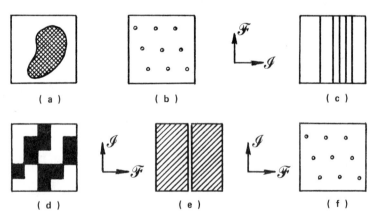

FIG. 8.5-1. Sequence of operations for space-variant 2-D processing via 2-D to 1-D mapping: (a) input distribution, (b) skewed sample mask, (c) result of vertical transform, (d) processing mask, (e) slit, (f) sampling mask for output.

and output distributions by sample arrays numbering 3000 pixels or less. This number may be impractically small for many applications.

B. Implementation by Spatial Frequency Coding

A second method for general 2-D space-variant processing is based on a spatial frequency coding of sample cells of the input distribution. We consider first the case of a simple geometrical transformation. Assume that the input complex wave amplitude distribution $f(x, y)$, representing the input function, consists of an array of nonoverlapping spatial pulses; i.e.,

$$f(x, y) = \sum_{m,n} f_{mn} p(x - ms, y - ns), \tag{8.5-6}$$

where $p(x, y)$ is the basic pulse function and s is the pulse spacing. The corresponding intensity distribution is given by

$$|f(x, y)|^2 = \sum_{m,n} |f_{mn}|^2 |p(x - ms, y - ns)|^2, \tag{8.5-7}$$

where we have used the fact that the pulses are nonoverlapping. Now let $f(x, y)$ illuminate a mask that "tags" each spatial pulse with a spatial frequency carrier of the form $\exp[j2\pi s(kx + ly)]$, k and l being integers. The result is the distribution

$$f'(x, y) = \sum_{m,n} f_{mn} p(x - ms, y - ns) e^{j2\pi s(kx + ly)}, \tag{8.5-8}$$

where k and l are functions of m and n. If this distribution is Fourier-transformed in x and y, the result is the distribution

$$F'(u, v) = \sum_{m,n} [f_{mn} e^{j2\pi s^2(mk + nl)}] P(u - ks, v - ls) e^{-j2\pi s(mu + nv)}, \tag{8.5-9}$$

where $P(u, v) = \mathscr{F}\{p(x, y)\}$, the pulse transform. To the extent that overlap of adjacent pulse transforms can be ignored,† the corresponding intensity distribution is given by

$$|F'(u, v)|^2 = \sum_{m,n} |f_{mn}|^2 |P(u - ks, v - ls)|^2. \tag{8.5-10}$$

Since k and l are functions of m and n, we see that this intensity distribution corresponds to a geometrical mapping of the input intensity distribution. A pulse of relative strength $|f_{mn}|^2$ at coordinates $x = ms$, $y = ns$ is mapped into a pulse of the same relative strength at coordinates $u = ks$, $v = ls$. So long as the sampling conditions for space-variant systems are satisfied [18], this

---

† Obviously, since the spatial pulse sequence was assumed nonoverlapping, there must be some overlap in the sequence $P(u - ks, v - ls)$. However, we assume that the cross-terms are small compared to the self-product terms and therefore can be ignored.

operation can represent a general geometrical transform of a spatially continuous input intensity distribution. If suitable phase compensation masks are used in the Fourier transform plane, it can also represent a general geometrical transformation on the input wave amplitude distribution.

In order for such a mapping to be effective, it is necessary that the spatial pulses making up $f'(x, y)$ and $F'(u, v)$ be essentially nonoverlapping. Therein lies the principal limitation of this scheme, for the space–bandwidth products of input and output distributions are, as a consequence, severely limited. To obtain a feel for this limitation, we assume that $p(x, y)$ is a 2-D Gaussian pulse:

$$p(x, y) = \exp\left[-\pi(x^2 + y^2)\right], \qquad (8.5\text{-}11)$$

a choice motivated by the observation that the uncertainty product is a minimum for the Gaussian function. We take an optimistic view and assume the pulse separation s to equal unity. There is actually significant overlap in this case; however, computations are simplified, and we only seek a rough estimate of the processor space–bandwidth product. Within a cell width, then, the number of periods of the highest frequency spatial carrier is given by the maximum value assumed by k (or l). The total number of cells in a row or column must therefore be limited such that the number of cells times the maximum possible frequency does not exceed the 1-D space–bandwidth product of the processor. Generalizing to two dimensions, we conclude that the inequality

$$(N \times M)(K \times L) \leq 2\text{-D space–bandwidth product} \qquad (8.5\text{-}12)$$

must be satisfied, where $N \times M$ and $K \times L$ denote the size of the cell (pulse) arrays in the input and output planes, respectively. As an example, a general geometrical transformation of a 50×50 cell input into a 50×50 cell output would require a processor with a space–bandwidth product (2-D) of approximately 6.25×10^6.

It is not necessary in effecting a geometrical transformation to operate on a spatially discretized input. For example, Bryngdahl has described a scheme [19] where an input transparency is placed in contact with a mapping element, either refractive or diffractive (e.g., a hologram), that multiplies the input distribution locally by a spatially varying carrier frequency term. To the extent that the input and the carrier frequency vary sufficiently slowly, the result is a fully continuous version of the scheme described above. Limitations on the space–bandwidth product capabilities of such a processor are similar.

Recently, Goodman [2] has described an extension of Bryngdahl's technique that is capable of effecting general 2-D space-variant operations, as opposed to geometrical transformations. In this scheme, the input—again assumed slowly varying—is considered to consist of an array of small squares, or pixels, each with its own value of amplitude transmittance. Behind each pixel is placed a small hologram that generates in a nearby subsequent plane (perhaps following a Fourier transform lens) a distribution specific to that hologram. Let the input distribution lie in the ξ-η plane, and the output in the x-y plane. If the holograms are constructed such that the hologram at input coordinates (ξ, η) produces output plane distribution $h_1(x, y; \xi, \eta)$, then the superposition of all cell responses, weighted by the input pixel amplitudes, is the desired output.

C. Coordinate Transformation Processing

Certain space-variant 2-D processing operations can be performed using conventional (i.e., space-invariant) coherent spatial filtering systems if the input to the processor first undergoes an appropriate distortion or coordinate transformation and the output of the processor a complementary "inverse" distortion. Such operations, first used in digital image processing [20, 21], have been investigated extensively in optical processing by Casasent and colleagues [22], who have emphasized their application in pattern recognition.

We illustrate the basic idea of coordinate transformation processing by considering the restoration of imagery blurred by the camera rotation discussed in Section 8.3. This type of motion blur is space-variant; normally, subsequent deblurring would entail a space-variant processing operation. However, it is possible to make the deblurring operation space-*invariant* if we first distort the blurred image in such a way that the blur has the *same* form and scale throughout the entire (distorted) image. Whereas the original image $i(x, y)$ is described by the general superposition integral

$$i(x, y) = \int_{-\infty}^{\infty} \int_{-\infty}^{\infty} o(\xi, \eta) h_1(x, y; \xi, \eta) \, d\xi \, d\eta, \qquad (8.5\text{-}13)$$

where $o(x, y)$ is the object distribution, the distorted image $i_d(x', y')$ is given by a *convolution*:

$$i_d(x', y') = o_d(x', y') * h_d(x', y'). \qquad (8.5\text{-}14)$$

In this latter equation, i_d and o_d are distorted versions of the original object and image distributions, respectively, and h_d is the corresponding space-invariant blur function; x' and y' are Cartesian coordinates. For rotation blur, the required distortion is the polar-to-rectangular-coordinate mapping depicted in Fig. 8.5-2. Whereas in the undistorted distribution the two-point object responses have different lengths and amplitudes, after the coordinate transformation they are the same, being simply displaced from one another. In this new representation, the blur function $h_d(x', y')$ is given by

$$h_d(x', y') = \delta(x') \, \text{rect}[(y' - \Delta\theta/2)/\Delta\theta]. \qquad (8.5\text{-}15)$$

If $i_d(x', y')$ is convolved with a deblurring function $s(x', y')$, the result is

$$\hat{i}_d(x', y') = o_d(x', y') * [h_d(x', y') * s(x', y')]. \qquad (8.5\text{-}16)$$

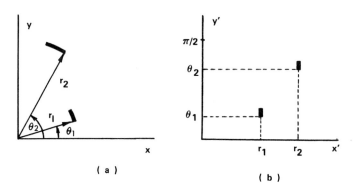

FIG. 8.5-2. Polar-to-rectangular coordinate transformation: (a) input, (b) output.

The result of this convolution is of course still a distorted representation of the image, but this can be corrected by subjecting i_d to the inverse of the original geometrical transformation.

The basic scheme illustrated by this simple example is characteristic of coordinate transformation deblurring (and other spatial filtering operations) generally: The blurred image is subjected to a geometrical distortion (the coordinate transformation) that leaves the blur function the *same* throughout the image. The distorted image then undergoes linear space-invariant processing, with the resultant output being transformed so as to undo the distortion. In some cases the specific form of the geometrical transformation needed to make the processing operation space-invariant can be determined by visualizing a rubber sheet distortion that makes the blur function the same throughout the entire image distribution. More generally, the analytical methods described in Ref. [23] can be used to determine the transformation, given the form of $h_1(x, y; \xi, \eta)$.

The distortion operations themselves cannot as a rule be performed optically with any great practicality. However, in many cases of interest (e.g., rotation blur, coma blur, and the diffusion blur of Section 8.3), the coordinate transformation can be introduced by scanning the input image and simultaneously writing the scan signal on a transparency in accord with the desired mapping. The polar-to-Cartesian distortion of Fig. 8.5-2, for example, can be effected in this manner if an arc scan in the x-y representation is mapped into a vertical line scan in the x'-y' representation. Casasent and Psaltis have employed logarithmic scans in x and y to transform two functions differing in scale, $f(x, y)$ and $g(x, y) = f(ax, ay)$, into two functions that are related by a simple shift: $g_d(x', y') = f_d[x' - \log(a), y' - \log(a)]$ [24]. Such an operation, essentially a 2-D version of the Mellin transform [4], can be used in scale-invariant optical processing [22].

Several practical limitations of coordinate transform processing are noted. First, in order to implement the scheme, the desired distortion operation must be known. In some cases (e.g., atmospheric turbulence blur), this will not be possible. And even if the coordinate transformation is known, it may be impossible to implement for the entire image. For example, the polar-to-Cartesian coordinate mapping becomes more and more difficult as r approaches zero. Also, there is once again a penalty paid in terms of space–bandwidth product. The smallest structural detail of the distorted distribution input to the linear shift-invariant optical processor must be resolvable by the processor. Yet much detail of the input distribution is stretched by the coordinate transformation. The conclusion is that the space–bandwidth product of the original input must be smaller than the space–bandwidth product of the optical processor itself, how much depending on the specific transformation applied. As a final limitation of coordinate transformation processing, we note that processing for noise reduction may in fact be made more difficult by a coordinate transformation, for noise processes that are statistically stationary (space-invariant) in the original image will be non-stationary (space-variant) in the distorted image distribution.

D. Holographic Multiplexing Techniques

Three techniques for general 2-D space-variant optical processing have been proposed that exploit one or another characteristic of holographic data storage for multiplexing of information, in this case the information being the different responses of the space-variant optical processor to input source points at different locations. The first characteristic so exploited is Bragg selectivity [25]. Figure 8.5-3a shows an optical system for recording a multiple exposure Fourier transform hologram in a thick recording medium. For the nth hologram recording, a plane reference wave is produced by a point source at location (ξ_n, η_n) in the input plane; the object distribution is the 2-D Fourier transform (with respect to x and y) of $h_1(x, y; \xi_n, \eta_n)$. This multiple exposure hologram serves as the spatial filter transparency in the coherent spatial filtering system shown in Fig. 8.5-3b. When light of amplitude $f(\xi_n, \eta_n)$ from input point (ξ_n, η_n) illuminates the hologram with the original nth reference wave, the response $h_1(x, y; \xi_n, \eta_n)$ is reconstructed in the output plane. To the extent that Bragg selectivity can prevent all but the nth hologram from responding to light from (ξ_n, η_n), the composite output plane response to a sampled input distribution represents the desired superposition.

The problem is that, although Bragg selectivity can do an excellent job of suppressing cross-talk between responses to input points separated (for this particular geometry) in the ξ-direction, it does not work well in the η-direction. This characteristic is unavoidable and, as a consequence, the

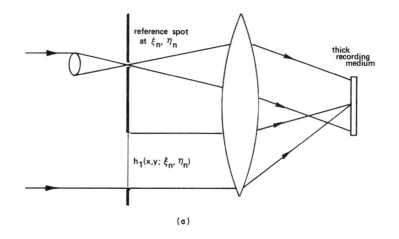

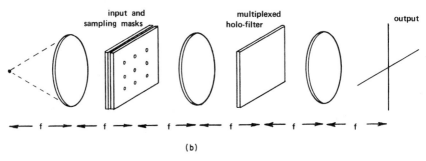

FIG. 8.5-3. Space-variant processing using thick hologram: (a) recording geometry, (b) hologram used as space-variant spatial filter.

Bragg effect method is best suited for effecting superposition operations of the form

$$g(x, y) = \int_{-\infty}^{\infty} \int_{-\infty}^{\infty} f(\xi, \eta) h(x - \xi, y; \eta) \, d\xi \, d\eta, \qquad (8.5\text{-}17)$$

i.e., space-invariant in one direction only.

An alternative holographic multiplexing method [26] that overcomes this limitation on 2-D space-variant processing exploits the associative memory characteristics of holograms. As before, a multiple-exposure hologram is recorded with Fourier transforms of $t(x, y; \xi_n, \eta_n)$ as the object distributions, but this time the recording medium is thin and the reference wave, instead of being a plane wave, is a random wave distribution produced by passing light from point (ξ_n, η_n) through a diffuser. In the subsequent spatial filtering operation, light from input point (ξ_n, η_n) again illuminates the diffuser to recreate the nth reference wave. Through the associative memory characteristics, this elicits the response $h(x, y; \xi_n, \eta_n)$ as desired. Ideally, a wave from any other input point, say (ξ_i, η_i), elicits only its own associative memory response $h(x, y; \xi_i, \eta_i)$ and not $h(x, y; \xi_n, \eta_n)$. Unfortunately, cross-talk cannot be completely avoided in practice, and the output of the processor will contain an undesirable background distribution that increases with the number of input points and with the number of multiplexed holograms. A further restriction on the scheme is imposed by the limited dynamic range of the recording medium, which limits the total number of holograms that can be multiplexed with acceptable diffraction efficiency.

Finally, we mention a scheme based on a time integration synthesis of the desired output distribution [27]. A holographic recording system is used to record a *multiple exposure* hologram, where the nth contributing recording is of the distribution $f(\xi_n, \eta_n)h_1(x, y; \xi_n, \eta_n)$. The reference wave is the same for all recordings. When the multiple exposure hologram is reconstructed, a wave distribution of the form $\sum_n f(\xi_n, \eta_n)h_1(x, y; \xi_n, \eta_n)$ is produced, which can be smoothed to yield the final output $g(x, y)$. Two drawbacks of this approach are evident: A new multiple exposure hologram must be recorded for each new input to the processor, and bias buildup may severely limit the diffraction efficiency of the hologram. The seriousness of these limitations has not been fully explored.

8.6 CONCLUDING REMARKS

A variety of coherent-optical methods for realizing space-variant super-position integrals have been presented in this chapter. Systems for processing 1-D signals were treated in particular depth, since they are relatively easily implemented, are fully parallel, and their analytical characteristics have been studied in great detail. Techniques for 2-D space-invariant processing are still evolving; limitations, particularly with respect to space–bandwidth product, currently restrict their applicability. The most easily applied method is probably the geometrical transformation method. In general, however, it requires scanning operations and is therefore not fully parallel in its oper-ation. Perhaps one of the potentially most important methods for space-variant optical processing is based on ultrahigh speed matrix operations realized with incoherent-optical processors. However, this method does not involve the optical Fourier transform in any way and is as a consequence not discussed here. The interested reader is referred to Ref. [2] for a discussion. In addition to this reference, Refs. [3, 22] provide additional insight into general space-variant processing operations.

ACKNOWLEDGMENTS

The support of the National Science Foundation and the U.S. Army Research Office for work in space-variant optical processing is gratefully acknowledged.

REFERENCES

[1] J. D. Gaskill (1958). "Linear Systems, Fourier Transforms, and Optics," p. 56. Wiley, New York.

[2] J. W. Goodman (1981). Linear space-variant optical data processing, in "Optical Information Processing" (S. Lee, ed.). Springer Publ., New York.

[3] J. F. Walkup (1980). Space-variant coherent optical processing, Opt. Eng. 19, 339–346.

[4] R. N. Bracewell (1965). "The Fourier Transform and its Applications," Chap. 17. McGraw-Hill, New York.

[5] L. J. Cutrona, E. N. Leith, C. J. Palermo, and L. J. Porcello (1960). Optical data processing and filtering systems, IRE Trans. Inf. Theory IT-6, 386–400.

[6] L. J. Cutrona (1965). Recent developments in coherent optical technology, in "Optical and Electro-Optical Information Processing" (J. T. Tippett, D. A. Berkowitz, L. C. Clapp, C. J. Koester, and A. Vanderburgh, eds.), pp. 83–124. MIT Press, Cambridge, Massachusetts.

[7] R. J. Marks and S. V. Bell (1978). Astigmatic coherent processor analysis, Opt. Eng. 17, 167–169.

[8] B. J. Pernick (1980). Optical systems for combined 1-D image-orthogonal Fourier transform processing, Appl. Opt. 19, 754–760.

[9] J. W. Goodman, P. Kellman, and E. W. Hansen (1977). Linear space-variant optical processing of 1-D signals, Appl. Opt. 16, 733–738.

[10] R. J. Marks, J. F. Walkup, M. O. Hagler, and T. F. Krile (1977). Space-variant processing of 1-D signals, Appl. Opt. 16, 739–745.

[11] J. M. Florence (1979). Frequency variant optical signal processing, Ph.D. Thesis, Georgia Institute of Technology, Atlanta (Univ. Microfilm No. 79-13794).

[12] W. T. Rhodes (1976). Log-frequency variable resolution optical spectrum analysis using holographic mapping techniques, Opt. Commun. 18, 492–495.

[13] W. T. Rhodes and J. M. Florence (1976). Frequency variant optical signal analysis, Appl. Opt. 15, 3073–3079.

[14] A. Papoulis (1968). "Systems and Transforms with Applications in Optics," p. 38. McGraw-Hill, New York.

[15] D. Casasent and E. Klimas (1978). "Multichannel optical correlator for radar signal processing, Appl. Opt. 17, 2058–2063.

[16] R. A. K. Said and D. C. Cooper (1973). Crosspath real-time optical correlator and ambiguity function processor, Proc. IEEE 120, 423–428.

[17] R. J. Marks, J. F. Walkup, and T. F. Krile (1977). Ambiguity function display: An improved coherent processor, Appl. Opt. 16, 746–750.

[18] R. J. Marks, J. F. Walkup, and M. O. Hagler (1976). A sampling theorem for space-variant systems, J. Opt. Soc. Am. 66, 918–921.

[19] O. Bryngdahl (1974). Geometrical transformations in optics, J. Opt. Soc. Am. 64, 1092–1099.

[20] G. M. Robbins and T. S. Huang (1972). Inverse filtering for linear shift-variant imaging systems, Proc. IEEE 60, 862–872.

[21] A. A. Sawchuk (1972). Space-variant image motion degradation and restoration, *Proc. IEEE* **60**, 854–861.

[22] D. Casasent and D. Psaltis (1978). Deformation invariant, space-variant optical pattern recognition, *Prog. Opt.* **16**, 289–356.

[23] D. Psaltis and D. Casasent (1977). Deformation invariant optical processors using coordinate transformations, *Appl. Opt.* **16**, 2288–2292.

[24] D. Casasent and D. Psaltis (1976). Scale invariant optical transform, *Opt. Eng.* **15**, 258–261.

[25] L. M. Deen, J. F. Walkup, and M. O. Hagler (1975). Representations of space-variant optical systems using volume holograms, *Appl. Opt.* **14**, 2438–2446.

[26] T. F. Krile, R. J. Marks, J. F. Walkup, and M. O. Hagler (1977). Holographic representations of space-variant systems using phase-coded reference beams. *Appl. Opt.* **16**, 3131–3135.

[27] R. J. Marks (1979). Two-dimensional coherent space-variant processing using temporal holography: Processor theory, *Appl. Opt.* **18**, 3670–3674.

Chapter 9

Fourier Optics in Nonlinear Signal Processing

A. A. SAWCHUK AND T. C. STRAND

IMAGE PROCESSING INSTITUTE
DEPARTMENT OF ELECTRICAL ENGINEERING
UNIVERSITY OF SOUTHERN CALIFORNIA
LOS ANGELES, CALIFORNIA

9.1 INTRODUCTION

A useful measure of the complexity of a signal is its time–bandwidth or space–bandwidth product. There is a great general need for systems that can perform fast, parallel, multidimensional operations on signals with large time–bandwidth and space–bandwidth products. In many cases traditional analog or digital electronic systems are overburdened or simply inadequate for processing these signals. The need for great processing capability arises in guidance, control, image processing [1, 2], radar signal processing [3], image pattern recognition [4, 5], and machine perception.

In the earlier chapters of this book, optical processing systems were invariably taken from the set of linear space-invariant (LSI) systems. W. T. Rhodes enlarged the set by including space-variant systems in his discussion in Chapter 8. However, as we shall see below, nonlinear systems can also be usefully applied in optical processing. Moreover, in some applications, the only option is to deal with nonlinear systems.

The parallel nature of optical systems and their inherently large space–bandwidth product has led to the development of many systems and techniques for optical information processing [6]. A fundamental difficulty with optical processing has been the limited range of operational software available [7, 8]. Thus, general nonlinear operations such as logarithms, power laws, and limiters have been very hard to implement, while linear operations such as correlation, convolution, and Fourier filtering have been relatively

371

easy. Many new techniques of signal processing and pattern recognition require nonlinear functions as part of their operation, and these functions have been achieved digitally, although in serial form [2]. In this chapter we describe a large number of techniques for achieving these and many other nonlinear operations in optical systems. Many of the techniques utilize real-time optical input transducers which can convert electronic or image information into a form suitable for input to an optical processor [9]. With most of these techniques, the processing is performed almost exclusively in an analog fashion. Recently, several concepts for binary [10–21] or residue [22–28] numerical optical processing have been developed in which the signals are processed as discrete levels within the system. This new approach holds much promise for the future if real-time processing speed, accuracy, and flexibility can be maintained. We show in this chapter how some of these operations can be treated as a form of nonlinearity.

In the systems to be described here, the Fourier transform generally enters in one of two ways. In many of the systems Fourier transform processing is integral to the implementation of the nonlinearity. This is the case for halftone techniques (Section 9.4,A) and spatial frequency modulation techniques (Section 9.4,B). In other systems, Fourier transform processing is an important adjunct to the nonlinearity in the operation of the overall system. This is the case for the systems described as composite nonlinear systems in Section 9.5.

There are a few nonlinear processing systems that do not involve Fourier transform processing. Such systems have also been included in this review for completeness, in recognition of their significance in Fourier optics in the broad sense. These systems are described in Section 9.4,C.

Although we have attempted a general overview of the subject of non-linear processing, the emphasis in the description of systems and experimental results is on work performed by the authors and their colleagues. The reason for this biased emphasis is that this is the work we feel most qualified to discuss and the only work we feel qualified to discuss in detail. Although by no means exhaustive, the references cited in the text along with the brief discussions of the cited works should give a more general view of the breadth and depth of research in the area of nonlinear optical processing.

9.2 CHARACTERISTICS OF NONLINEAR SYSTEMS

In this section we present the concept of nonlinear systems as applied to optical processing. The section defines nonlinear systems and describes what types of nonlinear systems will be discussed in this chapter. This latter function is perhaps the most important, because no monograph of this length can address all aspects of such a broad and ill-defined subject as nonlinear systems. The final purpose of this section is to provide background information which will serve as a framework for later discussions of specific systems.

A. Linear Space-Invariant Imaging Systems

Nonlinear systems can only be defined in term of what they are not; that is, to state the obvious, the set of nonlinear systems comprises all systems which are not linear. Therefore, before continuing, it is useful to specify what a linear system is. A system defined by the operator $\mathscr{L}$ produces an output $g_2(x_2, y_2)$ when given an input $g_1(x_1, y_1)$:

$$g_2(x_2, y_2) = \mathscr{L}[g_1(x_1, y_1)]. \tag{9.2-1}$$

The system $\mathscr{L}$ is linear if the superposition rule

$$\mathscr{L}[af(x, y) + bg(x, y)] = a\mathscr{L}[f(x, y)] + b\mathscr{L}[g(x, y)] \tag{9.2-2}$$

holds for all inputs f and g and all constants a and b. Such linear systems can be represented mathematically by an integral operator

$$g_2(x_2, y_2) = \int_{-\infty}^{\infty} \int_{-\infty}^{\infty} g_1(x_1, y_1)h(x_1, y_1; x_2, y_2) \, dx_1 \, dy_1, \tag{9.2-3}$$

where h is the impulse response of the system. Given the above criterion for establishing linearity or, conversely, nonlinearity of a system, we add the further restriction of space invariance which applies throughout this chapter to both linear and nonlinear systems. Space-variant systems represent an area of considerable research interest, particularly with regard to linear systems. Space-variant systems are the subject of Chapter 8 of this volume and will not be discussed further here. A linear system $\mathscr{L}$ is space-invariant if the impulse response has the form

$$h(x_1, y_1; x_2, y_2) = h(x_2 - x_1, y_2 - y_1). \tag{9.2-4}$$

In this case Eq. (9.2-3) reduces to a simple convolution expression

$$g_2(x_2, y_2) = \int_{-\infty}^{\infty} \int_{-\infty}^{\infty} g_1(x_1, y_1)h(x_2 - x_1, y_2 - y_1) \, dx_1 \, dy_1$$

$$= g(x_1, y_1) * h(x_1, y_1). \tag{9.2-5}$$

In the following sections, we will represent the linear space-invariant system schematically as shown in Fig. 9.2-1.

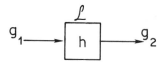

FIG. 9.2-1. Schematic representation of a linear space-invariant system characterized by the impulse response h with input g_1 and output g_2.

The real importance of space invariance is that it has a very simple description in the Fourier transform domain. Letting $G_1(u, v)$, $G_2(u, v)$, and $H(u, v)$ be the Fourier transforms of $g_1(x, y)$, $g_2(x, y)$, and $h(x, y)$, respectively, then

$$G_2(u, v) = G_1(u, v)H(u, v). \qquad (9.2\text{-}6)$$

In words, the Fourier transform G_2 of the output is equal to the product of the Fourier transform G_1 of the input and the Fourier transform H of the system impulse response. Here $H(u, v)$ is called the system transfer function. From the above we see that a linear space-invariant system can be described as three discrete operations on the input: Fourier transformation, multiplication with the transfer function, and inverse Fourier transformation, in that order. This is shown schematically in Fig. 9.2-2. This breakdown of the linear system will be useful in our classification of nonlinear systems in the next section.

B. Classification of Nonlinear Systems

Because it is impossible to treat all nonlinear systems, we must identify a clearly defined subset of nonlinear systems for analysis. Not only should this subset be small enough to be manageable, but it should be general enough to include as much of the work in nonlinear optical processing as possible and it should have a straightforward and logical description.

As a first step in defining the subset of nonlinear systems which we will discuss, we consider point nonlinearities, i.e., nonlinearities where the output g_2 at a given point (x_2, y_2) is a nonlinear function $\mathcal{N}$ of the input g_1 at a corresponding point (x_1, y_1) and is independent of the value of the input at all other points. The process of recording a light distribution on an ideal (infinite resolution) photographic film is an example of a point nonlinearity.

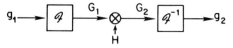

FIG. 9.2-2. Schematic representation of a linear space-invariant system in terms of Fourier transformations and the transfer function H. Here $\mathcal{F}$ and $\mathcal{F}^{-1}$ represent the Fourier transform operation and its inverse, respectively.

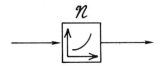

FIG. 9.2-3. Schematic representation of a point nonlinearity $\mathcal{N}$.

Although such point nonlinearities can obviously not be achieved in the strict sense, for practical purposes, many physical processes can be suitably modeled as point nonlinearities when the extent of the impulse response is small compared to the finest detail in the object. We will schematically represent a point nonlinearity as shown in Fig. 9.2-3.

Although the point nonlinearity by itself is of considerable interest, there are many cases of interest which cannot be described as simple point nonlinearities. However, if we combine a point nonlinearity with a linear system, we obtain a flexible model which can describe most nonlinear optical processing systems. This is the model which we will use in the following treatment.

C. Input–Output Plane Nonlinearity

The simplest combination of a point nonlinearity with a linear system is placement of the point nonlinearity just before or after the linear system, as shown in Fig. 9.2-4. These combinations will be referred to as input plane nonlinearities and output plane nonlinearities, respectively.

The simplest case of an output nonlinearity is observation of the image intensity in a coherent imaging system. The ideal coherent imaging system is linear in the complex amplitude of the input. However, the process of detecting the image involves the inherent nonlinearity of taking the square of the absolute value of the amplitude. This relation between the image intensity

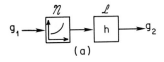

(a)

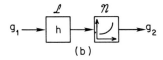

(b)

FIG. 9.2-4. Input–output plane nonlinearities. (a) The system consists of a point non-linearity $\mathcal{N}$ applied to the input, followed by a linear system $\mathcal{L}$. (b) Here the system has a linear component $\mathcal{L}$ with the point nonlinearity $\mathcal{N}$ applied to the output.

I_2 and the object amplitude U_1 is expressed by

$$I_2(x_2, y_2) = |U_2(x_2, y_2)|^2$$

$$= \left| \int_{-\infty}^{\infty} \int_{-\infty}^{\infty} U_1(x_1, y_1) h(x_2 - x_1, y_2 - y_1) \, dx_1 \, dy_1 \right|^2, \quad (9.2\text{-}7)$$

where h is the coherent impulse response of the imaging system, U_1 is the complex amplitude in the input plane, and U_2 and I_2 are the complex amplitude and intensity, respectively, in the image plane.

With minor modifications this model can be applied to many practical systems which are partially coherent, including in particular space-invariant systems with temporally coherent illumination. Several authors have dealt extensively with such systems (see, for example, Refs. [29, 30] and references cited therein), and we will not go into the subject here.

Independent of the inherent nonlinearity relating detected image intensity to the complex amplitude, most imaging systems are still subject to detector response nonlinearity at the output. The most common detectors of images, the retina, photographic film, and vidicons, exhibit strong nonlinearities which play an important role in the detection process. Thus, we see that the output nonlinearity model depicted in Fig. 9.2-4b is fundamental to many imaging systems.

The second general type of nonlinear system to be discussed is just the reverse of output nonlinearity. In this case, which we refer to as an input nonlinearity, the nonlinear element operates on the input before the signal is passed through the linear system. A model of this system is shown in Fig. 9.2-4a. A common example of an input nonlinearity is radiography, where the quantity of interest, the x-ray absorption, is proportional to the logarithm of the transmitted x-ray intensity.

Although it is obvious that the nonlinear element operator and the linear system operator do not in general commute, there is one trivial case when the input and output nonlinearity models are equivalent. This occurs if the impulse response of the linear system is a delta function ($H \equiv 1$). In terms of imaging systems, if the imaging process is perfect, it makes no difference to the overall response whether the nonlinear element is at the input or the output. The overall response of such systems is simply a point nonlinearity described by the nonlinear element. The purpose of most of the systems to be discussed in this chapter is to implement such a point nonlinearity. Although many of these systems include optical imaging systems where the transfer function definitely does not fulfill the perfect imaging requirement, $H \equiv 1$, the overall system will be modeled as a simple point nonlinearity. The consequences of not meeting the perfect imaging condition must be carefully studied in these cases.

D. Fourier Plane Nonlinearities

Applying a point nonlinearity to either the input or the output of a linear imaging system gives us models which are adequate for describing much of the work done in nonlinear optical processing. However, these models still represent only a very limited subset of the class of systems which can be obtained by combining a nonlinear element with an otherwise linear system. Having exhausted the two possibilities for combining a nonlinear element with the linear system model of Fig. 9.2-1, we now consider the additional possibilities evident when the linear system is described as in Fig. 9.2-2. The explicit Fourier transform representation of Fig. 9.2-2 is especially well suited to coherent imaging systems where the Fourier plane signals G_1, G_2, and H can be physically realized. Thus, it is a simple matter to apply a point nonlinearity to either the Fourier transform G_1 of the input, the Fourier transform G_2 of the filtered input, or the Fourier transform H of the impulse response of the linear system. These possibilities, which will be referred to as Fourier plane nonlinearities, are schematically represented in Fig. 9.2-5. Again we point out that, for an ideal transfer function, $H \equiv 1$, these systems simplify considerably. For this special case the systems of Fig. 9.2-5a and b are equivalent nonlinear systems and Fig. 9.2-5c reduces to a linear system. Although important examples of each type of Fourier plane nonlinearity of Fig. 9.2-5 can be found in the literature, and will be discussed in a subsequent section of this chapter, they are not as numerous as the input–output type of nonlinear system. This is perhaps due in part to the fact that the function of the Fourier plane nonlinearity is less transparent than that of the input–

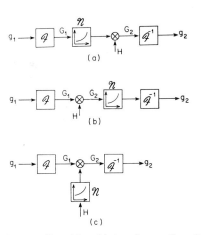

FIG. 9.2-5. Fourier plane nonlinearities. (a) A point nonlinearity $\mathcal{N}$ applied to G_1, the Fourier transform of the input. (b) A point nonlinearity applied to G_2, the filtered input. (c) A point nonlinearity applied to H, the filter function.

output nonlinearity. However, as will be seen later, the applications that have been found for this type of nonlinear system indicate a great potential for future research in this area.

9.3   APPLICATIONS OF NONLINEARITIES

In this section we briefly address the question of applications of nonlinear processing. Just as the theory of nonlinear processing is not as straightforward as that of linear processing, it is similarly not as straightforward to define large classes of operations such as the linear operations of convolution and filtering which can be immediately applied to a wide range of problems. Here we will discuss applications which have been demonstrated and attempt to define some general types of operations that have a wide range of applicability. Specific examples of the various applications are discussed in more detail in subsequent sections, along with descriptions of special applications not covered in this overview section.

A. Point Nonlinearities

As discussed above in Section 9.2,B, the simplest nonlinear system from a theoretical standpoint is one designed to produce a point nonlinearity. This is also the nonlinearity which has found the widest range of application.

The most straightforward applications of point nonlinearities are those associated with gray scale manipulation. These applications include increasing and decreasing contrast, linearizing the response of a system, extending the dynamic range or alternatively implementing clipping or thresholding of the input signal. One gray scale manipulation of particular interest is inversion or the formation of a negative. This allows optical division to be performed by equivalently multiplying by the inverse of a function [31]. Because of their utility and simplicity these gray scale operations are some of the most commonly used in practical processing systems.

Several other operations which are not so straightforward are also achieved with point nonlinearities. An example of this is the level slice where all gray levels are suppressed except in one predetermined range. Once this function has been obtained, it finds wide application and can be readily extended to many other areas. Equidensitometry, for example, consists of producing a series of narrow level slices in an image producing equidensity contours [32, 33]. The concept of multiple level slices can also be extended to the concept of analog-to-digital (A/D) conversion via the generation of bit planes [34, 35]. Once one has begun to consider digital applications, many other possible avenues become apparent. In particular, it is possible to implement any combinatorial logic operation by means of a point nonlinearity [36]. A final example of a useful function which can be obtained with a point nonlinearity is histogram equalization [33], where the gray scale is altered in such a way as to produce an image where all gray levels occur with approximately the same frequency.

B. Composite Nonlinear Systems

Extending the concept of nonlinear systems to include those which can be described as a point nonlinearity inserted in a linear system, we obtain what will be referred to as composite nonlinear systems. As discussed in the last section, a point nonlinearity can be utilized in the input plane, the output plane, or the Fourier plane of a linear system to produce an overall response which is no longer linear, nor is it adequately described as a point nonlinearity. In the following paragraphs we will first discuss some general operations which can be achieved by utilizing point nonlinear operations in the input and/or output planes of an otherwise linear system. Then we will mention some of the general applications of introducing point nonlinearities in the Fourier plane.

Probably the most common application of input–output plane nonlinearities is in homomorphic filtering [37, 38]. In homomorphic filtering for multiplied signals, a logarithmic nonlinearity is applied to the input. The resultant image is then filtered by a linear system. The filtered output can then be retransformed by an exponential nonlinearity (see Fig. 9.3-1). Such filtering operations are particularly well suited to the problem of filtering signals with multiplicative noise. The logarithmic operation converts the problem to one of additive signals which can then be handled by normal filtering procedures.

Input nonlinearities can in general be useful whenever certain intensity ranges of the input need to be emphasized or suppressed prior to spatial filtering. Such might be the case if one has an image with a constant noise level and one wishes to suppress regions where the signal-to-noise ratio is low. Several specific examples of input–output nonlinearities will be discussed in Section 9.5,A.

The most common application of point nonlinearities in the Fourier plane is in obtaining a filter which is difficult to produce directly but which can be implemented exactly, or at least approximately, by applying a point

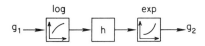

FIG. 9.3-1. A homomorphic filtering system for multiplied signals consists of a logarithmic nonlinearity applied to the input, followed by a linear filter and finally an exponential nonlinearity if desired.

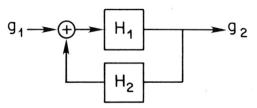

FIG. 9.3-2. A general feedback system consists of a feedforward leg with a transfer function H_1 and a feedback leg with a transfer function H_2. The equivalent transfer function is given by $H = H_1/(1 - H_1 H_2)$.

nonlinearity to an existing filter. This is the essence of feedback systems and several other systems which will be described in Section 9.5,B. Feedback systems are important enough to warrant a brief discussion here. A block diagram of a general feedback system is depicted in Fig. 9.3-2. It is easily shown that this system is equivalent to a simple linear system with the transfer function H given by

$$H = H_1/(1 - H_1 H_2). \tag{9.3-1}$$

where H_1 is the transfer function of the feedforward leg of the system and H_2 is the transfer function associated with the feedback leg. In describing the feedback in this manner we are assuming that H_1 and H_2 incorporate any gain or loss factors associated with the system so that H_1, H_2, and H are not normalized quantities. Since the denominator in Eq. (9.3-1) can be made very small, filters with a large effective dynamic range, such as those required for inverse filtering, can be implemented from filters which themselves may have a very small dynamic range. In practical applications of this procedure, close attention must be paid to noise and stability problems of course, but the technique is still very powerful. Several other systems in which a point nonlinearity is directly applied in the Fourier plane of an optical processor will be discussed in the next section.

9.4   POINT NONLINEAR SYSTEMS

The next several sections describe specific systems which directly perform point nonlinear operations on two-dimensional input functions. There are three major categories of techniques: (1) Halftone processing, (2) intensity-to-spatial frequency conversion, and (3) direct nonlinear processing. The best developed technique is halftone processing, and this section describes theory, degrading effects, compensation, experimental results, and generalizations such as pseudocolor. The next section describes theta modulation and a relatively recent nonlinear processing method relying on variable grating mode liquid crystal devices. This technique uses the intensity-to-spatial frequency conversion characteristics of these devices to implement nonlinear functions. The last section describes electro-optical systems which directly implement nonlinear functions by utilizing inherent characteristics of devices without the need for Fourier transforms.

A. Halftone Processing

The halftone method of nonlinear processing is a two-step procedure. The first step converts the continuous level two-dimensional input signal into a pulse-width-modulated (ideally) binary input. This operation is exactly the halftone procedure used in the graphic arts to represent an image containing gray tones as a binary picture. The halftoning step uses a mask transparency called a halftone screen, along with a high contrast (ideally a sharp threshold) photographic material or real-time coherent-optical input transducer.

For both photographic and real-time halftoning, the second step of the process is Fourier filtering and recombination of diffraction components in a coherent-optical system. The many variations of halftone screens and filtering procedures permit great flexibility in the nonlinear functions that can be achieved.

This discussion of halftone nonlinear processing is divided into several sections. The remainder of this section is a qualitative discussion of the process. The next sections contain mathematical details of halftone screens and the filtering process. Section 9.4,A,2 describes theoretical effects of using recording materials (film or real-time devices) which are not ideally binary. Some techniques for compensating for these effects are given. Section 9.4,A,3 shows experimental results for several specific kinds of nonlinear functions. Examples include photographic preprocessing and real-time implementation with a liquid crystal light valve. With this type of real-time device, halftoning has been accomplished at television rates (approximately 30 image frames per second). The last section describes extensions of the technique, including pseudocolor nonlinear optical processing.

The halftone process has been used for some time in the graphic arts industry to control density transfer characteristics in photoreproduction. Halftone screens generally consist of a one-dimensional or two-dimensional periodic array of identical continuous amplitude transmittance profiles, each varying continuously from opaque (amplitude transmittance $t = 0$ or density $D = \infty$) to transparent ($t = 1$ or $D = 0$) [39–44]. Figure 9.4-1 is a diagram of the preprocessing steps for a one-dimensional input. In Fig. 9.4-1a the one-dimensional continuous input $D_p(x)$ is plotted logarithmically as a photographic density. A typical halftone screen periodic density profile is shown as $D_s(x)$ in Fig. 9.4-1b. In the preprocessing step, the continuous

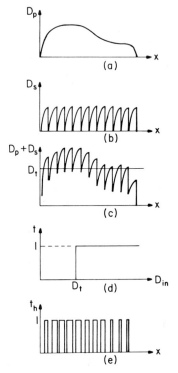

FIG. 9.4-1. The halftone encoding process. A continuous input density (a) adds with the halftone screen density (b) to produce a modulated density (c). This is recorded with a hard-clipping film whose characteristic is shown in (d). The resultant halftoned image is shown in (e). (From Armand [44].)

input and halftone screen are placed together and are photographically copied (by imaging or contact printing) onto a high contrast recording medium that ideally has a sharp threshold response, as shown in Fig. 9.4-1d. This figure shows the ideal transmittance of a high contrast reversal photographic material versus an input scale D_{in}, where D_{in} is proportional to the logarithm of exposure through the sandwich of input and halftone screen. The D_t shown in Fig. 9.4-1c and d is a clip level and is the maximum density through which the recording medium can be exposed. The clip level is logarithmically proportional to the controllable uniform illumination used; thus the effect of varying the illumination is to move the D_t value up and down on the axis of Fig. 9.4-1c.

The transmission of the recording medium is ideally either 1 or 0, because of its high contrast characteristics. The halftoned version of a continuous density distribution such as that in Fig. 9.4-1a will appear as shown in Fig.

9.4-1e. All values of x for which the density is less than the clip level turn black, hence transmit no light. All values of x for which the density is greater than the clip level expose the medium, resulting in unity transmission.

It is this halftoned picture, shown in profile in Fig. 9.4-1e, which is capable of yielding a nonlinearity when placed in a coherent-optical system as shown in Fig. 9.4-2. When this halftoned input picture is made with a two-dimensional halftone screen, the Fourier transform plane is a two-dimensional array of points of light which are ideally the centers of distinct spectral islands, each of which contains complete image information. This assumes that the picture is sampled at an adequately high rate. If the sampling rate is not high enough, the spectral islands will not be separable, and aliasing will occur. One of these diffraction orders is selected by a spatial filter and retransformed by the second lens to yield the demodulated output. As shown in this work, the output will be a nonlinear version of the original picture, where the nonlinearity depends on the halftone screen and the diffraction order chosen.

Section 9.4,A,1 considers in detail the dependence of the system transfer function on the halftone screen and diffraction order used. Input–output relationships are established as a function of system parameters.

1. Theory of Operation

In this section we present a formulation of the halftone process, which considers halftone screens with cells of any shape and recording media with general characteristic curves [44]. The input–output curves for ideally binary recording media are then derived.

For a general recording medium the binary characteristic curve of Fig. 9.4-1d is replaced by a curve such as that shown in Fig. 9.4-3. This curve will be denoted subsequently by $g(\log E)$, where $g(\cdot)$ is a general nonlinear function and the proportionality between D_{in} and $\log E$ is included in the definition of the function. This causes the amplitude transmittance of the halftoned picture to consist of pulses that are no longer rectangular, as

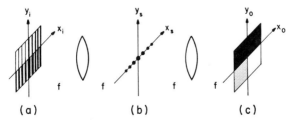

FIG. 9.4-2. Demodulation of the halftoned picture in a coherent-optical system: (a) halftoned picture, (b) spatial filter, and (c) desampled output. (From Armand [44].)

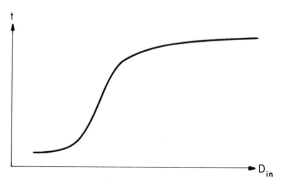

FIG. 9.4-3. Transmittance versus input density for a general nonideal recording medium. (From Armand [44].)

shown in Fig. 9.4-4. The amplitude, width, and shape of these pulses depend on the input picture density levels, the halftone screen density profile, and the shape of the characteristic curve of the recording medium. Each group of pulses corresponds to a constant intensity (or density) subregion in the input picture. The period L (Fig. 9.4-4) of the halftone screen is chosen to be small in comparison with the period of the highest spatial frequency component in the input picture, so any local region of the amplitude transmittance of the halftoned transparency is approximately a periodic sequence of pulses. In the analysis that folows, I_{in} denotes the local input picture intensity that produces this pulse sequence. We will now find the output intensity I_0 as a function of I_{in}, assuming I_{in} varies slowly compared to the spatial sampling rate of the halftone screen as stated. With this assumption, the inherent dependence of I_{in} and I_0 on spatial position in the input image can be dropped. The amplitude transmittance of the halftoned input is denoted by t_h. It can be expanded

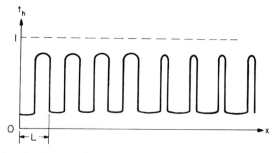

FIG. 9.4-4. Amplitude transmittance of a halftoned transparency made with a nonideal recording medium. (From Armand [44].)

in a complex Fourier series:

$$t_h(x) = \sum_{k=-\infty}^{+\infty} B_k \exp(-j2\pi kx/L), \qquad (9.4\text{-}1)$$

where

$$B_k = \frac{1}{L} \int_0^L t_h(x) \exp(j2\pi kx/L)\, dx. \qquad (9.4\text{-}2)$$

In the above sum each term represents a grating order, and when we take the Fourier transform of the halftoned picture in the coherent-optical processor, these orders appear in the Fourier plane as isolated spectral islands. The spatial filter in this plane then selects a single order. We omit wavelength and lens geometrical scale factors for clarity, and the resulting intensity distribution at the processor output is

$$I_0(I_{in}, k) = |B_k|^2 = \left| \frac{1}{L} \int_0^L t_h(x) \exp(j2\pi kx/L)\, dx \right|^2, \qquad (9.4\text{-}3)$$

which relates the intensity at a locally constant region of the output picture to the amplitude transmittance of the halftoned picture and the selected order.

The halftone periodic transmittance $t_h(x)$ can be related to the input intensity I_{in} as follows. If the density variations of one period of the halftone screen are represented by $f(x)$ [this is the function plotted as $D_s(x)$ in Fig. 9.4-1b], then the intensity transmitted by the halftone screen is $I_{in}10^{-f(x)}$ [6]. Using the amplitude transmittance versus log exposure curve of the recording medium as described by $g(\log E)$, we can then write

$$t_h(x) = g\{\log[I_{in}10^{-f(x)}]\} = g[\log I_{in} - f(x)], \qquad (9.4\text{-}4)$$

and replacing it in Eq. (9.4-3), we have

$$I_0(I_{in}, k) = \left| \frac{1}{L} \int_0^L g[\log I_{in} - f(x)] \exp(j2\pi kx/L)\, dx \right|^2, \qquad (9.4\text{-}5)$$

which relates the intensity at a locally constant region in the output picture to the intensity of the corresponding region in the input picture through a nonlinear integral relationship. When the specific forms of $g(\log E)$ and $f(x)$ are substituted in this relationship and the integral is solved, the overall relation between the output intensity and the input intensity is nonlinear. This nonlinearity depends on $g(\log E)$, $f(x)$, and the value of the order selected. Note also that the process of desampling the halftoned input in the Fourier plane produces a spatially continuous output.

The above formulation could be performed in terms of the intensity transmittance $\tau(x)$ of the halftone screen rather than its density profile. In this case the intensity transmittance-versus-exposure curve of the recording medium $T(E)$ is needed rather than $g(\log E)$. After a similar derivation as above we obtain

$$I_0(I_{\text{in}}, k) = \left| \frac{1}{L} \int_0^L T[I_{\text{in}}\tau(x)] \exp(j2\pi kx/L) \, dx \right|^2. \tag{9.4-6}$$

a. *Binary Recording Medium* A model for the characteristic curve of an idealized binary recording medium is shown in Fig. 9.4-5. Ideally, $a = 0$ and $b = 1$. Note that this form of characteristic curve is applicable to a positive transparency (reversal photographic medium). We could also consider the more familiar negative transparency, although the basic results remain the same. We will choose the positive transparency curve because these curves are more similar to the characteristic curves for real-time devices. We now simplify the general relationship of Eq. (9.4-5) using the characteristic curve of Fig. 9.4-5.

b. *Zero Order* When only the zero order of the Fourier-transformed picture is used, $k = 0$ in Eq. (9.4-5) and we have

$$I_0(I_{\text{in}}, 0) = \left\{ \frac{1}{L} \int_0^L g[\log I_{\text{in}} - f(x)] \, dx \right\}^2. \tag{9.4-7}$$

Considering Fig. 9.4-5 we can write

$$\text{if} \quad \log I_{\text{in}} - f(x) < \log I_r, \quad \text{then} \quad g(\log E) = a \tag{9.4-8}$$

and

$$\text{if} \quad \log I_{\text{in}} - f(x) \geq \log I_r, \quad \text{then} \quad g(\log E) = b. \tag{9.4-9}$$

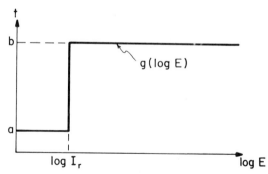

FIG. 9.4-5. Characteristic curve of a binary recording medium. (From Armand [44].)

Assuming $f(x)$ to be a monotonically increasing function we have

$$\text{if} \quad x > f^{-1}[\log(I_{in}/I_r)], \qquad \text{then} \quad g(\log E) = a \qquad (9.4\text{-}10)$$

and

$$\text{if} \quad x \le f^{-1}[\log(I_{in}/I_r)], \qquad \text{then} \quad g(\log E) = b, \qquad (9.4\text{-}11)$$

where $f^{-1}(\cdot)$ denotes the inverse function of $f(x)$. Substituting Eqs. (9.4-10) and (9.4-11) in (9.4-7) gives

$$I_0(I_{in}, 0) = \{a + [(b - a)/L]f^{-1}[(\log(I_{in}/I_r)]\}^2 \qquad (9.4\text{-}12)$$

after some simplification. This result is a generalization of previous work for the case $a = 0$, $b = 1$ [43]. Because f and f^{-1} are monotonic functions, $I_0(I_{in}, 0)$ is also a monotonic function. By appropriate choice of $f(x)$, a variety of monotonic input–output curves can be synthesized in the zero order.

c. *Nonzero Order*   For $k \neq 0$, the preceding expressions for the characteristic curve can be used in Eq. (9.4-5) to obtain

$$I_0(I_{in}, k) = \frac{(b - a)^2}{\pi^2 k^2} \sin^2\left[\frac{\pi k}{L} f^{-1}\left(\log \frac{I_{in}}{I_r}\right)\right] \qquad \text{for } k \neq 0. \quad (9.4\text{-}13)$$

Note that the $\sin^2$ dependence in Eq. (9.4-13) allows nonmonotonic nonlinear functions to be achieved if nonzero orders are used in the Fourier filtering step. In general, the diffraction order required is proportional to the number of sign changes in the slope of the desired nonlinear function.

d. *Halftoning Screen Density for Some Useful Nonlinearities*   The halftone screen density function for different nonlinearities with a general recording medium characteristic curve has been determined. Here we illustrate the procedure by finding the halftone screen density functions for several specific nonlinearities assuming a binary recording medium.

Logarithmic Transformation   This transformation is monotonic and can be obtained in the zero order. The desired logarithmic transformation is

$$I_0(I_{in}, 0) = K \log(I_{in}/I_r), \qquad (9.4\text{-}14)$$

where K is a constant. Combining Eqs. (9.4-12) and (9.4-14), we obtain

$$K \log(I_{in}/I_r) = \{a + [(b - a)/L]f^{-1}[\log(I_{in}/I_r)]\}^2. \qquad (9.4\text{-}15)$$

If we represent $\log(I_{in}/I_r)$ by $f(x)$ and note that $f^{-1}[f(x)] = x$, then Eq. (9.4-15) can be written as

$$f(x) = (1/K)[a + [(b - a)/L]x]^2, \qquad (9.4\text{-}16)$$

which gives the density profile of the halftone screen for a logarithmic transformation.

Exponential Transformation   This transformation can also be obtained in the zero order. We want

$$I_0(I_{in}, 0) = \alpha \cdot \beta^{(I_{in}/I_r)}, \qquad (9.4\text{-}17)$$

where α and β are positive constants and $\beta > 1$. Combining Eq. (9.4-12) and Eq. (9.4-17) we obtain

$$\alpha(\beta)^{I_{in}/I_r} = \{a + [(b - a)/L]f^{-1}[\log(I_{in}/I_r)\}^2. \qquad (9.4\text{-}18)$$

Representing $\log(I_{in}/I_r)$ by $f(x)$ as before and simplifying, we obtain

$$f(x) = \log(2\log\{a + [(b - a)/L]x\} - \log\alpha) - \log(\log\beta). \qquad (9.4\text{-}19)$$

This gives the density profile of the halftone screen for an exponential transformation. For brevity, we omit here certain additional constraints on α and β that ensure the existence and nonnegativity of $f(x)$.

Power Transformation   This is another example of a transformation possible in the zero order. We want

$$I_0(I_{in}, 0) = \gamma(I_{in}/I_r)^{\lambda}, \qquad (9.4\text{-}20)$$

where γ and λ are positive constants. Combining this with Eq. (9.4-12) we have

$$\gamma(I_{in}/I_r)^{\lambda} = \{a + [(b - a)/L]f^{-1}[\log(I_{in}/I_r)]\}^2, \qquad (9.4\text{-}21)$$

and with the usual substitutions and simplifications we obtain

$$f(x) = (2/\lambda)\log\{a + [(b - a)/L]x\} - (1/\lambda)\log\gamma. \qquad (9.4\text{-}22)$$

This gives the density profile of the halftone screen for a power transformation. As before, additional constraints on λ and γ that ensure the existence of $f(x)$ are omitted.

Level Slice Transformation   In this case the desired input and output intensities are

$$I_0 = \begin{cases} 0 & \text{for } I_{in} < I_1, \\ c^2 & \text{for } I_1 \le I_{in} < I_2, \\ 0 & \text{for } I_2 \le I_{in}, \end{cases} \qquad (9.4\text{-}23)$$

and this function is shown in Fig. 9.4-6. This transformation is nonmonotonic and can be achieved in the first order; i.e., $k = 1$ in Eq. (9.4-13). We have

$$I_0(I_{in}, 1) = [(b - a)^2/\pi^2]\sin^2\{(\pi/L)f^{-1}[\log(I_{in}/I_r)]\}. \qquad (9.4\text{-}24)$$

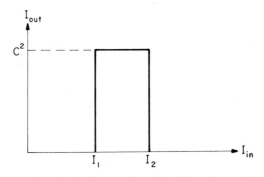

FIG. 9.4-6. Level slice transformation. (From Armand [44].)

Combining Eq. (9.4.23) and Eq. (9.4-24) we can see that

$$f^{-1}[\log(I_{in}/I_r)] = 0 \qquad \text{for} \quad I_{in} < I_1, \qquad (9.4\text{-}25)$$

$$f^{-1}[\log(I_{in}/I_r)] = (L/\pi)\sin^{-1}[c\pi/(b-a)] \qquad \text{for} \quad I_1 \le I_{in} < I_2, \qquad (9.4\text{-}26)$$

and

$$f^{-1}[\log(I_{in}/I_r)] = L \qquad \text{for} \quad I_{in} \ge I_2. \qquad (9.4\text{-}27)$$

These are summarized in Fig. 9.4-7, where

$$r = (L/\pi)\sin^{-1}[c\pi/(b-a)]. \qquad (9.4\text{-}28)$$

Replotting the data in Fig. 9.4-7 we obtain Fig. 9.4-8, which gives the density profile of the halftone screen for the level slice transformation as shown in Fig. 9.4-6. Note that the halftone screen for this function is a Ronchi ruling of the specified duty cycle and contrast.

Some experimental results for these nonlinear functions are given in Section 9.4.A,3. Additional variations in the halftone screens are possible;

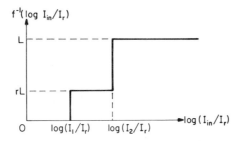

FIG. 9.4-7. Inverse function of the density profile of the halftone screen which performs the level slice transformation. (From Armand [44].)

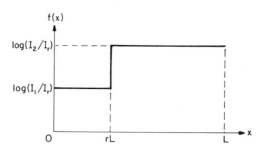

FIG. 9.4-8. Density profile of the halftone screen which performs the level slice transformation. (From Armand [44].)

using nonmonotonic cell profile functions $f(x)$ allows a nonlinearity with many slope changes to be achieved in the first diffraction order [45]. Related design algorithms have also been given by Matsumoto and Liu [46].

2. Degrading Effects and Compensation

As shown in previous sections, the copy medium used in the pulse-width modulation step of nonlinear halftone processing must ideally have a sharp threshold characteristic. Although some photographic materials closely approach this ideal, they have the disadvantage of slow, clumsy operation. Recently, many different types of real-time optical input modulators have appeared, which can convert electronic or image information to a form for input to a coherent-optical processing system [9, 47–50]. Many different technologies are utilized in these devices, but at present most of them have the smooth linear transfer characteristics of photographic film used for ordinary continuous tone application. Devices which operate at television rates (on the order of 30 frames/sec) and perform sharp thresholding remain unavailable. Details of some of these devices and experimental results for nonlinear processing are given later.

The general analysis of the halftone process as expressed by Eq. (9.4-5) can be used to predict the effects of the copying medium given a screen profile and diffraction order. Several analyses of the problem have used computer simulation to predict degradation of the input–output curves from the ideal. The results of this work are extensive [44, 51]; only a sample is given here.

For smooth monotonic nonlinear functions such as the logarithm, the main source of output degradation is the linear section of the t-versus-D_{in} characteristic curve as shown in Fig. 9.4-3. The saturation regions where the slope of the curve is zero for low and high values of D_{in} have less effect on the performance. Figure 9.4-9 shows these effects. Figure 9.4-9a is the ideal two-decade logarithmic response, where the horizontal axis is plotted on a normalized scale. Figure 9.4-9b shows the degraded response for a copying

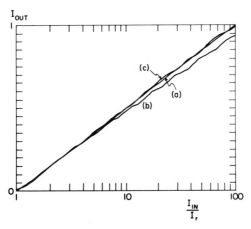

FIG. 9.4-9. Logarithmic function for a piecewise linear model of a recording medium with gamma = 3.0. (a) Ideal. (b) Degraded. (c) Optimized. (From Armand [44].)

medium having a photographic gamma (slope of the linear part of the density-versus-log exposure curve) of 3.0. The curve showing the degradation was computed using a piecewise linear model for the curve $g(\cdot)$ of the copy medium, although any particular measured response curve can be used in Eq. (9.4-5). In Fig. 9.4-9, the I_{out} response tends to fall below ideal for high values of normalized I_{in}.

For nonlinearities with sharp slope changes and sharp corners, such as the level slice function, the sharp rising and falling transitions have a reduced slope, and the sharp corners tend to become rounded. The reason for this is that sharp threshold functions rely completely on the thresholding characteristic of the copy medium to attain their sharp slope. This effect can be seen by comparing the ideal level slice shown in Fig. 9.4-10a with the degraded results shown in Fig. 9.4-10b. Given a copy medium with a finite gamma greater than 1, it is possible to increase the effective gamma by making a copy of the first halftoned image. The overall gamma of the process will increase, and the threshold will be sharper. However, this procedure is clumsy and impractical for real-time implementation.

Given a nonideal recording medium described by $g(\cdot)$, and a desired input–output curve, two methods are available for designing an optimum halftone screen which produces the best fit to the desired nonlinear response. Both techniques begin with Eq. (9.4-5) relating these functions [44]. The first method analytically inverts the integral equation of Eq. (9.4-5). Although this method is exact, it is currently restricted to certain monotonic non-linearities such as power law and exponential transformations, and it is also restricted to a piecewise linear model for the recording medium characteristic

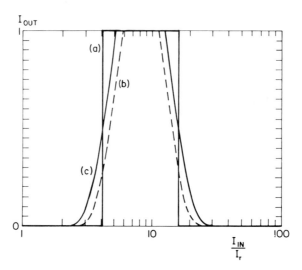

FIG. 9.4-10. Level slice function for a piecewise linear model of a recording medium with gamma = 3.0. (a) Ideal. (b) Degraded. (c) Optimized. (From Armand [44].)

curve. Solution for more complicated models may be possible but is analytically very difficult.

In an alternative method, the halftone screen density profile to be derived is assumed to be quantized. Optimum quantized values are found by minimizing the difference between desired and degraded outputs in the least squares sense. This method can be used for any form of the recording medium characteristic curve and any type of nonlinearity. In application, the halftone screen may be made on a plotting device with discrete density levels; thus this procedure gives a good practical solution.

The results of computer simulation of compensation for a logarithmic function are shown in Fig. 9.4-9c. In this simulation, 30 discrete points in the halftone screen density profile and screen density values from 0 to 2 are assumed. The optimized output curve is seen to approximate the ideal result in Fig. 9.4-9a with much less error than before. Similar results for a level slice function are shown in Fig. 9.4-10c. Here the optimization procedure is successful in improving the fit to the ideal but cannot increase the finite slope of the sharp transitions at the boundaries of the degraded response. Details of these procedures and many additional simulation results are given in Refs. [44–51].

3. Experimental Results

This section describes experimental results of nonlinear processing by halftoning. The first section summarizes results from traditional photographic methods; the second section describes recent real-time implementation with coherent input transducers.

a. *Photographic Implementation* The use of halftoning to modify gray level transfer characteristics dates back to the late 1800s with the development of the modern graphic arts and photographic industry. As early as 1893 Abbe [52] experimentally noted the diffraction orders present in the Fourier spectrum of a sampled image. More recently, Marquet [53] and later Marquet and Tsujiuchi [54] noted that demodulating a halftoned photographic image in a coherent-optical system produced various monotonic and nonmonotonic nonlinearities depending on the diffraction order chosen in the Fourier transform plane. Similar experiments and the basic idea of halftoning for nonlinear processing were described by Delingat [55]. Other halftone experiments were reported by Pappu *et al.* [56], and Roychoudhuri and Malacara [57] explored contrast reversal with halftoning.

Kato and Goodman [39, 40] experimentally found that a commercially available halftone screen performed a logarithmic nonlinearity over 2 log units of exposure. They demonstrated the use of the technique for the separation of multiplied signals by homomorphic filtering [37, 38]. In homomorphic filtering the logarithm of two signals which have been multiplied is taken first. This log operation converts the signals to additive form so that traditional linear filtering can be used. Homomorphic filtering is effective for separating signals with a multiplicative noise component. Kato and Goodman demonstrated its application in separating periodic multiplicative noise from an image, as shown in Fig. 9.4-11. Part (a) of this figure shows a face and a grating multiplicatively combined. Part (b) shows the homomorphically filtered result in which a halftone screen has been used to

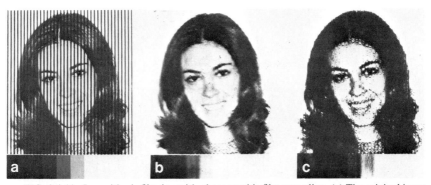

FIG. 9.4-11. Logarithmic filtering with photographic film recording. (a) The original input pattern is a girl's face multiplied with a grating. The dynamic range is from 0 to 2 in density. (b) Logarithmic filtering to remove the grating. (c) Linear filtering to remove the grating. (From Kato and Goodman [39, 40].)

convert the input to additively combined signals which can be easily separated with Fourier filtering. Figure 9.4-11c shows the result of pure linear filtering without the halftone process to remove the grating. These workers describe additional experimental results for suppressing speckle noise and removing the screen grid from radiographs. Dashiell and Sawchuk [43] also made logarithmic halftone screens on a plotting microdensitometer. These screens also gave a good approximation to a logarithmic response over two decades of input dynamic range. These workers also experimentally performed exponentiation having an output dynamic range greater than three decades.

Dashiell and Sawchuk have experimentally demonstrated level slicing using the first Fourier diffraction order as described in Section 9.4,A.1 [43, 58]. Their halftone screen was a low contrast copy of a Ronchi ruling. Figure 9.4-12 shows their experimental results; part (a) is the original continuous tone scene, and part (b) shows the isodensity contours of level slicing.

Much early work on halftone nonlinear processing was concerned with multiple isodensity or multiple isophote level slicing. Schwider and Burow [59] described a technique for filtering an image hologram to obtain isophotes. Delingat [60] showed that passing high diffraction orders of a halftoned image produced isophotes. Schneider [61] and Schneider et al. [62] described the design of special halftone screens and filtering systems to obtain isophote and isodensity contours. Liu et al. [63] have also demonstrated multiple isophote contours by filtering a halftoned image. Figure 9.4-13, taken from Strand [41], shows an example of multiple isophote processing. Figure 9.4-13a shows the input image, and Fig. 9.4-13b shows three isophotes, each representing a doubling of the input amplitude.

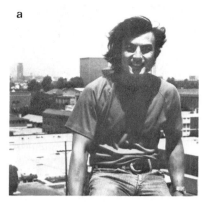

FIG. 9.4-12. Level slice function with photographic film recording. (a) Continuous tone input. (b) Level slice output. (From Sawchuk and Dashiell [58]. © 1975 IEEE.)

 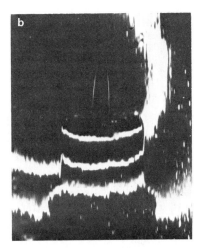

FIG. 9.4.13. Isophote generation with photographic film recording. (a) Input image. (b) Isophote output. Each isophote represents a doubling of input amplitude. (From Strand [41].)

Dashiell and Sawchuk [45] have demonstrated optical image quantization and intensity notch filtering by halftoning. They have described a synthesis procedure for halftone cell profiles that are not monotonic. Using such a screen, they can achieve an arbitrary number of sign changes in the input–output curve using the first diffraction order.

Lohmann and Strand [34] and Liu [64] have performed analog-to-digital conversion by halftoning using photographic film. Figure 9.4-14 shows a three-bit A/D output [34]. The results of the conversion appear serially as bit planes, each of which displays the information of a particular significant bit of the digitized image. The solid lines plotted on the I_{in} axis of the figure show the nonlinear transfer characteristic needed to produce the bit planes of the three-bit reflected binary or Gray code. The bottom curve showing I_{03} versus I_{in} is the least significant output bit. The halftone process was used, and the output results are shown scaled and superimposed on the drawing. The next most significant bit, I_{02}, has the same characteristic curve as I_{03} except that the horizontal axis is expanded by a factor of 2. This expansion is achieved experimentally by attenuating the input by 2 and repeating the experiment. Similarly, I_{01} is the most significant bit, and it is obtained by attenuating by a factor of 4. In this way, each bit can be obtained sequentially by halftoning, and any analog input on the scale from 0 to 8 gives a unique three-bit quantized and digitized representation. Results of A/D conversion for a two-dimensional image are shown in Figs. 9.4-15 and 9.4-16 [65]. Figure 9.4-15 is the input image. Figure 9.4-16 shows the Gray code bit planes. The images on the left side of Fig. 9.4-16 were generated by the

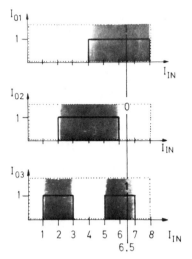

FIG. 9.4-14. First three bit planes of a Gray code A/D conversion performed on a linear intensity wedge. (From Lohmann and Strand [34].)

optical A/D conversion. Bit planes produced by a scanning microdensitometer are shown on the right side of Fig. 9.4-16 for comparison. The discrepancies near the perimeter of the image in the least significant bit are attributable to nonuniform illumination in the optical processor.

Liu [64] has designed a halftone screen which produces several bit planes with only one halftoned photograph. The different bit planes in decreasing order of significance are obtained by passing higher order diffraction terms in the Fourier filtering step. Liu [66] has also described another halftone procedure for A/D conversion, which relies on an electronic logic array to

FIG. 9.4-15. Input image for optical A/D conversion.

produce the digital output at each image point. A photographic halftone procedure is used first as quantizer to give a discrete input to the logic system.

Real-Time Implementation   There has been very little experimental work on real-time nonlinear halftone processing because of the lack of real-time

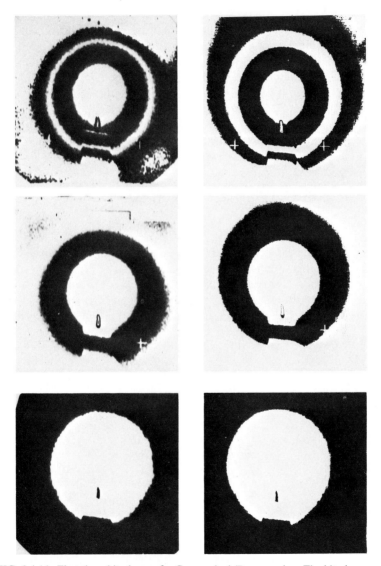

FIG. 9.4-16. First three bit planes of a Gray code A/D conversion. The bit planes on the left-hand side were generated optically. The bit planes on the right-hand side were produced digitally after scanning with a microdensitometer.

optical input transducers with a threshold characteristic. Logarithmic and level slice nonlinear processing has been attempted with a standard Hughes liquid crystal light valve (LCLV) suitable for linear incoherent-to-coherent optical conversion.

The LCLV exists in both electronically and optically controlled versions. The optical LCLV serves as a real-time incoherent-to-coherent converter. Incoherent light impinges on a photoconductor, which in turn changes the local electric field across a liquid crystal layer. The change in electric field alters the local birefringence of the liquid crystal material. This birefringence pattern can be read out as a spatial amplitude modulation by placing an analyzer in the output beam oriented orthogonal to the initial polarization of the readout beam. The readout illumination can be spatially coherent or incoherent but must be temporally narrowband. Additional variations in the device response are made possible by introducing a twist in the alignment of the liquid crystal molecules during device assembly. The LCLV device operates at television frame rates (approximately 30 msec cycle time) and generally is designed to have a linear response over two decades of dynamic range. Many references that discuss construction and operational details are available [47, 48, 50].

To obtain a real-time logarithmic nonlinearity [44], the characteristic curves of a Hughes 45° twisted nematic LCLV were measured. Figure 9.4-17a is the original characteristic curve of the LCLV, showing that the device has a smooth curve approximately that of photographic film with a gamma of 2–3. With these data, an optimum compensated discrete halftone screen was made using the procedure outlined in Section 9.4,A,1. This half-

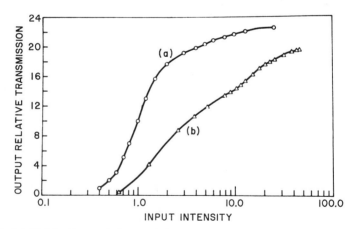

FIG. 9.4-17. Liquid crystal characteristic curve. (a) Original transfer function without a halftone screen. (b) Generated logarithmic transfer function with a halftone screen designed to give a logarithmic response over two decades. (From Armand *et al.* [73]. © 1978 IEEE.)

tone screen had a fundamental spatial frequency of 3 cycles/mm and was designed to work in the zero diffraction order.

The experimental setup used is shown in Fig. 9.4-18. The continuous level input is placed in contact with the halftone screen and imaged onto the control surface of the LCLV with incoherent illumination. The LCLV is read out with coherent light between crossed polarizers, and a low-pass spatial filter sufficient to pass the image bandwidth is placed in the zero order in the Fourier plane of the system. Figure 9.4-17b shows the response curve of the system with the halftone screen in place. The result approximates a logarithmic transfer function with less than 5% error over one decade and less than 10% error over two decades.

To test the effectiveness of the logarithmic filtering system in another experiment, two crossed multiplicatively combined Ronchi rulings were used as an input picture for the experimental setup of Fig. 9.4-18. The period of these rulings was approximately 3 mm, much longer than the halftone screen period of 0.33 mm. The spectrum in the filter plane is shown in Fig. 9.4-19a. Next, the logarithm halftone screen was placed in contact with the rulings. The filter plane spectrum is shown in Fig. 9.4-19b. The difference in Fourier spectra between multiplicatively combined gratings and additively combined gratings obtained by real-time logarithmic filtering is as follows: The additive spectrum components lie only on the x- and y-axis around the zeroth diffraction order in the frequency domain, while the multiplicative spectrum contains cross-term off-axis components. Figure 9.4-19 also shows the higher diffraction orders that arise because of the halftone screen. For simple logarithmic processing, these higher orders would be eliminated by spatial filtering.

The system has also been applied to an image filtering problem. For this experiment a picture with multiplicative noise was generated. The goal was to eliminate the multiplicative noise by homomorphic filtering, i.e., a sequence of a logarithmic transformation followed by linear filtering. Ideally, this

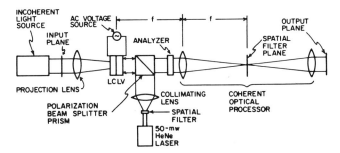

FIG. 9.4-18. Experimental setup for real-time halftone processing with a LCLV. (From Armand *et al.* [73]. © 1978 IEEE.)

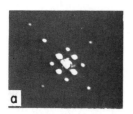

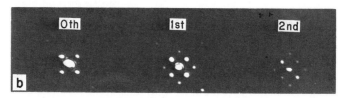

FIG. 9.4-19. Real-time logarithmic processing with a halftone screen and a LCLV. (a) Fourier transform of two crossed gratings imaged on the LCLV with no halftone screen. The gratings are oriented at 45° and 135° with respect to a horizontal axis. The grating spectra are convolved with one another. (b) Fourier transform of crossed gratings with the halftone screen. In the zeroth order, the off-axis terms have been eliminated, indicating the grating spectra are not convolved with one another. (From Armand *et al.* [73]. © 1978 IEEE.)

would be followed by an exponential transformation, but this is not essential in demonstrating noise reduction [44]. The exponentiation was not included in the following experiments. The noise generated for this experiment models that of a pushbroom scanner that scans six lines at a time with six independent detectors. Any variation in the sensitivity or gain along the row of detectors gives rise to a periodic six-bar noise structure across the image. This effect is shown in Fig. 9.4-20a. The density range of the image with the simulated scanner noise was 2.0*D*. This was chosen to match the 100:1 operating range of the halftone screen.

The noisy image was sandwiched with the halftone screen and imaged with incoherent illumination onto the LCLV. The resultant image was read out with a helium–neon laser. A low-pass spatial filter with bandwidth sufficient to pass the image was placed in the Fourier plane of the system to select a single (zeroth) diffraction order in the halftone spectrum. Since the screen effectively performs a logarithmic transformation, the pattern in the zeroth diffraction order of the Fourier plane consisted of the sum of the image diffraction pattern and the pushbroom scanner diffraction pattern. The scanner diffraction pattern consisted of a series of isolated diffraction orders which could be filtered out without significantly degrading the image. Thus in the output plane the image is reconstructed without the multiplicative noise. This is shown in Fig. 9.4-20b. Without the logarithmic transformation,

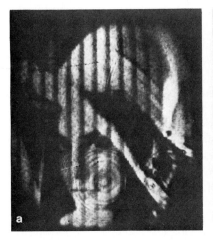

FIG. 9.4-20. Real-time homomorphic filtering. (a) Input image is a face with simulated pushbroom scanner noise. Overall density range is 2.0D. (b) Homomorphic filtered output.

the noise and image spectra would have been convolved with one another, making them impossible to separate.

These experiments demonstrate the feasibility of performing real-time nonlinear filtering for smooth functions using an LCLV or some other real-time device. A real-time level slice experiment has also been performed with a standard linear 45° twisted nematic LCLV [67], and the results are comparable to the level slice work shown in the simulations of Fig. 9.4-10b. Until sharp threshold real-time devices become available, halftone nonlinear processing will be limited to smooth nonlinear functions.

4. Extensions and Conclusions

Several extensions of nonlinear processing useful in image manipulations have been reported experimentally. Pseudocoloring is the process of associating colors with different gray levels in a monochrome image so that the image is enhanced for display. One motivation for the use of pseudocolor is that the human eye is much more sensitive to small changes in color than to small changes in gray level. The pseudocoloring system described by Liu and Goodman [68] begins by halftoning the continuous tone input using a screen designed to produce multiple isophotes [63]. The halftoned image is then placed in a coherent processor as shown in Fig. 9.4-2 and is sequentially illuminated by red, green, and blue monochromatic collimated plane waves. Different diffraction orders may be selected at each wavelength, and the resulting images for each color are summed photographically on color film at the output. The result contains different combinations of the three primary colors at each point as a function of input level.

Indebetouw [69] has described a pseudocoloring system which combines halftoning with elements of the theta modulation procedure discussed in Section 9.4,B. A one-dimensional halftone screen designed to produce a level slice is used in contact with the continuous tone input to expose a high contrast film as usual. This process is repeated with the exposure changed, so that the effective position of the level slice on the I_{in}-axis is shifted. In addition, the one-dimensional screen is rotated through a small angle between exposures. In effect, the various level slices are encoded as directional information on the binary copy film. The copy film is then placed in a coherent processor, except that a white light point source is used in front of the collimating lens to produce white light plane wave illumination. In the Fourier plane, level slice information recorded at a particular range of intensity values appears as a line of diffraction spots located at a unique angle. This pattern is modified with an angular array of color filters. After inverse transforming, the desampled image has various colors associated with different bands of gray levels. An advantage of this method is that the colored result is not formed sequentially and need not be summed photographically on color film.

The preceding theory and experimental results are a summary of the present state of halftone nonlinear processing. The procedure allows a large variety of nonlinear functions to be performed but also has limitations. Two major limitations are that accurate halftone screens with sufficient gray level resolution must be available, and that a sharp threshold is needed on the copy medium (photographic film or a real-time device) to ensure a binary input to the processor. The copy medium must also have a spatial resolution sufficient to record the finest detail in a sampled halftoned image of the original. Closely related to this is the tradeoff between spatial resolution in the copy medium and gray scale accuracy in the final output due to the pulse width modulation nature of the process. In short, the overall performance of nonlinear halftone processing depends strongly on the binary, high resolution character of the copy medium. These properties can generally be achieved in photographic processes but are still unavailable in a real-time optical input transducer. A final limitation is noise in the form of speckle, interference fringes, and other artifacts that appear in coherent systems. Careful precautions must be taken to avoid these effects. Finally, nonlinear processing using halftoning and incoherent spatial filtering is possible, in exactly the same way as the eye views a halftoned photograph. This procedure eliminates coherent noise problems but is limited to monotonic nonlinear transformations of intensity. Achieving nonmonotonic functions requires the interference properties of coherent light.

B. Intensity-to-Spatial Frequency Conversion

One very convenient method of obtaining point nonlinearities is through intensity-to-spatial frequency conversion. The idea is to encode each resolution element of an image with a grating structure where the period and/or the orientation of the grating is a function of the image intensity at the point in question. If certain sampling requirements are met, each intensity level of interest is uniquely assigned to a different point in Fourier space and all points with a given intensity in the image are assigned to the same point in Fourier space (assuming space-invariant operation is desired). Then a pure amplitude spatial filter can alter the relative intensity levels in an arbitrary way. The overall transformation can be found graphically, as in Fig. 9.4-21 which shows the implementation of a level slice. In this figure, the quantity s is used as a generalized spatial frequency coordinate. In terms of polar spatial frequency coordinates, s typically varies either along an arc of constant radius ρ or along a radius of constant azimuth θ (see Fig. 9.4-22b and c). The first case is referred to as theta modulation, and the latter case is referred to by the somewhat ambiguous term, frequency modulation. In both theta modulation and frequency modulation methods there is a certain connection between the number of intensity levels to be distinguished, the bandwidth of the object, and the spatial frequency required for the modulated grating. For the general case of intensity-to-spatial frequency conversion, the spatial frequency is modulated along some arbitrary curve s in the spatial frequency plane (see Fig. 9.4-22a). A grating with a given fundamental frequency (ρ, θ)

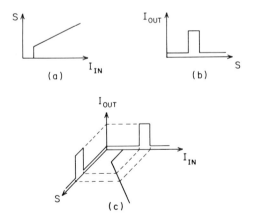

FIG. 9.4-21. Intensity-to-spatial frequency conversion for nonlinear processing. (a) A device is used to convert input intensity to a varying spatial frequency s. (b) A Fourier plane filter selectively attenuates the spatial frequency component s. (c) The I_{out}-versus-I_{in} characteristic of the overall filtering system can be found graphically by tracing through the two characteristics of (a) and (b).

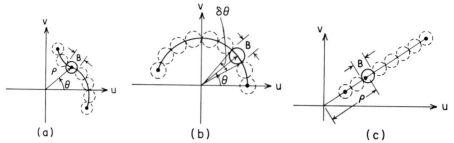

FIG. 9.4-22. Intensity-to-spatial frequency conversion (a) General case. (b) Theta modulation. (c) Frequency modulation.

in polar coordinates, corresponding to a specific input intensity, produces a diffraction order at (ρ, θ) surrounded by a diffraction spot whose dimension is roughly equal to the bandwidth B of the object. If one wants to process N distinct intensity levels, then the curve s must be chosen so that N nonoverlapping object spectra can be placed along s as indicated in Fig. 9.4-22. For the case of theta modulation, where $\rho = \rho_0$, the curve s is restricted to an arc of π radians. If the number N of intensity levels to be encoded is large, we can assume that the angular width $\delta\theta$ of the object spectrum is

$$\delta\theta \cong B/\rho_0. \tag{9.4-29}$$

If we want N distinct levels, then we must satisfy the condition

$$N\delta\theta = NB/\rho_0 \equiv \Delta\theta \leq \pi \tag{9.4-30}$$

or

$$\rho_0 \geq NB/\pi. \tag{9.4-31}$$

For frequency modulation, the useful modulation range is limited by the fact that the grating will generally have multiple diffraction orders. If the lowest fundamental grating frequency is ρ_0 and its first significant harmonic is ρ_1, then the usable modulation range $\Delta\rho$ is

$$\Delta\rho = \rho_1 - \rho_0. \tag{9.4-32}$$

If we assume there are no missing orders in the grating spectrum, then

$$\Delta\rho = \rho_0. \tag{9.4-33}$$

Furthermore we have the equality for the spectral width,

$$\delta\rho = B. \tag{9.4-34}$$

In this case we must satisfy the condition

$$N\delta\rho = NB \leq \rho_0 = \Delta\rho \tag{9.4-35}$$

or

$$\rho_0 \geq NB. \tag{9.4-36}$$

Since the bandwidth B is equal to two times the highest spatial frequency in the object, Eqs. (9.4-31) and (9.4-36) give us the relationship between the required sampling frequency ρ_0 and the largest object spatial frequency. Obviously, if N is large, very high spatial frequency gratings will be required, which is a limiting factor in these techniques. This would particularly be true if the optical system had to be capable of imaging these gratings. Fortunately, since spatial filtering of the grating spectra for selecting a single diffraction order is an integral part of the process, the space–bandwidth product requirements on the optical system are not so severe.

Of these two intensity-to-spatial frequency conversion techniques, theta modulation was the first to be discussed and demonstrated [70, 71]. A practical problem associated with theta modulation, as well as with frequency modulation, is how to obtain the spatial frequency encoding. One solution to this problem for theta modulation which has been demonstrated involves the use of special halftone screens [62]. The screen profile is designed to produce halftone dots which have one edge whose orientation is a function of the intensity in the original image. The theta modulation technique is especially well suited to white light image processing, since the spectra at different orientations do not overlap as they do in general for a pure frequency modulation scheme. The pseudocolor system of Indebetouw [69], described in the previous section, also uses halftone techniques to achieve theta modulation.

Lee has recently described a system which utilizes both theta modulation and frequency modulation [72]. A scanning interferometric system is used to encode each picture element of an image as a small grating segment whose frequency and orientation can be changed in accordance with input intensity levels. If real-time processing is not required, this system has the advantage of being able to utilize more fully the Fourier plane than either theta modulation or frequency modulation.

Frequency modulation has recently become a potentially viable means of achieving point nonlinearities in real-time with the advent of a device referred to as a variable grating mode (VGM) liquid crystal device [73, 74]. The VGM device consists of a thin (<12-μm) nematic liquid crystal layer sandwiched with a photoconductor between two electrodes (see Fig. 9.4-23). When a dc voltage is applied across the liquid crystal, a phase grating structure is produced in the liquid crystal material. As the voltage is varied, the period of the grating structure is altered, as shown in Fig. 9.4-24. As seen in this figure, the fundamental frequency of the grating structure can typically be varied between 200 and 600 line pairs/mm. Since the voltage across the

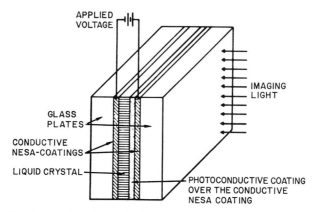

FIG. 9.4-23. Construction of an optically activated VGM real-time device. (From Armand *et al.* [73]. © 1978 IEEE.)

liquid crystal can be locally controlled by projecting an image onto the photoconductor, the device functions as an intensity-to-spatial frequency converter. The attractive aspects of this device are that it can operate in real-time and that it is a parallel processing device. Furthermore, many useful nonlinearities can be implemented with very simple filters. For example, a level slice operation is obtained by placing a slit in the Fourier plane passing only a narrow spatial frequency band corresponding to a given input intensity range. The level slice can be continuously varied by merely translating the slit filter. The level slice function has been demonstrated with the VGM device [75].

The VGM device is still very much in the development stage. There are many problems to be solved before it can be considered a practical element of an optical processing system. One of the principal problems involves defects in the grating structure, as seen in Fig. 9.4-25. These defects cause the diffraction orders of the grating to smear. As can be easily deduced from the earlier analysis, any smearing of the diffraction orders reduces the number of distinct intensity levels which can be processed. Therefore, it is necessary to keep the defects at a minimum if one wishes to process an image with a large number of intensity levels. Another important question which remains to be answered concerns the speed of the VGM device. Although current devices are slow and, because of their nature, it is unlikely that high speeds will ever be attained, the goal of attaining speeds approaching television frame rates would provide an adequate processing rate because of the parallel operation of the device. Other areas that are being studied concern such issues as device lifetime, uniformity, and reproducibility. It is expected that a better understanding of the underlying physics of the VGM effect will have a significant impact on all the problem areas discussed.

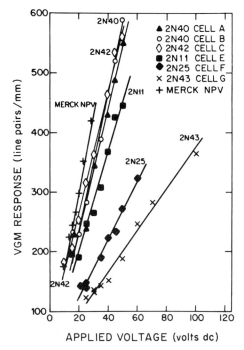

FIG. 9.4-24. VGM spatial frequency as a function of applied voltage.

FIG. 9.4-25. The phase grating structure of the VGM device viewed through a phase-contrast microscope. (Reprinted with permission from Armand *et al.* [73]. © 1978 IEEE.)

One application area of nonlinear processing which is of great general interest and which is of particular interest in work with current VGM devices is the implementation of optical logic. Figure 9.4-26 shows how logic operations can be viewed as simple point nonlinearity operations. This figure assumes that two binary inputs are superimposed on a device and indicates the nonlinear response required to obtain various combinatorial logic functions. It is seen that the required nonlinearities are simple binary functions. These functions are particularly well-suited to VGM devices in principle, since the desired nonlinearities can be obtained by simple slit apertures in the Fourier plane of the VGM device. The logic functions are also well-suited to the practical limitations of current devices, because there are only a small number of intensity levels that need to be distinguished in a logic system. This in turn implies that defect-broadened diffraction orders can be tolerated.

Of course the interest in performing logic operations optically is again due to the processing advantage of a parallel processing system. Each resolution element of the processing system functions as an independent logic gate, and all resolution elements are processed simultaneously. Thus, even an inherently slow device can provide a high effective data processing rate. Furthermore, if the data to be processed are initially in the form of an image, having a parallel optical processor can obviate the need for scanning and serializing the data. Given an analog image, one still needs to perform an analog-to-digital conversion, but this can also be done in parallel optically, either with a VGM device or with other nonlinear techniques discussed in this chapter [34, 35].

Figure 9.4-27 shows how a VGM device can be used to perform two-input combinatorial logic functions. The two binary inputs are imaged onto a transparent photoconductor. This may be a completely incoherent imaging

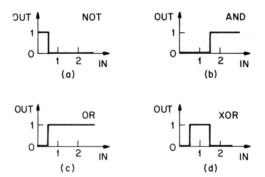

FIG. 9.4-26. Logic functions as binary nonlinearities. Given an input consisting of the sum of two binary inputs, different logical operations can be effected on these inputs by means of the depicted nonlinearities. (a) NOT, (b) AND, (c) OR, (d) XOR. (From Soffer et al. [75].)

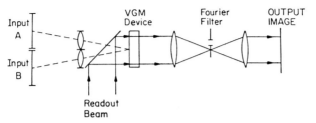

FIG. 9.4-27. Experimental arrangement for performing logical operations on two-dimensional binary inputs with a VGM device. The two binary input images are superimposed on the photoconductor. The device is read out in transmission. Simple slit filters can be used to achieve the desired logic operations. (From Soffer *et al.* [75].)

process. A spatially coherent readout beam is passed through the device. The read and write beams in this setup should be such that they can be separated either with a spatial filter or a color filter. A slit aperture placed in the Fourier plane is used to implement the desired logic operation. Experiments were conducted using two inputs which were chosen such that the output would be in the format of a truth table. The results are shown in Fig. 9.4-28. In these experiments the image area was approximately 1 cm². The images were written with a filtered mercury arc lamp. An optical bias in the form of uniform background illumination was included so that the zero input condition would correspond to a nonzero VGM frequency. This offset is

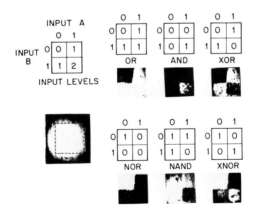

FIG. 9.4-28. VGM logic results. Two binary images were superimposed on the photoconductor to produce the input intensities as shown in the upper left corner. Without filter, the output is ideally a uniform field (logical 1). The output field with no filter is shown in the lower left corner with the image area of interest indicated by the dotted lines. The truth tables for the various logic functions are shown on the right above their corresponding experimental output. (From Soffer *et al.* [75].)

necessary for implementing the NOT function or any other function requiring a nonzero output for a zero input. The image was read out with a collimated helium–neon laser beam.

Not only can one implement the basic combinatorial logic functions with this system, but one can also obtain any commutative Boolean function with a single pass through the system. Thus, for example, a full adder can be implemented directly with a single device. Indeed, since the VGM device produces a symmetric spectrum, one can filter the positive diffraction orders to obtain the sum bit of a full adder and simultaneously filter the negative orders to obtain the carry bit. The experimental setup for realizing a full adder which simultaneously operates on all points of one-image bit plane is shown in Fig. 9.4-29.

By adding feedback to the system, it is possible to extend this technique to sequential logic. Furthermore, if a means can be devised of interconnecting the pixels on a given device, one can construct entire circuits on a single VGM device.

It should be noted in closing this discussion that, even if the VGM device itself does not prove to be a practical optical processing component, the concept of having an element which performs a two-dimensional intensity-to-spatial frequency conversion in real time is so powerful that it warrants research into other possible means of achieving this function.

C. Direct Nonlinearities

Direct nonlinear optical functions are achieved using the inherent nonlinear transfer characteristics of an optical recording medium or real-time image transducer. With this type of nonlinear processing, there is no pulse width modulation, intensity-to-spatial frequency conversion, or other type of intermediate mechanism. Although few of these particular techniques rely on Fourier plane filtering, they are included in this chapter to complete a broad-based view of the overall field of nonlinear optical processing.

Photographic film has been used since its invention to perform nonlinear

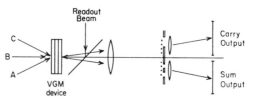

FIG. 9.4-29. Implementing sum and carry bit planes of a full adder with one VGM device. The inputs are bit planes from images A and B and the carry bit plane C. Whenever two or more of the three inputs are 1, the carry output should be 1. The sum bit should be 1 if an odd number of inputs are 1.

image transformations. Although most photography uses the linear part of the density-versus-log exposure curve to ensure linear recording, the limiting characteristics of the saturation region of the film curve and the nonlinear curve (toe) rising to the linear region have been used for artistic and scientific effects. Tai *et al.* [76] have obtained a direct logarithm by carefully choosing operating points on the characteristic curves of film. A two-film procedure with a copying step is used, and a dynamic range of two decades is obtained without the need for a halftone screen. They have applied this procedure in homomorphic filtering of multiplied signals, with results comparable to those shown in Fig. 9.4-11.

Agfa [77] has produced a special photographic product called Agfa-contour containing a sandwich of positive and negative emulsions. This combination gives the film a direct level slice characteristic which produces an isodensity contour output.

For real-time direct nonlinear processing, television systems and solid state sensor arrays combined with electronic scanning and nonlinear processing circuitry have been used. The characteristic transfer curves can be electronically shaped so that the complete system, consisting of scanned image sensor, processing, and display, performs a desired nonlinearity. Because of the scanned nature of these systems however, the parallel processing capabilities of optics are lost.

A real-time optical transducer which has been used for nonlinear processing is the Itek Pockels readout optical modulator (PROM) [49]. This device is based on a bismuth silicon oxide ($Bi_{12}SiO_{20}$) crystal which is both birefringent and photoconductive. The PROM is sensitive at blue wavelengths and is read out nondestructively at red wavelengths through crossed polarizers. By altering bias voltages in a particular sequence with the application of a control image, various linear operations such as contrast inversion and real-time spatial wavefront modulation can be performed. Using a different control sequence produces an approximate real-time level slice effect. PROM devices are still in the research and development stage, but their flexibility in performing linear and nonlinear optical real-time processing makes them interesting candidates for future work.

The Hughes liquid crystal light valve described in Section 9.4.A,3 is another real-time device that has been applied to real-time direct linear and nonlinear processing [48, 50]. The particular characteristic curves of such devices can be used for direct nonlinear processing in real time in exactly the same way as photographic film, with a choice of positive or negative characteristic curves. An example of this is a real-time incoherent image subtractor which has been built using two LCLVs [78].

A direct implementation of A/D conversion in real time using a special type of LCLV has been described by Armand *et al.* [79]. The system performs the sampling, quantization, and digitization of a two-dimensional data array without scanning. The method uses the nonlinear device characteristics of a special LCLV having uniform parallel alignment of the liquid crystal material. In this configuration, the device exhibits a pure birefringent effect (i.e., no optical activity) that varies with the local electric field controlled through a photoconducting layer. The special LCLV is called a multiple-period device; the overall relationship between the intensity transmittance of the device and the incident intensity at any point is given ideally by the sinusoidal curve shown with dashes in Fig. 9.4-30a.

The digital results of A/D conversion at each image point may be output serially as a bit sequence or in parallel as bit planes like those shown in Fig. 9.4-16. The solid line curves of Fig. 9.4-30 show the nonlinear transfer characteristic needed to produce the bit planes of the three-bit reflected binary or Gray code and their relationship to the dashed curves of sinusoidal device characteristics. When the output of Fig. 9.4-30a is thresholded at $\frac{1}{2}$, a 1 output is produced above threshold and a 0 output below, as shown by the curves

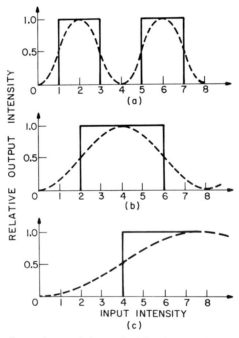

FIG. 9.4-30. Nonlinear characteristic required for the three-bit Gray code. Solid curves are the desired characteristics for the bit plane outputs. Dotted curves are the ideal sinusoidal response of a linear birefringent device. Parts (a) through (c) represent increasingly significant output bits. (Reprinted with permission from Armand *et al.* [73]. © 1978 IEEE.)

with solid lines. This thresholding can be done electronically following light detection by a parallel array of sensors. The threshold output in Fig. 9.4-30a is the least significant bit of the three-bit Gray code. The other two bits are obtained by attenuating the input intensity to rescale the horizontal axis effectively. Use of the full dynamic range (0–8) gives the least significant bit. Attenuating the input by a factor of $\frac{1}{2}$ (to the range 0–4) gives the first cycle of the characteristic curve shown in Fig. 9.4-30b. The last (most significant) bit is obtained by using an attenuation of $\frac{1}{4}$ so that the curves of Fig. 9.4-30c result. Note that any continuous input between 0 and 8 gives a unique quantized three-bit output. Although the outputs in Fig. 9.4-30 are the three bits of the Gray code, other A/D code conversions, such as the usual straight binary code, can be achieved by translating these curves left or right along the horizontal axis. This can be done by introducing phase retardation plates with different delays along orthogonal axes into the crossed-polarizer system.

The system can produce these bits in parallel when an array of three periodically repeated attenuating strips is placed over the write surface of the liquid crystal device, as shown schematically in Fig. 9.4-31. The strips have attenuation factors of 1, $\frac{1}{2}$, and $\frac{1}{4}$, and the image of the strips is in register with a parallel photodetector array with electronic thresholding in the output plane. All three bits are sensed in parallel in this way. The period of the strips should be much smaller than the inverse of the maximum spatial frequency of the input picture to avoid aliasing. A two-dimensional array of attenuating spots with a corresponding detector array can also be used instead of the linear strip array. Simpler but slower operation can be achieved by using only one detector array and sequentially uniformly attenuating the

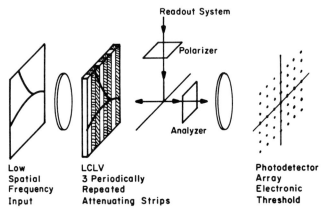

FIG. 9.4-31. System for parallel A/D conversion. (Reprinted with permission from Armand *et al.* [73]. © 1978 IEEE.)

entire input array. Ideas such as this have been used for electro-optic A/D conversion, but none have used an optically controllable real-time device.

The experimental system shown in Fig. 9.4-32 was used to obtain the input–output characteristic of a pure birefringent LCLV and to demonstrate the concept of A/D conversion. The incoherent source illuminating the input plane is a mercury arc lamp. A fixed and rotatable polarizer pair in the input light beam is used to vary the input light intensity. The real-time device is a Hughes parallel aligned nematic LCLV with a cadmium sulfide (CdS) photoconductor. The short-wavelength cutoff filter eliminates wavelengths shorter than 493 nm to make sure that the write beam wavelength is matched to the sensitivity range of the CdS photoconductor. The read light source is a xenon arc lamp. Because of the dispersion of birefringence in the liquid crystal material, the read light should have a narrow spectral bandwidth. An interference filter with a peak wavelength of 434.7 nm and a bandwidth of 18.4 nm is used to meet this requirement for the read light. With no picture in the input plane, the output intensity varies in a quasi-sinusoidal fashion with increasing input illumination because of the changing birefringence. If the amount of birefringence varied linearly as a function of the write beam intensity, a strictly sinusoidal variation of the output intensity would be expected. However, a number of factors, including the optical properties of the liquid crystal and the photoconductor characteristic properties, affect the output curve and produce an approximately sinusoidal output whose frequency varies (monotonically) with input intensity. The experimental response curve obtained is shown in Fig. 9.4-33.

Although the theory behind the A/D conversion assumes a strictly periodic response characteristic, it is possible to produce the desired bit planes by using the quasi-periodic response curves of the actual device. The

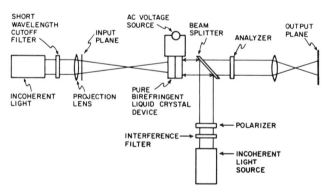

FIG. 9.4-32. Experimental setup for real-time parallel A/D conversion [79].

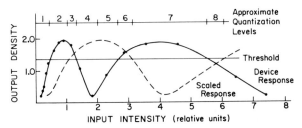

FIG. 9.4-33. Response curve of the liquid crystal device used for the three-bit A/D conversion. The solid curve is the measured response. The dotted curve represents the same response with a fixed attenuation of the input [79].

tradeoff is that nonuniform quantization results. The quantization levels obtained in this experiment are shown in Fig. 9.4-33.

There are no attenuating strips on the liquid crystal device used in this experiment. Instead, the bit planes were generated serially. Also, the output was recorded on hard-clipping film rather than a thresholding detector array. A test target was generated that consisted of a tablet with eight gray level steps. The gray levels were chosen to match the quantization levels shown in Fig. 9.4-33. This test object was imaged onto the liquid crystal device, and the output was photographically hard-clipped to produce the least significant bit plane of a three-bit A/D conversion. Next the write illumination intensity was decreased, effectively rescaling the response curve of the device to generate the next bit plane. The last bit plane (the most significant bit) was obtained by attenuating the write intensity again and photographing the output. The input and the three bit planes generated are shown in Fig. 9.4-34.

Although the output contains some noise, the experiment illustrates the principle of real-time parallel incoherent-optical A/D conversion. It was found later that the computer-generated gray scale was somewhat noisy because of the grain of the high contrast film used. It is possible that future experiments with clearer inputs and improved periodic light valves will produce better experimental results and more bits of quantization.

The potential A/D conversion rate can be estimated from typical parameters of currently available devices. The important parameters are device resolution (typically 40 cycles/mm), device size (typically 50×50 mm), and speed (generally 30 frames/sec). Multiplying all these parameters together and dividing by 3 for the attenuating strips implies an A/D conversion rate of 4×10^7 points/sec. A fully parallel system with one light valve and detector array for each bit plane can achieve 1.2×10^8 points/sec.

An advantage of this technique is that it operates with incoherent input. The requirements on the spatial and temporal coherence of the readout illumination are sufficiently relaxed that noise problems associated with

FIG. 9.4-34. Direct A/D conversion. The eight-level analog input is shown at the top. Below is the binary-coded output in the form of three bit planes of the Gray code. (Reprinted with permission from Armand *et al.* [73]. © 1978 IEEE.)

coherent spatial filtering or transforming techniques are avoided. The technique also minimizes the spatial frequency requirements of the real-time device because the sharp edges of the binary dots in halftoning do not have to be maintained.

Although these initial results are encouraging, further application of the technique must await improved real-time devices. The aperiodic nature of the LCLV results in unequal quantization intervals and limits the number of bits. The LCLV device used in this experiment is inherently aperiodic, because of the nonlinear response of the photoconductor and the nonlinear relationship between applied voltage and effective birefringence. Further developments in device technology may improve the periodicity and overall performance of the technique.

In a related application of a Hughes birefringent LCLV, Collins *et al.* [80] implemented binary combinatorial logic operations by directly using the characteristic curves of the device. they achieved two input logic functions, such as AND, NOR, and XOR, by viewing logic as a nonlinearity, similar to the procedure followed in the VGM logic described in Section 9.4,B. Their LCLV device was a parallel aligned pure birefringent cell similar to that used by Armand *et al.* [79] for A/D conversion. By appropriate choice of operating point on the device curves, shown as dashed lines in Fig. 9.4-30, different functions were achieved. These workers used various rotations and combinations of polarizer–analyzer pairs to translate the curves left and right to match the desired characteristics. Their experimental results demonstrate several of the standard binary combinatorial logic functions.

Lee and co-workers have developed a system for obtaining logic operations using a liquid crystal optically addressed through a photoconductor matrix [72, 81]. Each element of the array can perform simple combinatorial logic operations on two binary inputs.

As noted in Section 9.4,A, real-time implementation of halftoning is limited by the lack of an optical input transducer with a threshold characteristic. The threshold itself is useful as a direct nonlinear function; for example, the nonlinear curves needed for optical logic can be expressed as a sequence of thresholds. An attempt to improve the threshold characteristics of the Hughes LCLV has been described by Michaelson [82]. The feedback arrangement shown in Fig. 9.4-35 was used. As shown in this figure, a portion of the output light was directed back to the input of the light valve via a combination of beam splitters and lenses. The light was summed with the input illumination at the surface of the light valve to produce positive feedback. Both the read and the write illumination were derived from an argon ion laser operating at 514.5 nm. It should be noted that, although coherent light was used in the experimental procedure, the feedback arrangement and the characteristics of the light valve are such that incoherent illumination would have worked as well. To avoid unwanted interference between the coherent input light and the feedback light, the input illumination was configured with its polarization orthogonal to the feedback component. In this manner, the incident light on the light valve was simply the sum of the intensities of the two components. In the initial state, with no input illumination and the feedback component set to zero by momentarily blocking the feedback path, the device remains in the off state, resulting in zero output intensity. As the input illumination is increased, the device remains off until the threshold level of the light valve is reached. At this point, the device begins to turn on and, as a result of the optical gain characteristic of the light valve and the positive feedback, regeneration occurs. The device is switched on with no further increase in input illumination. The regenerative process continues until a point is reached on the light valve transfer curve

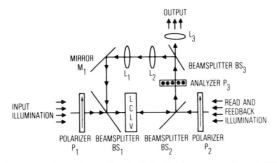

FIG. 9.4-35. Experimental optical feedback system using a LCLV.

where the loop gain (the device gain less the feedback losses) drops to unity. In practice this point is reached before the saturation level of the light valve is reached. As a result, the desired binary transfer function is not entirely achieved, but rather a soft shoulder is obtained with a further increase in input illumination. Experimental results are shown in Fig. 9.4-36 together with the light valve transfer function for comparison. As seen in this figure, the feedback system provides a very sharp threshold characteristic as well as a marked improvement in overall gamma.

The light valve used to obtain the data in Fig. 9.4-36 did not exhibit a particularly high gamma. Thus the output range over which regeneration occurred was somewhat limited. Further, device nonuniformities over the aperture precluded the possibility of testing the feedback arrangement with two-dimensional input images. To be effective, the device would have to be sufficiently uniform to maintain the input threshold point within a reasonably narrow band over the area of the input image. Improved devices exhibiting both the necessary uniformity and increased initial gamma could render the feedback system a viable means of obtaining real-time sharp thresholding operation from the light valves.

In a related system, Sengupta *et al.* [83] have used feedback around a LCLV to implement an optical flip-flop. Pairs of image elements on the LCLV are coupled by optical feedback so that latching at one of two stable states occurs. A parallel array of flip-flops is achieved with one LCLV by arranging the output of half the active image area to control the opposite half, and vice versa. These authors present stability theory for predicting the stable states and show experimental confirmation.

Another method of direct nonlinear processing uses several optically and electrically controlled liquid crystal light valves. This system has been

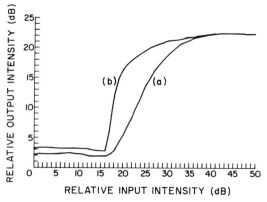

FIG. 9.4-36. LCLV feedback system response: (a) open loop transfer function and (b) closed loop transfer function.

described in detail by Michaelson and Sawchuk [84] and is called the multiple light valve system (MLVS). The system performs nonlinear processing in real time and has the advantages of incoherent operation and electronic programming of arbitrary nonlinearities.

A functional schematic of the MLVS is shown in Fig. 9.4-37. The first element in the system maps the intensity variations of the two-dimensional input image into two-dimensional outputs with constant magnitude and temporal separation. This step is accomplished by a time-scanning level slice produced with a LCLV. The constant magnitude outputs are then weighted as a function of time by another LCLV to give the desired nonlinear transformation. The weighted outputs are then integrated over an appropriate time interval to give a nonlinearity-transformed output image. The operation is shown schematically in Fig. 9.4-37 for a simple three-intensity-level, two-dimensional input image. The intensity-to-time converter maps input intensities over the range of 0 to I_{max} into the time interval $[0, T]$. Thus, for a linear mapping, the output planes for the intensity-to-time converter will remain dark during the time interval $0 < t < t_1$, since there are no intensity components in the input image between $I = 0$ and $I = I_1$. At time t_1 corresponding to the input intensity I_1, the output plane will have a constant intensity response over all portions of the input image for which $I = I_1$. Similar output responses occur at times t_2 and t_3, as shown in the figure. For all other times $0 < t < T$, with $t \neq t_1, t_2, t_3$, the output intensity remains at zero. The temporal intensity weighter is simply an electro-optic attenuator which weights the constant intensity, time-sequential outputs of the intensity-to-time converter in the desired nonlinear manner. The time-sequential weighted responses are then summed over the interval $[0, T]$ in the integrator. Thus at time $t = T$, a nonlinearly transformed input image

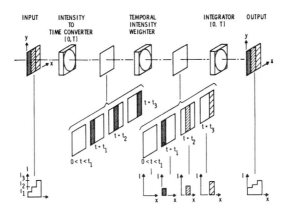

FIG. 9.4-37. Schematic diagram of the multiple light valve system (MLVS) for nonlinear optical processing. (From Michaelson and Sawchuk [84].)

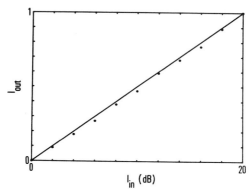

FIG. 9.4-38. Response of the MLVS programmed for a logarithmic characteristic. (From Michaelson and Sawchuk [84].)

will be present at the integrator output. A simple logarithmic compression is depicted in Fig. 9.4-38. In practice, the integrator can be a television system, photographic film, or another real-time image transducer.

One of the key features of the system is the ability to arbitrarily change the form of the nonlinear function in real time. The nonlinearity is introduced by applying a nonlinearly shaped voltage waveform into the temporal intensity weighter during the time interval $[0, T]$. In the system being evaluated, the waveform is produced by a microprocessor–D/A converter and can arbitrarily be changed to effect numerous transformations including logarithm, exponentiation, level slice, and A/D conversion. Details of implementation and experimental measurements are given in Refs. [82, 84].

Figures 9.4-38 and 9.4-39 show operation of the system with a logarithmic and exponential characteristic. Figure 9.4-40b shows the experimental

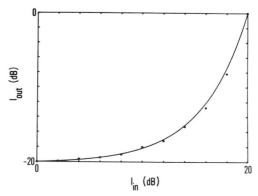

FIG. 9.4-39. Response of the MLVS programmed for an exponential characteristic. (From Michaelson and Sawchuk [84].)

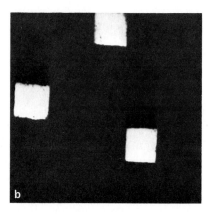

FIG. 9.4-40. Level slice operation performed with the MLVS: (a) input image and (b) level slice output. (From Michaelson and Sawchuk [84].)

results of a level slice operation performed on the variable intensity input image of Fig. 9.4-40a.

Farhat [31] has discussed the possibility of using a photochromic material to implement a direct nonlinearity. In particular, he suggests the possibility of performing division optically by placing a transparency in contact with a photochromic material having a negative response characteristic. In his paper, Farhat also suggests several variations of composite nonlinear systems obtained by placing a photochromic nonlinear element at various places in an optical processing system.

A final technique for direct point nonlinearities has been developed by Santamaria *et al.* [85] for providing a square root nonlinearity. The method does not fit into any of the three rubrics listed above under point nonlinearities. The technique consists of doing a time-sequential reconstruction of an image by imaging through a filter that transmits the zero order and one other spatial frequency component. The second component is selected by scanning an aperture in the Fourier plane.

9.5 COMPOSITE NONLINEAR SYSTEMS

In this section we describe several nonlinear processing systems whose overall function cannot be modeled as a simple point nonlinearity but rather consists of a point nonlinear operation injected at some point in what is otherwise a linear filtering system. The nonlinearity combined with an impulse response of finite extent form what is analogous to a nonlinear system with a memory in electronics. The optical system differs from its electronic analog in that the signal is two-dimensional and the system is noncausal.

Although many of the systems described above in the section on point nonlinearities may be more complex in their execution and consist of more processing stages than some of the systems described in this section, these systems represent a step up in terms of their functional complexity according to the organizational scheme we have adopted to describe nonlinear systems.

A. Input–Output Plane Nonlinearities

As mentioned in Section 9.3, the best example of the application of input–output plane nonlinearities is homomorphic filtering where typically a logarithmic nonlinearity is applied to an input in order to convert multiplicatively combined signals to additive signals. The additive signals are then filtered by a linear system, and the output is processed by an exponentiation, the inverse of the logarithmic transformation. These systems were discussed in the last section on point nonlinearities, since they were the same as other systems discussed there and because they only implemented the initial logarithmic nonlinearity before the filter; i.e., only an input plane nonlinearity was involved.

Several systems utilize a point nonlinearity in the output plane or image plane. Of course any system with a nonlinear detector, as most detectors are to some extent, technically has an output plane nonlinearity. In many cases this nonlinearity actually enhances the output by increasing the contrast, softening highlights, shifting color balance, or some other effect. Although many examples of this type of nonlinearity can be found, the enhancement is, as often as not, a chance occurrence of secondary importance. We will not attempt to list these systems here. One of the earliest and perhaps most widely used examples where an output plane nonlinearity is an essential design feature is the "sharpness meter" or focus detection scheme used in many autofocus systems [86–88]. This type of focus detection depends upon the fact that any defocus blur will reduce the effective contrast in the image. If one integrates the image intensity with a linear detector, there will be no difference between the in-focus and the defocused images. However, if the image is integrated with a nonlinear detector, there will be an extremum in the output signal when the image is in focus. This is due to the fact that the operations of integration and of applying a point nonlinearity do not commute. Since most detectors are inherently nonlinear, this focus detection technique is simple to implement with a single large-area detector and a focusing servomechanism.

Another use of a nonlinearity applied to the output plane of a system is in improvement of the signal-to-noise ratio in an image. An example of this was described by Palermo et al. [89]. They showed that the holographic image of a high contrast object could be improved by detecting the image with a detector with a limiter function which could suppress the low-intensity twin image of an on-axis hologram. This technique can be used in other situations where the image information and the noise to be suppressed have sufficiently different intensity histograms.

There are at least two systems which have been described in the literature where a linear filtering system was designed to be used with a hard-clipping output nonlinearity to encode certain information about the input. In the first of these, Schneider et al. [62] recorded a filtered image with a halftone process. The screens they used were designed to provide pictographic representation of the output levels; that is, a symbol was produced at each pixel and the symbol depended upon the incident intensity. Depending upon the type of filtering that was performed in the first stage of the system, the final display encoded either density or spatial frequency distribution information for the input image. This type of processing was found to be particularly useful in aiding image evaluation by a human observer.

A second system which depends on threshold detection of the output is a system described by Braunecker et al. [90] for hybrid character recognition in incoherent light. Their system is based on decomposing the characters to be recognized into a set of principal components for a given input. A set of computer-generated filters are then used to detect the values of these principal components. The filters are further designed to produce binary outputs under ideal conditions, so that the output of the system should be a binary code for the input character. By using a threshold detector in the output plane, a properly binary output is still obtained in the presence of small disturbances in the form of noise or input irregularities. Thus the nonlinearity here again provides a certain amount of noise immunity.

B. Fourier Plane Nonlinearities

After the introduction of the off-axis holograms of Leith and Upatnieks [91] one of the early problems of holography was the intermodulation noise generated by nonlinear recording media. Although there are straightforward techniques for mediating this noise source, they are generally implemented at the expense of signal strength. However, Ghandeharian [92] has recently described a means of designing the nonlinear characteristic of the recording medium into a Fourier plane holography system. His technique allows one to exploit the full dynamic range of the recording material and still avoid the intermodulation noise associated with a simple power law type of nonlinearity. The method consists of recording a Fourier hologram with two or more reference beams. The mixing of the reference beams caused by the recording nonlinearity results in reconstructions at several spatially distinct locations. By proper selection from these locations, an image free of intermodulation noise is obtained.

Another use of nonlinear mixing in the Fourier plane of an optical system was described by Lee and Stalker in 1972 [93–95]. They used an input consisting of two separate images g_1 and g_2 and a reference wave. A nonlinear medium placed in the Fourier plane of the system generated cross-product terms between the Fourier transforms G_1 and G_2 of the input images. Thus, after a second Fourier transform, the output contains terms corresponding to the correlation and the convolution of the input images. A schematic of this system is shown in Fig. 9.5-1.

As mentioned earlier in this chapter, a number of optical processing systems have been developed which are essentially linear but which use a point nonlinearity to generate a desired Fourier plane filter. Chief among these systems are those that rely on feedback. Feedback systems constitute an area that has seen considerable research work because of its wide range of applications in filtering, in nonlinearity implementations, and in space-variant processing. The subject of feedback is too broad to cover in this chapter. Cederquist and Lee have recently written a good review of the subject [96], and Bartholomew and Lee have provided a recent description of a feedback system applied specifically to the problem of nonlinear optical processing [97] (see also Refs. [72, 81, 95]). Finally, a different type of feedback system which again performs a linear process with a nonlinearly derived filter is the optical implementation of Gerchberg's [98] iterative algorithm for signal extrapolation [99]. The interested reader is referred to these articles and the references cited therein for information on feedback systems.

Feedback is a very elegant means of introducing a nonlinearity into the generation of a filter function for a linear filtering system. There have also

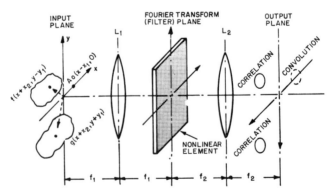

FIG. 9.5-1. Correlation and convolution obtained by using a nonlinear element in the Fourier plane. (From Lee [95].)

been several more direct applications of nonlinearities to the production of spatial filters. In 1968, two groups reported on the advantages of nonlinear recording of matched filters [100, 101]. In particular they noted that discrimination in character recognition could be improved by suppressing the low frequency components in the filter which carry most of the energy but little of the information. One simple means they found for suppressing the low frequency information was to record the matched filter with a beam ratio that was matched to unity for some high spatial frequency. With this approach the low spatial frequencies produced very low contrast fringes which could be totally eliminated by overexposure for low spatial frequency components. In a similar vein, Vander Lugt and Rotz [102] have described a "transposed processing" scheme where the Fourier transform of the input signal with its attendant noise is recorded on film and the resultant is used like a matched filter where the reference signal becomes the input. The underlying idea is to use the film nonlinearity, in particular its saturation characteristic, to suppress the signal in the parts of the spectrum dominated by noise. In a different vein, Dallas et al. [103] have shown that a film with a logarithm nonlinearity can be used to record an approximate inverse filter directly from the desired reference function. Thus the nonlinearity provides a greatly simplified means of obtaining inverse filters. Vijaya Kumar and Casasent [104] have also discussed the possible advantages of nonlinear recording of matched filters for use in optical correlators.

C. Combination Systems

There are a few optical processing systems where nonlinearities are introduced in several places in the system. In particular, Lohmann has suggested optical implementation of a transmitting system that utilizes nonlinear processing [105]. A monotonic nonlinearity is applied at the input of a communications channel for purposes of histogram equalization or signal-to-noise ratio improvement. After transmission the original signal is reconstructed by an iterative process that includes the nonlinearity as part of each iteration.

Finally, several hybrid systems have been proposed by Hunt [106–108], which involve multiple nonlinear stages. The first of these systems was designed for MAP-type image restoration [106]. It utilized coherent-optical processing in a feedback loop with analog electronic elements to provide the desired nonlinearities. Hunt then described a possible optical implementation of interpolated differential pulse code modulation for bandwidth compression of imagery [107]. Although some of the systems he described require coherent feedback, the main system is incoherent, requires no feedback, and involves only a quantizer type of nonlinearity. The most recent in this line of contributions is also directed at bandwidth compression [108]. Here the aim is to achieve adaptive compression by turning off the compression algorithms in the vicinity of edges. The main nonlinearities involved are again a quantizer and a level slice used in the edge detection process.

9.6 SUMMARY

It has been the intent of this chapter to offer an overview of the topic of nonlinear processing and its relationship to Fourier transform processing. We have introduced a classification scheme based on a Fourier transform description of a linear system with point nonlinearities inserted at different points in the system. It is hoped that this classification may assist the reader in understanding the nonlinear systems that have been developed and provide some insight for possible future work in the area.

Several approaches in implementing point nonlinearities have been described. The two main approaches have been halftone techniques and spatial frequency modulation. Real-time implementations of both approaches were discussed in the chapter along with application areas.

Finally, a variety of composite nonlinear systems were also covered. These systems combine point nonlinearities with Fourier transform processing to obtain an overall function which is no longer describable as a simple point nonlinearity. Feedback was also discussed here briefly as a special case, although most feedback systems actually perform linear processing on the input. The composite nonlinear systems were typically developed with a very specific application in mind, and these applications were mentioned.

ACKNOWLEDGMENTS

The authors wish to acknowledge the support of the Air Force Office of Scientific Research, Electronics and Solid State Sciences Division, under grant AFOSR-77-3285, the National Science Foundation, under grants ENG76-15318 and ENG78-11368, and the Joint Services Electronics Program (AFSC) at USC.

The authors are grateful to A. Armand, P. Chavel, S. Dashiell, J. Michaelson, B. H. Soffer, and A. R. Tanguay, Jr., for their contributions to the research described in this chapter, and to R. Keller for his careful reading of the manuscript.

REFERENCES

[1] H. C. Andrews, A. G. Tescher, and R. P. Kruger (1972). *IEEE Spectrum* **9**(7), 20–32.
[2] B. R. Hunt (1975). *Proc. IEEE* **63**, 693–708.
[3] D. Casasent and W. Sterling (1975). *IEEE Trans. Comput.* **C-24**, 348–358.

[4] D. Casasent and D. Psaltis (1980). *Opt. Lett.* **5**, 395–397.

[5] C. F. Hester and D. Casasent (1980). *Appl. Opt.* **19**, 1758–1761.

[6] J. W. Goodman (1968). "Introduction to Fourier Optics." McGraw-Hill, New York.

[7] S. H. Lee (1974). *Opt. Eng.* **13**, 196–207.

[8] J. W. Goodman (1977). *Proc. IEEE* **65**, 29–38.

[9] D. Casasent (1977). *Proc. IEEE* **65**, 143–157.

[10] R. Landauer (1976). *In* "Optical Information Processing" (Y. E. Nesterikhin, G. W. Stroke, and W. E. Kock, eds.), pp. 219–253. Plenum, New York.

[11] H. F. Taylor (1978). *Appl. Opt.* **17**, 1493–1498.

[12] L. Goldberg and S. H. Lee (1979). *Appl. Opt.* **18**, 2045–2051.

[13] D. H. Schaefer and J. P. Strong, III (1977). *Proc. IEEE* **65**, 129–138.

[14] R. A. Athale and S. H. Lee (1979). *Opt. Eng.* **18**, 513–517.

[15] K. Preston, Jr. (1972). "Coherent Optical Computers," Chap. 8. McGraw-Hill, New York.

[16] C. C. Guest and T. K. Gaylord (1979). *SPIE Proc. Opt. Process. Syst.* **185**, 42–49.

[17] J. Cederquist and S. H. Lee (1979). *Appl. Phys.* **18**, 311–319.

[18] R. P. Akins, R. A. Athale, and S. H. Lee (1980). *Opt. Eng.* **19**, 347–358.

[19] S. A. Collins, Jr., V. H. Gerlach, and Z. M. Zakman (1979). *SPIE Proc. Opt. Process. Syst.* **185**, 36–41.

[20] E. Garmire, J. H. Marburger, and S. D. Allen (1978). *Appl. Phys. Lett.* **32**, 320.

[21] B. Kruse (1978). *AFIPS Conf. Proc.* **47**, 1015–1024.

[22] A. Huang (1973). *IEEE Trans. Comput.* **C-22**, 14–18.

[23] A. Huang, Y. Tsunoda, J. W. Goodman, and S. Ishihara (1979). *Appl. Opt.* **18**, 149–162.

[24] D. Psaltis and D. Casasent (1979). *Appl. Opt.* **18**, 163–171.

[25] F. A. Horrigan and W. W. Stoner (1979). *SPIE Proc. Opt. Process. Syst.* **185**, 19–27.

[26] A. Tai, I. Cindrich, J. R. Fienup, and C. C. Aleksoff (1979). *Appl. Opt.* **18**, 2812–2823.

[27] S. A. Collins, Jr. (1977). *SPIE Proc. Eff. Util. Opt. Radar Syst.* **128**, 313–319.

[28] S. A. Collins, Jr., J. Ambuel, and E. K. Damon (1978). *Proc. Int. Comm. Opt. Conf., Madrid* **ICO-11**, 311–314.

[29] B. J. Thompson (1969). *In* "Progress in Optics—VII" (E. Wolf, ed.), pp. 169–230. North-Holland, Amsterdam.

[30] M. J. Beran and G. B. Parrent, Jr. (1964). "Theory of Partial Coherence." Prentice-Hall, Englewood Cliffs, New Jersey.

[31] N. H. Farhat (1975). *IEEE Trans. Comput.* **C-24**, 443–448.

[32] E. Lau and W. Krug (1968). "Equidensitometry." Focal Press, New York.

[33] W. K. Pratt (1978). "Digital Image Processing." Wiley (Interscience), New York.

[34] A. Lohmann and T. C. Strand (1975). *Proc. Electro-Opt. Syst. Des., Anaheim, Calif.* pp. 16–21.

[35] A. Armand, A. A. Sawchuk, T. C. Strand, D. Boswell, and B. H. Soffer (1980). *Opt. Lett.* **5**, 129–131.

[36] B. H. Soffer, D. Boswell, A. M. Lackner, P. Chavel, A. A. Sawchuk, T. C. Strand, and A. R. Tanguay, Jr. (1980). *SPIE Proc. Int. Opt. Comput. Conf., Washington, D.C.* **232**, Book II, 128–136.

[37] A. V. Oppenheim (1967). *Inf. Control* **11**, 528–536.

[38] A. V. Oppenheim, R. W. Schafer, and T. G. Stockham, Jr. (1968). *Proc. IEEE* **56**, 1264–1291.

[39] H. Kato and J. W. Goodman (1973). *Opt. Commun.* **8**, 378–381.

[40] H. Kato and J. W. Goodman (1975). *Appl. Opt.* **14**, 1813–1824.

[41] T. C. Strand (1975). *Opt. Commun.* **15**, 60–65.

[42] S. R. Dashiell and A. A. Sawchuk (1975). *Opt. Commun.* **15**, 66–70.

[43] S. R. Dashiell and A. A. Sawchuk (1977). *Appl. Opt.* **16**, 1009–1025.

[44] A. Armand (1979). Ph.D. Thesis, Dep. Electr. Eng., Univ. of Southern California, Los Angeles (USCIPI Rep. No. 880).

[45] S. R. Dashiell and A. A. Sawchuk (1977). *Appl. Opt.* **16**, 1936–1943.

[46] S. Matsumoto and B. Liu (1979). *Appl. Opt.* **18**, 2792–2802.

[47] T. D. Beard, W. P. Bleha, and S.-Y. Wong (1974). *Appl. Phys. Lett.* **22**, 90.

[48] J. Grinberg, A. Jacobson, W. Bleha, L. Miller, L. Fraas, D. Boswell, and D. Myer (1975). *Opt. Eng.* **14**, 217–225.

[49] S. Iwasa and J. Feinleib (1974). *Opt. Eng.* **13**, 225.

[50] W. Bleha, L. Lipton, E. Wiener-Avnear, J. Grinberg, P. Reif, D. Casasent, H. B. Brown, and B. Markevitch (1978). *Opt. Eng.* **17**, 371–384.

[51] S. R. Dashiell and A. A. Sawchuk (1977). *Appl. Opt.* **16**, 2279–2287.

[52] E. Abbe (1893). *Arch. Mikrosk. Anat.* **9**, 413.

[53] M. Marquet (1959). *Opt. Acta* **6**, 404–405.

[54] M. Marquet and J. Tsujiuchi (1961). *Opt. Acta* **8**, 267–277.

[55] E. Delingat (1973). *Optik* **37**, 82–90.

[56] S. V. Pappu, C. A. Kumar, and S. D. Mehta (1978). *Curr. Sci.* **47**(1), 1–6.

[57] C. Roychoudhuri and D. Malacara (1975). *Appl. Opt.* **14**, 1683–1689.

[58] A. A. Sawchuk and S. R. Dashiell (1975). *Proc. IEEE Int. Opt. Comput. Conf., Washington, D.C.* 73–76.

[59] J. Schwider and R. Burow (1970). *J. Opt. Soc. Am.* **60**, 1421.

[60] E. Delingat (1972). *Optik* **34**, 433–441.

[61] W. Schneider (1974). *Opt. Acta* **21**, 563–576.

[62] W. Schneider, F. Fink, and H. Van Der Piepen (1975). *Opt. Commun.* **14**, 42–45.

[63] H.-K. Liu, J. W. Goodman, and J. Chan (1976). *Appl. Opt.* **15**, 2394–2399.

[64] H.-K. Liu (1978). *Appl. Opt.* **17**, 2181–2185.

[65] T. C. Strand (1976). Ph.D. Thesis, Dep. Appl. Phys. Inf. Sci., Univ. of California-San Diego, La Jolla.

[66] H.-K. Liu (1978). *Opt. Lett.* **3**, 244–246.

[67] A. A. Sawchuk and S. R. Dashiell (1976). *SPIE Proc. Opt. Inf. Process.* **83**, 130–136.

[68] H.-K. Liu and J. W. Goodman (1976). *Nouv. Rev. Opt.* **7**, 285–289.

[69] G. Indebetouw (1977). *Appl. Opt.* **16**, 1951–1954.

[70] B. Morgenstern and A. W. Lohmann (1963). *Optik* **20**, 450–455.

[71] J. D. Armitage and A. W. Lohmann (1965). *Appl. Opt.* **4**, 399–403.

[72] S. H. Lee (1981). *In* "Optical Information Processing Fundamentals" (S. H. Lee, ed.), Chap. 7. Springer-Verlag, Berlin and New York.

[73] A. Armand, A. A. Sawchuk, T. C. Strand, D. Boswell, and B. H. Soffer (1978). *Proc. Int. Opt. Comput. Conf., London* pp. 153–158.

[74] A. Armand, A. A. Sawchuk, T. C. Strand, D. Boswell, and B. H. Soffer (1978). *Proc. Int. Comm. Opt. Conf., Madrid* **ICO-11**, 253–255.

[75] B. H. Soffer, D. Boswell, A. M. Lackner, A. R. Tanguay, Jr., T. C. Strand, and A. A. Sawchuk (1980). *SPIE Proc. Devices Syst. Opt. Signal Process.* **218**, 81–87.

[76] A. Tai, T. Cheng, and F. T. S. Yu (1977). *Appl. Opt.* **16**, 2559–2564.

[77] Agfa-Gevaert, Inc., 275 North Street, Teterboro, New Jersey.

[78] J. Grinberg and E. Marom (1977). *SPIE Proc. Opt. Signal Image Process.* **118**, 66–74.

[79] A. Armand, A. A. Sawchuk, T. C. Strand, D. Boswell, and B. H. Soffer (1980). *Opt. Lett.* **5**, 129–131.

[80] S. A. Collins, Jr., M. T. Fatehi, and K. C. Wasmundt (1980). *SPIE Proc. Int. Opt. Comput. Conf.* **232**, Book 2, 168–173.

[81] S. H. Lee (1978). *In* "Optical Information Processing" (E. S. Barrekette, G. W. Stroke, V. E. Nesterikhin, and W. E. Kock, eds.), Vol. 2, pp. 171–191. Plenum, New York.

[82] J. Michaelson (1979). Ph.D. Thesis, Dep. Electr. Eng., Univ. of Southern California, Los Angeles (USCIPI Rep. No. 930).

[83] U. K. Sengupta, U. H. Gerlach, and S. A. Collins, Jr. (1978). *Opt. Lett.* **3**, 199–201.

[84] J. Michaelson and A. A. Sawchuk (1980). *SPIE Proc. Devices Syst. Opt. Signal Process.* **218**, 107–115.

[85] J. Santamaria, J. Ojeda-Castaneda, and L. R. Berriel-Valdes (1980). *In* "Optics in Four Dimensions" (ICO, Ensenada) (M. A. Machado and L. M. Narducci, eds.), pp. 154–159. Amer. Inst. Phys., New York.

[86] D. R. Craig (1961). *Photogr. Sci. Eng.* **5**, 337–342.

[87] A. Erteza (1976). *Appl. Opt.* **15**, 877–881.

[88] R. J. Gagnon (1978). *J. Opt. Soc. Am.* **68**, 1309–1318.

[89] C. J. Palermo, E. N. Leith, R. O. Harger, and W. A. Loucka (1970). *Appl. Opt.* **9**, 2813–2814.

[90] B. Braunecker, R. Hauck, and A. W. Lohmann (1977). *Photogr. Sci. Eng.* **21**, 278–281.

[91] E. N. Leith and J. Upatnieks (1962). *J. Opt. Soc. Am.* **52**, 1123–1130.

[92] H. Ghandeharian (1980). *J. Opt. Soc. Am.* **70**, 835–842.

[93] S. H. Lee and K. T. Stalker (1972). *J. Opt. Soc. Am.* **62**, 1366A.

[94] K. T. Stalker and S. H. Lee (1974). *J. Opt. Soc. Am.* **64**, 545A.

[95] S. H. Lee (1976). *In* "Optical Information Processing" (Y. E. Nesterikhin, G. W. Stroke, and W. E. Kock, eds.), pp. 255–279. Plenum, New York.

[96] J. Cederquist and S. H. Lee (1979). *Appl. Phys.* **18**, 311–319.

[97] B. J. Bartholomew and S. H. Lee (1980). *Appl. Opt.* **19**, 201–206.

[98] R. W. Gerchberg (1974). *Opt. Acta* **21**, 709–720.

[99] R. J. Marks, II (1980). *Appl. Opt.* **19**, 1670–1672.

[100] D. Raso (1968). *J. Opt. Soc. Am.* **58**, 432–433.

[101] R. A. Binns, A. Dickinson, and B. M. Watrasiewicz (1968). *Appl. Opt.* **7**, 1047–1051.

[102] A. Vander Lugt and F. B. Rotz (1970). *Appl. Opt.* **9**, 215–222.

[103] W. J. Dallas, R. Linde, and H. Weiss (1978). *Opt. Lett.* **3**, 247–249.

[104] B. V. K. Vijaya Kumar and D. Casasent (1980). *Opt. Commun.* **34**, 4–6.

[105] A. W. Lohmann (1977). *Opt. Commun.* **22**, 165–168.

[106] B. R. Hunt (1975). *Proc. Int. Opt. Comput. Conf., Washington, D.C.* pp. 11–13.

[107] B. R. Hunt (1978). *Appl. Opt.* **17**, 2944–2951.

[108] B. R. Hunt and S. D. Cabrera (1980). *Proc. Int. Opt. Comput. Conf., Washington, D.C.* pp. 210–215.

Chapter 10

Optical Information Processing and the Human Visual System

BAHAA E. A. SALEH

DEPARTMENT OF ELECTRICAL AND COMPUTER ENGINEERING
UNIVERSITY OF WISCONSIN
MADISON, WISCONSIN

For several decades it has been realized that Fourier analysis and linear systems theory, as well as the concepts of communication and information theory, provide a foundation on which a new theory of image formation can be constructed [1–4]. Fourier analysis has today become an indispensible tool for the description of optical systems, as this book well demonstrates.

The application of Fourier methods in physical optics has been paralleled by a similar application in physiological optics [6–8]. This is not surprising, for the eye is an image-forming system. More recently, however, it was realized that Fourier techniques could also be successfully used to describe the transmission of information through the retina–brain part of the visual system [7, 8]. This was an important advance, since it provided a unified approach for describing the flow of spatial data in a cascade of optical and neural systems. Another recent significant advance was the discovery of psychophysical and neurophysiological evidence [9, 10] indicating the existence of neural channels which are selectively sensitive to the Fourier components of a viewed spatial pattern. This suggests that the visual system performs the operation of two-dimensional Fourier transform or acts as a spatial spectrum analyzer.

It is apparent that present efforts to understand the processing of spatial data in the visual system would benefit considerably from theoretical and technical advances in the fields of coherent and Fourier optics and image processing. Likewise, research in image processing should benefit from

431

knowledge of the properties, capabilities, and organization of the visual system. This chapter reports on some aspects of mutual interaction between the two disciplines of vision and optical information processing.

In Section 10.1, the use of Fourier methods in describing the transmission of spatial information through the visual system is reviewed.

Sections 10.2–10.4 present applications of coherent-optical techniques in vision research.

When illuminated by coherent light, the eye lens can act as a Fourier transform lens which images on the retina a two-dimensional Fourier transform of the viewed pattern. By using a two-point pattern, a sinusoidal grating is created on the retina. Such a pattern can be used to test the visual pathway system. The technique is described in Section 10.2.

When the eye forms an image of *diffused* coherent light on the retina, it transforms a random phase, uniform intensity light into a random intensity pattern (speckle) which is viewed by the subject. Motion of the speckle pattern that results from moving the diffuser is a good indicator of the refractive errors of the eye. The use of the speckle method for eye refraction is introduced and discussed in Section 10.3.

In Section 10.4 we explore the potential uses of coherent-optical processors in computing two-dimensional Fourier transforms or performing spatial filtering operations on complex patterns used in vision research.

Section 10.5 addresses the problem of picture preemphasis in the compensation of visual distortions. It discusses some of the efforts of image processing engineers to include properties of the visual system in their model.

Other applications of coherent optics in vision research, which are not discussed in this chapter, include the use of holography and holographic interferometry in observation of the eye [11–17], the use of light beating spectroscopy for studying blood flow within the eye and for studying lens cataracts [18, 19], and applications which utilize the high power densities attainable with laser sources.

10.1   SPATIAL INFORMATION PROCESSING
IN THE VISUAL SYSTEM

The spatial information contained in a viewed picture undergoes three stages of processing: optical, neural, and psychophysical [20]. Optical processing involves formation of an image of the picture on the retina. Like any other optical imaging system, the performance of the eye is limited by distorting effects such as diffraction, which results from the finiteness of the system's aperture, the pupil; aberration, which results from deviations of the imaging surfaces from ideal curvature; and defocusing or refractive error; in addition to other special effects such as intraocular scattering. All these effects can be perfectly described within the framework of physical optics.

The second stage of spatial information processing is neural. It starts at the retina where the image is detected by a mosaic of photoreceptors, from which signals are transmitted through a neural network of ganglion cells to the geniculate fibers and ultimately to neurons of the visual cortex. The nature of neural processing is only partially understood. Examples of effects that lead to distortions in this processing stage are the effect of sampling the image by a finite array of detectors, the shot noise that results from the quantum nature of light, and the structure of the neural network itself, which includes effects such as mutual interaction between the signals generated by groups of neighboring receptors in the form of spatial summation or lateral inhibition.

The third stage is that of the final perception of the picture, which is psychophysical in nature and therefore not amenable to the usual quantitative analysis.

In an attempt to describe the relationship between the original picture and the perceived picture, we realize immediately that the perceived picture cannot be directly measured and may not be completely describable by a numerical function. We can only assume the existence of a hypothetical numerical perceived picture brightness as function of position. An equivalence between the hypothetical picture and the actual perceived picture cannot, in general, be established. However, for the equivalence to hold, all judgments based on the perceived picture, regarding for instance the absence or presence of a certain feature, should be derived correctly from the hypothetical quantitative picture by using consistent rules. We can now use a system's approach to the study of transmission of information by defining a system whose input is the original picture and whose output is the hypothetical perceived picture. Many efforts have been dedicated to a characterization that includes all known physical, neurophysiological, and psychophysical information about the visual system. The simplest model [21] is shown in Fig. 10.1-1. It is composed of three cascaded systems.

(i)   A *linear* system which represents the optics of the eye.

(ii)   A logarithmic *point nonlinearity* which represents the response of the photoreceptors. It includes the dynamics of dark and light adaptation.

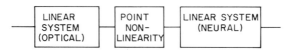

FIG. 10.1-1. Simple model of the visual system.

(iii)   A *linear* system which represents neural processing. It includes the effects of spatial summation and lateral inhibition.

The overall system is, of course, nonlinear.

A. Linearized Model: The Modulation Transfer Function and the Contrast Sensitivity Function

The complications that arise from the nonlinearity of the visual system can be avoided by using a perturbation approach in which the viewed picture has low contrast, i.e., contains one mean value light level and small fluctuations around it. The system then becomes approximately linear and can be completely characterized by its impulse response. The response of the visual system to an impulse depends on the location of the impulse; i.e., the system is space-variant or nonisoplanatic. This is clear from the known variation in visual acuity, spatial summation, and receptor density with eccentricity. It has been argued, however, that the system is approximately locally isoplanatic over regions within which structural inhomogeneity is effectively self-homogenizing [22]. When a system is linear and space-invariant, then it can be completely characterized by its response to sinusoidal stimuli. The response to a sinusoidal input is also sinusoidal, having the same frequency but suffering from an amplitude attenuation and a phase shift. The attenuation as a function of frequency defines the *modulation transfer function* (MTF). The phase shift as a function of frequency defines the *phase transfer function*. The combination of the modulation and phase transfer functions determines the system's *transfer function*.

The response of the system to a stimulus of any complex form is the superposition of the responses to each of the sinusoidal Fourier components of the stimulus.

The transfer function of a linear system is ordinarily determined by measuring the attenuation and phase shift of sinusoidal stimuli of different frequencies. To apply this technique to the visual system, sinusoidal spatial gratings

$$I(x, y) = I_0\{1 + m \cos[2\pi f_0(x \cos \theta - y \sin \theta) + \phi]\}$$

of contrast m, frequency f_0, and orientation θ, such as that shown in Fig. 10.1-2, are used. However, there is no way of directly measuring the amplitude or the contrast of the perceived sinusoidal pattern. This led Schade [5], who first applied Fourier techniques to vision, to the idea of measuring the *contrast sensitivity function* (CSF). The contrast of a spatial sine wave grating displayed on an oscilloscope is reduced, keeping its mean value fixed, until it reaches its threshold, i.e., until the viewer sees a uniform pattern. The reciprocal of this contrast level is then the contrast sensitivity. The measure-

FIG. 10.1-2. Sine wave grating at an angle $\theta = 45$ deg.

ment is repeated at a number of spatial frequencies, and the contrast sensitivity function $C(f)$ is determined. To determine the relationship between the CSF and the MTF of the linearized visual system, a mechanism of threshold detection must be postulated. The simplest mechanism (as illustrated in Fig. 10.1-3) is that of comparing the perceived small contrast to a fixed threshold level δ which when exceeded the pattern is declared to be seen. The measured contrast threshold of the original pattern $[C(f)]^{-1}$ should, when reduced by the MTF, $H(f)$, of the visual system be equal to the threshold level δ. Therefore, $[C(f)]^{-1}H(f) = \delta$, or

$$H(f) = \delta C(f); \tag{10.1-1}$$

i.e., the MTF is proportional to the CSF.

Extensive research has been dedicated to measurement of the CSF

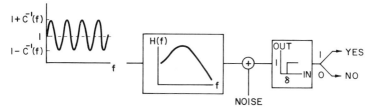

FIG. 10.1-3. Single-channel model.

[10, 23–34]. Much effort has also been given to separating the optical and neural contributions to the CSF. This was accomplished by creating sinusoidal patterns of varying known frequency and contrast on the retina, as will be discussed in Section 10.2. The optical contribution is the MTF of the eye's imaging system. It characterizes the optical performance of the eye [35–39].

An example of experimental data [36] for the MTF of the optics of the eye and the CSF of the visual pathway appears in Fig. 10.1-4, which shows that the optics of the eye acts as a low-pass spatial filter, whereas the retina–brain system acts as a bandpass filter attenuating both very low and very high spatial frequencies.

The spatial frequency of maximum sensitivity is typically between 3.0 and 4.5 cycles/deg (degree of arc subtended in the viewer's field of vision). The sensitivity also depends on θ, being a maximum for $\theta = 0$ or 90 deg and a minimum at $\theta = 45$ deg.

B. Theory of Fourier Channels

Responses of the visual systems of cats and monkeys to sinusoidal spatial patterns have been measured at the ganglion cells that transmit the retinal

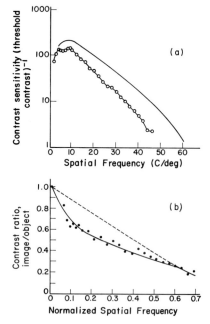

FIG. 10.1-4. (a) ○, contrast sensitivity function of the overall visual system (2-mm pupil); ——, contrast sensitivity function of the retina–brain system. (b) ——, MTF of the optics of the eye derived from (a) and (dotted line) diffraction-limited MTF of a system of a 2-mm pupil. (Adapted from Campbell and Green [36].)

signals to the geniculate body [40], at the geniculate body units [41], and at the striate visual cortex cells [41, 42]. Cells were found that were tuned to limited bands of spatial frequency. Cortical neurons were found that were tuned sensitively to different orientations of the grating [43].

This suggests that the visual system of humans is similarly organized in channels that are sensitive to the frequency and orientation of the Fourier components of the spatial pattern. The detection model should then be organized in multiple channels as illustrated in Fig. 10.1-5 instead of in a single channel as in Fig. 10.1-3.

Evidence that supports the Fourier channel theory is reviewed below.

(i) Campbell and Robson [9, 10] measured the contrast threshold for a number of periodic patterns. By resolving each pattern into its Fourier components, they found that a pattern was detectible whenever the contrast of any of its components exceeded the independent contrast threshold for that component. Thus the contrast threshold for the periodic pattern

$$I(x, y) = I_0 \left[1 + \sum_{n=1}^{\infty} m_n \cos(2\pi n f_0 x + \phi_n) \right]$$

is reached whenever

$$m_n > C(n f_0) \qquad \text{for any} \quad n,$$

where $C(f)$ is the contrast sensitivity function.

They also found that one pattern was distinguished from another whenever the contrast of any one distinguishing frequency component exceeded its own threshold. For example, a square wave grating

$$I_1(x, y) = I_0 \left[1 + \frac{4}{\pi} m \sum_{n=1,3,\dots}^{\infty} \frac{1}{n} \sin(2\pi n f_0 x) \right]$$

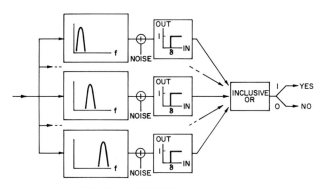

FIG. 10.1-5. Multiple-channel model.

is distinguished from a sine wave grating

$$I_2(x, y) = I_0[1 + (4/\pi)m \sin(2\pi f_0 x)]$$

whenever any of the harmonics in $I(x, y)$ reaches its own threshold.

They concluded that the different frequency components were detected independently, as illustrated in Fig. 10.1-5.

Statistical independence of the detection channels was tested by observing the detection threshold of the mixture of two sinusoidal frequencies [44, 45]. It was confirmed that the threshold of the mixture is a probabilistic summation of the thresholds of the individual components, which verifies the assumption that the detection noises in the separate channels (Fig. 10.1-5) are statistically independent. The results were not explainable within the single-channel model of Fig. 10.1-3.

In both detection and identification tasks, it was shown [45] that the data became inconsistent with independent processing only when the important spatial frequency components were so closely spaced that they activated common channels. The channel bandwidth was estimated at plus or minus one octave around the center frequency of the channel (i.e., $\pm\frac{1}{2}$ the center frequency). In the angular direction it was estimated as ± 10 deg.

(ii)   Campbell and Kulikowski [46] used the technique of masking a low contrast test grating with a high contrast grating of the same orientation and spatial frequency. They found that, unless the two gratings had the same orientation and spatial frequency, the masking effect was considerably less. A similar technique [47] is to adapt to a grating and then inspect another. It was necessary for the two gratings to match in frequency and orientation in order for the adaptation to have an effect. It was also observed [48] that, when a subject adapted to a square wave grating, the threshold for detecting both the fundamental and the third harmonics was elevated.

(iii)   Evoked potential response to a viewed pattern was recorded from the visual area of the scalp. It was confirmed [49, 50] that, at the level where the evoked potential arises, there is a mechanism generating an electric signal that is selectively sensitive to spatial frequency and orientation.

(iv)   Some patients with neurological disorders were found to suffer from a reduction in their contrast sensitivity function, which was particularly severe at certain spatial frequency bands [51].

All these different experiments support the theory that the CSF is an envelope of a set of constituent-independent sensitivity functions, each tuned around a central spatial frequency and ascribed to an independent channel.

C. Suprathreshold Vision

Threshold vision is concerned with the classification of what is visible and what is not. It is relatively much simpler to assess, since it involves binary decisions. Suprathreshold vision (or vision above threshold) is much more complex because it involves the appearance of viewed pictures.

As we mentioned earlier, when the viewed picture is of low contrast, the system that determines the perceived picture is approximately linear and can be described completely by its response to sinusoidal gratings. We have argued that the MTF of the linearized system is proportional to the CSF [Eq. (10.1-1)]. This was based on the assumption that sine wave patterns are attenuated by the system's MTF and then compared to a fixed threshold δ to determine detectability. What if δ were a function of frequency? The MTF

$$H(f) = \delta(f)C(f) \qquad (10.1\text{-}2)$$

would then be different from the CSF.

Indeed, measurements of the suprathreshold MTF, which were performed by matching sinusoidal gratings of different spatial frequency, resulted in a shape different from that of the CSF [28, 30, 52–58]. As shown in Fig. 10.1-4, a typical CSF falls off at high and low spatial frequencies. The suprathreshold MTF was found to be considerably flatter. The matter is complicated further by the effect of nonlinearity, which cannot be neglected as the contrast increases. It has been suggested [57, 59, 60] that neural processing compensates for optical degradation and thus achieves a dramatic deblurring of the image. Another interesting phenomenon of suprathreshold vision is that observed by Campbell et al. [61–63]. They compared the detection threshold and the suprathreshold appearance of the square wave pattern

$$I_1(x, y) = I_0 \left[1 + \frac{4m}{\pi} \sum_{n=1,3,5,\ldots} \frac{1}{n} \sin(2\pi n f_0 x) \right] \qquad (10.1\text{-}3)$$

and a similar square wave pattern without the fundamental

$$I_2(x, y) = I_0 \left[1 + \frac{4m}{\pi} \sum_{n=3,5,\ldots} \frac{1}{n} \sin(2\pi n f_0 x) \right]. \qquad (10.1\text{-}4)$$

When the frequency f_0 is sufficiently small ($f_0 < 1$ cycles/deg) and the contrast m is not so large, it is possible to have

$$m < C(f_0);$$

i.e., the fundamental component is below its detection threshold, and simultaneously

$$m(1/n) > C(n f_0), \qquad n = 3, 5, \ldots;$$

i.e., the first few harmonics exceed their thresholds. They found that then the two patterns could not be distinguished. This is not surprising for, after all, they only differ in the fundamental which is not detectable on its own. This is consistent with the theory of Fourier channels. What is surprising is that both patterns are perceived as though they both had a square wave luminance distribution instead of some other profile. It seems that the missing fundamental in $I_2(x, y)$ has been "filled in" and the periodicity f_0 is visualized. Campbell *et al.* [63] suggest that the visual system responds "as though hardwired to detect square gratings and edges by means of quasi-Fourier analysis."

It seems that our understanding of suprathreshold spatial vision is still very limited.

10.2 MEASUREMENT OF THE CONTRAST SENSITIVITY FUNCTION OF THE VISUAL PATHWAY BY GENERATION OF SINUSOIDAL PATTERNS ON THE RETINA

Properties of the retina–brain part of the human visual system can be measured, in isolation of the influence of the dioptrics, by generating known test patterns on the retina.

As previously mentioned, the retina–brain part of the visual system is extremely difficult to understand or characterize completely. Its most prominent and relatively simple characteristic is the contrast sensitivity function. The CSF of the visual pathway can be measured by creating sinusoidal patterns of known spatial frequencies and known adjustable contrast on the retina and measuring the threshold contrast for their detection. Because the sinusoidal pattern is the Fourier transform of a two-point pattern, the human eye can be used as a Fourier transform lens to transform coherent light from two points of adjustable separation into a sinusoidal retinal pattern of adjustable frequency. The sinusoidal pattern can also be regarded as interference fringes resulting from a two-point source, as in Young's experiment. The technique was invented by LeGrand [64] and used by Arnulf and Dupuy [7] and Westheimer [8] who used tungsten or high pressure mercury vapor light sources. Since Campbell and Green [36] used a helium–neon laser for this purpose, the technique has been established as a powerful research tool [26, 27, 37, 65–71].

A. Generation of a Sinusoidal Pattern on the Retina

A sinusoidal pattern can be generated on the retina by allowing the eye lens to produce an optical Fourier transform [72] of a two-point object distribution as illustrated in Fig. 10.2-1. An object consisting of two points separated by a distance Δ,

$$f(x, y) \propto [\delta(x - \tfrac{1}{2}\Delta) + \delta(x + \tfrac{1}{2}\Delta)]\delta(y), \qquad (10.2\text{-}1)$$

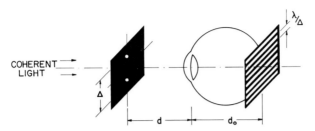

FIG. 10.2-1. Formation of the Fourier transform of a two-point light distribution on the retina.

has a Fourier transform of magnitude

$$|F(\omega_x, \omega_y)| \propto \cos[(\tfrac{1}{2}\Delta)\omega_x], \tag{10.2-2}$$

which when obtained with a lens of focal length f and a coherent wave of wavelength λ corresponds to an intensity distribution

$$I(x, y) \propto \cos^2[(\pi\Delta/\lambda f)x] = \tfrac{1}{2} + \tfrac{1}{2}\cos[(2\pi\Delta/\lambda f)x] \tag{10.2-3}$$

in the Fourier plane (see Chapter 1). When the eye is focused at infinity, i.e., $d_0 = f$, the Fourier plane coincides with the retina, and the above distribution corresponds to a sinusoidal pattern of frequency $\Delta/\lambda f$ cycles/mm (or Δ/λ cycles/radian) and unit contrast. The pattern is of course the known Young's interference fringes. The frequency of the pattern can be varied by changing the separation of the points Δ. In order to vary the contrast, an *incoherent* uniform illumination of varying intensity I_{in} can be added to the sinusoidal pattern, resulting in

$$I(x, y) = \tfrac{1}{2}I_c\{1 + \cos[(2\pi\Delta/\lambda f)x]\} + I_{in}, \tag{10.2-4}$$

where I_c is the peak intensity of the coherent pattern. The contrast becomes $I_c/(I_c + 2I_{in})$. It remains now to show that the frequency and contrast of the sinusoidal retinal pattern is independent of (or insensitive to) the optics of the eye.

B. Effect of the Optics of the Eye on the Frequency and Contrast of the Retinal Sinusoidal Pattern

(i) *Effect of refractive error* In the previous analysis, we assumed that the eye acts as a Fourier transform lens which transforms a two-point object distribution into a sinusoidal image distribution. We assumed that the eye is focused at infinity, thus satisfying the condition for Fourier transformation irrespective of the distance of the object plane from the eye (see Chapter 1). What influence does the varying accommodation of the eye have on the frequency, contrast, and perhaps the shape of the retinal pattern?

Consider Fig. 10.2-1. The eye is focused at a distance D but is viewing a *coherent* two-point object located at a distance d. To determine the pattern in the image (retinal) plane, we recall that the system has an impulse response (in the x-direction) [72]

$$h(x, x_1) \propto \exp[j(\pi/\lambda)(x_1^2/d_0 + x^2/d)]\exp[-j(\pi/\lambda\epsilon)(x/d + x_1/d_0)^2], \quad (10.2\text{-}5)$$

where

$$\epsilon = 1/d_0 + 1/d - 1/f = 1/d - 1/D \quad (10.2\text{-}6)$$

represents the focusing error.

Therefore the response to the object $[\delta(x - \frac{1}{2}\Delta) + \delta(x + \frac{1}{2}\Delta)]$ is

$$I(x) = |h(x, \tfrac{1}{2}\Delta) + h(x, -\tfrac{1}{2}\Delta)|^2 \propto \cos^2[(\pi\Delta/\lambda\epsilon dd_0)x]$$
$$= \tfrac{1}{2} + \tfrac{1}{2}\cos[(2\pi\Delta/\lambda d_0)x/(1 - d/D)], \quad (10.2\text{-}7)$$

which is a sinusoidal pattern of frequency

$$f_s = \frac{\Delta}{\lambda}\frac{1}{1 - d/D} \quad \text{cycles/radian.} \quad (10.2\text{-}8)$$

This result was obtained by Campbell *et al.* [65] using a different approach. When $d \ll D$, then $f_s \simeq \Delta/\lambda$, as was previously obtained. Therefore, we conclude that, when d is sufficiently small, the state of accommodation of the eye does not affect the fringe spacing. The best choice of d is, of course, $d = 0$, i.e., making the object plane coincide with the eye's nodal plane.

(ii) *Effect of diffraction* When $d = 0$, the highest frequency sinusoidal pattern that can be generated on the retina using the above method is that obtained by having the separation between the two points Δ equal to the diameter of the pupil Δ_p. This highest frequency is then given by

$$f_{s,\max} = \Delta_p/\lambda \quad \text{cycles/radian.}$$

When $\Delta < \Delta_p$, the finiteness of the pupil, which ordinarily results in diffraction errors, has no effect on the frequency or the contrast of the created sinusoidal pattern. For a pupil of diameter 4 mm and for $\lambda = 0.6\ \mu m$, $f_{s,\max} = 116$ cycles/deg.

(iii) *Effect of aberrations* Whereas aberration could, for large apertures, play an important role in determining the MTF of the optical part of the visual system [35], ideally it has no effect on the contrast and frequency of retinal sinusoidal patterns generated by the above-discussed interference method when $d = 0$. Its effect is merely a shift of the created pattern. Because aberration causes only a phase shift between the waves at the two object points, it results in only a phase shift of their interference fringes.

(iv) *Effect of lens opacities* (*cataract*)   A cataract causes a loss of lens transparency, occurring as part of the general aging process, as a result of ocular or systemic disease, or from traumatic or perforating injury to the globe. Lens opacities result in loss of vision. Moreover, they often make it difficult to use ophthalmoscopic examinations to detect further macular abnormalities. Lens opacities are usually not homogeneous. Even if the whole lens appears opaque, microscopic holes and regions of relatively higher transparency are present.

Coherent-optical Fourier transform techniques allow us to form, from light that crosses the lens at only two pinpoints, a fine sinusoidal pattern on the retina. Such a task is impossible when incoherent light is used.

Sinusoidal gratings can therefore be formed on the retina of cataractous eyes by searching for pairs of relatively transparent holes in the lens opacity through which coherent light can enter unscattered and interfere to form a fine testing pattern at the retina. This technique has been successfully demonstrated by Green [68, 69], who has suggested that determining acuity to retinal gratings can be used for assessing the potential for visual improvement in cataract patients before their eye lenses are surgically removed.

(v)   *Other effects*   Scattering from time-varying inhomogeneities in the humor aqueous and from nonuniform distribution of tears may also result in the appearance of false patterns which have random orientations and which move slowly. Such effects create a noisey background that makes the task of detecting the true pattern more difficult. It has been reported, however, [71] that these extraneous patterns disappear after subjects shake their heads from side to side and close their eyes for a few minutes.

C. Practical Systems

Several optical techniques can be used to create two-point light distributions near or in the eye's nodal plane. Westheimer [8] and Mitchell *et al.* [66] obtained the diffraction pattern (Fourier transform) of a Ronchi ruling and used a mask to separate two points, as shown in Fig. 10.2-2. Campbell and Green [36] employed a beam splitter and a mirror to separate a laser beam into two beams focused by a lens onto two points in the nodal plane of the eye (Fig. 10.2-3). Ohzu [70] used two dove prisms adhered together at their bases with a semitransparent thin film coating to split a laser beam into two beams and to mix with them an incoherent beam for adjusting the contrast (Fig. 10.2-4). In all the above methods contrast is changed by varying the ratio of the intensities of the coherent and incoherent beams by the use of orthogonal polarizers P_1 and P_2.

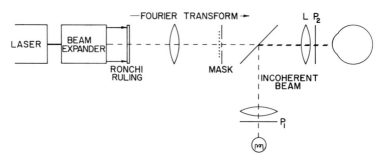

FIG. 10.2-2. Practical system for creation of a sine wave grating on the retina using a Ronchi ruling.

D. Results

Several vision researchers measured the contrast sensitivity function of the retina–brain part of the visual system using the technique of creating sinusoidal patterns on the retina. Typical results for a normal visual system are shown in Fig. 10.1-4. The CSF of the overall system of the same subject is also shown. From the two functions, the reduction in contrast due to the optics of the eye is calculated, which determines the MTF of the optical part of the visual system.

This technique has been used to test the performance of the eye [35–39], to determine the effect of orientation on visual resolution [66, 67], and to evaluate the capability of the visual system of cataract patients [68, 69] and keratoconus patients [70].

E. Creation of Other Test Patterns on the Retina

Just as a two-point light distribution is Fourier-transformed by the eye lens into a sinusoidal distribution on the retina, other distributions can be similarly created [73]. For example, a periodic square wave of frequency f_0

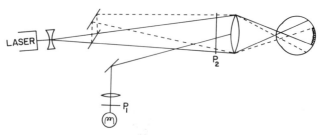

FIG. 10.2-3. Practical system for creation of a sine wave grating on the retina using a beam splitter.

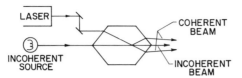

FIG. 10.2-4. System for combining two coherent beams and an incoherent beam using two dove prisms.

cycles/deg and 100% contrast can be generated by use of the field distribution

$$f(x) = \delta(x) + \frac{2j}{\pi} \sum_{n=1,3,5,\ldots} \frac{1}{n} \left[\delta\left(x - n\frac{\Delta}{2}\right) - \delta\left(x + n\frac{\Delta}{2}\right) \right],$$

where $\Delta = \lambda f_0$. This is a symmetric set of points located at $x = \pm n(\Delta/2)$, whose intensities drop as $1/n^2$ as shown in Fig. 10.2-5. The pattern can be made by obtaining on a hologram of an optical Fourier transform of a square wave grating, or by introducing 90° phase shift by use of retardation plates.

FIG. 10.2-5. Pattern whose Fourier transform is a square wave grating.

To create more complex patterns on the retina, we may be required to Fourier-transform functions that are complex. These can be synthesized by the use of holographic techniques.

10.3 MEASUREMENT OF THE OPTICS OF THE EYE BY GENERATION OF A SPECKLE PATTERN ON THE RETINA

A. Basic Properties of Speckle

When a coherent uniform light beam is reflected from (or transmitted through) a rough surface of uniform reflectance (or transmittance), it undergoes different random phase shifts at different points on the surface. When an image of the surface is formed on a screen by an *ideal* imaging system, as in Fig. 10.3-1a, it has uniform intensity. This is because each point Q on the screen receives light from only one conjugate point P on the rough surface. The intensity of the light at Q is affected only by the intensity of the light at P and not by whatever phase shift the surface roughness introduces at P.

On the other hand, if an image of the coherently illuminated rough surface is formed by a nonideal imaging system such as an unfocused lens, as in Fig. 10.3-1b, or by simple radiation through free space, then light that reaches a screen point Q originates from a patch of points surrounding P on the rough surface. The random phase shifts introduced by the rough surface at

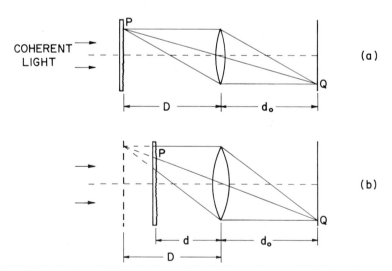

FIG. 10.3-1. Imaging of diffused coherent light: (a) focused system, (b) defocused system.

different points of this patch result in their random interference, hence in a random intensity at Q. If the patch around P contains N independent identical subareas, each producing an intensity I at Q, then the total intensity at Q due to the interference of light from all the subareas will have a random value ranging anywhere from 0 to N^2I. Different points on the screen have different random intensities, and the image of the rough surface has the random grainy structure known as speckle.

The formation, properties, and applications of speckle have been extensively investigated in recent years [74, 75].

B. Dynamics of Speckle Due to a Moving Rough Surface

Consider Fig. 10.3-1 again. In the absence of the imaging lens, the optical system is space-invariant (or isoplanatic), and therefore a translation of the rough surface in the object plane results in an identical translation of the speckle pattern in the image plane. If the rough surface moves with a velocity v, the speckle pattern will move with the same velocity and in the same direction.

In the presence of the lens, however, the imaging system is space-variant. Motion of the rough surface with a certain velocity does not necessarily result in accompanying motion of the speckle with the same velocity or in the same direction. It turns out that the velocity (magnitude and direction) of the speckle pattern depends rather sharply on the state of focus of the imaging system.

In order to determine the dynamics of a random time-changing pattern such as that of speckle generated by a moving rough surface, we have to determine the cross-correlation between the intensity of the pattern $I(x, t)$ at the two space–time points (x, t) and $(x + \xi, t + \tau)$. If this cross-correlation is peaked sharply at all $\xi = v'\tau$ and is very small otherwise, then we can say that the random pattern moves unchanged with velocity v'. To determine the cross-correlation function $\langle I(x, t)I(x + \xi, t + \tau)\rangle$ or the cross-covariance function $\langle \Delta I(x, t)\Delta I(x + \xi, t + \tau)\rangle$, where $\Delta I(x, t) = I(x, t) - \langle I(x, t)\rangle$, we write the intensity in terms of the field at the image plane,

$$I(x, t) = |V(x, t)|^2, \qquad (10.3\text{-}1)$$

where $V(x, t)$ is the complex amplitude at position x and time t in the image plane. Under some rather general conditions, the field $V(x, t)$ has complex circularly symmetric Gaussian statistics [76] which enable us to relate the intensity cross-covariance to the complex amplitude covariance as

$$\langle \Delta I(x, t)\Delta I(x + \xi, t + \tau)\rangle = |\langle V^*(x, t)V(x + \xi, t + \tau)\rangle|^2. \quad (10.3\text{-}2)$$

Now we can relate the field $V(x, t)$ in the image plane to the field $V_0(x, t)$

radiated from the rough surface in the object plane by

$$V(x, t) = \int_{-\infty}^{\infty} h(x, u)V_0(u)\, du, \tag{10.3-3}$$

where $h(x, u)$ is the impulse response of the imaging system. Because we know that the object field moves with a velocity v in the x-direction, we can write

$$\langle V_0^*(x, t)V_0(x + \xi, t + \tau)\rangle \propto \delta(\xi - v\tau). \tag{10.3-4}$$

Substituting in Eqs. (10.3-1)–(10.3-4) we obtain

$$\langle \Delta I(x, t)\Delta I(x + \xi, t + \tau)\rangle \propto \left| \int_{-\infty}^{\infty} h^*(x, u)h(x + \xi, u + v\tau)\, du \right|^2. \tag{10.3-5}$$

It now remains to substitute the appropriate expression of $h(x, u)$ in Eq. (10.3-5) to determine the desired speckle cross-covariance function. For the simple defocused imaging system shown in Fig. 10.3-1b and neglecting the effect of finiteness of aperture, we have [72]

$$h(x, u) = \exp[j(\pi/\lambda)(x^2/d + u^2/d_0) - j(\pi/\lambda\epsilon)(x/d + u/d_0)^2], \tag{10.3-6}$$

where

$$\epsilon = 1/d - 1/D. \tag{10.3-7}$$

Substitution in Eq. (10.3-5) results in

$$\langle \Delta I(x, t)\Delta I(x + \xi, t + \tau)\rangle \propto \delta(\xi + v'\tau), \tag{10.3-8}$$

where

$$v' = -(D/d_0)\eta v$$

and

$$\eta = 1 + (1 + d_0/D)(1 - d/D). \tag{10.3-9}$$

This demonstrates that the speckle pattern indeed moves unchanged with a velocity v'. The speckle velocity v' is equal to the rough surface velocity v scaled by the magnification factor $-D/d_0$ and multiplied by a factor η. This factor decreases linearly with an increase in d. When the system is focused, $d = D, \eta = 1, v' = -(D/d_0)v$, and the speckle moves with the same velocity as an image of the rough surface.

When the imaging system is focused such that $d < D$ and $\eta > 1$, then the speckle moves faster than an image of the rough surface and in the same direction. When $d > D$ and $\eta < 1$, then the speckle slows down. When $d > s_0 > D$, where

$$s_0 \equiv D(2 + d_0/D)/(1 + d_0/D), \tag{10.3-10}$$

the speckle reverses direction.

At $d = s_0$ the speckle has zero velocity. Instead of having the appearance of a solid pattern that is moving, it has instead the appearance of a continuously changing pattern that is statistically stationary. It appears to be "boiling" instead of flowing.

The finiteness of the aperture of the imaging system, which was ignored in the previous analysis, results in a broadening of the cross-correlation function of Eq. (10.3-8) while maintaining its peak in the same location [77].

C. An Astigmatic Imaging System

The above derivation of the velocity of speckle was based on a one-dimensional imaging model. It can be extended easily to a two-dimensional model if the imaging system is spherical. The analysis can also be generalized to astigmatic systems. For example, if the x- and y-axes coincide with the principal axes of the imaging system, and if v_x and v_y are the x- and y-components of the rough surface velocity, then the components of the speckle velocity are given by a generalization of Eq. (10.3-9),

$$v'_x = -(D_x/d_0)\eta_x v_x, \qquad v'_y = -(D_y/d_0)\eta_y v_y, \qquad (10.3\text{-}11)$$

where

$$\eta_x = 1 + (1 + d_0/D_x)(1 - d/D_x), \qquad \eta_y = 1 + (1 + d_0/D_y)(1 - d/D_y), \qquad (10.3\text{-}12)$$

and D_x and D_y are the distances of the planes that are conjugate to the image plane for the x- and y-components, respectively. It is therefore apparent that the motion of speckle could be in a direction different from that of the rough surface. An example is the case in which $v_x = v_y$, i.e., the velocity vector $\mathbf{V}$ makes a 45-deg angle with the principal x-axis. If D_x and D_y are such that η_x is negative but η_y is positive, then $\mathbf{V}'$ is oriented at an angle between 270 and 360 deg. Such a change in direction is indicative of the astigmatic nature of an imaging system.

D. Clinical and Vision Research Applications of Speckle

We have demonstrated so far that the motion of a speckle pattern resulting from an image of a moving rough surface changes dramatically with the focusing state of the imaging system. This phenomenon can be utilized in measuring the refractive state of the human eye [78–80]. The basic idea is to replace the imaging system of Fig. 10.3-1 with the eye together with lens L_1 (as shown in Fig. 10.3-2). The eye is accommodated at a distance D by viewing a target at this distance. The subject views the motion of the speckle formed on his retina as the rough surface moves. The examiner changes the power of the lens, by using a trial set of lenses, until the subject views a stationary boiling speckle. By using Eq. (10.3-10) and knowing the power of the lens

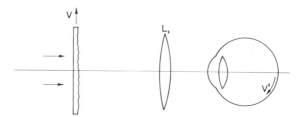

FIG. 10.3-2. Imaging coherent light diffused by a moving diffuser on the retina.

L_1, we can compute the refractive power of the eye. A practical implementation of the system which incorporates the Badal principle which is illustrated in Fig. (10.3-3). An additional lens L_2 called a Badal lens is introduced in front of the eye so that the refractive power of the combination L_1, L_2 is changed by displacing L_1 relative to L_2 instead of by replacing L_1. The result is a laser speckle optometer [81–88]. This method of eye examination changes the task of the subject from that of judging which lens is optimal for reading the letters of an eye chart to that of deciding whether the speckle pattern moves to the right or to the left or remains stationary. The latter judgment is usually easier and more accurate. The speckle technique of eye refraction has been used to perform measurements of refraction, meridional measurements, measurement of chromatic aberration, measurement of astigmatic aberration, and determination of the resting point of eye accommodation.

10.4 APPLICATIONS OF TWO-DIMENSIONAL SPATIAL TRANSFORMATIONS IN VISION RESEARCH

In recent years the Fourier transform property of the lens has been successfully used as the main building block of the coherent-optical processor. As demonstrated in this book, use of the optical processor to perform trans-

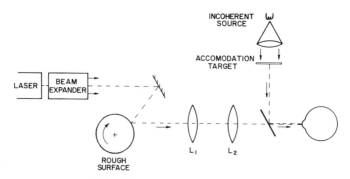

FIG. 10.3-3. Laser speckle optometer.

formations on two-dimensional signals (pictures) has been established as an efficient and reliable technique.

The use of optical (and digital) two-dimensional image processing as an important tool in vision research has only recently been realized. In visual studies, it is often useful to determine the two-dimensional Fourier transform of a picture or a test pattern. It is also frequently important to know the way pictures look after being operated on by other transformations suggested by optical, neurophysiological, or psychophysical effects in visual systems [89]. In this section we discuss some of the applications of Fourier transforms and spatial filtering in vision.

A. Two-Dimensional Fourier Transform

(i)   As discussed in Section 10.1, evidence has been advanced of the existence of channels in the visual system that are selectively sensitive to specific spatial frequencies and orientations. A Fourier theory of spatial vision has been formulated and is the subject of current extensive investigations. Such investigations necessitate the computation of two-dimensional Fourier transforms of the used test patterns in order to determine their expected detection thresholds.

Comparison of the thresholds of detection of sine wave gratings and other periodic one-dimensional patterns having known Fourier transforms—such as square wave gratings–has shed some light on the mechanism of detection and indeed has provided evidence supporting the channel theory. Other simple two-dimensional periodic patterns have also been used [90]. Their Fourier transforms can be computed analytically. As more complex patterns are introduced, especially aperiodic realistic patterns such as those that surround us in daily life, optical or numerical two-dimensional Fourier transform computations will be needed.

(ii)   The need to compute two-dimensional Fourier transforms of pictures also arises in testing whether certain illusions related to the perception of spatial patterns are explainable by the spatial filtering characteristics of the visual system. An example, which was the subject of recent discussions [91, 92], is the "pincushion illusion."

When the pincushion grid pattern shown in Fig. 10.4-1 is observed with the dark lines horizontal and vertical, an illusion of crisscrossing white diagonal lines extending between the points of the pincushion is observed. When the grid is rotated 45 deg, the white lines appear vertically and horizontally and the illusion is intensified.

In order to explain this and other similar illusions, the first step is to check whether they simply result from the two-dimensional spatial filtering of the visual system. To do this, the two-dimensional Fourier transform of

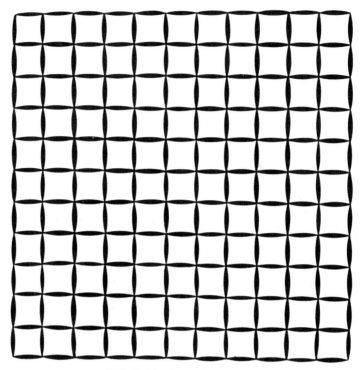

FIG. 10.4-1. Pincushion pattern.

the pattern should be determined. An optical Fourier transform of the pin-cushion grid has been obtained [92], and it contains horizontal, vertical, and diagonal components. The diagonal components obviously correspond to the illusion. The actual mechanism by which the lines are seen is not clear; however, the existence of the diagonal components of the Fourier transform suggests that the illusionary lines are physically present because of the two-dimensional spatial filtering characteristics of the visual system [92].

(iii)   Knowledge of the Fourier transform of typical pictures from the visual environment is necessary for the analysis of possible interactions between frequency-sensitive influences of the visual environment and development of the visual system. An example is the work of Switkes *et al.* [93] in which optical Fourier transforms of photographic samples of visual environments (indoor carpentered, outdoor carpentered, and pastoral) were measured (Fig. 10.4-2). They provided a systematic comparison of the anisotropy of the spatial frequency content of these environments and the anisotropy of visual acuity. Their work aimed at testing the assumption that anisotropy of the visual acuity of Western adults, which is higher in the vertical and horizontal direction than in the diagonal direction, results from the rectilinear vertical–horizontal structures present in the carpentered environment. A study of the radial distribution of the spatial spectra of typical environments should also allow the investigation of a possible correlation between the spatial structure of information presented to the visual system and the shape of the contrast sensitivity function [94].

FIG. 10.4-2. Sample pictures from outdoor carpentered and pastoral environments and their power spectra. (Reprinted with permission from *Vision Research* **18,** E. Switkes, M. Mayer, and J. Sloan. Spatial frequency analysis of the visual environment: Anisotropy and the carpentered environment hypothesis, Copyright 1978, Pergamon Press, Ltd.)

(iv)   Studies on texture descrimination [95] often require determination of the Fourier transforms of texture patterns.

B. Two-Dimensional Spatial Filtering Analysis

A linear system is characterized completely by its transfer function. A response of the system to any stimulus can be determined by analyzing the stimulus in terms of its Fourier components and superposing the responses due to each of these components. For two-dimensional systems, such as optical and visual systems, the computations are very lengthy. Coherent-optical processors can be efficiently used to perform such computations. A typical optical processor capable of performing any linear space-invariant spatial filtering operation is illustrated in Fig. 10.4-3, in which lenses L_1 and L_2 have focal lengths f. The mask has a transmittance

$$t(x, y) = \mathscr{H}(x/\lambda f, y/\lambda f),$$

where $\mathscr{H}(f_x, f_y)$ is the transfer function of the spatial filtering system to be simulated and λ is the wavelength of the laser. The input picture is placed in plane P_1, and the output picture is recorded in piane P_2. The signal processing properties of such systems is discussed in Chapters 1 and 2.

Several examples of applications of spatial filtering analysis in vision research are discussed below.

(i)   *Simulation of the optical filtering properties of the eye*   If the MTF of the optical part of the visual system is known (e.g., through measurements such as those discussed in Section 10.2), then we should be able to compute the pattern that appears on the retina when the eye views any arbitrary scene. Such a facility would be very useful in evaluating the effect of optical deficiencies in image perception. The eye dioptrics can of course be simulated by the use of an artificial eye. However, effects such as aberration and random

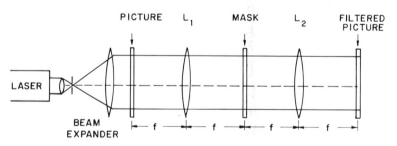

FIG. 10.4-3. Spatial filtering using a coherent-optical processor.

intraocular scattering may be difficult to reproduce with reasonable accuracy. If the MTF is measured, the system can be simulated using a coherent-optical processor. An example is an eye with a cataract. It is known that a cataract results in a disproportionate optical attenuation of low spatial frequencies, which has a very debilitating visual effect. In order to appreciate the nature of the visual world that cataract patients experience, it is important to see how they perceive complex visual scenes. Hess and Woo [96] obtained samples of complex scenes after propagation through a cataract, by optical simulations. They simulated a cataract by using a layer of scattering spherical particles in a liquid medium. Given the measured MTF of a cataractous eye, a coherent-optical (or digital) processor should prove useful in computing the response to any scene.

(ii) *Simulation of the spatial filtering characteristics of visual systems* If the MTF of a linearized model of the overall visual system is determined, e.g., by measuring the CSF, then we can compute the response of the modeled system to any scene.

Some especially interesting scenes are those that generate certain geometric or contrast illusions. An example is the illusory contour perceived in the Kanizsa triangle shown in Fig. 10.4-4. A black triangle is perceived as a unitary form separated in depth over a white triangle. The question is whether the perception of a triangle can be explained within the linear spatial filtering model of the visual system. Ginsburg [97] has attempted to answer this question by actually filtering this picture using a simulated system. He used a transfer function identical to that of the MTF of the visual system; i.e., he attenuated the different spatial frequencies according to the attenuation

FIG. 10.4-4. The Kanizsa triangle.

characteristics of the visual system. The filtered picture was smoothed and lost contrast; it did not contain an actual physical triangle, but it retained the illusory triangle. He then removed three-fourths of the higher spatial frequencies contained in the filtered pictures; i.e., he examined the signal contained in the low spatial frequency channels. The result was the appearance of a generally homogeneous dark triangle of sharp boundary line. The filtering enhanced pattern information (the illusory triangle) that was implicit, but not readily apparent, in the original pattern features.

(iii) *Relation between visual acuity and contrast sensitivity* The relationship between visual acuity and the spatial frequency components contained in the letters of an eye test chart can be established by computing the Fourier transform of each letter. An example is shown in Fig. 10.4-5. In order to determine the role played by high spatial frequencies in letter identification, we have used an optical processor to low-pass filter letters of the chart shown in Fig. 10.4-5a [98]. Letters of a row are smaller than those of the preceding row by a factor of $10^{0.1} \simeq 1.25$. Low-pass filtering at a certain cutoff frequency f_0 cycles/mm is progressively more severe as the letters become smaller. Figure 10.4-5b shows the filtered charts for $f_0 = 3$ cycles/mm. The row which has the smallest recognizable letters is the eighth. This corresponds to low-pass filtering at a spatial frequency of 1.64 cycles per vertical letter size and 1.32 cycles per horizontal letter size. These figures establish the spatial frequency components necessary to recognize the letters of the chart. To verify this claim, we have filtered again at $f_0 = 1.48$ cycles/mm. Results are shown in Fig. 10.4-5c from which we see that the lowest recognizable row is the fifth. This corresponds to low-pass filtering at 1.63 cycles per vertical letter size and 1.31 cycles per horizontal letter size; these are approximately the same figures obtained from Fig. 10.4-5b.

Ginsburg [9] addressed the same problem by computing the filtered

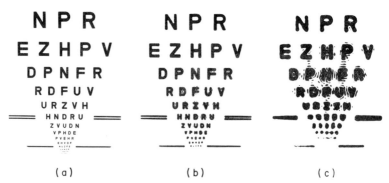

(a)   (b)   (c)

FIG. 10.4-5. (a) Visual acuity chart; (b) chart low-pass-filtered at 4.16 cycles/mm; (c) chart low-pass-filtered at 1.87 cycles/mm.

picture with a digital computer. His results show that a minimum spatial frequency of about 2.5 cycles per letter size is required for the detection and recognition of Snellen letters.

A relationship between the visual acuity and the contrast sensitivity function can be established by simulating a spatial filtering system whose MTF corresponds to the CSF of the subject. The letters of the test chart can then be filtered by this system and examined closely for recognition and determination of the visual acuity. Ginsburg [9] has succeeded in predicting the Snellen acuity of visually abnormal subjects (amblyopia and multiple sclerosis) from their contrast sensitivity functions.

(iv) *Spatial information content of different frequency channels* In an attempt to determine the spatial information in different frequency bands (channels), Ginsburg [100] filtered a portrait into seven different channels whose center frequencies were one octave apart (at 2^n, $n = 0, 1, \ldots, 6$ cycles/face width) with bandwidths of one and two octaves. The filtered pictures provided spatial information about the portrait ranging from basic shape and identification to texture and edges. Portraits filtered at one octave provided only minimal information about the portrait, whereas those filtered to two octaves were found to provide sufficient information for identification. This is consistent with Ginsburg's general conclusion [101] that we only require quite a narrow range of low spatial frequencies to recognize many objects.

(v) *Generation of clinical optotypes that possess certain spatial frequency properties* The spatial spectrum of ordinary clinical optotypes can be modified by filtering to create new optotypes which satisfy certain requirements.

An example is alphanumeric characters high-pass-filtered to remove the dc component [102]; i.e., the filtered characters have an average luminance equal to the background. This results in the character having identical detection and recognition thresholds; i.e., as the target is approached from a distance its presence will be detected, and at the same time it will be recognized and identified.

Such a test pattern may be of clinical value, for a patient need only report that the letter is seen instead of calling it out.

Because high-pass filtering of patterns that contain edges results in a ringing effect, Howland *et al.* [102] suggested the use of a black outline of the characters displayed on a gray background as shown in Fig. 10.4-6. The width of the black line is adjusted such that the black and white of a character sum to a gray equal to that of the background. This is a simple way of ensuring the absence of the dc spatial frequency component.

(vi) *Compensation of visual distortions* As described in Section 10.1, because of optical and neural effects, the visual system transmits different

FIG. 10.4-6. Letter "E" whose black and white segments average to equal the gray background. (From Howland *et al.* [102].)

spatial frequencies nonuniformly in accordance with its transfer function. This results in distortion which is an inherent part of normal vision. In some abnormal conditions such as amblyopia, cerebral lesions [51], and cataracts [96], spatial frequencies are suppressed nonuniformly.

Such linear distortions in the visual system may be partially compensated for by the use of spatial frequency preemphasis, i.e., by boosting spatial frequencies known to be suppressed by the visual system. A picture may be preemphasized by the use of either a digital or an optical spatial filtering processor. It is then displayed for viewing at the appropriate distance. A picture may be filtered using a coherent-optical processor and viewed directly in coherent light. An appropriate model for vision in coherent light must then be adopted [103].

When the visual distortion involves the phase, the compensation spatial filter must be complex. Holographic techniques can then be utilized. Phase aberration balancing techniques, using holography, have been used to see through *in vitro* human cataracts [104].

10.5 IMAGE SCIENCE AND VISION

In the previous sections, we described how vision research is benefiting from advances in coherent-optical techniques, as well as in methods of two-dimensional Fourier analysis. This section reviews briefly the influence of visual psychophysics and the physiology of human vision on research in image science.

When an image is prepared for transmission and display, it is useful to regard the visual system of the observer as a final stage of an overall image transmission system which is to be optimized in some prescribed manner. This helps in predicting how the performance of the transmission and display devices might be relaxed without resulting in appreciable overall degradation. Also, by deliberately distorting the image before it is transmitted, degradations due to transmission, to display, and possibly to the visual system itself might be compensated for, resulting in an improvement in image quality. The idea is to match the image transmission system to the visual system or place the image in a signal space maximally compatible with the human visual system. Knowledge of the structure of a visual system aids in establishing the appropriate fidelity criterion [105].

Many efforts have been dedicated to this approach since the work of Schade [5, 6]. Two ultimate conflicting limitations have guided such efforts— the validity of the used visual model and its simplicity. Very often simplicity has been opted for by considering only one feature of the complex visual system (e.g., spatial filtering properties, nonlinearity, masking, etc.).

Knowledge of the spatial filtering properties of the visual system has been helpful in determining the spatial bandwidth that ultimately limits the transmitted image. Reducing the transmission bandwidth to match that of the observer results in a reduction in noise. An image restoration filter may include partial restoration of the distortion that occurs in the visual system itself. For example, when images whose high frequency components had been attenuated through transmission were restored by inverse filtering, the quality was often improved beyond that of the "original" by enhancing the high frequencies to an extent substantially greater than the loss in the transmission process [106].

The nonlinear behavior of the visual system has also been utilized in image quantization, coding, enhancement, and restoration schemes. The logarithmic property of the visual system has motivated carrying out image restoration in the log luminance domain [107]. Homomorphic filters, which include logarithmic transformation followed by linear spatial filtering and exponential transformation, have resulted in very desirable qualities, under the assumption that the eye is a true logarithmic receiver. Indeed, impressive pictures have been obtained [105].

Another visual property which has been included in the optimization of image restoration filters and in image coding is masking. In the vicinity of sharp transitions in image intensity (edges), the contrast sensitivity of the visual system decreases considerably. The effect decreases exponentially as a function of the distance from the transition. The reduction in visibility of details near edges has suggested the adoption of a local fidelity criterion whereby low noise is allowed, noise which is later suppressed by the visual system. This allows the restoration filter to be strongly peaked, resulting in higher resolution [108–110].

Applications of visual models to image coding have also been very successful. The limited accuracy of the perception of the human observer makes certain perturbations or errors unnoticeable, or of less importance. These errors can be allowed in the encoded–decoded image presented to the observer. For a fixed image transmission rate, the optimum encoding method is that which introduces errors that are least noticeable by the observer. Thus the distortion introduced by the encoding–decoding system is matched to the human visual system or to a visual distortion criterion [111, 112].

Future advances in image processing are expected to be strongly influenced by our understanding of the visual system.

10.6 SUMMARY

Interaction between vision research and the fields of Fourier optics, optical information processing, and image processing has grown rapidly in recent years. Vision researchers are using coherent-optical techniques to probe the visual system. The following examples were reviewed in this chapter: use of speckle to measure the dioptrics of the eye, use of optical Fourier transforms to create on the retina fine patterns that can be used for measuring the transmission of spatial information in the retina–brain part of the visual system, and applications of optical spatial filtering techniques to pictures used in psychophysical studies. Moreover, vision researchers have successfully used the theoretical concepts of Fourier analysis, system theory, and information theory to characterize transmission of information in the visual system. On the other side, image processing engineeres are becoming more aware of the need to understand how images are processed by the human observer. This knowledge helps in improving the performance of image transmission and display devices. Some efficient image processing operations have even been copied from the visual system and implemented in image receivers.

ACKNOWLEDGMENT

The author acknowledges the support of the National Science Foundation, Grant No. ENG-7808674.

REFERENCES

[1] P. M. Duffieux (1946). "L'integrale de Fourier et Ses Applications a L'optique." Fac. Sci., Besançon, France.

[2] A. Maréchal and P. Groce (1953). *C. R. Hebd. Seances Acad. Sci.* **127**, 607–609.
[3] H. H. Hopkins (1953). *Proc. R. Soc. London, Ser. A* **217**, 408–432.
[4] E. L. O'Neill (1956). *IRE Trans. Inform. Theory* **IT-2**, 56–65.
[5] O. H. Schade (1948). *RCA Rev.* **9**, 5–37.
[6] O. H. Schade (1956). *J. Opt. Soc. Am.* **46**, 721–739.
[7] A. Arnulf and O. Dupuy (1960). *C. R. Hebd. Seances Acad. Sci.* **250**, 2727–2759.
[8] G. Westheimer (1960). *J. Physiol. (London)* **152**, 67–74.
[9] F. W. Campbell and J. G. Robson (1964). *J. Opt. Soc. Am.* **54**, 581A.
[10] F. W. Campbell and J. G. Robson (1968). *J. Physiol. (London)* **197**, 551–566.
[11] J. L. Calkins and C. D. Leonard (1970). *Invest. Ophthalmol.* **9**, 458–462.
[12] R. L. Wiggins, K. D. Vaughan, and G. B. Friedmann (1972). *Appl. Opt.* **11**, 179–181.
[13] J. L. Calkins (1975). *In* "Holography in Medicine" (P. Greguss, ed.), pp. 85–89. IPC Sci. Technol. Press, Guilford, Surrey, England.
[14] H. Ohzu (1975). *In* "Holography in Medicine" (P. Greguss, ed.), pp. 82–85. IPC Sci. Technol. Press, Guilford, Surrey, England.
[15] A. N. Rosen (1973). *Invest. Ophthalmol.* **12**, 786–788.
[16] T. Kawara and H. Ohzu (1977). *Jpn. J. Ophthalmol.* **21**, 287–296.
[17] T. Matsumoto, R. Nagata, M. Saishin, T. Matsuda, and S. Nakao (1978). *Appl. Opt.* **17**, 3538–3539.
[18] G. T. Feke and C. Riva (1978). *J. Opt. Soc. Am.* **68**, 526–531.
[19] G. B. Benedek (1971). *Appl. Opt.* **10**, 459–473.
[20] T. N. Cornsweet (1970). "Visual Perception." Academic Press, New York.
[21] C. F. Hall and E. L. Hall (1977). *IEEE Trans. Syst., Man Cybern.* **SMC-7**, 161–170.
[22] M. L. Davidson (1968). *J. Opt. Soc. Am.* **58**, 1300–1308.
[23] E. M. Lowry and J. J. Depalma (1961). *J. Opt. Soc. Am.* **51**, 740–746.
[24] E. M. Lowry and J. J. DePalma (1962). *J. Opt. Soc. Am.* **52**, 328–335.
[25] O. Bryngdahl (1964). *J. Opt. Soc. Am.* **54**, 1152–1160.
[26] D. G. Green and F. W. Campbell (1965). *J. Opt. Soc. Am.* **55**, 1154.
[27] F. W. Campbell (1968). *Proc. IEEE* **56**, 1009–1014.
[28] O. Bryngdahl (1966). *J. Opt. Soc. Am.* **56**, 811–821.
[29] E. L. Van Nes and M. A. Bauman (1967). *J. Opt. Soc. Am.* **57**, 401–406.
[30] A. Watanabe, T. Mori, S. Nagata, and K. Hiwatashi (1968). *Vision Res.* **8**, 1245–1263.
[31] G. Westheimer (1972). *In* "Handbook of Sensory Physiology" (D. Jameson and L. M. Hurvich, eds.), Vol. 7, Part 4, pp. 170–187. Springer-Verlag, Berlin and New York.
[32] D. H. Kelly and R. E. Savoie (1973). *Percept. Psychophys.* **14**, 313–318.
[33] R. L. Savoy and J. J. McCann (1975). *J. Opt. Soc. Am.* **65**, 343–350.
[34] D. H. Kelly (1977). *Opt. Acta* **24**, 107–129.
[35] A. Van Meeteren (1974). *Opt. Acta* **21**, 395–412.
[36] F. W. Campbell and D. G. Green (1965). *J. Physiol. (London)* **181**, 576–593.
[37] F. W. Campbell and R. W. Gubisch (1966). *J. Physiol. (London)* **186**, 558–578.
[38] R. W. Gubisch (1967). *J. Opt. Soc. Am.* **57**, 407–415.
[39] G. A. Fry (1970). *Prog. Opt.* **8**, 51–131.
[40] C. Enroth-Cugell and J. G. Robson (1960). *J. Physiol. (London)* **187**, 517–552.
[41] F. W. Campbell, G. F. Cooper, and C. Enroth-Cugell (1969). *J. Physiol. (London)* **203**, 223–235.
[42] F. W. Campbell, G. F. Cooper, J. G. Robson, and M. B. Sachs (1969). *J. Physiol. (London)* **204**, 120–121.
[43] F. W. Campbell, B. C. Cleland, G. F. Cooper, and C. Enroth-Cugell (1968). *J. Physiol. (London)* **198**, 237–250.
[44] M. B. Sachs, J. Nachmias, and J. G. Robson (1971). *J. Opt. Soc. Am.* **61**, 1176–1186.

[45] N. Graham and J. Nachmias (1971). *Vision Res.* **11**, 251–259.
[46] F. W. Campbell and J. J. Kulikowski (1966). *J. Physiol. (London)* **187**, 437–445.
[47] A. S. Gilinsky (1968). *J. Opt. Soc. Am.* **58**, 13–18.
[48] C. Blakemore and F. W. Campbell (1969). *J. Physiol. (London)* **203**, 237–260.
[49] L. Maffei and F. W. Campbell (1970). *Science* **167**, 386–387.
[50] F. W. Campbell and L. Maffie (1970). *J. Physiol. (London)* **207**, 635–652.
[51] I. Bodis-Wollner (1972). *Science* **178**, 769–771.
[52] O. Bryngdahl (1964). *J. Opt. Soc. Am.* **54**, 1152–1160.
[53] O. Bryngdahl (1965). *Kybernetic* **2**, 227–236.
[54] Y. Kohayakawa (1972). *J. Opt. Soc. Am.* **62**, 584–587.
[55] G. A. Hay and M. S. Chesters (1975). *J. Opt. Soc. Am.* **62**, 990–998.
[56] C. Blakemore, P. S. Muncy, and R. M. Ridley (1973). *Vision Res.* **13**, 1915–1931.
[57] M. A. Georgeson and G. D. Sullivan (1975). *J. Physiol. (London)* **252**, 627–656.
[58] J. J. Kulikowski (1976). *Vision Res.* **16**, 1419–1431.
[59] R. M. Springer (1978). *Vision Res.* **18**, 291–300.
[60] A. W. Snyder and M. V. Srinivasan (1979). *Biol. Cybern.* **32**, 9–17.
[61] F. W. Campbell, E. R. Howell, and J. G. Robson (1971). *J. Physiol. (London)* **217**, 17–18.
[62] C. S. Furchner, J. P. Thomas, and F. W. Campbell (1977). *Vision Res.* **17**, 827–836.
[63] F. W. Campbell, E. R. Howell, and J. R. Johnstone (1978). *J. Physiol. (London)* **284**, 189–201.
[64] Y. LeGrand (1935). *C. R. Hebd. Seances Acad. Sci.* **200**, 490.
[65] F. W. Campbell, J. J. Kulikowski, and J. Levinson (1966). *J. Physiol. (London)* **187**, 427–436.
[66] D. Mitchell, R. Freeman, and G. Westheimer (1967). *J. Opt. Soc. Am.* **57**, 246–249.
[67] D. G. Green (1970). *J. Physiol. (London)* **207**, 351.
[68] D. G. Green (1970). *Science* **168**, 1240–1242.
[69] D. G. Green (1971). *Trans. Am. Acad. Ophthalmol. Otolaryngol.* **75**, 629.
[70] H. Ohzu (1979). *Opt. Acta* **26**, 1089–1101.
[71] H. Ohzu (1978). *Mem. Sch. Sci. Eng., Waseda Univ.* No. 40.
[72] J. W. Goodman (1968). "Introduction to Fourier Optics," Chap. 5. McGraw-Hill, New York.
[73] B. E. A. Saleh (1978). *J. Opt. Soc. Am.* **68**, 1451(A).
[74] J. C. Dainty, ed. (1975). "Laser Speckle and Related Phenomena." Springer-Verlag, Berlin and New York.
[75] M. Françon (1979). "Laser Speckle and Applications in Optics." Academic Press, New York.
[76] B. E. A. Saleh (1978). "Photoelectron Statistics," Chaps. 2 and 3. Springer-Verlag, Berlin and New York.
[77] S. Komatsu, I. Yamaguchi, and H. Saito (1976). *Jpn. J. Appl. Phys.* **15**, 1715–1724.
[78] D. C. Sinclair (1965). *J. Opt. Soc. Am.* **55**, 575–576.
[79] H. A. Knoll (1966). *Am. J. Optom.* **43**, 532.
[80] E. Ingelstam and S. I. Ragnarsson (1972). *Vision Res.* **12**, 411–420.
[81] R. T. Hennessy and H. W. Leibowitz (1972). *Behav. Res. Methods Instrum.* **4**, 237.
[82] N. Mohon and A. Rodemann (1973). *Appl. Opt.* **12**, 783–787.
[83] W. O. Dwyer, D. Granata, R. Bossin, and S. R. Andreas (1973). *Am. J. Optom.* **50**, 222–225.
[84] D. Phillips, W. Sterling, and W. O. Dwyer (1974). *Am. J. Optom.* **51**, 260–263.
[85] W. N. Charman (1974). *Am. J. Optom. Physiol. Opt.* **51**, 832–837.
[86] H. W. Leibowitz and R. T. Hennessy (1975). *Am. Psychol.* **30**, 349–352.
[87] W. F. Long and C. L. Haine (1975). *Am. J. Optom. Physiol. Opt.* **52**, 582–586.

[88] D. W. Phillips, G. S. McCarter, and W. O. Dwyer (1976). *Am. J. Optom. Physiol. Opt.* **53,** 447–450.
[89] F. W. Campbell, R. H. S. Carpenter, and E. Switkes (1971). *J. Physiol. (London)* **217,** 18P–19P.
[90] D. H. Kelly (1975). *Vision Res.* **16,** 227–287.
[91] R. A. Schachar (1975). *Science* **192,** 389–390.
[92] A. P. Ginsburg and F. W. Campbell (1977). *Science* **198,** 961–962.
[93] E. Switkes, M. J. Mayer, and J. A. Sloan (1978). *Vision Res.* **18,** 1393–1399.
[94] F. W. Campbell (1974). *In* "The Neurosciences" (F. O. Schmitt and F. G. Worden, eds.), pp. 95–103. MIT Press, Cambridge, Massachusetts.
[95] J. E. W. Mayhew and J. P. Frisby (1978). *Nature (London)* **275,** 438–439.
[96] R. Hess and G. Woo (1978). *Invest. Ophthalmol. Visual Sci.* **17,** 428–435.
[97] A. P. Ginsburg (1975). *Nature (London)* **257,** 219–220.
[98] I. L. Bailey and J. E. Lovie (1976). *Am. J. Optom. Physiol. Opt..* **53,** 740–745.
[99] A. P. Ginsburg (1978). *J. Opt. Soc. Am.* **68,** 1455(A).
[100] A. Ginsburg (1975). *ARVO Ann. Meet., Sarasota, Fla.*
[101] A. P. Ginsburg (1978). *ARVO Ann. Meet., Sarasota, Fla.*
[102] B. Howland, A. Ginsburg, and F. Campbell (1978). *Vision Res.* **18,** 1063–1066.
[103] B. E. A. Saleh and W. C. Goeke (1979). *J. Opt. Soc. Am.* **69,** 1172–1177.
[104] G. O. Reynolds, J. L. Zuckerman, W. A. Dyes, and D. Miller (1973). *Opt. Eng.* **12,** 23–35.
[105] Z. L. Budrikis (1972). *Proc. IEEE* 771–779.
[106] W. F. Schreiber (1978). *Proc. IEEE* **66,** 1640–1651.
[107] T. G. Stockham, Jr. (1972). *Proc. IEEE* **60,** 828–842.
[108] A. N. Netravali and B. Prasada (1977). *Proc. IEEE* **65,** 536–548.
[109] G. L. Anderson and A. N. Netravali (1976). *IEEE Trans. Syst., Man. Cybern.* **SMC-6,** 845–853.
[110] T. Hentea and B. E. A. Saleh (1978). *IEEE Trans. Syst., Man Cybern.* **SMC-8,** 883–888.
[111] D. J. Sakrison (1977). *IEEE Trans. Commun.* **COM-25,** 1251–1267.
[112] D. J. Sakrison (1979). *In* "Image Transmission Techniques" (W. K. Pratt, ed.), pp. 21–71. Academic Press, New York.

Chapter 11

Statistical Pattern Recognition Using Optical Fourier Transform Features

HENRY STARK AND ROBERT O'TOOLE

DEPARTMENT OF ELECTRICAL, COMPUTER,
AND SYSTEMS ENGINEERING
RENSSELAER POLYTECHNIC INSTITUTE
TROY, NEW YORK

11.1 INTRODUCTION

In this chapter we return to pattern recognition by Fourier-optical methods, a discussion started by S. Almeida and G. Indebetouw in Chapter 2 and continued by others in subsequent chapters. There are many situations in pattern recognition where correlation is impractical because of non-negligible variations within the objects belonging to a given class. When this is the case, it may be necessary to depart from a purely optical approach and to use instead, a hybrid system such as the one discussed by J. R. Leger and S. H. Lee in Section 4.2 (see Fig. 4.2-1). In such a system what role should be assigned to the computer? This question is easily answered when one considers that statistical pattern recognition is often partitioned into three operations: data generation, feature selection, and classification. The last two operations—feature selection and classification—generally require many computations such as adding, subtracting, matrix multiplication and inversion, squaring, and square rooting. To date these operations cannot be done by an optical system at the level of accuracy or speed required for good performance. Hence they are best done by digital computer. Hybrid pattern recognition systems tend to share the same structure: raw data generation by the optical system and feature selection and classification by computer. This raises at least two questions. Can the optical system generate raw data with

465

enough imbedded features ultimately to allow classification? And, second, given its somewhat limited range of raw data options, how well can the hybrid system perform relative to all-digital systems with essentially unlimited options? These questions are the subject of this chapter. First we briefly review when correlation is or is not applicable. Then we discuss a canonical optical–digital computer (ODC) for pattern recognition based on optical Fourier transforms. Next we demonstrate how the optical–digital computer is used in pattern recognition of black lung disease (also called coal workers' pneumoconiosis or simply CWP) and multiclass problems involving arbitrary textures. We also compare the optical–digital approach with all-digital methods. Our main concern is with the pattern recognition of texture because of its ubiquitous nature: It finds application in radiography [1, 2] and in aerial and satellite image processing [3–6], as well as in other areas [7].

Readers of the technical literature will find that most optical pattern recognition schemes discussed there, as well as in this book, are based on correlation. Recently there have been some attempts to use other techniques in addition to correlation. For example, Mantock *et al.* [8] used pseudo-color encoding of local spatial frequencies to distinguish among cloud types. Caulfield *et al.* [9] discuss the use of moments and generalized matched filters for pattern recognition. Several other papers can be listed, but the fact remains that correlation techniques, i.e., matched filtering, is still pervasive in optical pattern recognition. In general, optical correlators are easy to implement and can be used in a number of different settings such as matched filtering of (i) stationary signals with stationary references (as in correlating an image with a fixed pattern), (ii) moving signals with stationary references (as in the use of acousto-optic signals with Fraunhofer or Fresnel diffraction masks), and (iii) moving signals with moving references (as in the SAW devices discussed by Das and Ayub in this book). What is the justification for using correlation? It is known from optimum receiver theory that correlation is the optimum linear signal processing scheme (in the sense of maximizing the signal-to-noise ratio at a given instant or point) for detecting a known signal in white noise. Perhaps less well known but more significant from the point of view of pattern recognition is the fact that correlation is the optimum (in the sense of Bayes) signal processing scheme when deciding which signal, from a known set of signals, is present in white, additive, independent Gaussian noise [10, p. 499]. The problem of character recognition falls into this category provided the character size, font, and orientation are known a priori. The efficacy of optical correlation rapidly decreases when scale, orientation, and fonts are allowed to vary even slightly. Nevertheless optical correlation remains a useful tool provided that it is judiciously applied, i.e., for the problems for which it was intended.

There are frequent situations where correlation is not applicable, and these include the many real-life pattern recognition situations where the user has no or little control over the patterns to be classified. In these situations the signals must be viewed as being fundamentally random, and one is better off talking about the statistics of the signal and not the signal itself, i.e., as a deterministic function of time, space, or frequency. Naturally occurring textures, for example, are signals of this kind and are best modeled as random processes. Any given texture sample is then treated as merely a realization of a random process and will generally not "look" like any other realization of the same texture except in a statistical sense. Since correlation is based on a signal searching for a duplicate reference, it cannot be efficiently used in this setting. For example, a human would have little trouble recognizing the letter "A" in italic, roman (old), roman (new), or Gothic type. However, it would be near impossible to design a correlator that could detect the letter "A" without specifying the font, size, orientation, etc.

Given then that different realizations of the same random process may not look sufficiently alike for correlation, we can still consider them as belonging to the same class if their underlying statistics are the same. The fact that samples of the *same* class may not look alike leads to the notion of within-class variation. The fact that samples for *different* classes do not look alike in some sense leads to the notion of between-class variation. Techniques for maximizing the ratio of suitable measures of between-class to within-class variations is the concern of *feature selection*. The problem of pattern recognition of textures can ultimately be broken down into the following components: (i) How can numerical data be generated from optical Fourier transforms? (ii) What are the best features that can be extracted from the numerial data? (iii) How should classification be made? Much of what follows in this chapter deals with these questions. However, we first describe the optical–digital texture computer in the next section.

11.2 AN OPTICAL–DIGITAL COMPUTER FOR TEXTURE ANALYSIS

The ODC is basically the coherent-optical spectrum analyzer (COSA) discussed in Chapter 1 with some minor modifications; it is shown in Fig. 11.2-1. A sample of a texture pattern is recorded on photographic film and inserted in a phase-matching fluid which negates unwanted phase modulation of the laser light that results from surface relief in the sample itself. We denote the amplitude transmittance (picture function) of the sample by $f(x, y)$. The sample is often a 35-mm high contrast film transparency. A lens is used to generate the spectral irradiance $|F(u, v)|^2$ of the picture function $f(x, y)$. The symbols (u, v) refer to rectangular spatial frequencies in the Fourier plane,

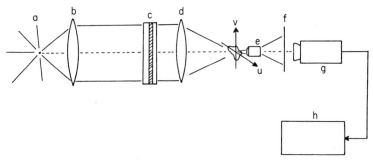

FIG. 11.2-1. Functional diagram of an optical–digital computer. Coherent light (a) is collimated by the lens (b). The sample (c) is held in a phase-matching oil and placed near the Fourier transform lens (d). The spectral irradiance in the uv plane is enlarged (e) and projected on a ground glass screen (f). A TV camera (g) scans the spectrum and transmits the information to a computer.

and $F(u, v)$ is the Fourier transform of $f(x, y)$. Vignetting is minimized by placing the sample against the Fourier transform lens. A microscope objective images an enlarged image of the Fourier irradiance onto a ground glass plate, and a spatial filter removes the zero-order aperture-diffracted light to protect the TV camera and allow the full dynamic range of the data to be recorded. The zero-order light is proportional to the square of the average value of the picture function and, since the latter is largely devoid of the texture detail, its removal does not represent a loss of information. The TV camera scans an enlarged image of the Fourier irradiance and generates a video signal that is branched to a monitor as well as to a sampler and digitizer. The digitized data are stored in a computer, whereupon they are converted to vectors, one datum vector for each texture sample. The organization of the Fourier irradiance into vectors is done with the aid of a computer-generated mask. The mask is simply a disjoint partitioning of the Fourier plane into L regions. The irradiance in each region is integrated numerically and normalized by the total integrated irradiance. Thus, if we denote by s_i the surface element associated with the ith partition, then the ith component of the datum vector $\mathbf{X}$ is simply

$$x_i = \iint_{s_i} |F(u, v)|^2 \, du \, dv \bigg/ \int_{-\infty}^{\infty} \int_{-\infty}^{\infty} |F(u, v)|^2 \, du \, dv, \quad i = 1, \ldots, L, \quad (11.2\text{-}1)$$

where

$$F(u, v) \equiv \int_{-\infty}^{\infty} \int_{-\infty}^{\infty} f(x, y) e^{-j2\pi(ux+vy)} \, dx \, dy. \quad (11.2\text{-}2)$$

Of course in practice Eq. (11.2-1) is realized by numerical integration, so that integration is replaced by summation. In general the partitions cover that portion of the Fourier plane over which significant irradiance is observed. Most often the partitions are annular regions, semiannular regions, wedges, or combinations thereof. The normalization expressed in Eq. (11.2-1) helps to eliminate factors that lead to errors, such as gray level variation from film to film and variations in laser light during the course of the experiment.

Once the datum vector X is produced from the irradiance spectrum, all subsequent operations are done digitally. Strictly speaking, since phase information is lost, the transformation from picture function to Fourier irradiance is not one-to-one. However, for all intents and purposes, it may be considered one-to-one for most textures encountered in practice.

The ODC discussed in this chapter is a concatenation of a COSA with a digital computer; the interface between the two is a TV camera. The COSA, of course, generates a Fourier transform at the speed of light. Moreover the time required to make this computation is independent of picture size and— given a lens of a decent size and of good quality—the resolution can be very high. There are two characteristics of the ODC approach that are worthy of note. First, the actual picture is never seen by the computer, hence is not processed by it. Second the data are generated primarily by the COSA and only minimally by the computer. As will be seen, these characteristics account for the fact that, when a large amount of imagery needs to be processed, it may be worthwhile to forego the versatility and simplicity of a digital computer for the gains of storage and significant decrease in computation time offered by the ODC.

11.3 FEATURE EXTRACTION AND CLASSIFICATION

A. Feature Extraction

The aim of feature extraction is to reduce the dimensionality of the raw datum vectors while finding a set of coordinates in which the different classes nicely separate. The reduction in dimensionality is quite important for at least two reasons: First, it leads to fewer computations, hence saves money and time; and second, it helps to avoid the pitfalls associated with the "curse of dimensionality." Meisel [11] suggests that to avoid apparently significant separations of identical classes, the ratio of the number of samples per class to the dimensionality of the feature space should not be less than 3. Without consideration of the "curse of dimensionality" one might expect classifier performance to improve as the number of variables is arbitrarily increased. However, this does not occur. Figure 11.3-1 exhibits the effects of dimensionality and sample size on the two-class Baye's classifier [12]. It is easy to

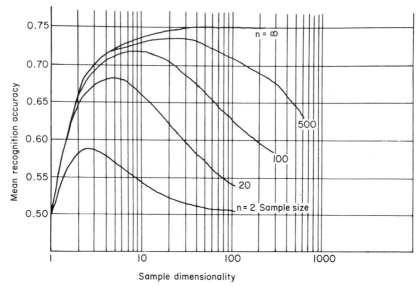

FIG. 11.3-1. Classifier accuracy versus dimensionality for the two-class Bayes classifier. (From Chen [12].)

see from the graphs that an arbitrary increase in sample dimensionality eventually reduces the classification accuracy.

The generation of feature vectors from datum vectors can be expressed mathematically as

$$\mathbf{Y} = \mathbf{AX}, \qquad (11.3\text{-}1)$$

where $\mathbf{X}$ is an $L \times 1$ datum vector, $\mathbf{Y}$ is an $M \times 1$ feature vector ($M \leq L$), and $\mathbf{A}$ is an $M \times L$ transformation matrix. We considered five standard feature selection algorithms and eventually settled on two—the Foley–Sammon (F–S) transform and the Hotelling trace transform. A brief description of these algorithms is given below.

1. Projection on One-Dimension (POD)

In the POD algorithm, a good component is one for which the ratio of the difference of class means to combined class scatter is large. The quality factor for the kth coordinate is given by

$$Q_k = \mu_{1k} - \mu_{2k} / \left[(1/N_1) \sum_{j=1}^{N_1} |x_{1jk} - \mu_{1k}| + (1/N_2) \sum_{j=1}^{N_2} |x_{2jk} - \mu_{2k}| \right], \qquad (11.3\text{-}2)$$

where μ_{ik} is the kth component of the sample mean $\boldsymbol{\mu}_i$, N_i is the number of samples in class i, and x_{ijk} is the kth component of the jth sample of the ith class ($i = 1, 2$). The Q values are computed for all N components in the original space and ranked so that $Q_{p(1)} \geq Q_{p(2)} \geq \cdots \geq Q_{p(M)} \geq \cdots$. The coordinates associated with the M largest Q values are chosen as the reduced M-dimensional feature space. Since the POD algorithm involves no rotation of the space and makes no use of the covariability of the data, its power as a feature selector is strongly impeded.

2. Karhunen–Loève (K–L) Transform

The K–L transform is widely discussed in the literature [13] because of its optimal properties in fitting and representing data. However, it is usually not optimum with respect to discriminating data, and its use in feature selection must be considered very cautiously. In our study, after initially furnishing poor results it was discarded.

3. Fukunaga–Koontz (F–K) Transform†

The F–K transform is an attempt to improve the performance of the K–L transform by ensuring that the eigenvectors which best fit one class are poorest for the other class. In this way Fukunaga and Koontz hoped to increase the likelihood that the transformed space would allow separability between the classes. The major problem with the F–K transform is that it does not readily exploit the information contained in the difference between the mean vectors. Furthermore, as Foley and Sammon [15] have shown in a classic example, the F–K transform does not always correctly order-rank the features that have good discriminating capabilities. For this reason it did not furnish good results in our examples and we discarded it in further work.

4. Foley–Sammon Transform‡

The F–S transform is specifically aimed at improving discrimination ability and therefore differs in a conceptual way from the K–L transform which is optimal with respect to data fitting. The optimality criterion for the F–S transform is the maximization, as a function of a vector $\mathbf{d}$, of the ratio of projected class mean differences to the sum of the projected within-class covariance along $\mathbf{d}$ (the modified Fisher ratio). The modified Fisher ratio is given by

$$R(\mathbf{d}) = [\mathbf{d}^T(\mathbf{\mu}_1 - \mathbf{\mu}_2)]^2 / \mathbf{d}^T(\mathbf{K}_1 + \mathbf{K}_2)\mathbf{d}, \qquad (11.3\text{-}3)$$

where $\mathbf{d}$ is an L-dimensional column vector on which the data are projected,

† Ref. [14].
‡ Ref. [15].

$\boldsymbol{\mu}_i$ is the sample mean of the ith class, and $\mathbf{K}_i$ is the sample covariance matrix of the ith class ($i = 1, 2$).

In finding the transformation to M-dimensional feature space we find the orthonormal set of vectors $\mathbf{d}_1, \mathbf{d}_2, \ldots, \mathbf{d}_M$ such that

$$R(\mathbf{d}_1) > R(\mathbf{d}_2) > \cdots > R(\mathbf{d}_M).$$

The matrix $\mathbf{A}$ in the transformation $\mathbf{Y} = \mathbf{AX}$ then becomes

$$\mathbf{A} = (\mathbf{d}_1 \mathbf{d}_2 \cdots \mathbf{d}_M)^{\mathrm{T}}. \qquad (11.3\text{-}4)$$

The procedure for computing the discriminant vectors is provided by Foley and Sammon in Ref. [15]. The Foley–Sammon transform, specifically designed for the separation of classes, is a useful technique when there is a significant difference in the class means and the underlying distributions are unimodal.

5. Hotelling Trace Criterion†

In this approach the criterion for class separability is the number J given by

$$J = \operatorname{tr} \mathbf{S}_2^{-1} \mathbf{S}_1, \qquad (11.3\text{-}5)$$

where tr denotes trace and, for the two-class case,

$$\mathbf{S}_1 \equiv \sum_{i=1}^{2} P_i(\boldsymbol{\mu}_i - \boldsymbol{\mu}_0)(\boldsymbol{\mu}_i - \boldsymbol{\mu}_0)^{\mathrm{T}}, \qquad (11.3\text{-}6)$$

$$\mathbf{S}_2 \equiv \sum_{i=1}^{2} P_i \mathbf{K}_i, \qquad (11.3\text{-}7)$$

$$\boldsymbol{\mu}_0 \equiv P_1 \boldsymbol{\mu}_1 + P_2 \boldsymbol{\mu}_2, \qquad (11.3\text{-}8)$$

$\mathbf{K}_i$ is the covariance of the ith class, P_i is the a priori probability of the ith class, and the other symbols were defined earlier. The matrix $\mathbf{S}_1$ is a measure of the between-class scatter, while $\mathbf{S}_2$ is a measure of the within-class scatter. To find the eigenvalues and eigenvectors of $\mathbf{S}_2^{-1} \mathbf{S}_1$, $\mathbf{S}_1$ and $\mathbf{S}_2$ must be diagonalized simultaneously. In practice this may require some nontrivial computations. The trace criterion algorithm gave excellent results.

The Hotelling trace criterion can be used for feature selection when there are more than two classes. In fact it can be used in two ways: the pooled mode and the pairwise mode. In the pooled mode, a single feature space is generated for all $N_c(N_c \geq 2)$ classes. Classification comparisons are then done in the same (single) feature space. In the N_c-class pooled mode Eqs. (11.3-6) to (11.3-8) are replaced by

† Ref. [16, p. 260].

$$\mathbf{S}_1 = \sum_{i=1}^{N_c} P_i(\boldsymbol{\mu}_i - \boldsymbol{\mu}_0)(\boldsymbol{\mu}_i - \boldsymbol{\mu}_0)^{\mathrm{T}}, \tag{11.3-9}$$

$$\mathbf{S}_2 = \sum_{i=1}^{N_c} P_i \mathbf{K}_i, \tag{11.3-10}$$

$$\boldsymbol{\mu}_0 = \sum_{i=1}^{N_c} P_i \boldsymbol{\mu}_i. \tag{11.3-11}$$

The optimum transformation $\mathbf{A}$, with respect to J, is then given by the first M eigenvectors ($M < L$) of $\mathbf{S}_2^{-1}\mathbf{S}_1$; i.e.,

$$\mathbf{A} = (\boldsymbol{\phi}_1\boldsymbol{\phi}_2\boldsymbol{\phi}_3 \ldots \boldsymbol{\phi}_M)^{\mathrm{T}}, \tag{11.3-12}$$

where T denotes transpose, $\boldsymbol{\phi}_i$ is the ith eigenvector of $\mathbf{S}_2^{-1}\mathbf{S}_1$, and λ_i is its corresponding eigenvalue and ordered such that

$$\lambda_1 > \lambda_2 > \lambda_3 > \cdots > \lambda_M.$$

This yields an M-dimensional feature space determined by the M largest eigenvectors of $\mathbf{S}_2^{-1}\mathbf{S}_1$.

In the pairwise mode, we create a different feature space for each class pair. For N_c classes there will be $N_c(N_c - 1)/2$ distinct class pairs to consider. For each class pair (C_i, C_j) we construct a transform according to Eq. (11.3-12), where $\mathbf{S}_2$ and $\mathbf{S}_1$ are now given by

$$\mathbf{S}_2 = P_i\mathbf{K}_i + P_j\mathbf{K}_j, \tag{11.3-13}$$

$$\mathbf{S}_1 = P_i(\mathbf{u}_i - \mathbf{u}_0)(\mathbf{u}_i - \mathbf{u}_0)^{\mathrm{T}} + P_j(\mathbf{u}_j - \mathbf{u}_0)(\mathbf{u}_j - \mathbf{u}_0)^{\mathrm{T}}, \tag{11.3-14}$$

where

$$\mathbf{u}_0 = P_i\mathbf{u}_i + P_j\mathbf{u}_j. \tag{11.3-15}$$

For all feature selection algorithms, the covariance matrix for samples of class i was computed according to

$$\mathbf{K}_i = 1/N_i \sum_{l=1}^{N_i} (\mathbf{X}_{il} - \mathbf{u}_i)(\mathbf{X}_{il} - \mathbf{u}_i)^{\mathrm{T}}, \tag{11.3-16}$$

where N_i is the number of samples in class C_i and $\mathbf{X}_{il}$ is the datum vector associated with the lth sample of the ith class.

The pairwise and pooled Hotelling trace criteria are identical in the two-class case. In the multiclass case, it may be more convenient to use the trace criterion in the pooled mode, since the feature space is then identical for all classes. However, this convenience is generally gained at the expense of class discrimination ability.

B. Classification

For classification we used an unweighted k-nearest-neighbor (KNN) decision rule. The KNN decision rule is applied in feature space and works as follows. Let Y_1, Y_2, ..., Y_n be feature vectors of known classification, and let Y be the feature vector to be classified. Assume there are two classes C_i and C_j. Among Y_1, Y_2, ..., Y_n, let Y_1^*, Y_2^*, ..., Y_k^* be the k nearest neighbors of Y as determined by an appropriate distance measure. Then Y is assigned the classification associated with the majority of its k nearest neighbors. In our case, we use the generalized (i.e., M-dimensional) Euclidean distance as the measure of closeness. The reason for using a KNN classifier is that, in addition to ease of implementation, its performance is near optimum in the Bayes sense.

In the multiclass case, the unknown test sample must be compared with respect to all $(N_c - 1)N_c/2$ class pairs (C_i, C_j). Thus to classify a test sample Y, say, in a four-class texture problem, we apply the KNN routine to each of the six pairs of classes:

$$[(C_i, C_j): (C_1, C_2), (C_1, C_3), (C_1, C_4), (C_2, C_3), (C_2, C_4), (C_3, C_4)].$$

We then use a majority vote scheme to arrive at a final decision. From the six classifications of the test sample, we assign Y to the class which received the most votes in the pairwise tests. In the case where we use the pooled four-class Hotelling trace feature extractor, the six class pair comparisons are all done in the same feature space. Note that, for the pairwise Hotelling trace and the Foley–Sammon feature extractors, each pairwise classification is performed in a separate and distinct feature space.

Because our sample sizes were small, we discarded the usual partitioning of the sample set into test and training sets. Rather we turned to a variation of the "leave-one-out" method designed by Fu [17]. In this approach the classification procedure consists of removing one sample point from the data set for subsequent use as a test point. The remaining samples serve as the prototypes. After classification, the test sample is returned to the sample pool, and a new test point is chosen. This procedure is repeated until every point in the sample has been classified.

It should be noted that, while samples play the dual role of test and prototype, at no time are a set of prototypes used to find separation surfaces which are then used to separate the prototypes, as is sometimes the case in linear discriminant analysis. In our use of the k-nearest-neighbor classifier, a test sample never influences the decision rule while it is being classified. For this reason, the KNN procedure is free from the bias that can result when training and test samples are the same.

Until now we have discussed the optical–digital computer and the mathematical machinery required to solve both two-class and multiclass problems. We now illustrate how the two-class (i.e., normal–abnormal) black lung disease problem can be classified by an ODC.

11.4 PATTERN RECOGNITION OF COAL WORKERS' PNEUMOCONIOSIS

A. Background

Coal and its derivative substances will play a major role in supplying our energy needs for the near-term future. Unfortunately most coal must be mined, and coal mining is a hazardous activity for humans. The most significant hazard results from the inhalation of coal dust particles, which eventually leads to a chronic condition known as coal workers' pneumoconiosis or simply black lung. The inhaled dust particles collect in the lungs and become sites for the formation of fibrous modules. As the disease progresses, fibrous tissue continually replaces elastic lung tissue, leading to loss of lung function. The latter is signaled by shortness of breath, wheezing, coughing, and difficulty in expectorating. Pneumoconiosis is incurable, and treatment is purely symptomatic. The disease is diagnosed from check x-rays in which the fibrous nodules appear as opacities in the affected regions of the lung field.

CWP is a disease that is endemic to the mining profession: NIOSH data [18] show that approximately 13% of active miners and as many as 28% of retired miners are afflicted with CWP. Recent government legislation now gives each coal worker the right to obtain a check x-ray at regular (<5 years) intervals for detecting the onset or progression of CWP. There are about 185 film readers actively screening for this disease. The levels of compensation furnished to miners depend in part on the diagnosed severity of CWP. To deal with both the problems of interreader variation and anticipated overburdening of the present manual techniques it would be highly desirable to develop machine techniques for mass screening of susceptible groups. Since the large majority of x-ray films apparently do not show signs of CWP, the machine techniques should be effective in screening out definite normals.

Kruger [18] attempted a computer-aided diagnosis of CWP using both all-digital and optical–digital techniques. His results were encouraging; the normal–abnormal classification accuracy was not less than 88%, while the comparable rate for physicians ranged from 83.0 to 97.9%. The false normal (missed detection) computer rate never exceeded 4%, while the comparable physician rate ranged from 1 to 7%.

The medical importance of the CWP problem has prompted additional recent research in this area [2, 19, 20]. The results point to the potential feasibility of using machine-assisted techniques for classifying CWP.

B. Procedure

We used the ODC shown in Fig. 11.2-1. Figures 11.4-1 and 11.4-2 show a normal and an abnormal chest radiogram, respectively. The monitor display of the Fourier irradiance generated by the normal x-ray film is shown in Fig. 11.4-3; that of the abnormal film is shown in Fig. 11.4-4. Immediately evident is the higher concentration of energy at high frequencies in the spectrum of the abnormal film. We can now see that the underlying reason for using the Fourier irradiance as the basis of a signature for CWP as manifest in radiograms is the fact that the texture of the abnormal x-ray film is different from that of the normal film [21]. The main radiographic CWP manifestation is a profusion of opacities in the affected zones, which add to or mask the normal lung vascularities. The presence of these radiopacities alters the visual impression of texture and, at least so it is argued, the qualitative changes in texture can be quantitatively translated through the Fourier irradiance spectrum.

In Figs. 11.4-3 and 11.4-4, the vertical waveform shows the video level

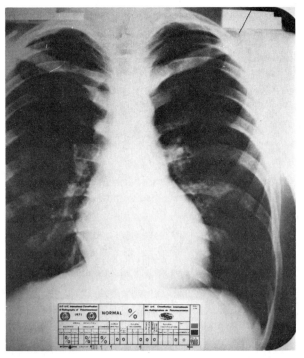

FIG. 11.4-1. Normal radiogram of chest. (From Stark and Lee [1].)

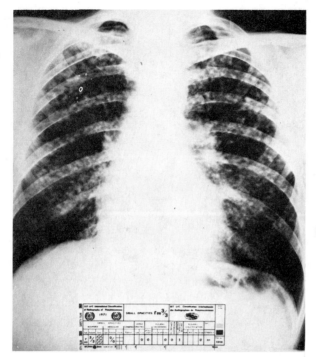

FIG. 11.4-2. Abnormal radiogram of chest. (From Stark and Lee [1].)

along the scan line in the center. The other two white lines on the left are reference lines for digitizing purposes. As explained in Section 11.2, once the data are in digitized form, the computer organizes the data into vectors, one for each sample. The mask used to do this is shown in Fig. 11.4-5; it consists of 16 wedges and 16 semiannular rings. The components obtained from the semiannular ring integrations give information about the rotationally symmetric portion of the spectrum. The components obtained from wedge integrations give information about the angularly distributed portion of the spectrum. The components x_i of the 32-component datum vector $\mathbf{X}$ are given by Eq. (11.2-1). We note that their sum

$$\sum_{i=1}^{32} x_i$$

is unity because of normalization. This negates noisy variations due to light fluctuations, etc.

The set of vectors $\{\mathbf{X}\}$ represents the preprocessed data base. When all these vectors have been computed, the problem of diagnosing CWP can be

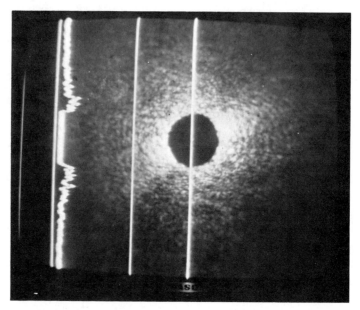

FIG. 11.4-3. Monitor display of Fourier irradiance generated by normal lung x-ray film. (From Stark and Lee [1].)

modeled as a classic two-stage pattern recognition problem involving feature selection and classification.

C. Description of the Data Base

The 1971 ILO U/C international classification of radiographs of pneumoconioses uses two figures for the profusion category and a letter to indicate the type of profusion. Other symbols for very large opacities and other significant disease manifestations are also in use [19]. However, for our purposes, the profusion category (number of opacities) and opacity size were of greatest interest. The profusion category is indicated by two numbers separated by a slash. The higher the number, the more advanced the disease. The first number is the actual category, and the number after the slash indicates that this alternative was seriously considered. Thus category 2/1 indicates a film in category 2, but that category 1 was seriously considered.

Our data set consisted of 64 samples. Of these, 27 were judged normal (0/, 0/0, 0/1) and 37 were judged abnormal (2/2, 2/3, 3/2, 3/3). Both rounded (p, q, r) and irregular (s, t, u) opacities were present. The area of examination roughly corresponded to a single lung zone, i.e., a third of a lung lobe.

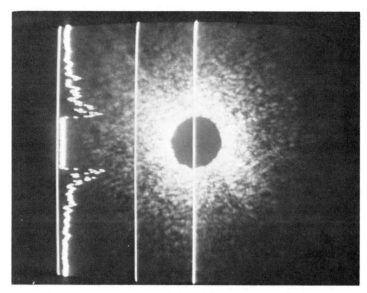

FIG. 11.4-4. Monitor display of Fourier irradiance generated by abnormal lung x-ray film. (From Stark and Lee [1].)

The films were made available by William Crawford, formerly of the Department of Radiology at Yale Medical School. About 20 of the films were international standards and furnished by the ILO. The remaining films were furnished by Alan H. Purdy of NIOSH. They were independently classified by five board-certified readers, including Dr. Crawford. The final classification was assigned according to majority rule.

D. Results

We found that the feature selection algorithms were critically important. Our results with the POD algorithm, the Karhunen–Loève transform, and the Fukunaga–Koontz transform were quite poor, as the correct classification percentage hovered near 70%. The missed detection (false normal) rate was about 10%, and the effective reduction in dimensionality was not clear. These poor results were not overly surprising in view of the properties of these feature selection algorithms.

On the other hand, the results obtained with the Foley–Sammon transform and the Hotelling trace criterion were very encouraging. For both the F–S transform and the trace criterion algorithm, the correct classification

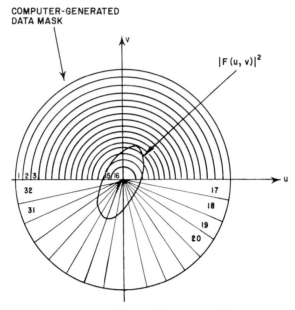

FIG. 11.4-5. Thirty-two sectors of integration generated by computer to produce raw datum vector **X** for film sample. (After Stark and Lee [1].)

rate was near 90%. In both cases, the reduction in dimensionality was essentially from 32 to 1. This suggests efficient exploitation of inherently strong correlations in the Fourier irradiance data. The results were robust vis-à-vis the value of k for the k-nearest-neighbor rule; i.e., for $3 \leq k \leq 10$, the classification performance remained the same.

The results of the classification algorithm when the F–S transform was used for feature selection are summarized in the following confusion matrix which is self-explanatory.

Assigned class	True class	
	Normal	Abnormal
Normal	22	2
Abnormal	5	35

Of 37 abnormals, only two were missed, giving a missed detection rate of 5.4%. The overall correct classification rate was 89%.

When the Hotelling trace criterion was used, the results were the following.

Assigned class	True class	
	Normal	Abnormal
Normal	22	1
Abnormal	5	36

Only one abnormal out of 37 was missed, giving a missed detection rate of only 2.7%. The overall correct classification rate was 90.7%.

E. Conclusions Drawn from CWP Pattern Recognition Study

Although our data base was smaller than we would have liked, the results of our study, especially when taken together with Kruger's results [18], suggest that the ODC technique is a promising method for the machine classification of CWP films into normal or abnormal classes. Our data base did not include any 1/0 and 1/1 cases, but 1/0 and 0/1 data are so borderline that they represent a source of confusion even to radiologists (hence the 1/0, 0/1 classification). In general, borderline data rated 1/- can be conservatively classified in a screening procedure by biasing the k-nearest-neighbor decision rule (or, for that matter, any other decision rule that is used).

It is appropriate to point out that our coherent-optical system employed standard components, and that we did not use either precision-grade lenses or specially designed so-called Fourier optics. No careful precautions were taken with respect to positioning the film samples in the optical system. Once the feature selection transformation matrix is determined, the only significant computation in processing a CWP film is in applying the k-nearest-neighbor algorithm. A special purpose dedicated computer can be used for this task.

In our next example, we consider a multiclass ($N_c = 4$) texture classification problem. We also try to deal with the question of how well optical Fourier transform techniques perform against all-digital techniques based on the second-order joint probability of picture intensity.

11.5 A FOUR-CLASS TEXTURE PROBLEM: THE ODC VERSUS THE ALL-DIGITAL APPROACH

A. The Textures: ODC Data Generation

For our comparative study, we selected four synthetic textures from Hornung [22]. The textures ranged from highly periodic and strongly directional to noiselike and isotropic and, although they were not samples of natural textures, they seemed to contain the nondeterministic quasi-repetitive structure evident in real-life scenes. While it was impossible to determine a priori how easily these textures could be correctly classified, we chose textures that were—to the human observer at least—highly disparate. We also felt that, in order to be useful, any noninteractive automatic texture recognition scheme should furnish a very high correct classification rate; a pattern recognition system yielding considerably less than 90–100% correct classification would, in most instances, not be considered practical.

The four texture classes are shown in Fig. 11.5-1. The Fourier irradiance of the four classes is shown in Fig. 11.5-2 where the dark region at the origin is an opaque stop to protect the TV camera from the zero-order aperture-diffracted light. The display system was limited to a maximum of four-bit resolution so that what is seen in these figures is in coarser gray level detail than the information actually stored in the computer.

The organization of the Fourier irradiance into vectors was done with the computer-generated mask shown in Fig. 11.5-3. As in the problem described in the previous section, the mask produces a 32-component datum vector $\mathbf{X} = (x_1, \ldots, x_{32})^{\mathrm{T}}$. However, in this case the texture mask does not contain any wedge partitions, since textures are not viewed in any preferred direction. Each component x_i represents the normalized, integrated irradiance as obtained by Eq. (11.2-1). The subscript i on the component x_i is the ring

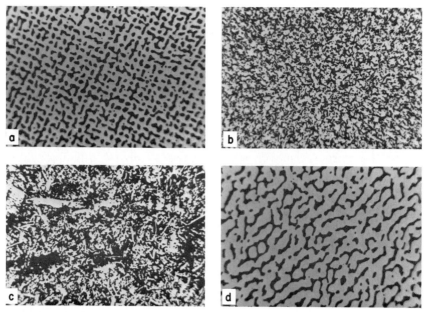

FIG. 11.5-1. The four texture classes: (a) class 1; (b) class 2; (c) class 3; (d) class 4. (From O'Toole and Stark [25].)

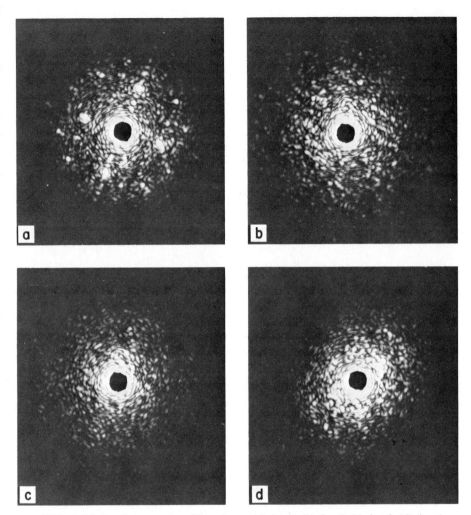

FIG. 11.5-2. Irradiance spectra of four classes: (a) class 1; (b) class 2; (c) class 3; (d) class 4. (From O'Toole and Stark [25].)

number and is proportional to the quantized spatial frequency $\omega = (u^2 + v^2)^{1/2}$. Figures 11.5-4a to 11.5-7a show the irradiance profiles of typical vectors from each of the four textures. The figures are actually graphs of percent spectral energy versus ring number moving outward from the origin. In general we observe that our textures have significant spectral differences mainly in the low to midspatial frequency region. Figures 11.5-4b to 11.5-7b show the intraclass variations in the spectra by superimposing several profiles from each class.

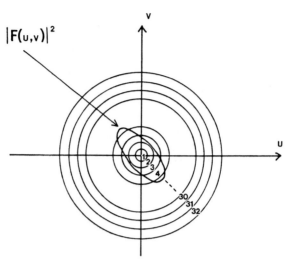

FIG. 11.5-3. Thirty-two rings of integration generated by computer to produce a 32-component raw datum vector **X** from Fourier irradiance. (From O'Toole and Stark [25].)

Despite there being a great deal of informal speculation about optical–digital versus digital methods, there have been very few actual studies comparing one method with the other. One problem that has been tried both ways is the CWP problem discussed earlier [2, 18]; roughly comparable results were reported. Weszka *et al.* [4] made a comparative multiclass study of texture measures for terrain classification and found that, on the whole, Fourier transform features were outperformed by features generated from the gradient distribution matrix (about which more will be said later). How-

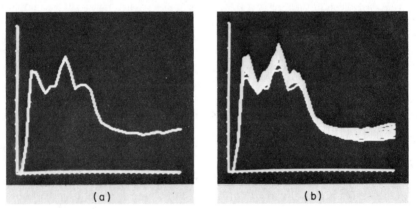

FIG. 11.5-4. Graph of irradiance vectors of class 1. (a) Single datum vector; (b) ensemble of data vectors superimposed to illustrate intraclass variation. (From O'Toole and Stark [25].)

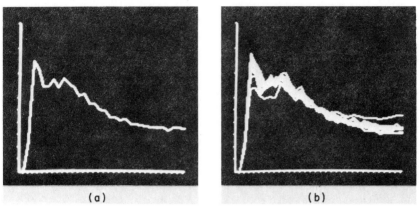

FIG. 11.5-5. Graph of irradiance vectors of class 2. (a) Single datum vector; (b) ensemble of data vectors superimposed to illustrate intraclass variation. (From O'Toole and Stark [25].)

ever, in Weszka's study, the Fourier transforms were computed digitally, at relatively low resolution. In our four-class study the Fourier transforms were obtained with a high resolution COSA. We also considered two powerful feature extraction algorithms which significantly reduced the dimensionality of the data.

B. Data Generation in the All-Digital Method

The digital approach dispenses with the optical preprocessor by using a vidicon camera to scan and digitize the *input texture directly*. No transfer to 35-mm slides is involved. Signatures are obtained from gray level spatial

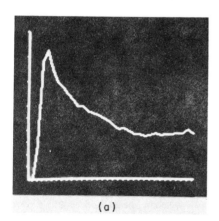
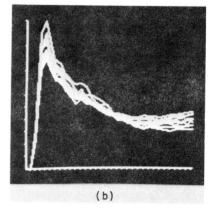

FIG. 11.5-6. Graph of irradiance vectors of class 3. (a) Single datum vector; (b) ensemble of data vectors superimposed to illustrate intraclass variation. (From O'Toole and Stark [25].)

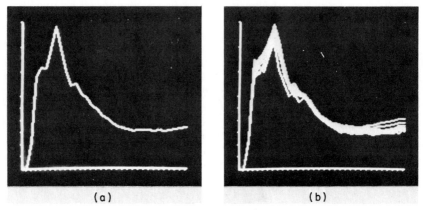

(a)   (b)

FIG. 11.5-7. Graph of irradiance vectors of class 4. (a) Single datum vector; (b) ensemble of data vectors superimposed to illustrate intraclass variation. (From O'Toole and Stark [25].)

dependence probability distribution matrices (gradient distribution matrices). These are obtained from measurements. First proposed by Haralick [23], these signatures have been used in other texture studies as well [3, 4, 24].

Given that a rectangular region of a texture pattern has been digitized into an array of pixels, we wish to consider the distribution of intensity differences existing between picture points as useful information. A gradient distribution matrix is constructed in order to preserve this information. Such a matrix $\mathbf{P}$ has elements p_{ij} which represent the joint probability of pixel pairs having intensities (i, j). An element p_{ij} is defined as the relative number of times a gray level pair (i, j) occurs when comparing pixels separated a distance d along an angle θ. The number of pixel pairs with intensities (i, j) is finally normalized by the total number of occurrences, giving a $\mathbf{P}$ matrix:

$$\mathbf{P} = \begin{bmatrix} p_{00} & p_{01} & p_{02} & \cdots & p_{0,N-1} \\ p_{10} & p_{11} & p_{12} & \cdots & p_{1,N-1} \\ \vdots & & & & \\ p_{N-2,0} & & & \cdots & p_{N-2,N-1} \\ p_{N-1,0} & p_{N-11} & & \cdots & p_{N-1,N-1} \end{bmatrix}, \tag{11.5-1}$$

$$p_{ij} = \frac{\text{No. of pixel pairs with intensities } (i, j)}{\text{Total no. of pairs considered}}, \tag{11.5-2}$$

where N is the number of gray levels. We note that $\mathbf{P}$ will be a $N \times N$ matrix symmetric about the diagonal ($p_{ij} = p_{ji}$). The dimension or size of $\mathbf{P}$ is thereby *independent of picture size*. Also, each $\mathbf{P}$ matrix is a function of the parameters d and θ.

Consider the following example [23]. Figure 11.5-8a represents a 4 × 4 pixel image quantized to four gray levels, 0–3. With $d = 1$ and $\theta = 0$ deg we are interested in horizontally adjacent pixel pairs. If, for example, we wish to calculate the matrix component p_{23}, we count the number of occurrences in which the first pixel of the pair has gray level 2 and the second has gray level 3. This happens just once in this example. Filling out the rest of the element values we obtain the resulting gradient distribution matrix shown in Fig. 11.5-8b. Finally, each element of the matrix is normalized by the total number of pairs considered, 24 in this case.†

From each **P** matrix the following four measurements, proposed by Haralick, are computed:

(a) *Angular second moment* (a measure of picture homogeneity):

$$f_1 = \sum_{i=0}^{N-1} \sum_{j=0}^{N-1} (p_{ij})^2. \qquad (11.5\text{-}3)$$

(b) *Contrast* (a measure of the amount of intensity variation):

$$f_2 = \sum_{n=0}^{N-1} n^2 \sum_{\substack{i=0 \ j=0 \\ |i-j|=n}}^{N-1 \ N-1} p_{ij}. \qquad (11.5\text{-}4)$$

(c) *Correlation* (measure of gray level dependencies):

$$f_3 = \sum_{i=0}^{N-1} \sum_{j=0}^{N-1} [(ij)p_{ij} - \hat{u}_x \hat{u}_y]/\hat{\sigma}_x \hat{\sigma}_y, \qquad (11.5\text{-}5a)$$

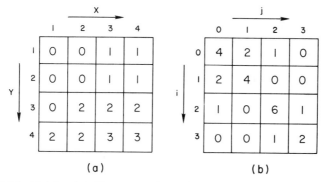

(a)   (b)

FIG. 11.5-8. (a) A 4 × 4 pixel image quantized to four gray levels, 0–3. (b) the corresponding unnormalized gradient distribution matrix. (From O'Toole and Stark [25].)

† The pixel pairs are of the form $\{(x_i, y_j), (x_{i+1}, y_j)\}$ or $\{(x_{i+1}, y_j), (x_i, y_j)\}$. Some of them are $\{(1, 1), (1, 2)\}$, $\{(1, 2), (1, 1)\}$, $\{(1, 2), (1, 3)\}$, etc.

where

$$\hat{u}_x = \sum_{i=0}^{N-1} \sum_{j=0}^{N-1} i p_{ij}, \tag{11.5-5b}$$

$$\hat{u}_y = \sum_{i=0}^{N-1} \sum_{j=0}^{N-1} j p_{ij}, \tag{11.5-5c}$$

$$\hat{\sigma}_x^2 = \sum_{i=0}^{N-1} \sum_{j=0}^{N-1} p_{ij}(i - \hat{u}_x)^2, \tag{11.5-5d}$$

$$\hat{\sigma}_y^2 = \sum_{i=0}^{N-1} \sum_{j=0}^{N-1} p_{ij}(j - \hat{u}_y)^2. \tag{11.5-5e}$$

(d) *Entropy* (another homogeneity measure):

$$f_4 = - \sum_{i=0}^{N-1} \sum_{j=0}^{N-1} p_{ij} \log_{10}(p_{ij} + \epsilon), \tag{11.5-6}$$

where ϵ is a small positive quantity (0.5×10^{-6} in our tests) inserted to avoid overflows when $p_{ij} = 0$.

The method of arriving at a datum vector from the gradient distribution matrices for a particular sample is as follows. First, a region of the texture sample is digitized into a 64×64 pixel array quantized to 16 levels. Next we generate **P** matrices for distances $d = 1, 2, 3$ and directions $\theta = 0, 45, 90,$ and 135 deg, a total of 12 matrices. From each matrix we calculate four function values, as defined in Eqs. (11.5-3) to (11.5-6), giving a total of 48 values. In order to reduce the effects of directional bias we take the average and the maximum spread of the function values, as they vary with θ, for each distance d. This in effect, tends to eliminate the variations due to the random alignments of texture samples from the same class during sampling. We are then left with 24 texture measurements. These values form the resultant 24-element sample datum vector shown in Fig. 11.5-9.

The two types of data vectors, one derived from optical Fourier transforms and the other derived from gradient distribution matrices, are extracted

$$
\begin{array}{ccc}
d = 1 & d = 2 \quad d = 3 \\
\overbrace{\qquad\qquad\qquad} & \overbrace{\ } \; \overbrace{\qquad}
\end{array}
$$

$$\mathbf{X} = (\bar{f}_1, \Delta f_1, \bar{f}_2, \Delta f_2, \bar{f}_3, \Delta f_3, \bar{f}_4, \Delta f_4 \; \vdots \; \cdots \; \vdots \; \cdots \Delta f_3, \bar{f}_4, \Delta f_4)^{\mathrm{T}}$$

$$x_1, x_2, x_3, x_4, \cdots \qquad\qquad\qquad \cdots x_{22}, x_{23}, x_{24}$$

FIG. 11.5-9. Twenty-four element sample datum vector. The first eight entries (x_1 to x_8) are for $d = 1$; the next eight entries are for $d = 2$; the last eight entries are for $d = 3$. The spread of f_i, i.e., $f_{i,\,\mathrm{max}} - f_{i,\,\mathrm{min}}$ is denoted by Δf_i. The overbar on the components means that they have been averaged over the four directions $\theta = 0, 45, 90,$ and 135°. (From O'Toole and Stark [25].)

from samples taken from each of the texture classes. Once the data bases are generated, the feature extraction and classification algorithms are invoked. Based upon our experience in the CWP experiment the only feature extraction algorithms we considered were the Hotelling trace criterion (both pairwise and pooled) and the Foley–Sammon transform.

C. Data Base and Classification

For each of the two texture measurement types, Fourier measures and gradient matrix measures, a data base consisting of 50 samples from each class, 200 total, was created. The KNN classification scheme, in conjunction with the various feature extractors, was then applied for several different values of the parameter k ($k = 1, 3, 5, 7, 9, 11$), and the results observed (recall that k is the number of nearest neighbors to the sample).

A computer classification run consists of removing one sample from the data collection and testing it against the remaining known samples. The test sample is then returned, and a new sample is chosen. This was done with 20 samples from each class for a total of 80 classification tests. Results were obtained in the form of tables that listed the overall classification accuracy as a function of which feature was used in the classification algorithm.

D. Results: Optical–Digital Approach

Texture samples for this experiment were in the form of 35-mm Panatomic-X black-and-white transparencies. Upon completion of the spectral irradiance ring integration procedure each sample was represented by a normalized 32-element datum vector $\mathbf{X}$. The entire ring sampling region included all spatial frequencies up to 350 cycles/mm.† Classification results using the different feature extraction algorithms are given in what follows. It was found that results were relatively stable over values of the parameter k in the classifier. Hence, for consistency, results are given for the same value of k, namely, $k = 9$.

1. Hotelling Trace Transform (As Used in the Pooled Mode)

Thirty-two features were extracted from each datum vector and ranked in descending order according to their respective eigenvalues. Classification runs were then made on the first 20 features, using *one feature at a time*. The tabulated results are in Table 11.5-1. It is evident that only the first three features are significant for discrimination purposes, with feature 1 giving 100% classification accuracy. This is not surprising in a four-class problem, since the between-class scatter matrix $\mathbf{S}_1$ has rank three and therefore the

† This figure refers to spectral information content in the 35-mm reduced slide.

rank of $S_2^{-1}S_1$ is at most three. Thus the matrix $S_2^{-1}S_1$ can have at most three nonzero eigenvalues. When the Hotelling trace approach is used in a pairwise mode, the rank of S_1 is one, hence we might expect only the first eigenvalue to be nonzero and only the first-ranked feature ($n = 1$) to be significant in classification. This is in fact what was observed. From Table 11.5-1 we see that classification accuracy decreases as we use lower-ranked features. This is to be expected, since the eigenvalues are a measure of the class discrimination power of their corresponding features.

When *more than one* of the top-ranked features were used, the accuracy remained at 100%. *A decrease in performance* was observed when a number of low-ranked features were included in the feature vector.

2. Hotelling Trace Transform (Used in Pairwise Mode)

In this mode we consider the four classes in a pairwise manner. This leads to 6 distinct class pairs and thus 6 distinct feature spaces. For each space we

TABLE 11.5-1

Single-Feature Classification Results for Optical–Digital Data and Hotelling Trace Feature Extractor in Pooled Mode

Rank of feature used	Number of samples correctly assigned to:				Total % accuracy
	Class 1	Class 2	Class 3	Class 4	
1	20	20	20	20	100
2	19	20	20	18	96
3	20	17	20	18	94
4	5	3	7	5	25
5	7	6	5	5	29
6	3	5	4	6	23
7	7	2	5	2	20
8	7	4	1	5	21
9	3	1	6	6	20
10	4	2	5	3	18
11	7	8	5	6	33
12	8	5	3	6	28
13	0	7	7	4	23
14	4	7	8	3	28
15	10	4	3	2	24
16	8	1	6	3	23
17	7	1	5	3	20
18	10	6	4	4	30
19	4	4	6	7	26
20	9	8	1	2	25

extract 32 features from each datum vector and rank them according to the magnitude of the corresponding eigenvalue. Results of classification runs using the first 20 features *one at a time* are listed in Table 11.5-2. We see that only the top-ranked feature is significant for classification, with performance dropping off considerably with higher-order ($n > 1$) features. As explained in the previous case, these results are not surprising in view of the fact that the rank of S_1 in this case is unity.

3. *Foley–Sammon Discriminant Vectors*

Again the texture classes are dealt with in a pairwise manner and a different feature space is required for each of the six pairwise tests. Separability of the classes is very strong with this feature extractor. Accurate recognition was accomplished with any of the first nine features. Table 11.5-3 gives the results of classification by single feature for all 20 features.

TABLE 11.5-2

Single-Feature Classification Results for Optical–Digital Data and Hotelling Trace Feature Extractor in Pairwise Mode

Rank of feature used	Number of samples correctly assigned to:				Total % accuracy
	Class 1	Class 2	Class 3	Class 4	
1	20	20	20	20	100
2	9	6	9	1	31
3	7	11	6	3	34
4	4	3	3	1	14
5	8	9	9	4	38
6	9	8	6	2	31
7	10	8	3	3	30
8	7	9	5	4	31
9	7	8	4	6	31
10	3	11	10	2	33
11	11	6	7	2	33
12	8	5	5	3	26
13	11	9	3	2	31
14	6	2	2	1	14
15	6	7	6	6	31
16	5	2	6	1	18
17	6	12	0	4	28
18	13	9	6	7	44
19	4	7	7	5	29
20	10	7	2	0	24

TABLE 11.5-3

Single-Feature Classification Results for Optical–Digital Data and Foley–Sammon Feature Extraction

Rank of feature used	Number of samples correctly assigned to:				Total % accuracy
	Class 1	Class 2	Class 3	Class 4	
1	20	20	20	20	100
2	20	20	20	20	100
3	20	20	20	20	100
4	20	20	20	20	100
5	20	20	20	20	100
6	20	20	20	20	100
7	20	20	20	20	100
8	20	20	20	20	100
9	20	19	19	20	97.5
10	20	19	14	20	81
11	20	16	20	20	95
12	8	12	19	20	74
13	20	13	17	19	86
14	18	16	19	19	90
15	20	19	20	19	97.5
16	20	20	20	13	91
17	20	13	20	8	76
18	20	17	20	8	81
19	20	19	19	13	89
20	20	16	18	10	80

E. Results: All-Digital Approach

After sampling and digitizing, a texture sample is stored in the form of a 24-element vector **X**. Presented below are classification results using each of the feature extraction methods. The same classification experiments, with $k = 9$ in the classifier, are performed as with optical–digital data.

1. Hotelling Trace Transform (Used in Pooled Mode)

Extracted from each datum vector were 24 features. They were ranked in descending order according to their corresponding eigenvalues. Classification runs were performed using each of the first 20 features, one at a time; the results are given in Table 11.5-4. None of these features taken singly yielded recognition accuracy better than 89%.

In addition to single-feature trials, classification runs employing multiple features were made in order to observe dimensionality effects on the classifier accuracy. Classification runs were made with the dimension of the feature space increasing with each successive run. A plot of overall classification

TABLE 11.5-4

Single-Feature Classification Results for Gradient Matrix Data and Hotelling Trace Feature Extractor in Pooled Mode

Rank of feature used	Number of samples correctly assigned to:				Total % accuracy
	Class 1	Class 2	Class 3	Class 4	
1	20	17	14	20	89
2	6	20	15	12	66
3	11	7	9	18	56
4	11	7	10	9	46
5	8	5	6	5	30
6	5	13	6	4	35
7	2	5	12	10	36
8	4	4	6	3	21
9	7	7	6	8	35
10	7	5	2	3	21
11	4	8	4	12	35
12	7	5	5	1	23
13	4	3	9	7	29
14	1	7	5	5	23
15	1	4	9	12	33
16	9	15	3	3	38
17	5	6	9	6	33
18	6	6	6	10	35
19	3	6	9	6	30
20	3	7	6	6	28

accuracy versus number of features used is shown in Fig. 11.5-10. The graph highlights the "curse of dimensionality": We see that performance increases to 99% correct classification when several features are added but declines from that figure when more than 13 features are used. The graph also demonstrates that using several good features jointly is often superior to using only one feature.

2. Hotelling Trace Transform (Used in Pairwise Mode)

Results of classification by a single feature are given in Table 11.5-5. Accuracy rapidly drops off from 100% with the first-ranked feature to an unacceptable 29% with the second-ranked feature. In effect all the features for $n > 1$ have no class discrimination capability.

3. Foley–Sammon Discriminant Vectors

The results of the single-feature classifier runs are given in Table 11.5-6. The first 16 features each furnished a four-class discrimination accuracy of

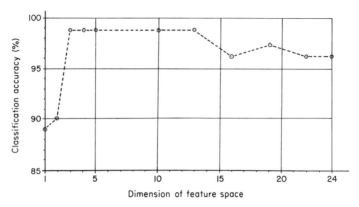

FIG. 11.5-10. Graph of classification accuracy versus feature vector length. Features generated from gradient matrix and Hotelling trace feature extractor in pooled mode. Classifier was KNN routine with $k = 9$. (From O'Toole and Stark [25].)

TABLE 11.5-5

Single-Feature Classification Results for Gradient Matrix Data and Hotelling Trace Feature Extractor in Pairwise Mode

Rank of feature used	Number of samples correctly assigned to:				Total % accuracy
	Class 1	Class 2	Class 3	Class 4	
1	20	20	20	20	100
2	9	4	4	2	24
3	7	8	7	4	33
4	8	7	3	2	25
5	8	7	2	6	29
6	5	1	7	1	18
7	9	5	3	0	21
8	3	4	2	4	16
9	5	10	4	2	26
10	6	7	4	6	29
11	8	2	7	1	23
12	9	12	5	0	33
13	9	8	0	7	30
14	13	6	4	2	31
15	12	2	6	2	28
16	8	5	4	2	24
17	10	4	5	2	26
18	9	3	3	0	19
19	5	8	3	2	23
20	8	14	8	2	40

TABLE 11.5-6

Single-Feature Classification Results for Gradient Matrix Data and Foley–Sammon Feature Extractor

Rank of feature used	Number of samples correctly assigned to:				Total % accuracy
	Class 1	Class 2	Class 3	Class 4	
1	20	20	20	20	100
2	20	20	20	20	100
3	20	20	20	20	100
4	20	20	20	20	100
5	20	20	20	20	100
6	20	20	20	20	100
7	20	20	20	20	100
8	20	20	18	20	97.5
9	20	20	19	20	98.8
10	20	20	18	20	97.5
11	20	20	18	20	97.5
12	20	20	18	20	97.5
13	20	20	18	20	97.5
14	20	20	19	20	98.8
15	20	20	19	20	98.8
16	20	20	18	20	97.5
17	20	14	11	20	81
18	20	19	18	20	96
19	16	20	18	17	89
20	20	20	19	20	98.8

97.5% or better. When top-ranked features were taken together, the classification accuracy remained at 100%. The Foley–Sammon feature extractor displayed exceptional robustness for both optical–digital and all-digital data.

F. Conclusions Drawn from ODC versus All-Digital Pattern Recognition of Texture

From the results given in the previous section we see that all the feature extractors performed adequately for both approaches, i.e., optical–digital and all-digital. They did not, however, perform equally well. In both approaches it is clear that the Foley–Sammon feature extractor was superior for the four textures chosen for this study. The Foley–Sammon feature extractor generated more features that led to successful classification. However, this is not a complete indication of the relative merits of the feature extractors. If we consider experiments employing just the first feature, we notice that in these cases both the pairwise Hotelling trace approach and the Foley–Sammon approach led to 100% recognition accuracy. Does this mean that these two feature extractors generated equally potent first-ranked features? Not necessarily. The merit of a feature extractor is best measured by how well it can cluster within-class samples while separating interclass samples in feature space. The relative separation of the texture classes is not directly apparent in these classification results. In order to compare the feature extractors more appropriately, we should choose four original textures which more closely resemble one another.

More central to the objective of this study is the comparison of performance using the two types of texture measures: the gradient matrix measures and the Fourier measures. *Neither measurement type led to consistently superior results as compared to the other for the textures chosen.* With the Foley–Sammon and Hotelling trace (pairwise mode) feature extractors the classification accuracy was almost identical for both measure types. Only when we observe results using the Hotelling trace algorithm in a pooled mode do we notice any differences. *Here the Fourier spectrum measures slightly outperform the gradient matrix measures.* Perfect classification was accomplished with just the first feature extracted from the Fourier data, while the first feature extracted from gradient matrix data could only yield 89% accuracy. This performance difference is significant and tends to indicate that, for the textures we have used, the Fourier measures are a superior alternative to gradient distribution matrix measures.

11.6 SUMMARY

Our aim in this chapter has been to show that coherent-optical systems, aided by a digital computer, can be used in pattern recognition problems that require more than matched filtering. Two examples were given: the two-class pattern recognition of CWP and the four-class pattern recognition of texture. In the latter we also attempted to compare the performance of optical Fourier transform techniques with all-digital gradient matrix measures. The results of the study indicate that the optical Fourier transform techniques performed at least as well as the sophisticated all-digital techniques.

In addition to performance, the physical limitations of the two texture measurement approaches should be considered. The digital scheme has an advantage here of flexibility, since its data generation operations are almost all performed by software. The system operator thereby can easily control the parameters which are involved in the process. The optical–digital approach on the other hand needs an optical bench and a coherent light source in addition to software control. Physical alignment and attention to optical cleanliness are thereby required. On the other hand, because of its parallel processing capabilities, the optical–digital approach should be carefully considered in situations requiring high speed processing of large amounts of texture data.

REFERENCES

[1] H. Stark and D. Lee (1976). *IEEE Trans. Syst., Man Cybern.* **SMC-6,** 788–793.
[2] E. L. Hall, R. P. Kruger, and A. F. Turner (1974). *Opt. Eng.* **13,** 250–257.
[3] H. Tamura, S. Mori, and T. Yamawaki (1977). *Proc. IEEE Conf. Pattern Recognition Image Process.* IEEE Cat. No. 77 CH1208-9C, pp. 289–298.
[4] J. Weszka, C. Dyer, and A. Rosenfeld (1976). *IEEE Trans. Syst., Man Cybern.* **SMC-6,** 269–285.
[5] H. Stark (1975). *IEEE Trans. Comput.* **C-24,** 340–347.
[6] D. Lee (1975). Ph.D. Thesis, Dept. Eng. Appl. Sci., Yale Univ., New Haven, Connecticut.
[7] G. Lendaris and G. Stanley (1970). *Proc. IEEE* **58,** 198–216.
[8] J. Mantock, A. A. Sawchuk, and T. C. Strand (1980). *Opt. Eng.* **19,** 180–185.
[9] H. J. Caulfield, R. Haimes, and D. Casasent (1980). *Opt. Eng.* **19,** 152–156.
[10] H. Stark and F. B. Tuteur (1979). "Modern Electrical Communications: Theory and Systems." Prentice-Hall, Englewood Cliffs, New Jersey.
[11] W. B. Meisel (1972). "Computer-Oriented Approaches to Pattern Recognition." p. 12. Academic Press, New York.
[12] C. Chen (1973). "Statistical Pattern Recognition," p. 69. Hayden Book Co., Rochelle Park, New Jersey.
[13] Y. T. Chien and K. S. Fu (1967). *IEEE Trans. Inf. Theory* **IT-13,** 518–520.
[14] K. Fukunaga and W. L. G. Koontz (1970). *IEEE Trans. Comput.* **C-19,** 311–318.
[15] D. H. Foley and J. W. Sammon, Jr. (1975). *IEEE Trans. Comput.* **C-24,** 281–289.
[16] K. Fukunaga (1972). "Introduction to Statistical Pattern Recognition." Academic Press, New York.
[17] K. S. Fu (1976). *IEEE Trans. Geosci. Electron.* **GE-15,** 10–18.
[18] R. P. Kruger, W. B. Thompson, and A. F. Turner (1974). *IEEE Trans. Syst., Man Cybern.* **SMC-4,** 40–49.
[19] E. L. Hall, W. O. Crawford, and F. E. Roberts (1975). *IEEE Trans. Biomed. Eng.* **BME-22,** 518–527.
[20] J. R. Jagoe and K. A. Paton (1976). *IEEE Trans. Comput.* **C-25,** 95–97.
[21] R. N. Sutton and E. L. Hall (1972). *IEEE Trans. Comput.* **C-21,** 667–676.
[22] C. Hornung (1976). "Background Patterns, Textures and Tints." Dover, New York.
[23] R. Haralick, K. Shanmugan, and I. Distein (1973). *IEEE Trans. Syst., Man Cybern.* **SMC-3,** 610–621.
[24] J. Tou and Y. Chang (1977). *Proc. IEEE Conf. Pattern Recognition Image Process.* IEEE Cat. No. 77 CH1208-9C, pp. 392–397.
[25] R. K. O'Toole and H. Stark (1980). *Appl. Opt.* **19,** 2496–2506.

Chapter 12

Incoherent-Optical Processing

H. BARTELT

PHYSIKALISCHES INSTITUT
  DER UNIVERSITÄT ERLANGEN-NÜRNBERG
ERLANGEN, FEDERAL REPUBLIC OF GERMANY

S. K. CASE

DEPARTMENT OF ELECTRICAL ENGINEERING
UNIVERSITY OF MINNESOTA
MINNEAPOLIS, MINNESOTA

R. HAUCK

FB7
UNIVERSITÄT ESSEN GHS
ESSEN, FEDERAL REPUBLIC OF GERMANY

12.1 INTRODUCTION

The previous chapters in this book have considered the use of optical Fourier transforms in spatially and temporally coherent-optical systems. In this chapter, we wish to consider optical systems in which we relax the coherence requirements in either the spatial or temporal domain and use the resulting gain in degrees of freedom to increase the utility of the system.

We start our discussion in Section 12.2 with the analysis of a generalized optical system containing an arbitrary source and calculate the optical properties of this system. For comparison with previous chapters, we then restrict our source spatially and temporally to show that our analysis reduces to that of a conventional coherent-optical processor. For this processor, we introduce the concepts of optical channels, information content, noise, redundancy, and multiplexing.

499

In Section 12.3, we then consider an optical system which is temporally coherent (quasi-monochromatic) but which is spatially incoherent. We show that such a system performs convolutions that are linear in intensity. We compare the operation of this system with that of a conventional coherent processor. In Section 12.4, we describe applications of this type of processor.

In Section 12.5, we consider an optical system which uses a polychromatic point source so that it is spatially coherent but temporally incoherent. We compare this system with a conventional optical processor. In Section 12.6, we describe several applications of this type of system.

12.2 GENERAL THEORETICAL ANALYSIS

A. Optical System Analysis

We consider a general, telecentric optical system as shown in Fig. 12.2-1. The system contains three lenses, each of focal length f, which are separated by distances $2f$. The four planes of interest are the source plane, S; the object plane, O; the pupil plane, P; and the image plane, I. The solid lines in the figure indicate how light from a point on the source would propagate through the optical system, while the dashed lines indicate how light from a point on the object propagates.

If we consider a general polychromatic source, one vector component of the electric field at one point on the source can be represented by the Fourier superposition

$$S(t) = \int_0^\infty s(v)e^{j[\phi(v) - 2\pi vt]}\, dv, \tag{12.2-1}$$

where $s(v)$ is the spectral amplitude at the temporal frequency v and $\phi(v)$ is the relative phase for this spectral component [1]. For an extended source, this becomes

$$S(\mathbf{r}_s, t) = \int_0^\infty s(\mathbf{r}_s, v)e^{j\phi(\mathbf{r}_s, v)}e^{-j2\pi vt}\, dv, \tag{12.2-2}$$

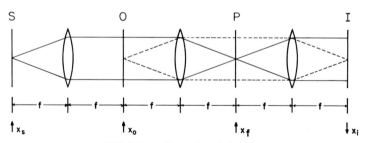

FIG. 12.2-1. General optical system.

where S is a function depending on source plane coordinates $\mathbf{r}_s = x_s\mathbf{i}_x + y_s\mathbf{i}_y$ and time t. We are interested in the temporal behavior of the signal because it will later be important for coherence studies or more specifically, for calculating the effect of eventually using a time-integrating detector. However, calculating the propagation of such complex time-varying signals is difficult. Hence, rather than immediately considering the signal $S(\mathbf{r}_s, t)$, which is made up of the entire spectrum of temporal frequencies, it is convenient to analyze first the propagation of one temporal frequency component (i.e., monochromatic light) through the optical system, since this propagation is well known, and later to perform the Fourier composition as in Eq. (12.2-2) to arrive at the final output signal. This approach is valid, since all operations that we will perform on the signal are linear, hence we can interchange the order of the spatial and temporal integrations.

For one temporal frequency, our source amplitude is represented by part of the integrand from Eq. (12.2-2):

$$U_s(\mathbf{r}_s, v) = s(\mathbf{r}_s, v)e^{j\phi(\mathbf{r}_s, v)}. \tag{12.2-3}$$

From this source component, the field amplitude in plane O, just before the object, is given by

$$U_o(\mathbf{r}_o, v) = \frac{1}{j\lambda f} \int_{-\infty}^{\infty} s(\mathbf{r}_s, v)e^{j\phi(\mathbf{r}_s, v)}e^{(-j2\pi/\lambda f)(\mathbf{r}_s \cdot \mathbf{r}_o)} \, d\mathbf{r}_s, \tag{12.2-4}$$

where we have integrated over all points in the source plane. We recognize the above equation as the spatial Fourier transform of the source distribution. We have written U_0 as a function of v to remind ourselves that this is the object illumination at a particular temporal frequency and have also used the associated wavelength $\lambda = c/v$.

We assume an object transparency with amplitude transmittance $O(\mathbf{r}_o, v)$ is located in the object plane. Here, the v term describes a frequency (wavelength)-dependent transmittance as may later be desirable for color film inputs. We also assume that the object does not change with time. The signal just after the transparency is the product

$$U_o(\mathbf{r}_o, v)O(\mathbf{r}_o, v). \tag{12.2-5}$$

The signal in the pupil (Fourier) plane, just before the pupil filter is then the spatial Fourier transform of Eq. (12.2-5):

$$U_f\left(\frac{\mathbf{r}_f}{\lambda f}, v\right) = \frac{1}{j\lambda f} \int_{-\infty}^{\infty} U_o(\mathbf{r}_o, v)O(\mathbf{r}_o, v)e^{(-j2\pi/\lambda f)(\mathbf{r}_o \cdot \mathbf{r}_f)} \, d\mathbf{r}_o. \tag{12.2-6}$$

The signal just after the pupil filter which has amplitude transmittance $\tilde{p}(\mathbf{r}_f/\lambda f, v)$ is then the product

$$U_f(\mathbf{r}_f/\lambda f, v)\tilde{p}(\mathbf{r}_f/\lambda f, v), \tag{12.2-7}$$

where we have again included frequency (wavelength)-dependent filter transmittance and the function $\tilde{p}$ has been written in terms of reduced coordinates.

In the image plane, the signal is then the spatial Fourier transform of the function in Eq. (12.2-7):

$$U_i(\mathbf{r}_i, v) = \frac{1}{j\lambda f} \int_{-\infty}^{\infty} U_f\left(\frac{\mathbf{r}_f}{\lambda f}, v\right) \tilde{p}\left(\frac{\mathbf{r}_f}{\lambda f}, v\right) e^{(j2\pi/\lambda f)(\mathbf{r}_f \cdot \mathbf{r}_i)} \, d\mathbf{r}_f. \qquad (12.2\text{-}8)$$

Using the convolution theorem with Eqs. (12.2-6) and (12.2-8), we may rewrite Eq. (12.2-8) as

$$U_i(\mathbf{r}_i, v) = \int_{-\infty}^{+\infty} U_o(\mathbf{r}_o, v)O(\mathbf{r}_o, v)P(\mathbf{r}_i - \mathbf{r}_o, v) \, d\mathbf{r}_o. \qquad (12.2\text{-}9)$$

We recognize this as the convolution of our object signal $U_o O$ with the Fourier transform P of our pupil function.

We have calculated our output amplitude for the temporal frequency v (i.e., monochromatic wavelength $\lambda = c/v$). The general temporal signal in the output plane is found by performing the Fourier superposition as given by Eq. (12.2-2):

$$U_i'(\mathbf{r}_i, t) = \int_0^{\infty} U_i(\mathbf{r}_i, v)e^{-j2\pi vt} \, dv \qquad (12.2\text{-}10)$$

$$= \int_0^{\infty} \int_{-\infty}^{\infty} U_o(\mathbf{r}_o, v)O(\mathbf{r}_o, v)P(\mathbf{r}_i - \mathbf{r}_o, v)e^{-j2\pi vt} \, d\mathbf{r}_o \, dv, \qquad (12.2\text{-}11)$$

where a prime is used to remind us when U_i is a function of time.

Finally, we consider the action of this signal on our image plane detector. Three effects we wish to include are that the detector responds to intensities, the detector integrates over a finite time period T, and the detector has a spectral response (i.e., λ or color sensitivity) given by the real variable $R(v)$. When the detector spectral response is included in Eq. (12.2-10), the exposure for a given small area around $\mathbf{r}_i$ on our detector is

$$E(\mathbf{r}_i) = \int_0^T \left| \int_0^{\infty} U_i(\mathbf{r}_i, v)R(v)e^{-2j\pi vt} \, dv \right|^2 dt. \qquad (12.2\text{-}12)$$

Inserting Eq. (12.2-9) into Eq. (12.2-12),

$$E(\mathbf{r}_i) = \int_0^T \int_0^{\infty} \int_0^{\infty} \int_{-\infty}^{\infty} \int_{-\infty}^{\infty} U_o(\mathbf{r}_o, v)O(\mathbf{r}_o, v)P(\mathbf{r}_i - \mathbf{r}_o, v)R(v)$$

$$\cdot e^{-j2\pi vt}U^*(\mathbf{r}_o', v')O^*(\mathbf{r}_o', v')P^*(\mathbf{r}_i - \mathbf{r}_o', v')R(v')$$

$$\cdot e^{j2\pi v't} \, d\mathbf{r}_o \, d\mathbf{r}_o' \, dv \, dv' \, dt, \qquad (12.2\text{-}13)$$

where we must integrate over three temporal and four spatial variables and we have used the fact that detectors are not phase-sensitive (i.e., R is real).

B. Quasi-Monochromatic Light Sources

Up to this point, our analysis has been completely general, since we have made no restrictive assumptions about our source or detector. A great deal of optical pattern recognition, however, uses quasi-monochromatic light sources, so that attention should be paid to them here. A quasi-monochromatic source produces a very narrow band of temporal frequencies such that $\Delta v/v \ll 1$. This assumption allows a number of simplifications. Since the transmittance of objects and pupil plane filters as well as the detector response usually vary relatively slowly as a function of temporal frequency (or wavelength), they can be considered constant over the frequency range Δv. Thus Eq. (12.2-13) can be rewritten

$$E(\mathbf{r}_i) = R^2 \int_{-\infty}^{\infty} \int_{-\infty}^{\infty} \int_0^T \int_0^{\infty} U_o(\mathbf{r}_o, v) e^{-j2\pi v t} \, dv \int_0^{\infty} U_o^*(\mathbf{r}_o', v') e^{+j2\pi v' t} \, dv'$$
$$\cdot O(\mathbf{r}_o) P(\mathbf{r}_i - \mathbf{r}_o) O^*(\mathbf{r}_o') P^*(\mathbf{r}_i - \mathbf{r}_o') \, d\mathbf{r}_o \, d\mathbf{r}_o' \, dt. \qquad (12.2\text{-}14)$$

The two integrals over v and v' are the temporal Fourier transforms of the source spectral distributions. Thus, performing these integrations, we have

$$E(\mathbf{r}_i) = R^2 \int_{-\infty}^{\infty} \int_{-\infty}^{\infty} \int_0^T U_o'(\mathbf{r}_o, t) U_o'^*(\mathbf{r}_o', t) \, dt$$
$$\cdot O(\mathbf{r}_o) P(\mathbf{r}_i - \mathbf{r}_o) O^*(\mathbf{r}_o') P^*(\mathbf{r}_i - \mathbf{r}_o') \, d\mathbf{r}_o \, d\mathbf{r}_o'. \qquad (12.2\text{-}15)$$

C. Coherence

The time-dependent part of the above integral is called the coherence function Γ [1], where

$$\Gamma(\mathbf{r}_o, \mathbf{r}_o') = \frac{1}{T} \int_0^T U_o'(\mathbf{r}_o, t) U_o'^*(\mathbf{r}_o', t) \, dt. \qquad (12.2\text{-}16)$$

The Γ function measures the spatial coherence of the light that illuminates the object. (Since the light is quasi-monochromatic, it already exhibits good temporal coherence.) Equation (12.2-16) can be seen to be the time average of the product of the illumination amplitudes at two different points in the object plane. Basically, if the relative phase between the two different object points $\mathbf{r}_o$ and $\mathbf{r}_o'$ is relatively constant during the detector integration time T, then the integral can have a large value and the light is described as spatially coherent. If the relative phase between the two points varies wildly during the integration time, the integral will have a value of approximately zero except when $\mathbf{r}_0 = \mathbf{r}_0'$ and the light is described as spatially incoherent. Situations between these two extremes are called partially coherent.

The coherence function is described in Eq. (12.2-16) as a measurement in the object plane. Our next task is to calculate the coherence function in terms of the structure of our source so that we can design sources for spatially coherent or incoherent processing. The signal $U_0'(\mathbf{r}_0, t)$ is given by the temporal Fourier transform of Eq. (12.2-4):

$$U_0'(\mathbf{r}_0, t) = \frac{1}{j\lambda f} \int_{-\infty}^{\infty} \int_0^{\infty} s(\mathbf{r}_s, v) e^{i\phi(\mathbf{r}_s, v)} e^{(-j2\pi/\lambda f)(\mathbf{r}_s \cdot \mathbf{r}_0)} e^{-j2\pi v t} \, d\mathbf{r}_s \, dv \quad (12.2\text{-}17)$$

$$= \frac{1}{j\lambda f} \int_{-\infty}^{\infty} S(\mathbf{r}_s, t) e^{(-j2\pi/\lambda f)(\mathbf{r}_s, \mathbf{r}_0)} \, d\mathbf{r}_s \quad (12.2\text{-}18)$$

by using Eq. (12.2-2). For a quasi-monochromatic source of mean frequency $\bar{v}$, the spectral amplitude $s(\mathbf{r}_s, v)$ is sharply peaked around the value $v = \bar{v}$ and is essentially zero elsewhere. Thus we expand the signal $S(\mathbf{r}_s, t)$ about the frequency $\bar{v}$ to have a form [2]

$$S(\mathbf{r}_s, t) = s'(\mathbf{r}_s, t) e^{j\phi'(\mathbf{r}_s, t)} e^{j2\pi \bar{v} t}, \quad (12.2\text{-}19)$$

where the high frequency oscillation is contained in the $e^{j2\pi\bar{v}t}$ term and the s' and ϕ' terms are the slowly varying envelope and phase of the wave train of mean frequency $\bar{v}$. We should note that the slow time variation in s' and ϕ' is present because of the finite band of frequencies present in our signal. For a strictly monochromatic signal (which is not physically realizable†), this time dependence would disappear.

Inserting Eq. (12.2-19) into Eq. (12.2-18) and then placing this in Eq. (12.2-16) enables us to write

$$\Gamma(\mathbf{r}_0, \mathbf{r}_0') = \frac{1}{T} \int_0^T \frac{1}{j\lambda f} \int_{-\infty}^{\infty} s'(\mathbf{r}_s, t) e^{j\phi'(\mathbf{r}_s, t)} e^{j2\pi\bar{v}t} e^{(-j2\pi/\lambda f)(\mathbf{r}_s \cdot \mathbf{r}_0)} \, d\mathbf{r}_s$$

$$\cdot \frac{1}{-j\lambda f} \int_{-\infty}^{\infty} s'(\mathbf{r}_s'', t) e^{-j\phi'(\mathbf{r}_s'', t)} e^{-j2\pi\bar{v}t} e^{(+j2\pi/\lambda f)(\mathbf{r}_s'' \cdot \mathbf{r}_0')} \, d\mathbf{r}_s'' \, dt$$

$$(12.2\text{-}20)$$

$$= \frac{1}{\lambda^2 f^2} \int_{-\infty}^{\infty} \int_{-\infty}^{\infty} \frac{1}{T} \int_0^T s'(\mathbf{r}_s, t) s'(\mathbf{r}_s'', t) e^{j[\phi'(\mathbf{r}_s, t) - \phi'(\mathbf{r}_s'', t)]} \, dt$$

$$\cdot e^{(-j2\pi/\lambda f)[(\mathbf{r}_s \cdot \mathbf{r}_0) - (\mathbf{r}_s'' \cdot \mathbf{r}_0')]} \, d\mathbf{r}_s \, d\mathbf{r}_s'', \quad (12.2\text{-}21)$$

where we have used the fact that s' and ϕ' are real functions. The different source elements at $\mathbf{r}_s$ and $\mathbf{r}_s''$ in our general extended source consist of different

† For one thing, the source would have to have been on from $t = -\infty$ to $t = +\infty$.

atomic radiators. Therefore the relative phase difference $\phi'(\mathbf{r}_s, t) - \phi'(\mathbf{r}''_s, t)$ between the two points will vary randomly during our detector integration time. The various source points can therefore be considered to be mutually incoherent, so that the integration over time in Eq. (12.2-21) has value zero unless $\mathbf{r}_s = \mathbf{r}''_s$. Therefore

$$\frac{1}{T} \int_0^T s'(\mathbf{r}_s, t)s'(\mathbf{r}''_s, t)e^{j[\phi'(\mathbf{r}_s, t) - \phi'(\mathbf{r}''_s, t)]} \, dt = \langle s'(\mathbf{r}_s, t)s'(\mathbf{r}''_s, t) \rangle \delta(\mathbf{r}_s - \mathbf{r}''_s), \quad (12.2\text{-}22)$$

where the brackets indicate time averaging and we have used the Dirac delta function [1]. Inserting Eq. (12.2-22) into Eq. (12.2-21) and integrating over $d\mathbf{r}''_s$ yields

$$\Gamma(\mathbf{r}_0, \mathbf{r}'_0) = \frac{1}{\lambda^2 f^2} \int_{-\infty}^{\infty} \langle s'(\mathbf{r}_s, t)^2 \rangle e^{(-j2\pi/\lambda f)[\mathbf{r}_s \cdot (\mathbf{r}_0 - \mathbf{r}'_0)]} \, d\mathbf{r}_s. \quad (12.2\text{-}23)$$

We see that the spatial coherence function for two points $\mathbf{r}_0$ and $\mathbf{r}'_0$ in the object plane is the spatial Fourier transform of the time-averaged intensity distribution in the source plane (van Cittert–Zernike theorem [1]). Inserting (12.2-23) into Eq. (12.2-15) results in

$$E(\mathbf{r}_i) = R^2 T \int_{-\infty}^{\infty} \int_{-\infty}^{\infty} \Gamma(\mathbf{r}_0, \mathbf{r}'_0) O(\mathbf{r}_0) P(\mathbf{r}_i - \mathbf{r}_0)$$
$$\cdot O^*(\mathbf{r}'_0) P^*(\mathbf{r}_i - \mathbf{r}'_0) \, d\mathbf{r}_0 \, d\mathbf{r}'_0. \quad (12.2\text{-}24)$$

D. Coherent-Optical Processing

Most of the previous chapters in this book have dealt with spatially and temporally coherent-optical processing. The light source for this type of processing is most often obtained by focusing the quasi-monochromatic light beam from a laser onto a pinhole of small diameter. For this situation, it is convenient and sufficiently accurate for our analysis to assume the light originates from a single point source:

$$\langle [s'(\mathbf{r}_s, t)]^2 \rangle = I_s \delta(\mathbf{r}_s), \quad (12.2\text{-}25)$$

where the source is assumed to be located on the optical axis and I_s is proportional to the source irradiance. In this case, our coherence function becomes

$$\Gamma(\mathbf{r}_0, \mathbf{r}'_0) = I_s/\lambda^2 f^2. \quad (12.2\text{-}26)$$

Since this function is independent of position in the object plane, the system is spatially coherent. Inserting Eq. (12.2-26) into Eq. (12.2-24) we have

$$E(\mathbf{r}_i) = \frac{I_s R^2 T}{\lambda^2 f^2} \int_{-\infty}^{\infty} \int_{-\infty}^{\infty} O(\mathbf{r}_o) P(\mathbf{r}_i - \mathbf{r}_o) O^*(\mathbf{r}_o') P^*(\mathbf{r}_i - \mathbf{r}_o') \, d\mathbf{r}_o \, d\mathbf{r}_o'$$

$$= \frac{I_s R^2 T}{\lambda^2 f^2} \int_{-\infty}^{\infty} O(\mathbf{r}_o) P(\mathbf{r}_i - \mathbf{r}_o) \, d\mathbf{r}_o \int_{-\infty}^{\infty} O^*(\mathbf{r}_o') P^*(\mathbf{r}_i - \mathbf{r}_o') \, d\mathbf{r}_o'$$

$$= \frac{I_s R^2 T}{\lambda^2 f^2} \left| \int_{-\infty}^{\infty} O(\mathbf{r}_o) P(\mathbf{r}_i - \mathbf{r}_o) \, d\mathbf{r}_o \right|^2 = \frac{I_s R^2 T}{\lambda^2 f^2} |O(\mathbf{r}_o) * P(\mathbf{r}_o)|^2, \quad (12.2\text{-}27)$$

where the output is written as a convolution.

Since this output depends on amplitudes, the optical system must be able to transfer and process complex information introduced by object and pupil transparencies. Figure 12.2-2 illustrates this point. The single-point source in Eq. (12.2-25) produces a plane wave of amplitude $\sqrt{I_s}/\lambda f$ which propagates along the optical axis. The amplitude and phase of this incident wave are altered by the object transparency. Because the relative phase of the wave that illuminates any two object points does not change with time, the relative phase between the light transmitted by these two object points (and any other object points) is preserved. Thus the object signal can be complex and, as shown in Fig. 12.2-2, the Fourier transform of the complex object signal can be displayed in the pupil (Fourier) plane. This, of course, makes it convenient to alter the object spatial frequency spectrum by inserting absorbing and phase-shifting filters in the Fourier plane, as has been described in previous chapters.

E. Optical Channels, Information Content, and Noise

Figure 12.2-2 also allows us to introduce the concept of optical channels. As can be seen from the figure, the entire object is illuminated by only one plane wave, so that we might say that the object information is carried on one optical channel.

The amount of spatial information carried by this channel is equal to the number of picture elements N in the object (pixels). This can be found by dividing the object area by the area of the smallest picture element. Alter-

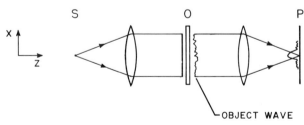

FIG. 12.2-2. Optical system with single-point source.

nately, the spatial information content, often called the space–bandwidth product, can be found by

$$N = \text{SBP} = A_o A_f / \lambda^2 f^2, \qquad (12.2\text{-}28)$$

where A_o is the object area and A_f is the area occupied by the object spectrum in the Fourier plane. As an example, commonly used 35-mm transparency film with a resolution of 25 μm has a SBP $> 10^6$.

The amount of information transmitted or processed by any communication channel is clearly maximized if the information is transmitted only once, as is the case for the single-channel optical system in Fig. 12.2-2. However, such a system is also the most susceptible to noise. Noise in the optical system can result from imperfect lenses, or very commonly from dust or other opaque particles that may collect on the input transparency, the lenses, or the pupil plane filters. As an example, one can easily see that an opaque spot on the surface of the Fourier transform lens in Fig. 12.2-2 could block the light emerging from a certain area on the object. Since all information is transferred through the optical system only once, the information from this object area would be distorted or lost. In addition, one would see the characteristic "bullseye" diffraction pattern of the opaque noise particle at the system output. The exact effect of noise on the system output signal depends on the exact location of the noise source and on the statistical nature of the noise signal. This subject is too lengthy to treat here, and readers are referred to other references on the subject [3, 4].

F. Redundancy and Multiplexing

To minimize noise, optical elements should be kept clean and free of defects. Additionally, information can be sent through several optical channels to be sure that the desired information will arrive via at least one of the channels (i.e., introduce redundancy into the system).

In Fig. 12.2-3, we show a two-channel optical system in which there are two equal-irradiance, mutually coherent point sources located at $\mathbf{r}_{s1}$ and $\mathbf{r}_{s2}$ in the source plane. Our time-averaged source amplitude can be written

$$S' = \sqrt{I_s}[\delta(\mathbf{r}_s - \mathbf{r}_{s1}) + \delta(\mathbf{r}_s - \mathbf{r}_{s2})e^{-j\phi}], \qquad (12.2\text{-}29)$$

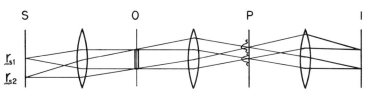

FIG. 12.2-3. Two-channel optical system. (A position vector is indicated by a wavy underline in figure and by bold type in text.)

where ϕ is the relative phase difference between the two point sources. The object illumination then consists of two coherent plane waves described by

$$U_o(\mathbf{r}_o) = \frac{\sqrt{I_s}}{\lambda f}(e^{(-j2\pi/\lambda f)(\mathbf{r}_{s1} \cdot \mathbf{r}_o)} + e^{-j[(2\pi/\lambda f)(\mathbf{r}_{s2} \cdot \mathbf{r}_o) + \phi]}). \quad (12.2\text{-}30)$$

Since the relative phase between the two source points does not change with time, the object illumination remains coherent and the system is still capable of processing complex data. Of course the object is illuminated with a sinusoidal interference pattern of spatial frequency $v = |\mathbf{r}_{s1} - \mathbf{r}_{s2}|/\lambda f$. In order to avoid any distortions of the output by the sinusoidal fringe pattern, the frequency v must be higher than the cutoff frequency of the detector (e.g., human eye, film) [5].

The advantage of the system is that, if an opaque spot is again present on the lens surface, light from a given point on the object may be blocked in one of the channels but transmitted through the other. The effect of the spot will now be present at two image points, but its effect will be reduced at each of them compared to the single-channel effect. Thus the noise will be less objectionable [6].

We also note that the object spectrum appears twice in the pupil plane. The two object spectra must not overlap if we want to place arbitrary filter masks in the pupil plane. Again, this is ensured if the object pixel size fulfills

$$\Delta x \geq \lambda f/|\mathbf{r}_{s1} - \mathbf{r}_{s2}|. \quad (12.2\text{-}31)$$

Placing a pair of filters in the pupil plane to intercept each of the pupil spectra in Fig. 12.2-3 results in noise reduction if the filters are identical, or in multiplexing capabilities if two different filters are used.

The disadvantage of the system is that twice the area is occupied in the pupil plane. Thus, in accordance with Eq. (12.2-28), the SBP of our transmission system must be doubled. If the SBP of the optical system is limited (because of finite lens size and resolution), this often means that we must reduce our object size or resolution.

In the next section, we consider the effect of making our multiple-point sources mutually incoherent so that the sinusoidal modulation in object illumination disappears [7]. Extending this principle to an optical system with N point sources results in N-channel redundancy for noise reduction.

12.3 SPATIALLY INCOHERENT PROCESSING

In this section, we will consider spatially incoherent-optical processing and will require a light source that is spatially incoherent, but quasi-monochromatic, so that the spectral bandwidth Δv is again small compared to the mean frequency $\bar{v}$:

$$\Delta v / \bar{v} \ll 1. \tag{12.3-1}$$

Spatially incoherent light cannot be totally monochromatic because the relative phase at the light emitted by two different points of the source must change in time to yield the spatial incoherence.

With the assumption of small spectral bandwidth, we obtained Eq. (12.2-24). For spatially incoherent illumination,

$$\Gamma(\mathbf{r}_o, \mathbf{r}_o') = I_o(\mathbf{r}_o) \, \delta(\mathbf{r}_o - \mathbf{r}_o'), \tag{12.3-2}$$

where I_o is proportional to the irradiance in the object plane. The output of the optical system is given by

$$
\begin{aligned}
E(\mathbf{r}_i) &= R^2 T \int_{-\infty}^{\infty} I_o(\mathbf{r}_o) O(\mathbf{r}_o) O^*(\mathbf{r}_o) P(\mathbf{r}_i - \mathbf{r}_o) P^*(\mathbf{r}_i - \mathbf{r}_o) \, d\mathbf{r}_o \\
&= I_o R^2 T |O(\mathbf{r}_o)|^2 * |P(\mathbf{r}_o)|^2,
\end{aligned}
\tag{12.3-3}
$$

assuming that the object is uniformly illuminated $[I_o(\mathbf{r}_o) = I_o]$.

The system performs a convolution of the intensity transmittance of the object and the point spread function. This is in contrast to the spatially coherent case where the convolution involved complex amplitudes. We also note that, in the general case of partially coherent illumination, there is no linear relation in terms of either amplitude or intensity between input and output.

A. Spatially Incoherent Light Sources

1. *Extended Spatially Incoherent Light Source*

In Section 12.2, we saw that the coherence function in the object plane was given by the spatial Fourier transform of the source irradiance when the source was spatially incoherent. If we assume a source with infinite extension and homogeneous irradiance, then the coherence function becomes a delta function. In this situation, it is instructive to derive the result of Eq. (12.3-3) in another manner. Consider first one point of the source and calculate the output distribution. Then add all distributions due to the multiple source points, as was done at the end of Section 12.2, except that our source points are now mutually incoherent. One source point at position $\mathbf{r}_s$ and with amplitude

$$s'(\mathbf{r}_s) \tag{12.3-4}$$

illuminates the object with a tilted plane wave:

$$[s'(\mathbf{r_s})/\lambda f]e^{(-j2\pi/\lambda f)(\mathbf{r_s}\cdot\mathbf{r_o})}. \qquad (12.3\text{-}5)$$

According to the coherent theory, the output field amplitude is given by

$$\frac{1}{\lambda f}\int_{-\infty}^{\infty} s'(\mathbf{r_s})O(\mathbf{r_o})e^{(-j2\pi/\lambda f)(\mathbf{r_s}\cdot\mathbf{r_o})}P(\mathbf{r_i} - \mathbf{r_o})\,d\mathbf{r_o}. \qquad (12.3\text{-}6)$$

The superposition in the output plane due to all source points is linear in intensity, because the light of separate source points does not lead to stationary interference patterns. We obtain the output by adding the intensities

$$E(\mathbf{r_i}) = \frac{R^2 T}{\lambda^2 f^2}\int_{-\infty}^{\infty}\int_{-\infty}^{\infty}\int_{-\infty}^{\infty} s'(\mathbf{r_s})s'^*(\mathbf{r_s})O(\mathbf{r_o})O^*(\mathbf{r_o'})e^{(-j2\pi/\lambda f)\mathbf{r_s}(\mathbf{r_o}-\mathbf{r_o'})}$$

$$\cdot P(\mathbf{r_i} - \mathbf{r_o})P^*(\mathbf{r_i} - \mathbf{r_o'})\,d\mathbf{r_o}\,d\mathbf{r_o'}\,d\mathbf{r_s}. \qquad (12.3\text{-}7)$$

The integral over $d\mathbf{r_s}$ yields $\delta(\mathbf{r_o} - \mathbf{r_o'})$ because of the exponential phase factor. Integration over $d\mathbf{r_o'}$ then leaves pairs of identical variables $O(\mathbf{r_o})$ and $P(\mathbf{r_i} - \mathbf{r_o})$ with the addtional assumption of uniform source irradiance $I_s = |s'|^2$, and Eq. (12.3-7) reduces to

$$E(\mathbf{r_i}) = \frac{I_s R^2 T}{\lambda^2 f^2}\int_{-\infty}^{\infty} |O(\mathbf{r_o})|^2 |P(\mathbf{r_i} - \mathbf{r_o})|^2\,d\mathbf{r_o}, \qquad (12.3\text{-}8)$$

which again is a convolution of intensities. To obtain the above result, we assumed that our source was infinite in extent. Such sources are prohibitively expensive (not to mention the storage problem). In realistic systems, the source can be a filtered incandescent source and must have an extension

$$\Delta S \geq \widetilde{\Delta O} + \widetilde{\Delta P}, \qquad (12.3\text{-}9)$$

where $\widetilde{\Delta O}$ is the width of the object spectrum and $\widetilde{\Delta P}$ is the width of the pupil as shown in Fig. 12.3-1. This source width ensures that all points in the object spectrum will illuminate all points in the pupil function.

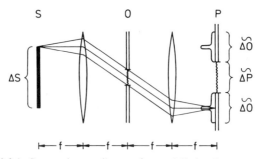

FIG. 12.3-1. Source size requirement for spatially incoherent processing.

2. Pupil Scanning

For pupil scanning, the object is illuminated with a plane wave whose incidence angle is scanned in time (e.g., by deflection of an expanded laser beam) [8, 9]. This corresponds to an extended light source in which one point after another is bright. The theory discussed in Section 12.3.A,1 applies, and the system is linear in intensity since sequentially illuminated light source points cannot interfere. The integration time of the detector must be sufficiently long that the entire pupil is scanned by the illuminating beams.

3. Spatially Incoherent Light Source in the Object Plane

There are several possibilities for this type of illumination. One is to place a filtered incandescent source in (or near) the object plane. Another useful situation is to have a self-luminous (e.g., CRT) or diffusely reflecting object [10–13].

To see why this setup produces correlations, we consider one object point with amplitude $O(\mathbf{r_o})$. This object point will produce a tilted plane wave in the Fourier plane, so that behind our pupil we will have the amplitude

$$O(\mathbf{r_o})e^{(-j2\pi/\lambda f)(\mathbf{r_o} \cdot \mathbf{r_f})}\tilde{p}(\mathbf{r_f}/\lambda f). \qquad (12.3\text{-}10)$$

In the output plane, we will then have

$$O(\mathbf{r_o})P(\mathbf{r_i} - \mathbf{r_o}). \qquad (12.3\text{-}11)$$

The superposition in the output plane is again linear in intensity, and our image irradiance can be found by squaring Eq. (12.3-11) and integrating over our source (object).

$$\int_{-\infty}^{\infty} |O(\mathbf{r_o})P(\mathbf{r_i} - \mathbf{r_o})|^2 \, d\mathbf{r_o} = \int_{-\infty}^{\infty} |O(\mathbf{r_o})|^2 |P(\mathbf{r_i} - \mathbf{r_o})|^2 \, d\mathbf{r_o}. \qquad (12.3\text{-}12)$$

We see that an intensity correlation in incoherent light is again obtained.

4. Object Scanning

In this case, one point after another in the object plane will be illuminated with a point source (e.g., a deflected, focused laser beam). The illumination cone must be sufficiently large (the illumination point must be sufficiently small) to illuminate a region of at least size $\Delta S = \widetilde{\Delta O} + \widetilde{\Delta P}$ in the pupil plane (see Fig. 12.3-1).

The image formation is the same as in Section 12.3,A,3, since the light of different object points cannot interact. The output convolution is linear in intensity.

5. Time-Varying Phase Function

Here we consider an illumination containing a time-varying phase function in the object plane. It is usually produced by illuminating a moving object plane diffuser [14]. For one object point as in Section 12.3,A,3, our object point amplitude is

$$e^{j\phi(\mathbf{r}_o,t)}O(\mathbf{r}_o),\tag{12.3-13}$$

which leads to an output plane amplitude

$$\int_{-\infty}^{\infty} e^{j\phi(\mathbf{r}_o,t)}O(\mathbf{r}_o)P(\mathbf{r}_i - \mathbf{r}_o)\,d\mathbf{r}_o\tag{12.3-14}$$

and image irradiance

$$\frac{1}{T}\int_{-\infty}^{\infty}\int_{-\infty}^{\infty}\int_{0}^{T} e^{j\phi(\mathbf{r}_o,t)}e^{-j\phi(\mathbf{r}_o',t)}O(\mathbf{r}_o)O^*(\mathbf{r}_o')P(\mathbf{r}_i - \mathbf{r}_o)$$

$$\cdot P^*(\mathbf{r}_i - \mathbf{r}_o')\,d\mathbf{r}_o\,d\mathbf{r}_o'\,dt.\tag{12.3-15}$$

The integral

$$\frac{1}{T}\int_{0}^{T} e^{j\phi(\mathbf{r}_o,t)}e^{-j\phi(\mathbf{r}_o',t)}\,dt \begin{cases} \equiv \hat{\phi}(\mathbf{r}_o, \mathbf{r}_o') & \text{in general,} & (12.3\text{-}16)\\ = \delta(\mathbf{r}_o - \mathbf{r}_o') & \text{ideally.} & (12.3\text{-}17) \end{cases}$$

For the ideal value of $\hat{\phi}$, Eq. (12.3-15) becomes

$$\int_{-\infty}^{\infty} |O(\mathbf{r}_o)|^2|P(\mathbf{r}_i - \mathbf{r}_o)|^2\,d\mathbf{r}_o.\tag{12.3-18}$$

If the phase function is produced by a moving diffuser with velocity $\mathbf{v}$,

$$e^{j\phi(\mathbf{r}_o,t)} = e^{j\phi(\mathbf{r}_o - \mathbf{v}t)},\tag{12.3-19}$$

and then

$$\hat{\phi}(\mathbf{r}_o, \mathbf{r}_o') = \text{AC}(e^{j\phi(\mathbf{r}_o)}).\tag{12.3-20}$$

The width of the autocorrelation (AC) corresponds to the size of the diffusing grain structure. In practice, the width should be

$$\Delta\hat{\phi} \leq 1/(\tilde{\Delta}O + \tilde{\Delta}P),\tag{12.3-21}$$

which means that the diffuser grain structure is sufficiently small to scatter light at a sufficiently large angle to shift the object spectrum completely over the pupil as shown in Fig. 12.3-1.

6. Comparison

The methods described so far produce convolutions of intensities. In Table 12.3-1 we compare the various practical methods of producing quasi-monochromatic incoherent light.

B. Comparison of Spatially Incoherent Processing with Completely Coherent Processing

The comparison will be carried out at the three important planes of a typical optical processor: input, pupil (filter), and output.

1. Input Plane

For spatially incoherent processing, we have seen that only object intensity $|O(\mathbf{r}_o)|^2$ is processed. There are several advantages to this: phase noise in the object is unimportant (e.g., from photographic film thickness variations); diffuse reflecting objects (e.g., printed matter) can be used as an input; self-luminous objects (e.g., CRTs) may be used for input, which makes it easy to implement hybrid (optical–digital) systems.

The disadvantage of the input plane is that only real, nonnegative functions can be processed, since phase information in the object is lost. In a number of situations, this limitation is not important, because one only has available or is only interested in object intensities.

TABLE 12.3-1

Comparison of Spatially Incoherent Quasi-Monochromatic Light Sources

Realization	Comments
Extended white light source with wavelength filter	Useful (e.g., CRT input)
	Wasteful in energy to produce good monochromaticity
Laser light and phase randomization	Commonly used
	Wasteful of energy if too fine a diffuser is used
	No complete incoherence possible because of nonideal diffusers
Laser light and deflection (pupil or object scan)	High technology required
	Long integration times necessary
	Loss of the parallel processing advantage

2. Filter Plane

The optical transfer function (OTF) of a linear space-invariant system is defined as the Fourier transform of the convolution kernel. For our system then

$$\text{OTF}(\mathbf{r}_f/\lambda f) = \text{FT}\{|P(\mathbf{r}_o)|^2\} = \text{FT}\{P(\mathbf{r}_o)\} \circledast \text{FT}\{P(\mathbf{r}_o)\}$$

$$= \tilde{p}(\mathbf{r}_f/\lambda f) \circledast \tilde{p}(\mathbf{r}_f/\lambda f), \qquad (12.3\text{-}22)$$

where FT means Fourier transform and ⊛ means autocorrelation [15]. Thus the OTF of a spatially incoherent system is given by the autocorrelation of the pupil function. This is in contrast to the completely coherent system where the OTF is the pupil function itself.

The disadvantage of this scheme is that not every desired OTF can be produced as the autocorrelation of a pupil function. The OTF is limited by the Lukosz bounds and will always have a low-pass characteristic, as shown in Fig. 12.3-2 [16, 17]. To overcome this limitation, many tricks have been employed (Section 12.3,C).

An advantage over the coherent case occurs in the filter plane with respect to filter positioning requirements. For the coherent case, we recall that the pupil plane filter must be centered on the optical axis and located one focal length behind the Fourier transform lens (Fig. 12.2-1). For the incoherent case, the pupil function need not be critically adjusted in the x-, y-, and z-directions. As shown in Table 12.3-2, a lateral shift in the filter by $\mathbf{r}_d$ causes a linear phase shift in the amplitude point spread function $P(\mathbf{r})$. A longitudinal shift in the filter by Δz causes a quadratic phase shift in $P(\mathbf{r})$. With incoherent illumination, however, only the intensity point spread function is important, so that the phase factors will drop out.

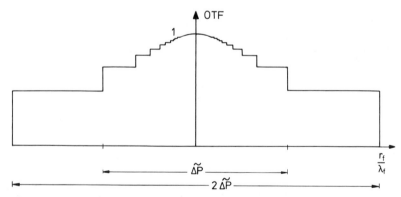

FIG. 12.3-2. Lukosz bounds. Upper limit to the normalized OTF with a nonnegative point spread function.

TABLE 12.3-2

Effects of Filter Displacement

Shift in pupil plane	Effect in output plane
$\tilde{p}\left(\dfrac{\mathbf{r}_f}{\lambda f} - \dfrac{\mathbf{r}_d}{\lambda f}\right)$	$P(\mathbf{r}_i)e^{(-j2\pi/\lambda f)(\mathbf{r}_d\ \mathbf{r}_i)}$
Lateral	
$\tilde{p}\left(\dfrac{\mathbf{r}_f}{\lambda f}, \Delta z\right)$	$P(\mathbf{r}_i)e^{(j\pi/\lambda f^2)\Delta z|\mathbf{r}_i|^2}$
Longitudinal	

If the convolution is obtained with a one-lens system (Fig. 12 3-3), the longitudinal displacement can be used to match the size of the point spread function to the object size. In Fig. 12.3-3 the system output is described by

$$\int_{-\infty}^{\infty} |O(\mathbf{r}_o)|^2 \left| P\left(\frac{\mathbf{r}_i - \mathbf{r}_o}{\lambda z}\right) e^{(j\pi/\lambda z)(\mathbf{r}_i - \mathbf{r}_o)^2}\right|^2 d\mathbf{r}_o. \qquad (12.3\text{-}23)$$

A final advantage of the incoherent system is with respect to pupil plane noise [3, 18]. Because all points in the object spectrum illuminate all points in the pupil plane (Fig. 12.3-1), a small dust particle in the filter plane will distort all frequencies of the object in the same way. Therefore there will be no noise artifacts (bullseyes), as in coherent systems, but only a decrease in contrast.

The increase in the system signal/noise ratio (SNR) with incoherent illumination can be understood in terms of redundancy as described in Section 12.2. To make a quantitative comparison, we must first find the number of optical channels in our incoherent system. If we refer to Fig. 12.2-3, where S is now a spatially incoherent source, we see that a small area on the source will produce one plane wave (one optical channel) for illuminating the object. If we consider an object with area A_o and width $\sqrt{A_o}$

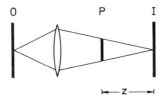

FIG. 12.3-3. Optical system for incoherent correlation with one lens.

to be the limiting aperture for the plane wave, the source size which could just be resolved through this aperture is given by

$$dS = \lambda f / \sqrt{A_o}. \tag{12.3-24}$$

Thus a source area

$$(dS)^2 = \lambda^2 f^2 / A_o \tag{12.3-25}$$

will produce one optical channel. If the total source area is A_s, then the number of optical channels is

$$N_c = A_s / (dS)^2 = A_s A_o / \lambda^2 f^2. \tag{12.3-26}$$

The gain in the signal-to-noise ratio for an incoherent-optical system in comparison to a coherent system is proportional to the square root of the number of optical channels [3] up to the limits where $N_c = N =$ the space–bandwidth product of our transmitted object [Eq. (12.2-28)]. If the number of channels N_c exceeds N, then the SNR gain for an incoherent system remains at $\sqrt{N}$.

From our source size requirement in Eq. (12.3-9), we see that the area of our source $A_s = (\Delta S)^2$ will always be greater than the area of our object spectrum $A_f = (\Delta \tilde{O})^2$. By comparing Eqs. (12.3-26) and (12.2-28), we see that we will have $N_c > N$, so that our SNR gain will be limited to $\sqrt{N}$. For an object SBP of 10^6, this yields a SNR increase of 1000, so that pupil noise is effectively eliminated with an incoherent system.

3. Output Plane

With an incoherent system, one can only add positive values obtained from convolution of the nonnegative object and point spread functions. Thus our output signal often contains a small modulation, the component of interest, on top of a large bias in which we have no interest. (This is not true for a coherent system which can have phase shifts for introducing negative amplitudes which can subtract the bias.) To record the output of the incoherent system therefore, we must have detectors that are linear over a very large dynamic range. This is often a limiting factor with incoherent systems.

C. OTF Synthesis

Having compared the performance of spatially coherent and incoherent systems, we now return to the problem of constructing the proper pupil filter to obtain a desired optical transfer function.

1. One-Channel Synthesis

If a certain OTF is desired, it is not very easy to construct directly a pupil function with the desired autocorrelation. Even if the inverse auto-correlation of the desired OTF exists, the pupil function is not unique, since many functions can give the same autocorrelation. One possible solution is as follows. From the desired OTF in Eq. (12.3-22), calculate the intensity point spread function $|P|^2$

$$|P(\mathbf{r}_o)|^2 = FT^{-1}\{\text{OTF}(\mathbf{r}_f/\lambda f)\}. \qquad (12.3\text{-}27)$$

Then calculate the amplitude point spread function

$$P(\mathbf{r}_o) = \sqrt{|P(\mathbf{r}_o)|^2}\, e^{j\phi(\mathbf{r}_o)}. \qquad (12.3\text{-}28)$$

The phase function $\phi(\mathbf{r}_o)$ is arbitrary, because only $|P(\mathbf{r}_o)|^2$ acts on our output. We next calculate the pupil function

$$\tilde{p}(\mathbf{r}_f/\lambda f) = FT\{P(\mathbf{r}_o)\} \qquad (12.3\text{-}29)$$

and plot it as a computer-generated hologram. One often uses a random or deterministic phase function $\phi(\mathbf{r}_o)$ to avoid a strong central peak in $\tilde{p}$ which would require excessive film dynamic range [19–22].

As an alternative to this technique, one could plot the intensity point spread function from Eq. (12.3-27), illuminate the plot with diffuse light, and produce an interferometric Fourier hologram.

2. Multiple-Channel Synthesis

Many ideas have evolved to overcome the restriction of being able to produce only nonnegative point spread functions. A survey of the various methods is given in Ref. [23]. The principal idea in the various solutions is to split up the desired complex (or bipolar) function into three (or two) positive components such that, when these components are combined with the proper phase relationship, they will produce the desired complex function. A filter for each of the positive functions is recorded on film as described in Section 12.3,C1. The convolution of the input function with each of the filters is performed separately, and the output images are then added together electronically with the proper phase relationships to give the desired complex OTF. The process is described mathematically in Section 12.6,A and will not be repeated here.

The practical problems involved with the system include registering the three output signals on the detectors, so that the proper pixels will be electronically added, and having a detector with sufficient dynamic range, since the desired final signal is often a small difference between several large output signals. Unfortunately, much of the advantage of improved SNR available with incoherent processing can easily be lost in the electronic detection process.

12.4 APPLICATIONS OF SPATIALLY INCOHERENT PROCESSING

By the use of electronic means (e.g., for modulating the input function and demodulating or subtracting the output functions) all operations and applications that are possible with coherent systems are, in principle, possible with incoherent systems. A number of experiments using spatially incoherent light have been described in the literature [8–14, 24–27].

In this section, we will show the results of one new experiment in which data are coded and later decoded to prevent unauthorized access. If the data are in the form of images (drawings, aerial photographs, medical images), the brightness of each pixel can, in general, be separately coded in accordance with a specific key known only to the transmitter and receiver. Such a coding is called cryptography [28, 29]. The experiments described here involve cryptoholography [30], in which a definite but secret interaction between the pixels is introduced. Specifically, the picture $|O(\mathbf{r})|^2$ is convolved with a secret two-dimensional encoding function $Q(\mathbf{r})$ to produce the coded image $C(\mathbf{r})$:

$$C(\mathbf{r}) = |O(\mathbf{r})|^2 * Q(\mathbf{r}). \qquad (12.4\text{-}1)$$

Here,

$$Q(\mathbf{r}) = |P(\mathbf{r})|^2, \qquad (12.4\text{-}2)$$

as in Eq. (12.3-8), since our convolution is done incoherently. The coded image can now be stored or transmitted. Decoding is accomplished by the convolution of $C(\mathbf{r})$ with the decoding function $Q'(\mathbf{r})$:

$$D(\mathbf{r}) = C(\mathbf{r}) * Q'(\mathbf{r}) \qquad (12.4\text{-}3)$$

$$= |O(\mathbf{r})|^2 * Q(\mathbf{r}) * Q'(\mathbf{r}). \qquad (12.4\text{-}4)$$

From Eq. (12.4-4), we see that $D(\mathbf{r})$ will be a replica of our object $|O(\mathbf{r})|^2$ if

$$Q(\mathbf{r}) * Q'(\mathbf{r}) = \delta(\mathbf{r}). \qquad (12.4\text{-}5)$$

The Fourier transform of Eq. (12.4-5) is

$$\tilde{q}(\mathbf{r}_f/\lambda f)\tilde{q}'(\mathbf{r}_f/\lambda f) = 1. \qquad (12.4\text{-}6)$$

From Eqs. (12.4-2) and (12.3-22), we see that

$$\tilde{q}(\mathbf{r}_f/\lambda f) = \mathrm{OTF}(\mathbf{r}_f/\lambda f),$$ (12.4-7)

hence we must choose

$$\tilde{q}'(\mathbf{r}_f/\lambda f) = [\mathrm{OTF}(\mathbf{r}_f/\lambda f)]^{-1}.$$ (12.4-8)

In order to be able to produce $\tilde{q}'$, which is a reciprocal, we must choose our original OTF such that no small magnitude values are present within the desired spatial frequency range.

In the experimental study, a line of text (Fig. 12.4-1a) was optically convolved with the point spread function shown within the box in Fig. 12.4-2. A wavelength-filtered incandescent lamp provided the spatially incoherent illumination for the object transparency. The encoded output image (Fig. 12.4-1b) was directly scanned by a computer-controlled microdensitometer to avoid film nonlinearities. In this study, the deconvolution was then performed digitally (Fig. 12.4-1c). With additional hard-clipping, a near perfect replica of the object was produced (Fig. 12.4-1d).

It should also be possible to perform the deconvolution optically. Using spatially incoherent light, a two-channel system would be necessary because the values of the deconvolution function are in general bipolar. Numerical simulations of this process disclosed that the final image would be obtained as a small difference between the two high biased images in the two output channels. Thus we could take advantage of parallel optical processing but would require a large detector dynamic range to produce good results.

FIG. 12.4-1. Cryptoholography experimental results. (a) Input object. (b) Incoherently encoded object. (c) Digitally decoded output image. (d) Hard-clipped version of (c).

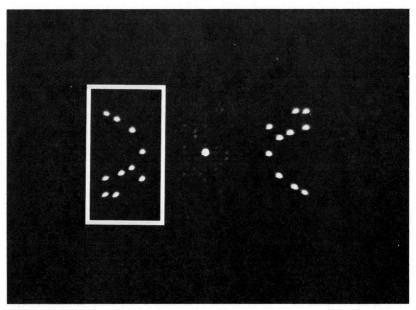

FIG. 12.4-2. Encoding point spread function (within box) obtained from a computer-generated hologram placed in the filter plane.

12.5 TEMPORALLY INCOHERENT PROCESSING

So far we have assumed spatially incoherent but monochromatic or quasi-monochromatic light. This means that the frequency range Δv is small in comparison to the mean frequency $\bar{v}$; i.e.,

$$\Delta v / \bar{v} \ll 1. \tag{12.5-1}$$

Now we will turn to polychromatic, i.e., temporally incoherent, light. We therefore introduce the parameters wavelength λ and temporal frequency v, which are connected by the equation

$$c = \lambda v, \tag{12.5-2}$$

where c is the speed of light. In order to discuss the coherence properties, we return to Eq. (12.2-13):

$$E(\mathbf{r}_i) = \int_{-\infty}^{\infty} \int_{-\infty}^{\infty} \int_{0}^{\infty} \int_{0}^{\infty} \int_{0}^{T} U_o(\mathbf{r}_o, v) O(\mathbf{r}_o, v) P(\mathbf{r}_i - \mathbf{r}_o, v) R(v)$$

$$\cdot e^{-2\pi j v t} U_0^*(\mathbf{r}_o', v') O^*(\mathbf{r}_o', v') P^*(\mathbf{r}_i - \mathbf{r}_o', v') R^*(v')$$

$$\cdot e^{2\pi j v' t} \, d\mathbf{r}_o \, d\mathbf{r}_o' \, dv \, dv' \, dt. \tag{12.5-3}$$

The time-dependent part is

$$\int_0^T e^{-2\pi j(v - v')t}\, dt = \int_{-\infty}^{\infty} \mathrm{rect}\left[\frac{t - (T/2)}{T}\right] e^{-2\pi j(v - v')t}\, dt. \quad (12.5\text{-}4)$$

This integral represents the temporal Fourier transformation of a rect function

$$\int_0^T e^{-2\pi j(v - v')t}\, dt = T\, \mathrm{sinc}[(v - v')T]e^{-2\pi j(v - v')T/2}. \quad (12.5\text{-}5)$$

The coherence properties depend on the integration time as well as on the frequency range used. If $(v - v')T$ is small, then

$$\mathrm{sinc}[(v - v')T]e^{-2\pi j(v - v')T/2} \approx 1. \quad (12.5\text{-}6)$$

If we insert this result in Eq. (12.5-3), we obtain

$$T\int_{-\infty}^{\infty}\int_{-\infty}^{\infty}\int_0^{\infty} U_o(\mathbf{r}_o, v)O(\mathbf{r}_o, v)P(\mathbf{r}_i - \mathbf{r}_o, v)R(v)\, dv \int_0^{\infty} U_o^*(\mathbf{r}_o', v')$$

$$\cdot O^*(\mathbf{r}_o', v')P^*(\mathbf{r}_i - \mathbf{r}_o', v')R^*(v')\, dv'\, d\mathbf{r}_o\, d\mathbf{r}_o'. \quad (12.5\text{-}7)$$

Hence frequency-dependent amplitudes are added if one specific point of the light source is considered. This light is called temporally coherent. The maximum integration time T, within which the light can be considered temporally coherent, signifies the coherence time. Coherence time and the corresponding frequency range are thus inversely related.

Now we assume $(v - v')T$ to be large, such that

$$T\, \mathrm{sinc}[(v - v')T]e^{-2\pi j(v - v')T/2} \approx \delta(v - v'). \quad (12.5\text{-}8)$$

This will always be the case for polychromatic light. Equation (12.5-3) then yields

$$\int_{-\infty}^{\infty}\int_{-\infty}^{\infty}\int_0^{\infty} U_o(\mathbf{r}_o, v)U_o^*(\mathbf{r}_o', v)O(\mathbf{r}_o, v)O^*(\mathbf{r}_o', v)P(\mathbf{r}_i - \mathbf{r}_o, v)$$

$$\cdot P^*(\mathbf{r}_i - \mathbf{r}_o', v)R(v)R^*(v)\, d\mathbf{r}_o\, d\mathbf{r}_o'\, dv. \quad (12.5\text{-}9)$$

Now intensities of light are added with respect to temporal frequency if one specific point of the light source is considered. This light is called temporally incoherent. It can be interpreted as emerging from mutually incoherent point sources of different wavelengths. We also note that the effective frequency range can be influenced by the spectral properties of the object and the filter. They can therefore change the coherence properties.

In the following, we will mainly use point sources. For a point source at

$\mathbf{r}_s$, Eq. (12.5-9) becomes

$$\int_{-\infty}^{\infty} \int_{-\infty}^{\infty} \int_{0}^{\infty} U_o(v) e^{(-j2\pi/\lambda f)(\mathbf{r}_s \cdot \mathbf{r}_o)} U_o^*(v)\, e^{(j2\pi/\lambda f)(\mathbf{r}_s \cdot \mathbf{r}_o')} O(\mathbf{r}_o, v) O^*(\mathbf{r}_o', v)$$

$$\cdot\, P(\mathbf{r}_i - \mathbf{r}_o, v) P^*(\mathbf{r}_i - \mathbf{r}_o', v) |R(v)|^2\, d\mathbf{r}_o\, d\mathbf{r}_o'\, dv$$

$$= \int_0^{\infty} |U_o(v)|^2 |O(\mathbf{r}_o, v) * P(\mathbf{r}_o, v)|^2 |R(v)|^2\, dv. \tag{12.5-10}$$

A. Temporally Incoherent Light and Sources

"Temporally incoherent light" is a very general term. Depending on the wavelength range and position of different point sources, many different distributions of temporally incoherent light are possible. We now look at several light sources.

The light emerging from a filament lamp resembles blackbody radiation (Fig. 12.5-1a). The wavelength spectrum is continuous and shows a maximum which depends on the filament temperature. Because of the finite size of the filament, the light does not radiate from a true point source. A point source must be achieved by imaging onto a pinhole. Consequently, the light effi-

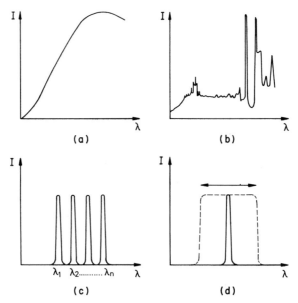

FIG. 12.5-1. Wavelength distribution of light sources. (a) Halogen lamp. (b) Xenon lamp. (c) Superimposed laser lines. (d) Tuned laser.

ciency will not be very high, even if a relatively large pinhole is used. For this source,† $U_0(\lambda) \simeq$ constant in Eq. (12.5-10) over a small range of v.

For greater light efficiency, gas discharge lamps are used. These sources have very high brightness, although it is generally considerably lower than that of a laser. In the visible range, a xenon lamp is preferable, as it shows a nearly flat visible spectrum and some intense UV peaks (Fig. 12.5-1b). For this source, $U_0(\lambda) \simeq$ constant for the entire visible spectrum. The light radiation is concentrated at the point of the gas discharge, which allows us to produce a point source with very high efficiency.

A temporally incoherent light source with completely different physical properties can be achieved by superimposing several laser beams from lasers radiating at different wavelengths. The superposition can be performed with beam splitters, for example. This source, as shown in Fig. 12.5-1c, will have an output spectrum which, to a first approximation, can be described by

$$U_0(\lambda) = \sum_{n=1}^{N} \delta(\lambda - \lambda_n). \tag{12.5-11}$$

A similar spectrum can also be produced by a laser that can radiate at several different visible wavelengths at the same time (e.g., argon ion lasers). With laser light of an extremely narrow linewidth, very good spatial coherence and brightness are present, making these very useful sources. The number of possible laser lines that can be added or produced, however, is restricted by practical considerations.

A different approach in producing light of this type is to use a dye laser that can radiate at different wavelengths. By tuning this laser in time, it is possible to scan the output wavelength. The scan has to be fast enough, compared to the time integration of the output device, for the output device to average and "see" white spatially coherent light (Fig. 12,5-1d). Here

$$U_0(\lambda) = \int_0^T \delta(\lambda - \lambda_0 + at) \, dt, \tag{12.5-12}$$

where $T = (\Delta\lambda/a)$, $\Delta\lambda$ is the range of the λ scan, and a is a constant. By rather simple means, this method produces many more wavelength channels than a combination of different lasers.

Up to this point, the relative position of the source points of different wavelengths has not been discussed. Usually, overlapping point sources are assumed. For special purposes, it can be of interest to produce light sources with a specific wavelength position relation. In the case of added laser lines,

---

† We take some license with our notation and replace the argument v with λ in $U_0(v)$. We do this because it is common to use wavelength rather than frequency in discussing the spectral distribution of a source.

it is easy to achieve a shift of the light sources in space. For all continuous light sources, we can achieve this effect by imaging the light source through a dispersive element such that

$$U_s(x_s, \lambda) = \delta(\lambda - c_1 x_s) U_s(\lambda). \qquad (12.5\text{-}13)$$

The value of c_1 may be constant as in a grating spectroscope, or wavelength-dependent as in a prism spectroscope.

Also, line sources shifted in the z-direction have been used for white light processing [31]. To this end, a white point source was imaged through a Fresnel zone plate, where the focal length f_m depends on the wavelength as

$$f_m(\lambda) = A^2/2\lambda|m| = A^2 v/2c|m|, \qquad (12.5\text{-}14)$$

where A is a (constant) length and $|m| = 1, 2, \ldots$ is the order number. The imaged light source was therefore of the type

$$U_s(z, v) = \delta(v - c_2 z) U_s(v). \qquad (12.5\text{-}15)$$

B. Comparison with Single-Wavelength Processing

1. Optical Channels

In comparison to single-wavelength processing, we gain a new degree of freedom with white light: wavelength λ. Therefore the number of information channels can be increased considerably. For monochromatic light, the number of independent information channels is characterized by the space–bandwidth product [Eq. (12.2-28)] or alternately by [32]

$$N = \Delta x_o \Delta y_o (\Delta x_f/\lambda f)(\Delta y_f/\lambda f). \qquad (12.5\text{-}16)$$

This represents the product of the area $\Delta x_o \Delta y_o$ in the object plane with the area $\Delta x_f \Delta y_f$ in the Fourier plane [multiplied by the scale factor $(\lambda f)^{-2}$]. For polychromatic light, the additional number of independent wavelength channels has to be considered. We will call two wavelengths independent (i.e., incoherent) if their wave trains do not produce stable interference fringes. This property, of course, depends on the light source as well as on the detector.

Two wave trains of frequency v and v' can be considered coherent if during the detector integration time T

$$(v - v')T < 1 \qquad (12.5\text{-}17)$$

[see also Eq. (12.5-6)]. If $T = 1$ sec, then $v - v' = \delta v \leq 1$ Hz. Therefore, each frequency shift of 1 Hz produces a new temporal optical channel. The total number of temporal channels N_t in our system has a maximum value $N_{t,\max}$. For a source with spectral width Δv, we have $N_{t,\max} = \Delta v/\delta v = \Delta v T$ temporal channels, so that our maximum space spectrum–bandwidth product becomes

$$N_{\max} = \Delta x_o \Delta y_o (\Delta x_f/\lambda f)(\Delta y_f/\lambda f)\, \Delta v T. \qquad (12.5\text{-}18)$$

The number of wavelength-dependent optical channels N_t that can actually be used (i.e., discriminated) in an optical system is much smaller than $N_{t,\max}$ because of the limited spectral resolution of optical components in the system. Therefore, to estimate the useful increase in the SBP, the wavelength dependence of each optical element now has to be considered. The most important effects are diffraction and refraction, which will be discussed in the following section. Other effects which change with wavelength are, for example, transmittance or absorption in color filters, transmittance or reflectance in interference filters, and scattering.

To discuss the wavelength dependence of diffraction, we assume an object with transmittance $O(x_o, y_o)$ to be illuminated by a spatially coherent white plane wave and transformed by a lens of focal length f. The resultant field amplitude at wavelength λ is described by (see Fig. 12.5-2 and [33])

$$\tilde{U}_f\left(\frac{x_f}{\lambda f}, \frac{y_f}{\lambda f}\right) = \frac{1}{j\lambda f} \int_{-\infty}^{\infty} \int_{-\infty}^{\infty} O(x_o, y_o)$$
$$\cdot e^{-j2\pi[(x_f/\lambda f)x_o + (y_f/\lambda f)y_o]}\, dx_o\, dy_o. \qquad (12.5\text{-}19)$$

The wavelength λ influences the result in two manners. The factor in front of the integral indicates a wavelength-dependent intensity of the diffraction pattern, but this weighting function is neglected in most cases. However, the scale factor $1/\lambda f$ in the exponent of the diffraction integral affects a far more important wavelength dependence. The different wavelengths produce diffraction patterns similar in form but different in size. This effect may be interpreted as a radial 1-D scaling process with respect to $r_f = (x_f^2 + y_f^2)^{1/2}$ and can be applied in spatial filtering to allow scale-invariant filtering (for scale searching) or wavelength-dependent filtering [34, 35].

The second wavelength-dependent effect to be discussed is dispersion $dn/d\lambda$, that is, the wavelength change in the refractive index n. A light ray

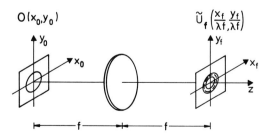

FIG. 12.5-2. Wavelength-dependent diffraction.

passing through a prism will have a wavelength-dependent deflection angle (Fig. 12.5-3). In the case of minimum deviation, the deviation angle ϵ is given by [36]

$$\epsilon(\lambda) = 2 \arcsin[n(\lambda) \sin(\alpha/2)] - \alpha, \qquad (12.5\text{-}20)$$

where α is the apex angle of the prism. If an object is imaged through the prism, dispersion will cause many laterally shifted images with shift

$$\Delta x(\lambda) = \epsilon(\lambda)f, \qquad (12.5\text{-}21)$$

where f is the lens focal length as shown in the figure. For an Amici prism, which actually consists of three adjoining prisms, the deviation angle is made zero for a middle wavelength λ, while only a lateral offset is produced. Similar effects can be achieved by introducing diffraction gratings into the optical system. Application of the above effects will be described in Section 12.6.

2. Redundancy and Multiplexing

In general, the new number of free channels available with multiple wavelengths can be used in two ways. Redundancy can be introduced in order to improve the signal-to-noise ratio. With N_t new parameters, the SNR can theoretically be improved by a factor of $\sqrt{N_t}$. This is only true if the noise is different in every channel. Polychromatic illumination will not reduce noise whose source is in the object or image plane but will reduce noise generated in the Fourier plane. To see this, consider an object of spatial frequency u_o that is illuminated with monochromatic light of wavelength λ, as in Fig. 12.5-4, to produce a focal spot at pupil plane position

$$x_f = \lambda f u_o. \qquad (12.5\text{-}22)$$

We assume that located at x_f we have an opaque defect of width δx. We next ask how much we must change the wavelength so that the diffracted light from the object will miss the opaque spot. The new shifted wavelength will constitute a new optical channel, since it will not "see" the same noise as the original wavelength. By taking the derivative of Eq. (12.5-22), we can see

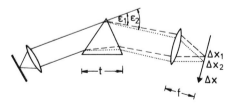

FIG. 12.5-3. Laterally displaced images produced by imaging through a prism.

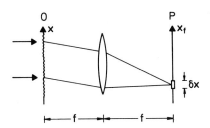

FIG. 12.5-4. Fourier plane noise reduction by using white light illumination.

that the wavelength must be changed by

$$\delta\lambda = \delta x / f u_o. \tag{12.5-23}$$

Hence, if our source emits light over a band $\Delta\lambda$, the new number of optical channels we will have for pupil plane noise suppression is

$$N_t = \Delta\lambda / \delta\lambda = \Delta\lambda f u_o / \delta x. \tag{12.5-24}$$

It can be seen from Eq. (12.5-24) that the SNR improvement is greatest for high object spatial frequencies u_o and for small pupil defects δx. A detailed discussion of noise characteristics in polychromatic optical systems is given in Ref. [4].

The second type of application for multichannel systems is multiplexing, which means independent use of different channels. Again the useful number of channels is restricted by practical considerations. For independent use, it is necessary to distinguish the different wavelength channels. The most effective systems for doing this are grating and prism spectroscopes, as previously mentioned [36].

The theoretical spectral resolution in a grating spectroscope depends on the order m used and the number of illuminated grating lines G:

$$N_t = \lambda / \delta\lambda = |m| G. \tag{12.5-25}$$

A prism spectroscope in the position of minimum deviation allows a resolution

$$N_t = \lambda / \delta\lambda = t |dn/d\lambda|, \tag{12.5-26}$$

where t is the effective width of the prism as shown in Fig. 12.5-3.

The actual number of optical channels that can be utilized depends on the exact system configuration. For instance, if several wavelength-encoded images are to be physically separated by a system such as that in Fig. 12.5-3, the allowed number of optical channels will depend on the physical size of the object as well as on the prism dispersion and wavelength range employed (in order to ensure that output images do not overlap).

Another consideration which affects the allowed object space–bandwidth product is that, in a practical spectroscope, the input slit has finite width. Thus the actual spectroscope output image consists of an ideal image convolved with the image of the input slit. This, of course, limits object resolution. These effects must be considered in any practical system.

12.6 APPLICATIONS OF TEMPORALLY INCOHERENT PROCESSING

Applications can be characterized by the physical principle used, e.g., the wavelength-dependent transmission of filters, the scale variance of diffraction, or the wavelength dependence of refraction. Another way of describing application possibilities can be based on an optical concept such as multichannel or parallel filtering, color encoding, or improvement of the SNR. We will use the second categorization and describe some examples of white light processing. Methods for storing wavelength-multiplexed information are not discussed. But it should be mentioned that there exist effects which allow a direct wavelength storage. There, the wavelength transmission or absorption of a recording material can be changed so that the output wavelength distribution represents the stored information [37].

A. Multichannel Filtering

Multichannel filtering means parallel spatial filtering with a set of different filters. In this case, the parameter for distinguishing these filters will be the wavelength transmission. Therefore, this method implies the use of a set of N colored spatial filters $\tilde{p}_n$ with a spectral transmittance $T_n(\lambda) \simeq \delta(\lambda - \lambda_n)$ such that

$$\tilde{p}\left(\frac{\mathbf{r}_f}{\lambda f}, \lambda\right) = \sum_{n=1}^{N} T_n(\lambda)\tilde{p}_n\left(\frac{\mathbf{r}_f}{\lambda f}\right). \tag{12.6-1}$$

In a filtering setup such as that in Fig. 12.2-1 with an input object $O(\mathbf{r}_o)$ illuminated with white light, we will obtain the output

$$|U_i(\mathbf{r}_i)|^2 = \sum_{n=1}^{N} |T_n(\lambda)|^2 |O(\mathbf{r}_o) * P_n(\mathbf{r}_o)|^2. \tag{12.6-2}$$

Often it is a problem to integrate the different filters into a single filter. In this case, the filters may be placed at positions $\mathbf{r}_f - y_{fn}$ next to one another without overlapping, such that

$$\tilde{p}(\mathbf{r}_f/\lambda f, \lambda) = \sum_{n=1}^{N} T_n(\lambda)\tilde{p}_n(\mathbf{r}_f/\lambda f - y_{fn}/\lambda f). \tag{12.6-3}$$

Then an additional phase factor will be introduced in the point spread function:

$$P(\mathbf{r}_i) = \sum_{n=1}^{N} T_n(\lambda) P_n(\mathbf{r}_i)\, e^{2\pi jyf_n y_i/\lambda f}. \qquad (12.6\text{-}4)$$

This phase factor will not be a problem if spatially incoherent illumination is used, although we would then be restricted to intensity correlations. Alternately, for spatially coherent processing, our light source could consist of an array of colored point sources that are laterally shifted in the source plane in accordance with the filter shift in the Fourier plane.

A solution between these two concepts is represented by interlacing different filters, for example, in a computer-generated hologram. Then the wavelength distinction has to be made by a color grating [38]. In any case, our goal is the result of Eq. (12.6-2), which shows an N-fold, parallel convolution.

The number N of achievable channels is mostly restricted by practical problems. Color film uses three different emulsions with slightly overlapping spectral responses and therefore is able to distinguish only three wavelength regions. This number can be sufficient for solving some problems of spatially incoherent filtering as follows. With monochromatic light in spatially incoherent systems, we saw in Sections 12.3 and 12.4 that we were restricted to producing positive point spread functions. With three independent positive functions, $\tilde{p}_1$, $\tilde{p}_2$, and $\tilde{p}_3$, complex filters can also be realized [39–41]:

$$\tilde{p}\left(\frac{\mathbf{r}_f}{\lambda f}\right) \text{complex} \rightarrow \tilde{p}\left(\frac{\mathbf{r}_f}{\lambda f}\right)$$

$$= \tilde{p}_1\left(\frac{\mathbf{r}_f}{\lambda f}\right) e^{-2\pi j/3} + \tilde{p}_2\left(\frac{\mathbf{r}_f}{\lambda f}\right) + \tilde{p}_3\left(\frac{\mathbf{r}_f}{\lambda f}\right) e^{+2\pi j/3}. \qquad (12.6\text{-}5)$$

There exist different possibilities for producing the positive $\tilde{p}_1$, $\tilde{p}_2$, and $\tilde{p}_3$. One solution is (omitting the arguments $\mathbf{r}_f/\lambda f$)

$$\tilde{p}_1 = -\tfrac{1}{3}\operatorname{Re}\tilde{p} - \tfrac{1}{3}\operatorname{Im}\tilde{p} - \tilde{p}_{\min}, \qquad (12.6\text{-}6\text{a})$$

$$\tilde{p}_2 = \tfrac{2}{3}\operatorname{Re}\tilde{p} - \tilde{p}_{\min}, \qquad (12.6\text{-}6\text{b})$$

$$\tilde{p}_3 = -\tfrac{1}{3}\operatorname{Re}\tilde{p} + \tfrac{1}{3}\operatorname{Im}\tilde{p} - \tilde{p}_{\min}, \qquad (12.6\text{-}6\text{c})$$

where Re and Im mean real and imaginary part, respectively, and $\tilde{p}_{\min}$ is the minimum of $\tilde{p}_1$, $\tilde{p}_2$, and $\tilde{p}_3$ (before subtracting $\tilde{p}_{\min}$). Each of the positive functions is recorded on a different emulsion in the color film. After filtering, the output of the different channels has to be added electronically and with the proper phase, which can be done in real time by a color TV system.

A simplified system of this type with two channels is sufficient for the realization of real, bipolar point spread functions:

$$\tilde{p}(\mathbf{r}_\mathrm{f}/\lambda f)\ \mathrm{real} \to \tilde{p}(\mathbf{r}_\mathrm{f}/\lambda f) = \tilde{p}_1(\mathbf{r}_\mathrm{f}/\lambda f) - \tilde{p}_2(\mathbf{r}_\mathrm{f}/\lambda f), \qquad (12.6\text{-}7)$$

where $\tilde{p}_1(\mathbf{r}_\mathrm{f}/\lambda f)$ and $\tilde{p}_2(\mathbf{r}_\mathrm{f}/\lambda f)$ are positive and

$$\tilde{p}_1(\mathbf{r}_\mathrm{f}/\lambda f) = \begin{cases} \tilde{p}(\mathbf{r}_\mathrm{f}/\lambda f) & \text{if } \tilde{p}(\mathbf{r}_\mathrm{f}/\lambda f) \geq 0 \\ 0 & \text{if } \tilde{p}(\mathbf{r}_\mathrm{f}/\lambda f) < 0, \end{cases} \qquad (12.6\text{-}8)$$

$$\tilde{p}_2(\mathbf{r}_\mathrm{f}/\lambda f) = \begin{cases} -\tilde{p}(\mathbf{r}_\mathrm{f}/\lambda f) & \text{if } \tilde{p}(\mathbf{r}_\mathrm{f}/\lambda f) \leq 0 \\ 0 & \text{if } \tilde{p}(\mathbf{r}_\mathrm{f}/\lambda f) > 0. \end{cases}$$

This represents a way to split a real function into two positive functions. The only electronic task to be done after the optical filtering is subtraction.

In a *spatially coherent* setup, every filter can perform an independent complex filtering, for instance, with a different scale or a different orientation of the point spread function. In the spatially incoherent case, we had to split a complex point spread function into three positive parts, requiring three filters. With the coherent method, one filter is sufficient for the same complex operation, so that with color film-type filters it is possible to realize a three-fold increase in the number of operations with a given number of filters.

Finally, if one wants to have greater wavelength selectivity in a film, in order to multiplex more filters into one film area, one can record filters as volume reflection holograms [42, 43]. As a result of the Bragg effect, each multiplexed filter responds only to a narrow band of wavelengths ($\delta\lambda \simeq 100$ Å), so that 20–30 filters can be multiplexed within the visible spectrum. Using computer-generated holograms, it has even been shown that it is possible to construct filters for readout at a wavelength other than that at which they were interferometrically recorded [43].

B. Scale-Multiplexed Filtering

We may often wish to perform optical pattern recognition on objects whose shape is known but which may appear in several scaled sizes. Biological cells at different stages of growth are an example of such objects. In the correlation setup of Fig. 12.6-1, we have an object whose shape is known but which can appear with different enlargements α so that our object function is written as

$$O(\mathbf{r}_\mathrm{o}/\alpha). \qquad (12.6\text{-}9)$$

The object is illuminated with white light and Fourier-transformed to yield

$$\tilde{o}(\alpha \mathbf{r}_\mathrm{f}/\lambda f), \qquad (12.6\text{-}10)$$

which is incident on a pupil plane filter (a computer-generated hologram,

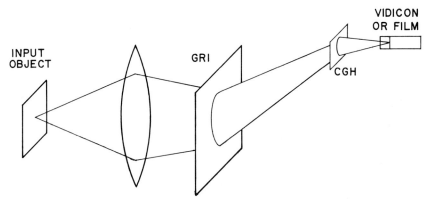

FIG. 12.6-1. Wavelength-multiplexed correlation.

CGH) constructed for the central wavelength of the spectrum λ_c such that the filter transmittance is

$$\tilde{p}(\mathbf{r}_f/\lambda f) = \tilde{\sigma}^*(\mathbf{r}_f/\lambda_c f). \tag{12.6-11}$$

For the wavelength $\lambda = \lambda_c/\alpha$, the filter and object transform are scale-matched and produce an intense correlation signal [34]. Thus the color (i.e., λ) of the correlation signal tells the scale of the object. Because the CGH filter is recorded on a carrier (grating), the output correlation signal will be dispersed, and it will be easy to find the peak wavelength. If one wishes to recognize the object but does not need to know the object scale, the system can be achromatized by adding an additional grating, GR1, as in Fig. 12.6-1 [43].

C. Color Production

Spatial structure has been used to produce color information [44]. In the method of theta modulation, each area (in a low spatial frequency object) that we desire to be of one color is modulated by a grating of specific spatial frequency and orientation. For example, if we wish a certain area near (x_o, y_o) on our object to ultimately have a color λ_n, then we record the grating G_n in this area such that

$$G_n(x_o, y_o) = \sin 2\pi(u_n x_o + v_n y_o), \tag{12.6-12}$$

where u_n and v_n are the grating spatial frequency components. When illuminated by white light, these gratings produce spectrally dispersed diffraction orders:

$$\tilde{g}_n(x_f/\lambda f, y_f/\lambda f) = C_{-1}\,\delta(x_f - u_n\lambda f, y_f - v_n\lambda f)$$
$$+ C_{+1}\,\delta(x_f + u_n\lambda f, y_f + v_n\lambda f) + C_o\,\delta(x_f, y_f), \tag{12.6-13}$$

so that at a given location $\mathbf{r}_f$ one particular λ is present. Then an appropriate wavelength or wavelength combination is selected by Fourier plane filtering with a filter that is opaque except for having holes at the proper positions, for example, at $\mathbf{r}_f = \pm(u_n\lambda_n f, v_n\lambda_n f)$, such that the transmitted light is given by

$$\tilde{g}_n\left(\frac{x_f}{\lambda f}, \frac{y_f}{\lambda f}\right)[\delta(x_f - u_n\lambda_n f, y_f - v_n\lambda_n f) + \delta(x_f + u_n\lambda_n f, y_f + v_n\lambda_n f)]$$

$$= C_{-1}\delta(\lambda - \lambda_n) + C_{+1}\delta(\lambda - \lambda_n). \qquad (12.6\text{-}14)$$

Thus, only the proper wavelengths are selected from the white spectrum for this portion of the object. An inverse transform then produces the properly colored object. Instead of sinusoidal gratings, Ronchi rulings may also be used to encode the information. The decoding mechanism is, in every case, wavelength-dependent diffraction.

D. Pseudocolor Encoding

While in the last section we overlaid gratings on objects in order eventually to produce colored images, in this section we will describe the production of colored images based on information already contained in the object. The process is called pseudocoloring and is often a desirable pre-processing operation. It can be especially important for human observers, since color discrimination is much better than gray level discrimination. The two main goals of pseudocoloring are processing in the space domain (translation of a gray tone into color) and processing in the frequency domain (translation of spatial frequency into color).

The second concept is well suited for optical implementation. The Fourier-transformed object $\tilde{o}(u, v)$ has to be filtered by a color-or wavelength-dependent filter [45]:

$$\tilde{o}\tilde{p} = \tilde{o}(x_f/\lambda f, y_f/\lambda f)\tilde{p}(x_f, y_f, \lambda). \qquad (12.6\text{-}15)$$

One problem with this method is the overlapping of different frequencies at a specific position in the Fourier plane because of the multispectral illumination. If an illumination spectrum of width $\Delta\lambda$ is used, a spatial frequency u' is spread over Δx_f:

$$\Delta x_f = x_{1f} - x_{2f} = u'\lambda_1 f - u'\lambda_2 f = u'f\,\Delta\lambda, \qquad (12.6\text{-}16)$$

so that there is no unique position where a color filter can be placed to act only on spatial frequency u'. Another problem arises with the production of wavelength-dependent filters. It is difficult to combine very many different color or interference filters into a desired arbitrary wavelength-dependent spatial filter.

Several of the above problems can be alleviated by the following optical arrangement which shows that only binary (black and white) filters are necessary for pseudocoloring if the object is 1-D or is made quasi-1-D by spatial filtering. As shown in Fig. 12.6-2, the object is illuminated with white light and Fourier-transformed. A slit is placed in the Fourier plane to favor the passing of the v_1 components of the transform. We also note that the transmitted Fourier transform is present in multiple scales because of the multispectral illumination. These multiple spectra are now dispersed by imaging through a prism such that

$$\tilde{o}(v, \lambda) \to \tilde{o}(v)\delta(u - c_1\lambda). \qquad (12.6\text{-}17)$$

A black and white filter $\tilde{p}(u, v)$ can be inserted, which independently operates on each different spatial frequency and wavelength:

$$\tilde{o}(v)\delta(u - c_1\lambda)\tilde{p}(u, v) = \tilde{o}(v)\tilde{p}(c_1\lambda, v), \qquad (12.6\text{-}18)$$

such that for a given spatial frequency v our filter transmits just the desired wavelength.

This method allows one to encode specific frequencies as well as specific frequency combinations (i.e., a spatial spectrum of patterns) with a specific wavelength or wavelength combinations [46]. Such pseudocoloring applied to pattern recognition is shown in Fig. 12.6-3. The input pattern consists of photographs of two different brick walls, as seen in Fig. 12.6-3a. A computer-generated filter was made which transmitted the red portion of the spectrum for all spatial frequencies present predominantly in the left half of the input image. Similarly, the filter transmitted the green portion of the spectrum for spatial frequencies predominantly present in the right half of the object. In Fig. 12.6-3b, the output image is photographed through a red filter which clearly shows that the left half of the output image has been pseudocolored red. Figure 12.6-3c shows a photograph of the output image taken through a green filter.

Finally, the first concept mentioned, pseudocoloring in the object domain, is more complicated to realize optically. All optical methods of this type are based on a preprocessing where gray levels are translated into spatial fre-

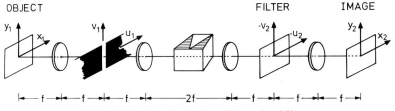

FIG. 12.6-2. Pseudocoloring with gray level filters.

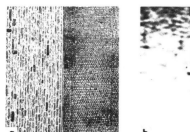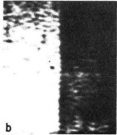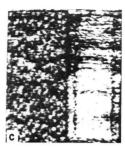

FIG. 12.6-3. Spatial frequency pseudocoloring. (a) Original object. (b) Red output image. (c) Green output image.

quencies, usually by halftone techniques [47]. These prepared inputs are then treated by one of the color selection methods already described [48, 49].

E. Improvement of the SNR

Instead of multiplexing, the additional degree of freedom λ can also be used to introduce redundancy into a processing system. If N_t independent channels are used, the signal-to-noise ratio can be improved by a factor of $\sqrt{N_t}$. Nevertheless until now such methods have rarely been implemented. One group has built an achromatic processor for noise reduction [4, 31]. An application for redundant information transmission through a fiber bundle has been published, which reduces errors caused by broken fibers [50]. A method for 1-D redundant spatial filtering is described in [51]. The principle corresponds to the system described in Fig. 12.6-2, except that instead of filtering with a different filter for each wavelength, identical filters are used, which are matched in scale to the different wavelengths. Thus no pseudocoloring is performed, but the effects of noise in the filter plane can be averaged out. Additionally, Rogers has long been an advocate of incoherent processing and has written a comprehensive book on the subject [52]. References [53–56] are very recent papers on white light processing.

12.7 CONCLUSION

In previous chapters, we have seen the application of optical Fourier transforms to completely coherent-optical systems. In this chapter, we have shown that optical systems can also be employed that use light that is either spatially or temporally incoherent, and we have outlined some of the trade-offs involved in using such systems. Fourier transform incoherent processing systems are relatively new compared to coherent processors, but they are developing at a rapid pace.

REFERENCES

[1] M. Born and E. Wolf (1975). "Principles of Optics." Chap. 10. Pergamon, New York.

[2] See Ref. [1], p. 496.

[3] P. Chavel and S. Lowenthal (1978). *J. Opt. Soc. Am.* **68**, 721–732.

[4] E. N. Leith and J. A. Roth (1979). *Appl. Opt.* **18**, 2803–2811.

[5] R. N. Bracewell (1956). *Aust. J. Phys.* **9**, 297.

[6] J. Upatnieks and R. W. Lewis (1973). *Appl. Opt.* **12**, 2162–2166.

[7] J. Upatnieks and B. J. Chang (1973). *Opt. Commun.* **9**, 348–349.

[8] P. Chavel and S. Lowenthal (1976). *J. Opt. Soc. Am.* **66**, 14–23.

[9] H.-E. Reinfelder (1979). Dissertation. Forschungsinst. für Informationsverarbeitung und Mustererkennung, Karlsruhe, West Germany.

[10] A. W. Lohmann (1968). *Appl. Opt.* **7**, 561–563.

[11] W. T. Maloney (1971). *Appl. Opt.* **10**, 2554–2555.

[12] S. Lowenthal and A. Werts (1968). *C. R. Hebd. Seances Acad. Sci., Ser. B* **266**, 542–545.

[13] B. Braunecker, R. Hauck, and A. W. Lohmann (1977). *Photogr. Sci. Eng.* **21**, 278–281.

[14] J. D. Armitage and A. W. Lohmann (1965). *Appl. Opt.* **4**, 451–467.

[15] P. M. Duffieux (1946). "L'integrale de Fourier et Ses Applications à L'optique." Fac. Sci., Besançon, France.

[16] W. Lukosz (1962). *Opt. Acta* **9**, 335–364.

[17] W. Lukosz (1962). *J. Opt. Soc. Am.* **52**, 827–829.

[18] B. Braunecker, R. Hauck, and W. T. Rhodes (1979). *Appl. Opt.* **18**, 44–51.

[19] E. N. Leith and J. W. Upatnieks (1964). *J. Opt. Soc. Am.* **54**, 1295–1301.

[20] C. B. Burkhardt (1970). *Appl. Opt.* **9**, 695–700.

[21] W. J. Dallas (1973). *Appl. Opt.* **12**, 1179–1187.

[22] Y. Nakayama and M. Kato (1979). *J. Opt. Soc. Am.* **69**, 1367–1372.

[23] A. W. Lohmann and W. T. Rhodes (1978). *Appl. Opt.* **17**, 1141–1151.

[24] A. W. Lohmann (1977). *Appl. Opt.* **16**, 261–263.

[25] W. Stoner (1978). *Appl. Opt.* **17**, 2454–2467.

[26] G. L. Rogers and J. G. Davies (1977). *Opt. Commun.* **21**, 311–317.

[27] D. Gorlitz and F. Lanzl (1977). *Opt. Commun.* **20**, 68–72.

[28] K.-D. Förster (1979). Diplomarbeit. Physikalisches Inst. Univ. Erlangen-Nürnberg, West Germany.

[29] K.-D. Förster and R. Hauck (1978). Annual Report—Applied Optics. Physikalisches Inst. Univ. Erlangen-Nürnberg, West Germany.

[30] A. W. Lohmann (1967). *IBM Tech. Discl. Bull.* **10**, 277.

[31] E. N. Leith and J. Roth (1977). *Appl. Opt.* **16**, 2565–2567.

[32] A. W. Lohmann (1966). *Proc. Summer Sch. Opt. Data Process., Univ. Michigan, Ann Arbor*; also IBM Res. Pap. RI 438 (1967).

[33] J. W. Goodman (1968). "Introduction to Fourier Optics." McGraw-Hill, New York.

[34] S. K. Case (1979). *Appl. Opt.* **18**, 1890–1894.

[35] S. P. Almeida, S. K. Case, and W. J. Dallas (1979). *Appl. Opt.* **18**, 4025–4029.

[36] M. Born and E. Wolf (1975). "Principles of Optics." Pergamon, New York.

[37] *Laser Focus* (1978). **9**, 30–34.

[38] H. O. Bartelt (1977). *Opt. Commun.* **23**, 203–206.

[39] P. Wiersma (1979). *Opt. Commun.* **28**, 280–282.

[40] D. Gùrlitz and F. Lanzl (1979). *Opt. Commun.* **28**, 283–286.

[41] D. Psaltis, D. Casasent, and M. Carlotto (1979). *Opt. Lett.* **4**, 348–350.

[42] S. K. Case (1976). Ph.D. Thesis, Univ. of Michigan, Ann Arbor.

[43] S. K. Case and W. J. Dallas (1978). *Appl. Opt.* **17**, 2537–2540.

[44] J. D. Armitage and A. W. Lohmann (1965). *Appl. Opt.* **4**, 399–403.
[45] J. Bescos and T. C. Strand (1978). *Appl. Opt.* **17**, 2524–2531.
[46] H. Bartelt (1981). *J. Opt.* **12**, 169.
[47] T. C. Strand (1975). *Opt. Commun.* **15**, 60–65.
[48] H.-K. Liu and J. W. Goodman (1976). *Nouv. Rev. Opt.* **7**, 285–289.
[49] G. Indebetouw (1978). *J. Opt.* **9**, 1–4.
[50] C. J. Koester (1968). *J. Opt. Soc. Am.* **58**, 63–70.
[51] J.-P. Goedgebuer and R. Gazeu (1978). *Opt. Commun.* **27**, 53–56.
[52] G. L. Rogers (1977). "Noncoherent Optical Processing." Wiley, New York.
[53] I. Glaser (1980). *J. Opt.* **11**, 215.
[54] D. Psaltis, O. Casasent, and M. Carlotto (1979). *Opt. Lett.* **4**, 348–350.
[55] G. M. Morris and N. George (1980). *Opt. Lett.* **5**, 446–448.
[56] G. M. Morris (1981), *Appl. Opt.* **20**, 2017–2025.

Index